TOWN AND COUNTRY PLANNING IN THE UK

Town and country planning has never been more important to the UK, nor more prominent in national debate. Planning generates great controversy: whether it's spending £80 million and four years' inquiry into Heathrow's Terminal 5, or the 200 proposed wind turbines in the Shetland Isles. On a smaller scale telecoms masts, takeaways, house extensions and even fences are often the subject of local conflict.

Town and Country Planning in the UK has been extensively revised by a new author group. The fifteenth edition incorporates the major changes to planning introduced by the Coalition government elected in 2010, particularly through the National Planning Policy Framework and associated practice guidance, and the Localism Act. It provides a critical discussion of the systems of planning, the procedures for managing development and land use change, and the mechanisms for implementing policy and proposals. It reviews current policy for sustainable development and the associated economic, social and environmental themes relevant to planning in both urban and rural contexts. Contemporary arrangements are explained with reference to their historical development, the influence of the European Union, the roles of central and local government, and developing social and economic demands for land use change.

Detailed consideration is given to:

- the nature of planning and its historical evolution;
- the role of the EU, central, regional and local government;
- mechanisms for developing policy and managing development;
- policies for guiding and delivering housing and economic development;
- sustainable development principles for planning, including pollution control;
- the importance of design in planning;
- conserving the heritage;
- community engagement in planning.

At the end of each chapter, suggestions for further reading are provided. Building on the work of Cullingworth and Nadin, this new edition will ensure that *Town and Country Planning in the UK* maintains its reputation as the 'bible' of British planning.

Barry Cullingworth was a Senior Research Fellow in the Department of Land Economy at the University of Cambridge, UK and Emeritus Professor of Urban Affairs and Public Policy at the University of Delaware, USA.

Vincent Nadin is Professor of Spatial Planning and Strategy at the Delft University of Technology, the Netherlands.

Trevor Hart is Visiting Research Fellow, **Simin Davoudi** is Professor of Environmental Policy and Planning, **John Pendlebury** is Professor and Head of School, **Geoff Vigar** is Professor of Urban Planning, **David Webb** is Lecturer in Planning and **Tim Townshend** is Head of Planning and Urban Design, all at the School of Architecture, Planning and Landscape at Newcastle University, UK.

TOWN AND COUNTRY PLANNING IN THE UK

Fifteenth edition

Barry Cullingworth, Vincent Nadin,
Trevor Hart, Simin Davoudi,
John Pendlebury, Geoff Vigar,
David Webb and Tim Townshend

Routledge
Taylor & Francis Group

LONDON AND NEW YORK

First published 1964
Fifteenth edition published 2015
by Routledge
2 Park Square, Milton Park, Abingdon, Oxon OX14 4RN

and in the USA and Canada by Routledge
711 Third Avenue, New York, NY 10017

Routledge is an imprint of the Taylor & Francis Group, an informa business

© Barry Cullingworth 1964, 1967, 1970, 1972, 1974, 1976, 1979, 1982, 1985, 1988

© Barry Cullingworth and Vincent Nadin 1994, 1997, 2002, 2006

© 2015 Barry Cullingworth, Vincent Nadin, Trevor Hart, Simin Davoudi, John
Pendlebury, Geoff Vigar, David Webb and Tim Townshend

British Library Cataloguing in Publication Data
A catalogue record for this book is available from the British Library

Library of Congress Cataloging in Publication Data
Cullingworth, J. B.
  Town and country planning in the UK / Barry Cullingworth, Vincent Nadin, Trevor
Hart, Simin Davoudi, John Pendlebury, Geoff Vigar, David Webb, and Tim
Townshend.—15th edition.
    pages cm
  Includes bibliographical references and index.
  1. City planning—Great Britain.   2. Regional planning—Great Britain.   I. Title.
  HT169.G7C8 2014
  307.1'2160941—dc23   2014025141

ISBN: 978-0-415-49227-0 (hbk)
ISBN: 978-0-415-49228-7 (pbk)
ISBN: 978-1-315-74226-7 (ebk)

Typeset in Garamond
by Keystroke, Station Road, Codsall, Wolverhampton
Printed in Great Britain by Ashford Colour Press Ltd,
Gosport, Hants

Contents

Plates

Figures

Tables

Boxes

Preface

It is fifty years since the first edition of this book was published. There have been many changes in the style and content of planning in the UK since then and this is reflected in the development of this book. It has grown significantly in size – a hardback edition from 1964 weighs in at half a kilo whilst a paperback of the fourteenth edition is edging towards three times that weight. This could be seen as a reflection of a number of factors – an apparent increase in complexity of the task of planning, a realisation that planners need to have a wider appreciation of what happens in other spheres of policy thereby extending the boundaries for planning and what planners need to be familiar with, and a seeming increasing propensity on the part of government to 'reform' planning. Such a list is certainly not complete. Comparing the two editions, there are about five times as many pages on plan making and the management of development in the most recent edition as in the first, a reflection of increasing complexity. But the priorities of planning have changed over fifty years. About a quarter of the material in the 1964 edition does not feature or have a high profile in the most recent edition: matters such as new and expanded towns, derelict land and regional planning have been replaced by coverage of environment and sustainable development, heritage and transport. This means that the issue of what to include and what to leave out has always been a consideration, and if we are to avoid producing a two-kilo book it seems even more pressing now than it was previously for Barry Cullingworth and Vincent Nadin.

As was the case for them, the team which took on this edition has had to make decisions which, at some points, have been to a degree personal. Whilst not everyone will agree with the choices we have made, we hope that we have maintained the traditional qualities of the book and that it continues to fulfil its role in providing a clear exposition of planning policies and tasks set within their historical context. We feel that the historical context has a particular value, not only because it shows how we reached where we now find ourselves, but also because it makes it possible to identify some key elements of consistency in planning, including those challenges it has yet to overcome, in spite of many years of practice.

Whilst our initial mission was merely to update, what is here is in fact extensively rewritten. This is not because of any perceived failings in the previous text, more because we found it easier to write in our own voices. This contributed to us deciding to add two new chapters, one on urban design and another on developing planning policies: the latter replaces the chapter in earlier editions focusing on 'land'. However, to avoid the book becoming ever larger, we have had to omit some items that have been included in previous editions. In some cases, these are items we felt were no longer as significant in the historical narrative, but the most significant change – in terms of the number of pages it has occupied – is the omission of the list of official publications. This is on the basis that what we see as the most relevant material is referred to in the text and is therefore included in the extensive bibliography, but it also reflects that much material is now available – or only available – on the Internet, and the UK government has been seeking to refine access to policy and consultation documents via its portal, gov.uk.

This edition has been written by a team of six people from the School of Architecture, Planning and

Landscape at Newcastle University. The fact that it now takes six people to complete a task that was for many years accomplished by Barry Cullingworth alone and then, for four editions, with the assistance of Vincent Nadin, highlights both the scale of their achievements and the growing scale of the task. The decision to use a team of people has allowed us to draw on individual enthusiasms and specialisms and we hope that this has yielded benefits to both individual chapters and to the book as a whole. The team involved were Simin Davoudi, John Pendlebury, Geoff Vigar, Dave Webb and Tim Townshend, with Trevor Hart taking the editorial role; the author(s) responsible for revising or writing individual sections are noted in the table of contents.

This text was largely completed by the spring of 2014 but, given the enthusiasm of recent governments to introduce changes to the planning system, there may well have been further changes introduced by the time you read this book. We have endeavoured to note significant proposals that were under consideration at the time of writing but readers will need to refer to government sources to ensure that they have pinned down an up-to-date picture of the ever-evolving UK planning system. However, this is a test we share with Barry Cullingworth, who in writing the first edition in 1963 was faced with anticipating the impact on planning of a London Government Bill and a Local Government Commission reviewing the structure and organisation of local government outside London. That its recommendations were subsequently abandoned did not ease his task.

Thanks are due to a number of people who offered advice or read and commented on drafts, including Jules Brand, Elizabeth Brooks, Jenny Crawford, Hannah Garrow, Graham Haughton, Neil Powe and Ernie Vickers; illustrations were prepared by Jenny Kynaston. Thanks are also due to the staff at Routledge and particularly to Andrew Mould and Sarah Gilkes, who offered an ideal blend of encouragement and prodding to get us to complete the text more or less on schedule.

Trevor Hart
School of Architecture, Planning and Landscape
Newcastle University

Barry Cullingworth 1929–2005

Barry Cullingworth died in February 2005 just before the fourteenth edition of *Town and Country Planning in the UK* was completed. He was particularly well known for this book but had a broad and distinguished academic record. As a researcher, consultant to government and prolific writer, he made an outstanding contribution to town and country planning and urban policy.

Cullingworth was born in Nottingham and started his higher education by taking a degree in music at Trinity College, London. He switched to sociology and took a degree at the University of London. In 1955 he was appointed as a research assistant at The University of Manchester and subsequently held lecturing and research appointments at Durham and Glasgow Universities. He published his first book in 1960, *Housing Needs and Planning Policy*, followed in 1963 by *Housing in Transition*. In 1966 he set up the Centre for Urban and Regional Studies at the University of Birmingham and in 1972 moved back to Scotland to set up the Planning Exchange.

While at Birmingham and Glasgow, Cullingworth chaired numerous government inquiries into housing and the new towns, the best known of which was on *Scotland's Older Houses*. The *Cullingworth Report*, as it is now known, revealed the parlous condition of private rented housing across the country and set the government on a path of radical reform. In later life he expressed disappointment with the relative lack of attention given to the quality and availability of affordable housing, especially in comparison to the priority given to protecting the countryside.

By the mid-1970s Cullingworth had published ten books, numerous official reports and undertaken consultancies at home and abroad, including reports for the OECD, WHO and United Nations. He was, therefore, the ideal candidate for appointment as Historian to the Cabinet Office to prepare the *Official History of Environmental Planning 1939–69*. With the late Gordon Cherry, he published the four volumes of the *History*, between 1975 and 1981. He explains in these volumes how 'a small group of visionaries in the civil service' reconstructed the government planning machinery intending 'to achieve a far greater degree of co-ordination and purposive action'. In many publications he was to advocate a positive role for planning as initiator of coordinated land use change.

In 1978, Cullingworth moved to North America, first as Chairman and Professor of Urban and Regional Planning at the University of Toronto and from 1983 as Unidel Professor of Urban Affairs and Public Policy at the University of Delaware. When he moved to Toronto this book was in its sixth edition and recognised as the 'leading review' in the field. He continued to publish in North America including *Urban and Regional Planning in Canada* and *Planning in the USA*, now in its fourth edition.

Cullingworth returned to Britain in 1994, working in an ambassadorial role for the University of Delaware, taking on a visiting position at Cambridge's Department of Land Economy and editing *British Planning: 50 Years of Urban and Regional Policy*. In recent years the writing of both the British and American textbooks has been shared with other authors. He was always an active partner, working energetically on the later editions until 2004. He was a generous co-writer too, with a willingness to update and change. His ability to digest vast quantities of information was matched only by his persistence in getting at the facts.

Cullingworth's publications reflect his energy, enthusiasm and commitment – and sheer capacity for work. They also owe something to the invaluable support of his wife Betty. He took a considered and meticulous approach to research and writing that lends authority to his publications. But he will be best remembered as an author who could draw out the significant from the routine and deliver his message in a meaningful and engaging style. He wrote with the intention of being understood and accessible.

Cullingworth's family remember him as a loving and funny man with a sense of mischief. He was, of course, usually surrounded by books, but it will be a surprise to many that he had a passion for DIY, finding time alongside the research and writing to work on renovating the many houses the family moved into. He was an accomplished pianist too, with a passion for music.

Cullingworth's publications have guided many thousands of students and practitioners over more than forty years. Despite this success, he was unpretentious and modest. While making great efforts to be comprehensive in his research he would never claim that the findings were exhaustive. He preferred instead to say that he was pointing the reader to some useful material. He did much more than that. Many more students will continue to benefit from his writing.

Barry Cullingworth devoted his life to his work and family. He is survived by his wife Betty, and his children, Wendy, Jane and Peter.

Vincent Nadin

Acronyms and abbreviations

Acronyms and abbreviations have been major growth areas in public policy. The following list includes all those used in the text and others that readers may come across in the planning literature. No claim is made for comprehensiveness.

1990 Act	The Town and Country Planning Act 1990
1991 Act	The Planning and Compensation Act 1991
2004 Act	The Planning and Compulsory Purchase Act 2004
2008 Act	The Planning Act 2008
AAP	area action plan
ACC	Association of County Councils
ACO	Association of Conservation Officers
ACOST	Advisory Council on Science and Technology
ACRE	Action with Communities in Rural England
ADAS	Agricultural Development and Advice Service
ADC	Association of District Councils
AESOP	Association of European Schools of Planning
ALA	Association of London Authorities (now ALG)
ALBPO	Association of London Borough Planning Officers
ALG	Association of London Government
ALNI	Association of Local Authorities in Northern Ireland
ALURE	alternative land use and rural economy
AMA	Association of Metropolitan Authorities
AMR	annual monitoring report
ANPA	Association of National Park Authorities
AONB	area of outstanding natural beauty
AOSP	areas of special protection (for birds)
APRS	Association for the Protection of Rural Scotland
AQMA	air quality management areas
AR	assessment report
ARC	Action Resource Centre
ASAC	area of special advertisement control
ASNW	area of semi-natural woodland
ASSI	area of special scientific interest (Northern Ireland)
ATB	Agricultural Training Board
BAA	British Airports Authority
BACMI	British Aggregate Construction Materials Industries
BANANA	build absolutely nothing anywhere near anything
BAR	buildings at risk
BAT	best available techniques
BATNEEC	best available techniques not entailing excessive cost
BFL	Building for Life (also BFL12)
BIC	Business in the Community
BID	business improvement district
BIS	Business, Innovation and Skills Department
BNFL	British Nuclear Fuels Ltd
BPEO	best practicable environmental option
BPF	British Property Federation

BPM	best practicable means	CEMR	Council of European Municipalities and Regions
BR	British Rail (now Network Rail)	CFC	chlorofluorocarbon
BRE	Building Research Establishment	CfIT	Commission for Integrated Transport
BRF	British Road Federation	CHP	combined heat and power
BRO	Belfast Regeneration Office	CIA	commercial improvement area
BSI	British Standards Institution	CIEH	Chartered Institute of Environmental Health
BTA	British Tourist Authority (now operating as VisitBritain)	CIL	Community Infrastructure Levy
BTC	British Transport Commission	CIPFA	Chartered Institute of Public Finance and Accountancy
BVPI	best value performance indicators		
BW	British Waterways	CIS	community involvement scheme (Wales)
BWB	British Waterways Board		
CA (1)	combined authority	CIT	Commission for Integrated Transport
CA (2)	Countryside Agency (formerly Countryside Commission)	CITES	Convention on International Trade in Endangered Species
CABE	Commission for Architecture and the Built Environment	CLA	Country Land and Business Association
Cadw	Not an acronym, but the Welsh name for the Welsh Historic Monuments Agency. The word means to keep, to preserve	CLES	Centre for Local Economic Strategies
		CLEUD	certificate of lawfulness of existing use or development
CAP	Common Agricultural Policy	CLOPUD	certificate of lawfulness of proposed use or development
CAT	City Action Team		
CBI	Confederation of British Industry	CLRAE	Conference of Local and Regional Authorities of Europe (Council of Europe)
CC	Countryside Commission (now Countryside Agency)		
CCRA	climate change risk assessment	CNCC	Council for Nature Conservation and Countryside (Northern Ireland)
CCS	Countryside Commission for Scotland (now Scottish Natural Heritage)	CNT	Commission for New Towns
		CO	Cabinet Office
CCT	compulsory competitive tendering	COBA	cost–benefit analysis
CCTV	closed circuit television	COE	Council of Europe
CCW	Countryside Council for Wales	COI	Central Office of Information (closed in 2011; remaining functions performed by Cabinet Office)
CDA	comprehensive development area		
CDC	city development company		
CDP	community development project	COPA	Control of Pollution Act 1974
CEC	Commission of the European Communities (European Commission)	COR	Committee of the Regions (EU)
		COREPER	Committee of Permanent Representatives
CEGB	Central Electricity Generating Board	CORINE	Community Information System on the State of the Environment (EU)
CEMAT	Conférence européene des ministres responsables de l'aménagement du territoire (European Conference of Ministers responsible for Regional Planning)	CoSIRA	Council for Small Industries in Rural Areas
		COSLA	Convention of Scottish Local Authorities

COTER	Commission for Territorial Cohesion (EU COR)	DCLG	Department for Communities and Local Government
CPO	compulsory purchase order	DCMS	Department for Culture, Media and Sport
CPOS	County Planning Officers' Society		
CPRE	Campaign to Protect Rural England (formerly Council for the Protection of Rural England)	DDA	Disability Discrimination Act 1995
		DEA	Department of Economic Affairs
		DECC	Department of Energy and Climate Change
CPRS	Central Policy Review Staff		
CPTED	crime prevention through environmental design	DEFRA	Department for Environment, Food and Rural Affairs
CPTUD	crime prevention through urban design	DETR	Department of Environment, Transport and the Regions
CPRW	Campaign (formerly Council) for the Protection of Rural Wales	DEVE	Committee on Development (EU COR)
CRBO	Community Right to Build Order	DfID	Department for International Development
CRE	Commission for Racial Equality (now part of the EHRC)	DfT	Department for Transport (formerly DoT)
CROW Act	Countryside and Rights of Way Act 2000	DG	Directorate General of the European Commission
CRP	city-region plan (Scotland)		
CRRAG	Countryside Recreation Research Advisory Group	DLG	derelict land grant
		DLR	Docklands Light Railway
CRT	Canal and River Trust	DLT	development land tax
CS	community strategy	DM	development management
CSD	Commission on Sustainable Development (UN)	DNH	Department of National Heritage
		DoE	Department of the Environment
CSERGE	Centre for Social and Economic Research on the Global Environment	DoENI	Department of the Environment for Northern Ireland
CSF	community support framework	DoH	Department of Health
CSR	Comprehensive Spending Review	DoT	Department of Transport (now DfT)
CWI	Controlled Waste Inspectorate	DP	development plan
DAFS	Department of Agriculture and Fisheries for Scotland	DPD	development plan document
		DPM	Deputy Prime Minister
DATAR	Délégation à l'aménagement du territoire et à l'action régionale (French national planning agency)	DPOS	District Planning Officers' Society
		DRIVE	dedicated road infrastructure for vehicle safety in Europe
DBFO	design, build, finance, and operate (roads by the private sector)	DSD	Department for Social Developme (NI)
DBRW	Development Board for Rural Wales	DTCPTF	Distressed Town Centre Prope Task Force
DC (1)	development control		
DC (2)	development corporation	DTI	Department of Trade and I
DC (3)	district council	DTLR	Department of Transport Government and the R (2000–2)
DCA	Department for Constitutional Affairs		
DCAN	development control advice note (NI)	DWI	Drinking Water Ins
DCC	Docklands Consultative Committee	EA	environmental asse

EAC	Environmental Audit Committee (House of Commons)
EAF	environmental action fund
EAFRD	European Agricultural Fund for Rural Development
EAGGF	European Agricultural Guidance and Guarantee Fund
EAP	environmental action programme
EAZ	education action zone
EBRD	European Bank for Reconstruction and Development
EC	European Community
ECMT	European Conference of Ministers of Transport
ECOSOC	Economic and Social Council (United Nations)
ECOTEC	emissions control optimisation technology
ECS	Economic and Social Committee (EU)
ECSC	European Coal and Steel Community
ECTP	European Council of Town Planners
EDC	economic development company
EDU	Equality and Diversity Unit (ODPM)
EEA (1)	European Economic Area (EU plus Iceland, Liechtenstein, Norway and Switzerland)
EEA (2)	European Environment Agency
EEC (1)	European Economic Community
EEC (2)	Energy Efficiency Commitment
EFS	England Forestry Strategy
EFTA	European Free Trade Association
EfW	energy from waste
EH	English Heritage
EHCS	English House Condition Survey
EHRC	Equality and Human Rights Commission
EIA	environmental impact assessment
EIB	European Investment Bank
EIP	examination in public
EIS	environmental impact statement
EMAS	eco-management and audit scheme
EMU	European Monetary Union
	English Nature
	English Partnerships
	educational priority area

EPA (2)	Environmental Protection Act 1990
EPC	Economic Planning Council
ERCF	Estates Renewal Challenge Fund
ERDF	European Regional Development Fund
ERP	electronic road pricing
ERRA	Enterprise and Regulatory Reform Act 2013
ES	environmental statement (UK)
ESA	environmentally sensitive area
ESDP	European Spatial Development Perspective
ESF	European Social Fund
ESPON	European Spatial Planning Observation Network
ESRC	Economic and Social Research Council
ETB	English Tourist Board
ETC	English Tourism Council
ETLLD	Scottish Executive Enterprise, Transport and Lifelong Learning Directorate
EU	European Union
EUCC	European Union for Coastal Conservation
EURATOM	European Atomic Energy Community
EUETS	EU Emissions Trading Scheme
EZ (1)	employment zone
EZ (2)	enterprise zone
FA	Forestry Authority
FC	Forestry Commission
FCGS	Farm and Conservation Grant Scheme
FEOGA	Fonds européen d'orientation et de garantie agricole (European Agricultural Guidance and Guarantee Fund)
FIFG	Financial Instrument for Fisheries Guidance
FIG	Financial Institutions Group
FMI	financial management initiative
FoE	Friends of the Earth
FoI	Freedom of Information
FPS	Fuel Poverty Strategy
FTA	Freight Transport Association

FUA	functional urban area		HIDB	Highlands and Islands Development Board (now HIE)
FWAG	Farming and Wildlife Advisory Group		HIE	Highlands and Islands Enterprise
FWGS	Farm Woodland Grant Scheme		HIP	housing investment programme
FWPS	Farm Woodland Premium Scheme		HL	House of Lords
GATT	General Agreement on Tariffs and Trade		HLC	Historic Landscape Characterisation
			HLCA	hill livestock compensatory allowances
GCR	Geological Conservation Review			
GDO	General Development Order		HLF	Heritage Lottery Fund
GDP	gross domestic product		HLW	high-level waste
GDPO	General Development Procedure Order		HMIP	Her Majesty's Inspectorate of Pollution
GEAR	Glasgow Eastern Area Renewal		HMIPI	Her Majesty's Industrial Pollution Inspectorate (Scotland)
GHG	greenhouse gases			
GI	green infrastructure		HMNII	Her Majesty's Nuclear Installation Inspectorate
GIA	general improvement area			
GIS	geographic information systems		HMO (1)	hedgerow management order
GLA	Greater London Authority		HMO (2)	house in multiple occupation
GLC	Greater London Council		HMR	Housing Market Renewal
GLDP	Greater London Development Plan		HMSO	Her Majesty's Stationery Office
GMCA	Greater Manchester Combined Authority		HMT	Her Majesty's Treasury
			HO	Home Office
GO	government office		HR	human resources
GOR	Government Offices for the Regions		HRF	Housing Research Foundation
GPDO	General Permitted Development Order		HSA	Hazardous Substances Authority
			HSE	Health and Safety Executive
GVA	gross value added		HWI	Hazardous Waste Inspectorate
HA	Highways Agency		IACGEC	Inter-Agency Committee on Global Environmental Change
HAA	housing action area			
HAG	housing association grant		IAEA	International Atomic Energy Agency
HAP	habitat action plan		IAPI	Industrial Air Pollution Inspectorate
HAT	housing action trust		IAPs	inner area programmes
HAZ	health action zone		IAS	inner area study
HBF	Home Builders' Federation		ICE	Institution of Civil Engineers
HBMC	Historic Buildings and Monuments Commission		ICNIRP	International Commission on Non-Ionising Radiation Protection
HC	House of Commons		ICOMOS	International Council on Monuments and Sites
HCA	Homes and Communities Agency			
HCiS	Housing Corporation in Scotland		ICT	information and communicatio⋅ technology
HER	Historic Environment Records			
HERS	Heritage Economic Regeneration Schemes (EH)		ICZM	integrated coastal zone mar
			IDC	industrial development cε
HHSRS	housing, health and safety ratings system		IDeA	Improvement and Deveˡ Agency
HIA	home improvement agency		IDP	infrastructure deliveˑ

IEEP	Institute for European Environmental Policy	LDDC	London Docklands Development Corporation
IEG	implementing electronic government	LDF	local development framework
IIA	industrial improvement area	LDO	local development order
ILD	Index of Local Deprivation	LDP	local development plan (Wales)
ILW	intermediate-level waste	LDS	local development scheme
IMP	Integrated Maritime Policy	LEADER	Liaison entre actions de développement de l'économie rurale
IMPEL	EU Network for the Implementation and Enforcement of Environmental Law	LEAP	local environmental agency plan
		LEC	local enterprise company (Scotland)
INTERREG	European Community initiative for transnational spatial planning	LEG-UP	local enterprise grants for urban projects (Scotland)
IPC (1)	Infrastructure Planning Commission	LEP	local enterprise partnership
IPC (2)	integrated pollution control	LETS	local exchange trading system
IPCC	Intergovernmental Panel on Climate Change	LFA	less favoured area (agriculture)
		LGA	Local Government Association
IPPC	integrated pollution, prevention and control	LGC	Local Government Commission for England
IRD	integrated rural development (Peak District)	LGF	local government finance
		LGMB	Local Government Management Board
ISOCARP	International Society of City and Regional Planners	LHS	local housing strategy (Scotland)
		LLW	low-level waste
ITA	integrated transport authority	LNP	local nature partnership
IUCN	World Conservation Union	LNR	local nature reserve
IWA	Inland Waterways Association	LOTS	living over the shop
IWAAC	Inland Waterways Amenity Advisory Committee	LPA	local planning authority
		LPAC	London Planning Advisory Committee
JNCC	Joint Nature Conservation Committee	LRT	Land Restoration Trust
JPL	*Journal of Planning and Environment Law*	LSC	Learning and Skills Council
		LSP	local strategic partnership
LA21	Local Agenda 21 (UNCED)	LSPU	London Strategic Policy Unit
LAA	local area agreement	LSTF	Local Sustainable Transport Fund
LAAPC	local authority air pollution control	LT	London Transport (now TfL)
LAQM	local air quality management	LTB	local transport board
LATS	landfill allowance trading scheme	LTP	local transport plan
LAW	Land Authority for Wales	LTS	local transport strategy (Scotland)
LAWDC	local authority waste disposal company	LUCS	Land Use Change Statistics
		LULU	locally unwanted land use
LBA	London Boroughs Association (now ALG)	LUTS	land use transportation studies
		LWRA	London Waste Registration Authority
LBAP	local biodiversity action plan		
LCO	landscape conservation order	MAFF	Ministry of Agriculture, Fisheries and Food
LDC	local development company		
LDD	local development document	MARS	Monuments at Risk Survey

MCC	metropolitan county council
MCZ	marine conservation zone
MEA	Manual of Environmental Assessment (for trunk roads)
MEGA	metropolitan European growth area
MEHRA	marine environmental high risk areas
MEP	Member of the European Parliament
MHLG	Ministry of Housing and Local Government
MLGP	Ministry of Local Government and Planning
MMG	marine minerals guidance note
MMO	Marine Management Organisation
MMS	multi-modal study
MNR	marine nature reserve
MOA	Mobile Operators Association
MoD	Ministry of Defence
MPA (1)	marine protected area
MPA (2)	mineral planning authority
MPG	minerals planning guidance note
MPP	Monuments Protection Programme
MPS	minerals policy statement
MSC	Manpower Services Commission
MSFD	Marine Strategy Framework Directive
MSP	maritime-spatial planning
MTAN	minerals technical advice note (Wales)
MTCP	Ministry of Town and Country Planning
MWMS	municipal waste management survey
NACRT	National Agricultural Centre Rural Trust
NAP	National Adaption Programme
NAO	National Audit Office
NARIS	National Roads Information System
NATA	New Approach to Appraisal (roads)
NAW	National Assembly for Wales
NBN	National Biodiversity Network
NCALO	Nature Conservation and Amenity Lands (Northern Ireland) Order 1985
NCC	Nature Conservancy Council
NCCI	National Committee for Commonwealth Immigrants

NCCS	Nature Conservancy Council for Scotland (now Scottish Natural Heritage)
NCVO	National Council of Voluntary Organisations
NDC	New Deal for Communities
NDO	Neighbourhood Development Order
NDP	Neighbourhood Development Plan
NDPB	non-departmental public body
NE	Natural England
NEC	noise exposure category
NEDC	National Economic Development Council
NEDO	National Economic Development Office
NEET	not in employment, education or training
NERC	National Environment Research Council
NETCEN	National Environmental Technology Centre
NFC	National Forest Company
NFFO	non-fossil fuel obligation
NGC	Northern Growth Corridor
NGO	non-governmental organisation
NHA	natural heritage area (Scotland)
NHB	New Homes Bonus
NHMF	National Heritage Memorial Fund
NHS	National Health Service
NIA	nature improvement area
NID	National Infrastructure Directorate
NII	Nuclear Installations Inspectorate
NIMBY	not in my back yard
NIO	Northern Ireland Office
NIREX	Nuclear Industries Radioactive Waste Executive
NLUD	National Land Use Database
NNR	national nature reserve
NPA	national park authority
NPCU	national planning casework unit
NPF (1)	National Planning Forum
NPF (2)	National Planning Framework (Scotland)
NPG	National Planning Guideline (Scotland)

NPPF	National Planning Policy Framework	PI	Planning Inspectorate (usually PINS)
NPPG	National Planning Policy Guideline (Scotland)	PIC	Planning Inquiry Commission
		PINS	Planning Inspectorate
NPS (1)	national policy statement	PIP	partnership investment programme
NPS (2)	noise policy statement	PIU	Performance and Innovation Unit
NR	Network Rail	PLI	public local inquiry
NRA	National Rivers Authority (now Environment Agency)	POS	Planning Officers' Society
		PPA (1)	planning performance agreement
NRF	Neighbourhood Renewal Fund	PPA (2)	priority partnership area (Scotland)
NRTF	national road traffic forecasts (GB)	PPC	Pollution, Prevention and Control Act 2000
NRU	Neighbourhood Renewal Unit		
NSA (1)	national scenic area (Scotland)	PPG	planning policy guidance note
NSA (2)	nitrate sensitive area	PPP (1)	polluter pays principle
NSIP	nationally significant infrastructure project	PPP (2)	public–private partnerships
		PPS (1)	planning policy statement (previously PPG)
NUTS	nomenclature of territorial units for statistics: designates levels of regional subdivision in the EU	PPS (2)	planning policy statement (NI)
		PPW	Planning Policy Wales
NVZ	nitrate vulnerable zone	PRIDE	Programmes for Rural Initiatives and Developments (Scotland)
NWDO	North West Development Office (NI)		
		PSA (1)	Property Services Agency
OBR	Office for Budget Responsibility	PSA (2)	public service agreement
ODPM	Office of the Deputy Prime Minister	PSI	Policy Studies Institute
OECD	Organisation for Economic Cooperation and Development	PSS	Planning Summer School (formerly TCPSS)
OEEC	Organisation for European Economic Cooperation	PTA	passenger transport authority
		PTE	passenger transport executive
OJ	*Official Journal of the European Communities*	PTRC	Planning and Transport Research and Computation
ONS	Office for National Statistics	PVC	polyvinyl chloride
OPCS	Office of Population Censuses and Surveys (now part of ONS)	QUANGO	quasi-autonomous non-governmental organisation
OPSR	Office of Public Services Reform	RA	renewal area
OS	Ordnance Survey	RB	regional body
PAG	Planning Advisory Group	RAC	Royal Automobile Club
PAN	planning advice note (Scotland)	RAWP	regional aggregates working parties
PAS	Planning Advisory Service	RCAHMS	Royal Commission on the Ancient and Historical Monuments of Scotland
PAT	policy action team		
PDG	Planning Delivery Grant		
PDL	previously developed land	RCC	rural community council
PDO (1)	permitted development order	RCEP	Royal Commission on Environmental Pollution
PDO (2)	potentially damaging operation (SSSI)		
		RCHME	Royal Commission on the Historical Monuments of England
PDR	permitted development right		
PFI	Private Finance Initiative	RCI	Radiochemical Inspectorate
PGS	planning gain supplement	RCU (1)	Regional Coordination Unit (ODPM)

RCU (2)	Road Construction Unit
RDA (1)	regional development agency
RDA (2)	rural development area
RDC	Rural Development Commission
RDG	regional development grant
RDO	Regional Development Office (NI)
RDP	rural development programme
RDPE	Rural Development Programme England
RDS	Regional Development Strategy Northern Ireland
REG	regional enterprise grant
RES (1)	race equality scheme
RES (2)	regional economic strategy
RGF	Regional Growth Fund
RHB	regional housing board
RHS	regional housing strategy
RIA	regulatory impact assessment
RIBA	Royal Institute of British Architects
RICS	Royal Institution of Chartered Surveyors
RIGS	regionally important geological/ geomorphological sites
ROI	regional output indicator
ROSCO	rolling stock operating company
RPB	regional planning body
RPG	regional planning guidance
RRAF	regional rural affairs forum
RS	regional strategy
RSA (1)	regional selective assistance
RSA (2)	Regional Studies Association
RSDF	regional sustainable development framework
RSL	registered social landlord
RSPB	Royal Society for the Protection of Birds
RSS	regional spatial strategy
RTB (1)	regional tourist board
RTB (2)	Right to Buy (public sector housing)
RTC	regional transport consortia (Wales)
RTP	regional transport partnership
RTPI	Royal Town Planning Institute
RTS	regional transport strategy
RUPP	road used as public path
RWMAC	Radioactive Waste Management Advisory Committee

SA	sustainability appraisal
SAC	special area of conservation (habitats)
SACTRA	Standing Advisory Committee on Trunk Road Assessment
SAGA	Sand and Gravel Association
SAP	species action plan
SAR	sustainability appraisal report
SC	standard charge
SCI	statement of community involvement
SCLSERP	Standing Conference on London and South East Regional Planning
SDA	Scottish Development Agency (now Scottish Enterprise)
SDC	Sustainable Development Commission
SDO	special development order
SDP	standard delivery plan (Scottish Housing)
SDS	Spatial Development Strategy (London)
SDU	Sustainable Development Unit
SE	Scottish Executive
SEA (1)	Single European Act 1987
SEA (2)	strategic environmental assessment
SEDD	Scottish Executive Development Department
SEEDA	South East England Development Agency
SEEDS	South East Economic Development Strategy
SEELLD	Scottish Executive Enterprise and Lifelong Learning Department
SEH	Survey of English Housing
SEHD	Scottish Executive Health Department
SEERAD	Scottish Executive Environment and Rural Affairs Department
SEM	Single European Market
SEPA	Scottish Environment Protection Agency
SERC	Science and Engineering Research Council
SERPLAN	London and South East Regional Planning Conference
SEU	Social Exclusion Unit
SFRA	strategic flood risk assessment

SHAC	Scottish Housing Advisory Committee	SPZ	simplified planning zone
SHG	social housing grant	SR	Spending Review
SHLAA	strategic housing land availability assessment	SRA	Strategic Rail Authority
		SRB	Single Regeneration Budget
SHMA	strategic housing market assessment	SSHA	Scottish Special Housing Association
SHQS	Scottish Housing Quality Standard	SSSI	site of special scientific interest
SI	statutory instrument	STB	Scottish Tourist Board
SIC	social inclusion partnerships (Scotland)	SUD	Committee on Spatial and Urban Development (EU)
SINC	site of importance for nature conservation	SUDS	sustainable urban drainage system
		SURF	Scottish Urban Regeneration Forum (Scotland)
SIP	social inclusion partnership (Scotland)	SURI	small urban regeneration inititive (Scotland)
SLF	Scottish Landowners Federation	TAN	technical advice notes (Wales)
SM	scheduled monument	TCPA	Town and Country Planning Association
SME	small and medium-sized enterprises		
SMR	sites and monuments records (counties)	TCPSS	Town and Country Planning Summer School (now PSS)
SNAP	Shelter Neighbourhood Action Project	TEC	training and enterprise council
		TEN	Trans-European Network(s)
SNH	Scottish Natural Heritage	TEST	Transport and Environment Studies
SO	Scottish Office		
SOAEFD	Scottish Office Agriculture, Environment and Fisheries Department	TEU	Treaty on European Union
		TfL	Transport for London
		THI	Townscape Heritage Initiative
SODD	Scottish Office Development Department	THORP	thermal oxide reprocessing plant
		TOC	train operating company
SOEnD	Scottish Office Environment Department (now SOAEFD)	TPI	Targeted Programme of Improvements (DfT)
SOID	Scottish Office Industry Department	TPO	tree preservation order
SOIRU	Scottish Office Inquiry Reporters Unit	TPPs	transport policies and programmes
		TRL	Transport Research Laboratory
SoS	Secretary of State	TSG	transport supplementary grant
SPA	special protection area (for birds) (EU)	TSO	The Stationery Office
		TUC	Trades Union Congress
		UA	unitary authority
SPAB	Society for the Protection of Ancient Buildings	UCO	Use Classes Order
		UDA	urban development area
SPD (1)	single programming document	UDC	urban development corporation
SPD (2)	supplementary planning document	UDG	urban development grant
SPG	supplementary planning guidance	UDP	unitary development plan
SPP	Scottish planning policy	UKAEA	United Kingdom Atomic Energy Authority
SPPS	strategic planning policy statement (Northern Ireland)		
		UKBAP	UK Biodiversity Action Plan
SPS	single payment scheme (CAP)	UKBG	UK Biodiversity Group

UNCED	United Nations Conference on Environment and Development (Earth Summit, Rio, 1992)
UNCSD	United Nations Commission on Sustainable Development
UNCTAD	United Nations Conference on Trade and Development
UNECE	United Nations Economic Commission for Europe
UNEP	United Nations Environment Programme
UNESCO	United Nations Educational, Scientific and Cultural Organisation
UNFCCC	United Nations Framework Convention on Climate Change
UP (1)	urban partnerships (Scotland)
UP (2)	Urban Programme
URA	Urban Regeneration Agency
URBAN	European Community initiative for urban regeneration
URC	urban regeneration company
UTF	Urban Task Force
VAT	value added tax
VDS	village design statement
VFM	value for money
VISEGRAD	four former communist countries: Poland, Czech Republic, Slovakia and Hungary
VOCS	volatile organic compounds
WAG	Welsh Assembly government
WCA	waste collection authority

WCED	World Commission on Environment and Development
WDA (1)	waste disposal authority
WDA (2)	Welsh Development Agency
WDP	waste disposal plan
WES	wildlife enhancement scheme
WFD	Water Framework Directive
WHO	World Health Organisation
WHS	World Heritage Site
WMEB	West Midlands Enterprise Board
WIC	Waste Infrastructure Credits
WIP	Waste Implementation Programme
WMO	World Meteorological Organisation
WO	Welsh Office
WOAD	Welsh Office Agriculture Department
WQO	water quality objectives
WRA	waste regulation authority
WRAP	Waste and Resources Action Programme
WRO	Wales Rural Observatory
WSP	Wales Spatial Plan
WTB	Welsh Tourist Board
WTO	World Trade Organisation
WWF	World Wide Fund for Nature (formerly World Wildlife Fund)
WWT	Wildfowl and Wetlands Trust

Encyclopedia refers to Malcolm Grant's *Encyclopedia of Planning Law and Practice*, London: Sweet and Maxwell, loose-leaf, regularly updated by supplements.

1 The nature of planning

If planning were judged by results, that is, by whether life followed the dictates of the plan, then planning has failed everywhere it has been tried. No one, it turns out, has the knowledge to predict sequences of actions and reactions across the realm of public policy, and no one has the power to compel obedience.

(Wildavsky 1987: 21)

Introduction

It is the purpose of this chapter to give a general introduction to the character and nature of planning. This may appear to be a philosophical or theoretical matter, and it is not the purpose of this book to review or engage with theory to any significant extent: this is more appropriately done elsewhere and some reading suggestions are given at the end of this chapter. However, if we are to engage successfully with the practical details of planning – an overarching purpose of this book – we at least need to know how the various elements of purpose and process connect to each other, so considering the framework within which they sit is an important foundation for making use of the rest of the contents. Therefore, in the next few pages we will consider what it is that planning is trying to do, the context in which it is trying to do it and the means it has developed to achieve its objectives. Having considered these matters, we then go on to indicate how the various specific elements of content in the rest of the book relate to these questions.

Whilst the evolution of planning is covered in detail in the following chapter, it is perhaps worth making a couple of points here about the development of planning as a professional activity, as a means of establishing that within a changing agenda there are some important consistencies. Planning has always been about 'making better places', to use the title of one of Patsy Healey's books. Writing on Christmas Day 1939, Thomas Sharp in the preface to his book *Town Planning* (1940) saw the product of planning as being 'a new and better way of life'. So, whilst these books were written seventy years apart, and Sharp's before the 1947 Planning Act launched town planning as we now know it in Britain, they agree on a purpose for planning, that of creating an improved environment for citizens. They both agree that the process of planning is likely to deliver a better outcome than a laissez-faire approach lacking in organisation and direction, or 'a dull and shifty opportunism' as Sharp (1945: 116) rather more colourfully puts it. That this organisation and direction needs to be part of a democratic process and not become a technocratic imposition on communities is also something on which there is general agreement. So, it would be accepted by everyone other than the most avid advocate of free market approaches that 'the idea of planning as an enterprise of collective action, of public policy, is linked to a belief that it is worth striving to improve the human condition' (Healey 2010: 118) and that this should be done in a way which allows and encourages the views of both public and 'experts' to be taken into account.

However, agreement on these fundamental items does not mean that an obvious and widely accepted solution always emerges from considering the process of planning. Politics, conflict and dispute are at the centre of land use planning. Conflict arises because of the competing demands for the use of land, because of the negative effects that can arise when the use of land changes, and because of the uneven distribution of costs and benefits which result from development. As Tewdwr-Jones (2012: 1) puts it, 'Planning as an activity that attempts to manage spatial change would not exist in any meaningful way if it was not for contention over the future use and development of the land.' Indeed, planning might usefully be defined as the process by which government resolves disputes about the use of land, and this very contention is also a constant.

However, whilst there are constants, the extent to which the context in which planning operates has changed makes it inevitable that changes of emphasis and focus have arisen in planning itself. At the dawn of what we might recognise as planning, the context was one of cities (and sometimes rural areas) characterised by unhealthy environments defined by poor physical fabric and living conditions. Later, in the period of post-war reconstruction, there was an imperative to address problems of acute housing need as part of a task of rebuilding towns and cities. At these times, planning was an almost evangelical activity, with the mission of creating better environments to the fore; then, the actions of planners tended to be widely supported. However, as the welfare state was rolled out and these clear and pressing physical problems began to be addressed, the mission of planning became wider, focusing on economic and social matters as well as improving the physical fabric. The activity of planning became more of a matter of debate and dispute where planners found themselves 'operating within a complex and often uncomfortable context, within which room for transformative manoeuvre seems slight' (Healey 1997: 8). Whilst much of this loss of a clear and relatively simple mission for planning could be placed at the door of contextual change, this was also reflected in a number of changes in the agenda for planning set by government

reviews of the planning system, which often cast doubt on the direction and process of planning. These are mapped and explored in the following chapters. At a number of points these changes have been prompted by what has been characterised by government as a failure on the part of planning to give sufficient importance to the role of facilitating and promoting economic growth. This points up what might be seen as a final constant, the nature of the relationship between planning and the market. Now, to a large extent, planning relies on the private sector to implement policies (Rydin 2011: 139), so how far it accommodates or seeks to adapt the working of the market is a matter which is overtly or covertly present in considering what the nature of planning can or should be.

An evolution in planning

The United Nations report *Planning Sustainable Cities* (2009: 10) identified socio-economic and institutional origins for modern town planning:

> 'Modern' urban planning emerged in the latter part of the 19th century, largely in response to rapidly growing, chaotic and polluted cities in Western Europe, brought about by the Industrial Revolution. The adoption of urban planning in this part of the world as a state function can be attributed to the rise of the modern interventionist state and Keynesian economics.

It goes on to point out that, at the outset, planning tended to be an exercise focused on physical planning and design, which was essentially the preserve of experts and was concerned with the production of some form of 'master plan'. However, such plans were often disconnected from the lives of those they served and proved ill equipped to adapt to contextual and institutional change. Some of the changes that planning has had to face include: the processes of globalisation and economic restructuring which have produced new challenges of inequality and societies which are more diverse than before; a growing concern about sustainability and the impact of climate change; an emerging distrust of technocratic approaches and a demand for more inclusive approaches to the task of planning; a widening planning agenda which gave new or increased prominence to matters such as economic change, equality and heritage; and a political disenchantment with the era of 'big government' coupled with a move towards a more fragmented institutional framework for the delivery of public services, including planning.

The move away from master planning led to greater emphasis being given to elements such as strategy and implementation within a more flexible planning framework. The emphasis now is on 'steering' rather than 'controlling', on seeking a future, not defining a singular idea of it (Healey 1997), with the general direction of travel indicated rather than trying, and failing, to meet a predetermined ideal (Hillier 2002).

Perhaps especially given this shift away from pre-defined end states, it can be quite hard to pin down a definition of terms such as 'strategy' or the qualities it imparts to the process of shaping and managing development (Shipley and Newkirk 1999). Like many other concepts in planning, it can be seen as being borrowed from elsewhere (Cooke 1983), in this case military and business spheres. An important component of the process of developing a strategy is that of 'making choices' – about what activities are carried out, how they are configured and how they relate to each other (Porter 1996). So we might expect a 'strategy' to relate to: some form of 'vision' for the future;[1] an awareness of context and relationships; some objectives; some guiding principles; and some indication of what might be developed where (Healey 2007; Roberts 1996). This offers a more extensive and varied menu of functions for planning than would be encompassed by physical master planning. The idea that planning is an essentially 'strategic' activity is not new: it has perhaps evolved over the last fifty years and it was not a feature of the first edition of this book. So, Healey's book *Urban Complexity and Spatial Strategies* (2007) points out that the approaches to strategy development popular in planning in the 1970s differed from what she feels is the position now in two important respects. First, the relatively systematic

ideas about the processes of strategy development associated with writers such as Etzioni[2] (1973) need to be replaced in an increasingly fragmented institutional landscape by more nuanced and subtle models where processes of discourse and influence assume greater importance and are an essential complement to an understanding of the physical environment which underpinned master planning. Second, in a world of greater mobility, the pattern of spatial relationships which characterised basic policy models of clear hierarchies of role and function for settlements needs to be replaced by an understanding rooted in relational rather than Cartesian geographies,[3] where planning needs to consider the determinants of the relationships between places and spaces rather than focus on a bounded analysis of the attributes of a place.

Planning has often been accused of paying insufficient regard to implementation of policy (Talen 1996), whilst some empirical research has suggested that 'plan implementation practice is generally poor' (Laurian *et al.* 2004: 573); but for planning as a public activity as the quotation at the start of this chapter suggests, now more than ever 'Promise must be dignified by performance' (Wildavsky 1973: 129). However, as Healey (2010: 230) notes, there is rarely a smooth transition from policy to action: 'instead, it involves a sustained struggle in the various arenas where place-management activity is performed, or major development projects nurtured from initiation to completion, or strategies converted into specific action programmes'. Forty years ago, Pressman and Wildavsky (1973) argued that a key contributor to implementation failure is that policymakers often do not understand the complexity and difficulty of coordinating activities and agencies involved in implementation; in the context of changing patterns of governance and a fragmenting state, the task of today's planners is certainly more complex than that faced by planners in earlier times when planning was very much about guiding the investments of the state.

Most planning policy is now implemented by the private sector, although in many cases interaction between public sector policymakers and private sector developers is important in achieving key outcomes such as area regeneration. Such interaction often takes place within some form of partnership and, according to Balloch and Taylor (2001: 1),

> partnership makes a lot of sense. At one level it is a rational response to divisions within and between government departments and local authorities, within and between professions, and between those who deliver services and those who use them. It is also a necessary response to the fragmentation of services that the introduction of markets brought with them.

However, whether we are talking about implementation by private sector developers or through some sort of partnership vehicle, the development of planning policy with an eye on implementation means that it cannot be a self-contained activity. In such a context, planning has to understand, and to some extent embrace, the aspirations and objectives of others, but it also has to take the consequences of limitations or reductions in the authority which it possessed in earlier times (Atkinson 1999a; Teisman and Klijn 2002).

The Planning and Compulsory Purchase Act 2004 formally introduced the concept of 'spatial planning', although a key document is the Royal Town Planning Institute's *New Vision*, produced in 2001, which advocates spatial planning as part of its future objectives for planning in the UK (RTPI 2001). Part of the stated logic of the move from 'planning' to 'spatial planning' is that of providing a more proactive coordinating role designed to bring together the increasingly diverse and fragmented agents of the state (Shaw and Lord 2009). However, pinning down an agreement on the nature – practical and philosophical – of spatial planning can seem a little difficult. As with all developments and changes in the profession, the move prompted some fierce debate. For some it was seen as a 'paradigm shift' (Morphet 2009: 393), but for others it was 'slippery, (Allmendinger and Haughton (2009b: 2547). In both these texts, it seems to be defined as much by what it isn't – planning as it was – as by what it is: for one it is the Promised Land, while for the other it is a mirage. However, if we move the focus from conflict to substance, maybe we can see

the introduction of the word 'spatial' as being a reminder to planners that at the heart of their discipline and profession lies the understanding of space and place and the importance of spatial relations. More concretely, spatial planning in an English context is aspiring to address some of the issues outlined above. Planning Policy Statement (PPS)1 *Delivering Sustainable Development*, published in 2005, identified cross-sectoral working, cross-boundary working and the integration of national, regional and local policy as among the attributes of successful spatial planning. Morphet (2009: 393) sees spatial planning as part of a wider process of local governance and as having as its role to 'deliver infrastructure within a local governance wide framework which comprises of a vision, objectives and shorter term delivery plans'. Ten years after the 2004 Act, spatial planning as a concept seems to have a somewhat lower profile but its formal introduction may have nudged planning as a profession towards some form of cultural change.

So, the nature of planning has evolved over time, but what is its mission now? The United Nations text quoted at the beginning of this section attempts to set out a definition of planning which it sees as being in tune with that identified by a network of twenty-five professional planning institutes from around the world, and this is reproduced in Box 1.1.

This picks up a number of items already discussed – the importance of strategy; the value of collective action, particularly in the context of a fragmenting state; although the word 'sustainability' is not used, it identifies that planning has to have a measured concern for the future. It also introduces the term 'ethical judgement', reminding us that planning should be aware of the range of values in play around any issue and have a concern for equality and social justice.

Distinctive features of the British planning system

Much of the above discussion could be applicable to a range of locations across the globe but, since the nature of a planning system is so much a product of culture and the different legal, political and administrative

BOX 1.1 A DEFINITION OF PLANNING

Definitions of planning have changed over time and are not the same in all parts of the world. Earlier views defined urban planning as physical design, enforced through land use control and centred in the state. Current perspectives recognise the institutional shift from government to governance (although in some parts of the world planning is still centred in the state), the necessarily wider scope of planning beyond land use, and the need to consider how plans are implemented.

Urban planning is therefore currently viewed as a self-conscious collective (societal) effort to imagine or reimagine a town, city, urban region or wider territory and to translate the result into priorities for area investment, conservation measures, new and upgraded areas of settlement, strategic infrastructure investments and principles of land use regulation. It is recognised that planning is not only undertaken by professional urban and regional planners (other professions and groupings are also involved); hence, it is appropriate to refer to the 'planning system' rather than just to the tasks undertaken by planners. Nonetheless, urban (and regional) planning has distinctive concerns that separate it from, for example, economic planning or health planning. At the core of urban planning is a concern with space (i.e. with 'the where of things', whether static or in movement; the protection of special 'places' and sites; the interrelations between different activities and networks in an area; and significant intersections and nodes that are physically co-located within an area).

Planning is also now viewed as a strategic rather than a comprehensive, activity. This implies selectivity, a focus on that which really makes a difference to the fortunes of an area over time. Planning also highlights a developmental movement from the past to the future. It implies that it is possible to decide between appropriate actions now in terms of their potential impact in shaping future socio-spatial relations. This future imagination is not merely a matter of short-term political expediency, but is expected to be able to project a transgenerational temporal scale, especially in relation to infrastructure investment, environmental management and quality of life.

The term 'planning' also implies a mode of governance (a form of politics) driven by the articulation of policies through some kind of deliberative process and the judgement of collective action in relation to these policies. Planning is not, therefore, a neutral technical exercise: it is shaped by values that must be made explicit, and planning itself is fundamentally concerned with making ethical judgements.

Source: UNHGR 2009: 19, for where it was adapted from Healey 2004a

approaches that this spawns, systems differ between countries. So, a quest to understand the British system[4] can be helped by comparing it with others, as it enables us to identify its distinctive features. However, it is also important to recognise that descriptions and analyses of systems will only take us so far in understanding what shapes planning outcomes, planning as experienced by citizens. As Lalenis (2010: 50) has stated, 'real planning, as opposed to that described in national planning legislation and documents, presents a wide range of variations, due to the co-existence of methods of action, more informal than formal, which are particular to each country'. Similarly, in considering a comparison between French and US planning, Cullingworth (1994: 165) observes: 'the formal system exists largely in law books, and the informal system makes it workable'.

In comparing planning systems, three features are of particular interest: first, the extent to which a planning system operates within a framework of constitutionally protected rights; second, the degree to which a system embodies discretion; and third, the importance of history and culture.

In many countries, the constitution limits governmental action in relation to land and property. The US Bill of Rights provides that 'no person shall . . . be deprived of life, liberty or property without due process of law; nor shall private property be taken without just compensation'. These words mean much more than is apparent to the casual (non-American) reader. Since land use regulations affect property rights, they are subject to constitutional challenge. They can be disputed not only on the basis of their effect on a particular property owner, but also in principle: a regulation can be challenged on the argument that, in itself, it violates the constitution. Moreover, the constitution protects against arbitrary government actions, and this further limits what can be done in the name of land use planning. No such restraints exist in the UK system.

Constitutions also often allocate powers to different tiers of government, which effectively ensures a minimum degree of autonomy for regional and local governments. Again, there is no such constitutional safeguard in the UK. As a result, the Thatcher government was able to abolish a whole tier of metropolitan local government in England and, in consequence, that part of the planning system that went with it. Similarly, when the Coalition government came to power in 2010, regional structures were abolished and a regional tier of planning disappeared. Such action would be inconceivable in most countries. In the United States, for example, there is little to compare with the central power which is exercised by the national government in Britain. Plan making and implementation are essentially local issues, even though the federal government has become active in highways, water and environmental matters and, in recent years, a number of states have become involved in land use planning. So local is the responsibility that even the decision on whether to operate land use controls is a local one, and many US local governments have only minimal systems so that, in contrast to the UK, it could not be said that there is a national planning system (Cullingworth 1994: 162). Similarly,

in much of Europe, regional and local government would not tolerate the extent of central government supervision (they might say interference) in local planning matters. But there is a point where decisions have to be made at a higher level because opposition from local decision-makers might mean that some nationally or internationally important developments never happen. Such a debate will be familiar to many readers through the controversy around what have come to be called nationally important infrastructure projects such as airport extensions, nuclear power installations and, most recently, the expansion of the rail network, High Speed 2.

Lack of constitutional constraint allows for a wide degree of discretion in the UK planning system. Describing the British planning system, Reade (1987: 11) noted that 'It rests on a high level of administrative discretion, where each piece of development requires permission: other counties are characterised by a greater closeness to a "rule of law" system.' In determining applications for planning permission, a local authority is mainly guided by the development plan, but other 'material considerations' can be taken into account. In most of the rest of the world, plans become legally binding documents. Indeed, they are part of the law and the act of giving a permit is no more than a certification that a proposal is in accordance with the plan. Plans in many other countries are different in character from those in the UK. The basis of regulation and planning in the US and many other jurisdictions is a system of zoning, based on the police powers of state and local governments. As Cullingworth and Caves state (2009: 63), 'much if not most of the land use planning in the United States is not planning but zoning and subdivision control'. Perceived advantages include relative effectiveness, ease of implementation, long-established legal precedent, and familiarity, but the dominant approach to zoning[5] has received criticism for its lack of the very flexibility which is seen as inherent to the British system.

This characteristic British discretion is further enlarged by the fact that the preparation of a local plan is carried out by the same local authority that implements it. This is so much a part of the tradition of British planning that no one comments on it. The

American situation is different, with great emphasis being placed on the separation of powers. (Typically the plan is prepared by the legislative body – the local authority – but administered by a separate board.) The British system has the advantage of relating policy and administration (and easily accommodating policy changes) but, to American eyes, 'this institutional framework blurs the distinction between policy making and policy applying, and so enlarges the role of the administrator who has to decide a specific case' (Mandelker 1962: 4). The Human Rights Convention also focuses attention on the separation of powers, since it provides for the right to appeal to an independent body against actions of government. While there is a limited right of appeal to the courts in the UK (which are independent) over planning procedures rather than substantive planning issues, most appeals are heard by the government or its representatives, in the form of the Planning Inspectorate.

Above all, in comparing planning systems, there are fundamental differences in the philosophy that underpins them. Thus, put simply (and therefore rather exaggeratedly), American planning is largely a matter of anticipating trends, while in the UK there is a conscious effort to bend them in publicly desirable directions. In France, *aménagement du territoire* deals with the planning of the activities of different government sectors to meet common social and economic goals, while in the UK town and country planning, even in the era of spatial planning, is about the management of land use, albeit taking into account social and economic concerns and the intentions of other agencies.

Planning systems are rooted in the particular historical, legal and physical conditions of individual countries and regions. In the UK, some of the many important factors which have shaped the system are the strong and long-established land preservation ethic and, in common with much of the rest of Europe, a growing conservationist ethic. In comparison, land in the United States has historically been a replaceable commodity that could and should be parcelled out for individual control and development. However, the history of early industrialisation in the UK, coupled with its small and densely developed nature, perhaps helps to explain these apparent differences.

However, a consideration of differences – with the principle one being the wide adoption of a system based on zoning compared with the British tradition of 'treating each case on its merits' – should not obscure a number of shared features across continents. These are to be found, particularly, in the realm of policy concerns. Although it may not always be expressed in the same language, the sometimes competing imperatives of economic competitiveness and sustainability are to be found as emerging agendas in most localities, whilst planning documents produced in many countries will espouse something which might be identified as some form of 'new urbanism'. Most countries will also have planning objectives which reflect a concern for the containment, management and regeneration of their urban communities and for the future of their rural communities, though, in the case of rural areas, the emphasis will vary depending on the degree of (e.g.) sparsity of population – for example, the concern with dying rural communities is much more prevalent in Australia than it is in Britain. This reference to socio-economic and geographical context is important, as policy should be a response to the nature of planning issues and be formed from an understanding of how communities 'work': unless such factors are consistent between countries then it is to be expected that, almost irrespective of the nature of the written planning system, the responses and outcomes will differ (van Leeuwen, 2010: 163–4).

Purpose and performance of planning in Britain

In legislation, for many years the stated purpose of planning in Britain was to 'regulate the development and use of land in the public interest'. From 2004, this was changed to 'contribute to the achievement of sustainable development'[6]. In 2012, the National Planning Policy Framework (NPPF) made a further change by introducing the notion that planning should be exercising a 'presumption in favour of sustainable development'. Like all policy statements, these have a very wide meaning, and one which is rather hard to pin down. This can lead to concern over just what the impact of adopting a particular guiding purpose for planning might be on the nature of development that takes place. Just what is the 'public interest' that guided planning? It assumed a consensus which maybe existed in the aftermath of the Second World War but which is far harder to pin down now. It would now be generally agreed (Taylor 1998: 34) that there is not a unitary public interest but rather multiple interests which may be in conflict over what planning should be trying to achieve and where priorities should be placed. The lack of concrete meaning for this term can be illustrated by the fact that government was able to pursue radically different emphases to policy in the period up to its being supplanted by the achievement of sustainable development as the purpose of planning.

Whilst sustainable development is a widely used term, that does not mean that there is a shared understanding of what it means for planning practice. The consequence is that 'different people interpret sustainable development in different ways' (Haughton and Counsell 2004: 214), so whilst many people would sign up to sustainable development as a guiding principle for planning, their commitment may be challenged when faced with its application to a particular development proposal affecting them and where they live. The elusive nature of some of the principles underpinning sustainable development – environmental capacity, environmental capital, economic benefits and distribution of environmental or social costs – means that they have to be translated into more concrete terms when they are related to particular localities and to particular issues with a local expression. The dilemma that planning has to face has been summed up by Susan Owens:

> Because land-use is so closely bound up with environmental change, land-use planning demands the translation of abstract principles of sustainability into operational policies and decisions. Paradoxically, this process is likely to expose the very conflicts that 'sustainable development' was meant to reconcile . . . The planning system is likely to remain a focus of attention because it is

frequently the forum in which these conflicts are first exposed

> (Owens 1995: 8)

The recent changes introduced by the NPPF included the introduction of a 'presumption in favour of sustainable development'. During the consultation phase starting in 2011, there were 11,000 responses, many of which expressed unease about just what this phrase might mean. Simon Jenkins, Chairman of the National Trust, felt that the content of the NPPF indicated that its proponents were 'in thrall to a few right wing nutters', perhaps reflecting a concern that this heralded a return to more laissez-faire approaches adopted in the 1980s: the guidance was more in favour of development than sustainability. This fear was reflected by the House of Commons Environmental Audit Committee in March 2011, when it highlighted that a lack of a statutory description for sustainable development in the guidance could be seen as running the risk that

> the principles of sustainable development – living with environmental limits, ensuring a strong, healthy and just society, achieving a sustainable economy, promoting good governance and using sound science responsibly – are unlikely to be adequately represented in the planning process.
> (Environmental Audit Committee 2011: 5)

However, it could be argued that each of these three attempts is more focused on the approach to be adopted than why we need planning in the first place, which might be thought to be closer to defining a purpose for planning. Reade (1987) felt that planning had largely avoided addressing this question because of 'premature legitimation' – planning achieved the status of a government activity before it had been properly established what it was supposed to do and why. It is relatively easy to track why this happened – regulation of development in the form of planning was introduced in response to the environmental and health problems produced by the absence of regulation – but if, as Rydin (2011: 12) puts it, 'planning is . . . a means by which society decides collectively what urban change should be like and tries to achieve that vision by a mix of means', there is still much room for debate about the nature and purpose of planning. Lack of agreement on a purpose is perhaps a significant reason why planning has been faced with so many challenges in Britain in recent years.

In his introduction to the NPPF, Greg Clark observed that 'Planning must not simply be about scrutiny. Planning must be a creative exercise in finding ways to enhance and improve the places in which we live our lives', perhaps seeking to reconnect planning to its more visionary role of former times. Without embracing a laissez-faire doctrine, it is possible to recognise a distinction between regulatory and enabling strands in planning. Janin Rivolin (2008: 182) distinguishes between what he terms 'conforming' and 'performing' roles for planning, concluding that 'in one case, implementation is intended as the capacity to "conform" development projects to a spatial strategy; in the other, implementation consists of promoting projects able to "perform" the strategy'. This is perhaps particularly relevant in the context of the greater attention being paid to the quality of outcomes achieved through the implementation of planning policies and strategies and highlighting potential differences between the nature of 'policy on the page' and the experience of 'development on the ground'. Should planning focus on moving towards an overall objective (however that may be understood) or should it be more concerned with tying new development to a set of 'rules'? Clearly, as it is an activity which has a legal basis, rules have to be followed if decisions are to be robust and defensible, but in doing so it is important not to lose sight of what planning/a plan is trying to achieve.

As has been pointed out above, one driver for change in planning has been institutional reviews of its purpose and performance. Perhaps the first of these was *The Future of Development Plans*, produced by the Planning Advisory Group (PAG) in 1965, and the most recent was the Conservative Party's 2010 Green Paper *Open Source Planning*. In his foreword to the first, Richard Crossman, the minister responsible for planning, noted that 'Planning is criticized on two main grounds: the delays it incurs and the quality of

its results.' This concern about delay/lack of speed in the planning system has been reflected in a number of other reports,[7] including *Open Source Planning*, which was concerned to get rid of 'Whole layers of bureaucracy, delay and centralised micro-management' (p. 2). It was also a notable element in the Labour government's Green Paper *Planning: Delivering a Fundamental Change* (DTLR 2001), which provided foundations for the 2004 Planning and Compensation Act and which sought a system which would come to 'robust decisions in sensible time frames' (para. 1.8). Whilst many planners would accept that unjustifiable delays can occur, others would question how far it is possible to achieve greater speed – in dealing with planning applications or producing a local plan – and at the same time ensure that better-quality decisions are made that better involve the public affected by them. However, each of the three reviews referred to above introduced significant changes to the structure of the planning system – the 1965 report was the precursor to a two-tier planning system of structure and local plans, the 2001 report led to the system of local development frameworks and an established role for regional planning, whilst the 2010 report removed the regional level and gave priority to planning at a local (neighbourhood) level.

Whilst performance as a concept clearly has a meaning in terms of just how quickly a plan or a planning decision is produced, it came to take on a wider meaning, that of how planning contributed to or inhibited national economic performance. Although it might not have been the first time that the issue was raised by Mrs Thatcher's administration, the White Paper *Lifting the Burden* (HM Government 1985) gave formal recognition to the assertion that planning could be damaging to national economic prospects and job creation, a precursor to the weakening of planning controls. The 2001 Green Paper noted that a 'successful planning system will promote economic prosperity' (para. 1.4) and this was followed by Kate Barker's two reports (2004; 2006) on the impact of planning on housing and the economy more generally. This heralded an emerging role for HM Treasury in shaping planning, with both reports being jointly sponsored by that department, as a part of Gordon Brown's approach to promoting national competitiveness. *Open Source Planning* continued the argument, noting that 'Without a transformed planning system, our chances of getting the investment and growth we need will be hampered and possibly crippled' and George Osborne has continued the precedent set by Gordon Brown in seeking a role in shaping planning and even announced in the 2014 Budget that new garden cities would be built.

In over thirty years there have been numerous initiatives attempting to move planning towards a position which is seen by their proponents to be more favourable towards economic growth and more market-friendly, but current political rhetoric suggests that more action is still needed. Is this because the British system of planning is hard to change, because the initiatives have been poorly founded, or is there some other reason? There are inevitable tensions between the objectives of planning and the market – planning looks long term and seeks to achieve results, some of which are hard (or impossible) to translate into monetary terms, whilst business tends to look short term and is focused on making a financial return. It could be argued that the resulting planning culture does not blend easily with a business culture, a position that is reinforced by lack of understanding on both sides. As the debate over the passage of the 1947 Planning Bill demonstrated, striking a balance between these interests by deciding how far business profits should 'pay for' wider social and environmental benefits is a contentious issue and may be one which will never be resolved to everyone's satisfaction, particularly in times when national economic growth and prosperity are seen as important overriding objectives. It does seem to be the case that it can be hard to radically alter the nature of British planning, partly because of the persistent nature of this tension, but also because the system has 'enough discretion and autonomy to allow local re-interpretation and resistance' (Allmendinger and Haughton 2013: 24) to change: such resistance can be nurtured by local public opposition to the idea of development. It also has to be said that many of the 'reforms' to planning have not been shaped by systematic research into the nature of the perceived 'problems';[8] rather they have been shaped to an appreciable extent by doctrine, but perhaps a significant weakness is that they have not been based on a clear and agreed articulation of just what it is planning should be trying to achieve. This suggests that debates and political initiatives will continue over the relationship between planning and the market. However, planning has to recognise that it has the power to guide and prevent, not initiate, development, which is initiated by market mechanisms, and that plans and policies which do not take cognisance of market mechanisms are unlikely to be put into practice. Therefore, planning policies and decisions *to some extent* have to reflect market preferences. The continuing debate is over where the balance should be struck between market objectives and broader planning concerns.

A further area where planning has been charged with underperformance is the engagement of the public. In spite of the fact that the importance of public participation was highlighted more than thirty years before by Seebohm (1968) and Skeffington (1969), the 2001 Green Paper felt able to state that the system 'often fails to engage communities. The result of all this is that the community feels disempowered' (para. 2.5). Subsequent response in legislation was primarily focused on structural change as the way to help address this problem. *Open Source Planning*, however, felt that these attempts had not worked and opined that 'To establish a successful democracy, we need participation and social engagement. But our present planning system is almost wholly negative and adversarial' (p. 1). Its approach to addressing the problem encompassed a 'localist' approach combined with incentives – the localist approach involving moving some decisions on planning policy closer to neighbourhoods and the incentive approach allowing communities to directly benefit from development as a 'real incentive for local people to welcome new homes and new businesses' (p. 2). However, whether such changes can better match the inclination and capacity of communities to become involved in the sometimes protracted and legalistic processes of planning is by no means certain: past experiences have raised elements of doubt. So, whilst the many adaptations of participation process which have followed the forty-five years since Skeffington (1969) have not proved to be a widely acclaimed success, this may only indicate that there are limits to what can be achieved. Should public participation levels be radically increased, it is quite possible that this would threaten two other areas where an improvement in performance was felt to be needed, delivering a faster planning system and one which was more 'market friendly'.

Challenges for planning in Britain

Whilst planning in Britain has experienced many changes since the passing of the 1947 Act – changes in the structure of organisations responsible for planning, changes in the form of plans, changes in the process of making plans, changes in policy priorities, an increase in the professionalisation of planning as an activity, amongst others – it still would espouse the mission of 'making better places'. It is tempting to think that many of the changes discussed above were stronger on rhetoric than impact, but over and above these challenges to its form and structure, planning as an activity continues to face a number of substantive challenges.

Because of the fact that planning, in deciding the use of land, has to mediate between many competing uses, possibly conflicting policy priorities and different interest groups – politicians serving various levels of government, the development industry, landowners 'the public' – it will always be operating in an area of conflicting opinions on what decisions it should take. Given the mix of public and private interests in play, there are many occasions when there are disagreements about the central concerns of the planning system – to determine what kind of development is appropriate, how much is desirable, where it should best be located and what it looks like.

Planning also has to operate in an environment characterised by uncertainty. Planning is, of necessity, about the future and the future is hard to predict, no matter how sophisticated the methods that are used in making forecasts. Whilst it is difficult to predict matters of direct concern to planners such as future numbers of jobs, numbers of people and households, and changes to travel patterns, the key controlling factor over the level and pace of development – the health of the economy and the state of the development industry's confidence – is, as experience over the last 100 years has shown, impossible to predict over the life of a plan. The fact that planning relies on the development industry for the implementation of many of its policies gives this a significant level of importance, but it is also important to consider the impact that planning has on the development industry. Several factors in the British planning system, including its inherent flexibility and the frequency of changes to the system, contribute to uncertainty for the development industry, which can act to discourage or slow a decision to invest in the development that may be essential to fulfil the ambitions of planning policy.

Complexity is also a two-pronged challenge for planning, in that it acts to increase the difficulty of the task of planning as well as making it more difficult for the public engagement now seen as essential for successful planning to happen. The growing size of this book over its fifteen editions is emblematic of the wider agenda that has developed for planning and has been reflected in the increasing volume of government guidance for planners. Social, economic and environmental matters have all grown in prominence; conservation, which merited a few pages in the first edition, now occupies a whole chapter; the range of professions and skills represented in a planning office compared with the years immediately following the 1947 Act – when admittedly there were few qualified planners – is greatly expanded, reflecting the widening range of knowledge necessary for making planning decisions. The growing range of skills is in part down to a growing sophistication in the methods and the range of data applied to making planning decisions, which itself may reflect that these decisions are in fact more complex than approaches taken in earlier times would suggest. There has also been a growing range of plans produced – with the most recent example being the neighbourhood-level plans introduced by the Localism Act 2011 – and a growing number of planning applications to be decided. As has been noted, this growing complexity makes it harder for members of the public to readily engage with planning, but the situation is made more difficult by the fragmentation of government: gone are the days when most things in a local area were the responsibility of 'the council'. At times it has been recognised by central government that local planning authorities need increased resources[9] to deal with this challenging agenda and at the same time achieve the level of performance sought, but maybe not consistently enough to satisfy all involved in delivering planning services.

The nature of these changes and challenges is reviewed in more detail in the following chapters. In particular, the longer-term evolution of planning is dealt with in Chapter 2 (but each chapter has a historical thread for its theme); the nature and the framework for developing strategies and policies in Chapters 4 and 6; decision-making and implementation in Chapters 5 and 6; reviews and questions of the performance of planning in Chapters 4 and 5; the relationship between planning and the market in Chapter 6; a focus on developing a 'quality of place' in Chapter 9; and sustainability, whilst featuring at many points, is particularly considered in Chapter 7.

A word about planning theory

Whilst this is not a book about, or rich in, theory, some of the issues discussed above in trying to pin down the nature of town planning illustrate that there are many debates on principle taking place in planning, so the value on giving some time to considering theory alongside the material in this book may be that it can 'provide some overall or *general* understanding of the nature of town planning' (Taylor 1998: v–vi). It can possibly help us to: understand the environment in which planners work; better understand the methods to be used in planning; and maybe help answer the question 'does it work?' along the lines implied by Wildavsky at the opening of this chapter. It may also help to justify the existence of the activity of 'town planning', particularly important given the number of challenges it seems to have faced in the last fifty years. However, typically practitioners are seen as ignoring

theory, on the grounds that it has no bearing or meaning for practice, but this does not sit easily with the RTPI's exhortation (2013: 5) that every member should be 'a reflective and analytical professional'.

A reading of a selection of the books suggested below will reveal that there is not a single 'planning theory'. Rather there are several bodies of theory which are relevant to planning. The view of their relative value and importance has changed over time but the fact that one group of approaches may now be receiving most attention does not mean that older theories do not have anything of value for us. There are several ways of categorising what we might call 'planning theories' and for many years the approach was to consider a simple bifurcation between what are called procedural and substantive theories. However, many would now see this as an outdated approach (for example, see Allmendinger 2009: 30–48) which neither accurately described nor included the full range of relevant theories. Procedural theories are concerned with how the activity of planning is carried out and can be particularly informative in considering the approaches to planning given in government guidance and what the implications might be for objectives of planning, such as public involvement and equality, of changes in the approach to be followed. Theories concerned with urban form – essentially physical planning – can help in considering historical aspects of planning but also new proposals such the reintroduction of the concept of garden cities. Theories which are concerned with the social, political and economic context of planning – a key element of the 'substantive' theories noted above – would help in considering matters such as the reach of planning and the nature of 'the public interest'.

An approach to planning common in the 1980s, but still of relevance now, which focused on 'getting things done' and the value of 'common sense' is associated with ideas of pragmatism. Some consideration of this body of theory can help us form a view on whether a focus on outcomes will lead to other important issues being ignored. Pragmatists often focus on the importance of reasoning. Judging a plan then becomes a process of accepting the reasoning behind the decisions and policies made and accepting that the knowledge useful in making a plan can come from a variety of sources. Such knowledge is often held within people – so called tacit knowledge – and the knowledge hosts might not even know they have it. Thus, for the planner to make a plan for a place is to somehow access this knowledge – through interactions, for example – and combine it with more 'codified' knowledge such as population projections, etc. A crude interpretation of this appeared strongly in UK planning in the 2000s as central government sought to accept plans and decisions on the basis that they had a 'sound evidence base'. This emphasis on reasoning and evidence, in the form of opinions, values and facts (insofar as these categories can be separated) is also strongly present in the discretionary decision-making processes of development management.

More recently, emerging bodies of theory have been concerned with questions such as how the quality and nature of the institutions responsible for planning and the network of actors and agents involved may shape the outcomes of planning, a matter which is clearly relevant to any consideration of the changing pattern of agencies and responsibilities introduced by different governments at different times. Finally, and perhaps the most elusive element of recent theory, is that of postmodernism. Whilst describing some-thing as 'elusive' might not be a great selling point, it does attempt to address a number of the questions which have been raised above, such as (amongst other things) complexity, uncertainty, power and meaning, with a focus on issues such as, in a complex constantly changing world, how planning needs some 'fixity' to plan or act but what implications may accompany such a 'fix', how might we understand the place of power in decision-making, and what shapes the behaviour of planners?

The list of further reading below tries to identify a range of texts which will help the reader to follow up on some of these ideas central to forming a reflective view of planning and changes in how we plan. Whilst this might not be an easy task – indeed, Allmendinger (2009: 47) identifies a regular question on planning theory as 'Why is planning theory writing so incomprehensible?' – there are many accessible texts and effort will be rewarded by a clearer insight into

the nature and implications of many of the matters considered in the following chapters.

Further reading

A good starting point on the nature of planning is Rydin (2011) *The Purpose of Planning*, which has been written to be as accessible as possible to people from all backgrounds. There are a number of very readable texts giving a critique of planning and Hall (1980) *Great Planning Disasters* is required reading for all planners, as well as for non-planners who want to know why planning is so difficult. A radical critique of the role of planning in society is provided in Ambrose (1986) *Whatever Happened to Planning?* but perhaps the classic critique of early approaches to urban planning is Jacobs (1961) *The Death and Life of the Great American City*, whilst a useful reflection on Jacobs' activities is provided by Flint (2009) *Wrestling with Moses*. An occasional collaborator with Jacobs and giving a similarly critical view of architecture and planning in Britain was Ian Nairn: Darley and McKie (2013) *Ian Nairn: Words in Place* gives some interesting reflections on his work.

Taylor's (1998) *Urban Planning Theory since 1945* provides an accessible overview of planning theory considered on a historical basis, while Allmendinger's (2009) *Planning Theory* offers a slightly more complex but more up-to-date review, arranged according to a typology. An alternative – and maybe dated but nonetheless interesting – typology of theory is offered by Yiftachel (1989) in 'Towards a new typology of urban planning theories'. This is reproduced in Hillier and Healey's (2008) *Critical Essays in Planning Theory*, which provides a large collection, spread over three volumes, of many of the significant and still relevant articles on various aspects of theory. A useful collection of early articles is contained in Faludi (1973) *A Reader in Planning Theory*, which includes a number of important articles not included in the more recent volume. Allmendinger and Tewdwr-Jones (2002) *Planning Futures: New Directions for Planning Theory* provides a selection of articles dealing with more recent developments in planning theory. Another older text still worth reading is Reade (1987) *British Town and*

Country Planning containing a number of critiques and arguments which the author sees as offering 'the only real hope of enabling a justification to be constructed for the planners' own existence' (p. xiii).

On politics and planning see Albrechts (2003) 'Reconstructing decision making'; for recent relations between political ideologies and planning see Tewdwr-Jones (2002) *The Planning Polity* and Allmendinger and Tewdwr-Jones (2000) 'New Labour, new planning?'. The challenge that Thatcherism and the market brought to ideas of planning has been addressed in many studies – notably Thornley (1993) *Urban Planning under Thatcherism: The Challenge of the Market*, Allmendinger and Thomas (1998) *Urban Planning and the British New Right* and Brindley *et al.* (1996) *Remaking Planning*. For a critical review of more recent trends in planning, see Allmendinger's (2011) *New Labour and planning: From New Right to New Left*, which provides interesting insights into the period from 1997 to 2010. A very good overview of the changes in the development plan system induced by legislation between 1947 and the end of the century is given by Poxon (2000) 'Solving the development plan puzzle in Britain: learning lessons from history'.

There is not a huge number of texts offering a comparison of planning systems but Cullingworth (1994) 'Alternate planning systems: is there anything to learn from abroad?' still provides interesting insights. A more recent and useful text is Lalenis (2010) 'A theoretical analysis on planning policies'. The rest of the study of which it is a part is also informative. National Housing and Planning Advice Unit (2009) *Review of European Planning Systems* is one of a small number of studies which provide a comparative study across Europe. For a comparative study of 'certainty and discretion' in planning, see Booth (1996) *Controlling Development*. Discretion is discussed at length (with a comparison of the UK and the USA) in Cullingworth (1993) *The Political Culture of Planning*. A broader discussion of the two countries is given by Vogel (1986) *National Styles of Regulation*.

The texts by Allmendinger, Healey and Hillier, and Taylor will provide insights into most aspects of planning theory but these can be supplemented by some focusing on particular themes. For physical planning many of the texts recommended in the following chapter on planning history will be useful, but Batty and Marshall (2009) 'The evolution of cities: Geddes, Abercrombie and the new physicalism' provides some interesting reflections on this strand of planning from a contemporary perspective. Reviews of initiatives such the New Towns programme can also be informative. A good example here is Ward (1993) *New Town, Home Town*. For procedural theory, a couple of texts from the heyday of such theory are Faludi (1973) *A Reader in Planning Theory* and McLoughlin (1969) *Urban and Regional Planning: A Systems Approach*. These can be profitably compared with recent guidance from government on plan preparation. For discussion of the relationship between planning and the market see Pennington (2002) *Liberating the Land* and Poulton (1991) 'The case for a positive theory of planning'. Allmendinger and Houghton (2013) 'The evolution and trajectories of English spatial governance: "neoliberal" episodes in planning' provides a systematic review of how planning was affected by neoliberal thinking between 1979 and 2010.

Communication, collaboration and networks dominated discussions about planning theory during the 1990s. Early contributions are Forester (1982) 'Planning in the face of power' and other papers brought together in Forester's book (1989) *Planning in the Face of Power*. Later contributions include Innes (1995) 'Planning theory's emerging paradigm: communicative action and interactive practice' and Healey (1997) *Collaborative Planning*, (1998) 'Collaborative planning in a stakeholder society' and (1992) 'Planning through debate: the communicative turn in planning theory'. Consequently, aspects of planning practice have been investigated using these ideas, for example in Healey (1993) 'The communicative work of development plans' and Tait and Campbell (2000) 'The politics of communication between planning officers and politicians: the exercise of power through discourse'. Campbell's work in general is good in discussing the public interest in planning and how the idea remains significant. For a critique of collaborative planning, see Tewdwr-Jones and Allmendinger (1998) 'Deconstructing communicative rationality: a critique of Habermasian collaborative planning'.

With regard to work on spatial planning, Vigar *et al.* (2000) *Planning, Governance and Spatial Strategy in Britain* argued the case for greater systemic attention to the visionary, rather than regulatory, side of planning, an idea that came to pass with mixed results in the 2000s. Much was subsequently written on spatial planning. From a large selection two interesting examples are Healey (2007) *Urban Complexity and Spatial Strategy* and Haughton *et al.* (2010) *The New Spatial Planning.*

Those who would like to gain an insight into postmodernism might try Miller (2002) *Postmodern Public Policy,* Gunder and Hillier (2009) *Planning in Ten Words or Less* and Allmendinger and Gunder (2005) 'Applying Lacanian insight and a dash of Derridian deconstruction to planning's "dark side"', which is not as forbidding as its title suggests. If you would like to get a broader view of postmodernism, Adamson and Pavitt (2011) *Postmodernism: Style and Subversion 1970–1990* provides context as well as some interesting contributions on architecture and design.

Notes

1   Healey (2007) uses the term 'inspirational vision' (p. 180), implying the engagement of wider groups, but the nature of stakeholder involvement can be particularly hard to pin down.
2   The thinking of Etzioni can be seen clearly reflected in the guidance given for development planning in the period, particularly in the idea that plans should focus on 'key issues'. Etzioni came from a sociology background to focus on processes of decision-making but he is recently better known for his espousal of 'communitarianism', thought by some to be an element in refashioning the British Labour Party into 'New Labour'. Other strands of thought in the 1960s and 1970s came from other disciplines, such as operational research, where texts such as Friend and Jessop (1969) and Friend and Hickling (1987) played a role in the development of ideas about how we plan.
3   A fuller and up-to-date discussion of this can be found in Davoudi (2012a). For some reflections on

the relationship between the work of Descartes and rational approaches to town planning, see Akkerman (2001).
4   Some would question whether there is there a UK 'national system' at all these days. A significant trend of the 2000s was the devolution of power to Northern Ireland, Scotland and Wales, such that a degree of policy divergence could be observed. That said, they all operate within an administrative tradition that is common and so can be considered as a whole, especially when compared with systems from very different traditions.
5   There are various approaches to zoning but the most prevalent in the USA is Euclidean Zoning, named after an approach to zoning and a subsequent legal case concerning the settlement of Euclid, Ohio. Euclidean zoning is characterised by the segregation of land uses into specified geographic districts and dimensional standards stipulating limitations on development activity within each type of district. The criticism it has received has led to the development of new systems of zoning. Such systems might focus on outcomes – for example, the provision of affordable housing – within a prescriptive system of zoning, allowing some discretion to be exercised depending on the range of desirable outcomes that a proposal might deliver. In other cases, 'design codes' might form a key component of planning control.
6   This 'purpose' was given by the Planning and Compulsory Purchase Act 2004, section 39.
7   If we were to include all changes to the planning system – changed guidance on aspects of plan making and development management, as well as review documents such as the PAG report – then we would have a list of considerable length. However, following PAG in 1965, we had the White Paper *Lifting the Burden,* which paralleled the Local Government Act 1985 introducing changes to the organisation of local government and the introduction of unitary development plans (UDPs). Then 1989 saw the consultation paper *The Future of Development Plans,* which seemingly addressed many of the same issues which exercised PAG, but subsequent legislation in 1990 and 1991 saw the shift to the 'plan-led' system, placing a higher value on policies in the development plan which some

might argue slowed the task of plan making. The 1998 consultation paper *Modernising Planning* almost in spite of its title again echoed the PAG concern with speeding up planning processes and was followed in 2001 by the Green Paper *Planning: Delivering a Fundamental Change*, which also included speed amongst its concerns. Whilst the language and emphases have changed, *Open Source Planning* continues the concern with the apparent slowness of planning processes, suggesting that planning is either fundamentally flawed or burdened with unrealistic expectations.

8   Whilst Kate Barker's published conclusions and recommendations were accompanied by significant volumes of research evidence, there was some feeling that the focus of the research was overly 'economistic'.

9   This issue was recognised in the 2000s by the introduction of a planning delivery grant which was used in the retention and recruitment of staff and increasing the use of ICT: resources were also made available to assist with the professional training of planners which, for a time, significantly increased numbers on university postgraduate courses.

2   The evolution of town and country planning

The first assumption that we have made is that national planning is intended to be a reality and a permanent feature of the internal affairs of this country.

(Uthwatt Report 1942)

Introduction

This chapter gives a brief introduction to the development of the planning system over the course of the last century or so. Its coverage is broadly chronological and concentrates on the development of the system, although giving some context on the major planning issues of the day. As the chapter comes closer to our present time, the discussion becomes briefer, as more detailed discussion of particular topics is picked up in the relevant thematic chapters.

The public health origins

Town and country planning as a task of government has developed from public health and housing policies. The nineteenth-century increase in population and, even more significant, the growth of towns led to public health problems which demanded a new role for government. Together with the growth of medical knowledge, the realisation that overcrowded insanitary urban areas resulted in an economic cost (which had to be borne at least in part by the local ratepayers) and the fear of social unrest, this new urban growth eventually resulted in an appreciation of the necessity for interfering with market forces and private property rights in the interest of social well-being.

Nineteenth-century public health legislation was directed at the creation of adequate sanitary conditions. Among the measures taken to achieve these were powers for local authorities to make and enforce building by-laws for controlling street widths and the height, structure and layout of buildings. Limited and defective though these powers proved to be, they represented a marked advance in social control and paved the way for more imaginative measures. The physical impact of by-law control on British towns is still very much in evidence, and it did not escape the attention of contemporary social reformers. In the words of Raymond Unwin:

> much good work has been done. In the ample supply of pure water, in the drainage and removal of waste matter, in the paving, lighting and cleansing of streets, and in many other such ways, probably our towns are as well served as, or even better than, those elsewhere. Moreover, by means of our much abused bye-laws, the worst excesses of overcrowding have been restrained; a certain minimum standard of air-space, light and ventilation has been secured; while in the more

modern parts of towns, a fairly high degree of sanitation, of immunity from fire, and general stability of construction have been maintained, the importance of which can hardly be exaggerated. We have, indeed, in all these matters laid a good foundation and have secured many of the necessary elements for a healthy condition of life; and yet the remarkable fact remains that there are growing up around our big towns vast districts, under these very bye-laws, which for dreariness and sheer ugliness it is difficult to match anywhere, and compared with which many of the old unhealthy slums are, from the point of view of picturesqueness and beauty, infinitely more attractive.

(Unwin 1909: 3)[1]

It was on this point that public health and architecture met. The enlightened experiments at Saltaire (1853), Bournville (1878), Port Sunlight (1887) and elsewhere had provided object lessons. Ebenezer Howard and the garden city movement were now exerting considerable influence on contemporary thought. In the commentary to the 2003 republication of Howard's book *To-morrow: A Peaceful Path to Real Reform* (originally published in 1898), it is described as 'almost without question the most important single work in the history of modern town planning'.[2] Howard's ideas about land reform and a 'socialist community' were initially influential, but it is his practical approach to the form of towns and how better towns could be created that was to have a more enduring impact, along with the creation of the first garden city at Letchworth from 1903 to the plan of Raymond Unwin and Barry Parker.

The National Housing Reform Council (later the National Housing and Town Planning Council) was campaigning for the introduction of town planning. Even more significant was a similar demand from local government and professional associations such as the Association of Municipal Corporations, the Royal Institute of British Architects, the Surveyors' Institute and the Association of Municipal and County Engineers. As Ashworth has pointed out:

the support of many of these bodies was particularly important because it showed that the demand for

town planning was arising not simply out of theoretical preoccupations but out of the everyday practical experience of local administration. The demand was coming in part from those who would be responsible for the execution of town planning if it were introduced.

(Ashworth 1954: 180)

The first Planning Act

The movement for the extension of sanitary policy into town planning combined with a wider reforming impetus gave considerable momentum in the early planning movement in the early years of the twentieth century. The garden city movement was growing and the teaching of town planning started with the foundation of the Liverpool School of Civic Design in 1909. The first legislation with 'town planning' in the title was enacted the same year. John Burns, President of the Local Government Board, in introducing the first legislation – the Housing, Town Planning, Etc. Act 1909 – spoke of its objectives as follows:

The object of the bill is to provide a domestic condition for the people in which their physical health, their morals, their character and their whole social condition can be improved by what we hope to secure in this bill. The bill aims in broad outline at, and hopes to secure, the home healthy, the house beautiful, the town pleasant, the city dignified and the suburb salubrious.

However, in reality though the legislation introduced the term 'town planning', its primary focus was upon housing. In the words of Booth and Huxley, the primary concerns were:

the sanitary and aesthetic improvement of working class dwellings – and the prevention of future slums in new peripheral suburban developments. The planning aspect of these concerns focuses in particular on extending regulatory control to the surroundings of the house.

(Booth and Huxley 2012: 268)

Thus the new powers provided by the Act were for the preparation of 'schemes' by local authorities for controlling the development of new housing areas. Though novel, these powers were effectively an extension of existing ones. It is significant that this first legislative acceptance of town planning came in an Act dealing with health and housing. The gradual development and the accumulated experience of public health and housing measures facilitated a general acceptance of the principles of town planning.

> Housing reform had gradually been conceived in terms of larger and larger units. Torrens' Act (Artisans and Labourers Dwellings Act, 1868) had made a beginning with individual houses; Cross's Act (Artisans and Labourers Dwellings Improvement Act, 1875) had introduced an element of town planning by concerning itself with the reconstruction of insanitary areas; the framing of bylaws in accordance with the Public Health Act of 1875 had accustomed local authorities to the imposition of at least a minimum of regulation on new building, and such a measure as the London Building Act of 1894 brought into the scope of public control the formation and widening of streets, the lines of buildings frontage, the extent of open space around buildings, and the height of buildings. Town planning was therefore not altogether a leap in the dark, but could be represented as a logical extension, in accordance with changing aims and conditions, of earlier legislation concerned with housing and public health.
>
> (Ashworth 1954: 181)

The 'changing conditions' were predominantly the rapid growth of suburban development: a factor which increased in importance in the following decades.

> In fifteen years 500,000 acres of land have been abstracted from the agricultural domain for houses, factories, workshops and railways . . . If we go on in the next fifteen years abstracting another half a million from the agricultural domain, and we go on rearing in green fields slums, in many respects, considering their situation, more squalid than

those which are found in Liverpool, London and Glasgow, posterity will blame us for not taking this matter in hand in a scientific spirit. Every two and a half years there is a County of London converted into urban life from rural conditions and agricultural land. It represents an enormous amount of building land which we have no right to allow to go unregulated.

> (Hansard, Parl Deb, 12 May 1908)

The emphasis was entirely on raising the standards of *new* development. The Act permitted local authorities (after obtaining the permission of the Local Government Board) to prepare town planning schemes with the general object of 'securing proper sanitary conditions, amenity and convenience', but only for land which was being developed or appeared likely to be developed.

Thus town planning in this legislation was largely concerned with the environment for new housing. It certainly did not include 'the remodelling of the existing town, the replanning of badly planned areas, the driving of new roads through old parts of a town – all these are beyond the scope of the new planning powers' (Aldridge 1915: 459). Nevertheless, the cumbersome administrative procedure devised by the Local Government Board (in order to give all interested parties 'full opportunity of considering all the proposals at all stages') might well have been intended to deter all but the most ardent of local authorities. The land taxes threatened by the Finance Act 1910, and then the First World War, added to the difficulties. It can be the occasion of no surprise that by the outbreak of the First World War only two schemes were approved under the 1909 Act, both in Birmingham, a city which had enthusiastically promoted planning. Nevertheless, the impetus for town planning was growing. The years 1910 and 1913 saw major town planning conferences and, in a reflection of the fact that planning was increasingly seen to demand a particular professional expertise, the Town Planning Institute was formed in 1914. And whilst meaningful statutory plans for town planning were some way off in the future, independent activities of civic surveying, following the inspiration of Patrick Geddes, and plan making were beginning to proceed apace, as Lucy Hewitt has described for London (2011, 2012).

Inter-war legislation

The first revision of town planning legislation which took place after the war (the Housing and Town Planning Act 1919) did little in practice to broaden the basis of town planning. The preparation of schemes was made obligatory on all borough and urban districts having a population of 20,000 or more, but the time limit (January 1926) was first extended (by the Housing Act 1923) and finally abolished (by the Town and Country Planning Act 1932). Some of the procedural difficulties were removed, but no change in concept appeared. Despite lip service to the idea of town planning, the major advances made at this time were in the field of housing rather than planning.

It was the 1919 Act which began what Bowley (1945: 15) has called 'the series of experiments in State intervention to increase the supply of working-class houses'. The 1919 Act accepted the principle of state subsidies for housing and thus began the nationwide growth of council house estates. Equally significant was the entirely new standard of working-class housing provided: the three-bedroom house with kitchen, bath and garden, built at the density recommended by the Tudor Walters Committee in 1918 of not more than twelve houses to the acre. At these new standards, development could generally take place only on virgin land on the periphery of towns, and municipal estates grew alongside the private suburbs: 'the basic social products of the twentieth century', as Briggs (1952, vol. 2: 228) has termed them.

This suburbanisation was greatly accelerated by rapid developments in road transport as well as the extension of public transport systems, such as the London Underground. The ideas of Howard and the garden city movement, of Geddes and of those who, like Warren and Davidge (1930), saw town planning not only as a technique for controlling the layout and design of residential areas, but also as part of a policy of national economic and social planning, were receiving increasing attention, but in practice

town planning typically meant little more than an extension of the old public health and housing controls, albeit at much lower densities, prompted by garden city ideas. One particular concern was urban sprawl and its impact on the countryside and rural England, already a major cultural concern for the nation's elite. The Council for the Preservation of Rural England (today, the Campaign to Protect Rural England (CPRE)) was formed in 1926 by Patrick Abercrombie amongst others and in 1928 Clough Williams-Ellis published his impassioned polemic against urban sprawl, *England and the Octopus*. Subregional plans, aimed at protecting the countryside, were supported by CPRE and undertaken by Abercrombie and others (Sheail 1981).

The next significant pieces of legislation were the Town and Country Planning Act 1932, which extended planning powers to almost any type of land, whether built-up or undeveloped, and, following the concerns of CPRE and Williams-Ellis, the Restriction of Ribbon Development Act 1935, which, as its name suggests, was designed to control the spread of development along major roads. But these and similar measures were largely ineffective. For instance, under the 1932 Act, planning schemes took about three years to prepare and pass through all their stages. Final approval had to be given by Parliament and schemes then had the force of law, as a result of which variations or amendments were not possible except by a repetition of the whole procedure. Interim development control operated during the time between the passing of a resolution to prepare a scheme and its date of operation (as approved by Parliament). This enabled, but did not require, developers to apply for planning permission. If they did not obtain planning permission, and the development was not in conformity with the scheme when approved, the planning authority could require the owner (without compensation) to remove or alter the development.

All too often, however, developers preferred to take a chance that no scheme would ever come into force, or that, if it did, no local authority would face pulling down existing buildings. The damage was therefore done before the planning authorities had a chance to intervene (Wood 1949: 45). Once a planning scheme was approved, on the other hand, the local authority ceased to have any planning control over individual developments. The scheme was in fact a zoning plan: land was zoned for particular uses such as residential or industrial, though provision could be made for such controls as limiting the number of buildings and the space around them. In fact, so long as developers did not try to introduce a non-conforming use, they were fairly safe. Furthermore, most schemes did little more than accept and ratify existing trends of development, since any attempt at a more radical solution would have involved the planning authority in compensation which they could not afford to pay. In most cases, the zones were so widely drawn as to place hardly more restriction on the developer than if there had been no scheme at all. Indeed, in the half of the country covered by draft planning schemes in 1937 there was sufficient land zoned for housing to accommodate 291 million people (Barlow Report 1940: para. 241).

A major weakness was the administrative structure itself. District and county borough councils were generally small and weak. They were unlikely to turn down proposals for development on locational grounds if compensation was involved or if they would thereby be deprived of rate income. The compensation paid either for planning restrictions or for compulsory acquisition, which had to be determined in relation to the most profitable use of the land, even if it was unlikely that the land would be so developed, and without regard to the fact that the prohibition of development on one site usually resulted in the development value (which had been purchased at high cost) shifting to another site. Consequently, in the words of the Uthwatt Committee,

> an examination of the town planning maps of some of our most important built-up areas reveals that in many cases they are little more than photographs of existing uses and existing lay-outs, which, to avoid the necessity of paying compensation, become perpetuated by incorporation in a statutory scheme irrespective of their suitability or desirability.

These problems increased as the housing boom of the 1930s developed; 2,700,000 houses were built in England and Wales between 1930 and 1940. At the outbreak of the Second World War, one-third of all the houses in England and Wales had been built since 1918. The implications for urbanisation were obvious, particularly in the London area. Between 1919 and 1939 the population of Greater London rose by about two million, of which three-quarters of a million was natural increase, and over one and a quarter million was migration (Abercrombie 1945). This growth of the metropolis was a force which existing powers were incapable of halting, despite the large body of opinion favouring some degree of control.

The depressed areas

The urban sprawl generating concern in London and the south of England was the result of increasing affluence for some in the country. However, elsewhere the picture was very different: whilst there was affluence and poverty across the country, the familiar north-south divide was becoming more established and deepening. Thus the intensifying pressure for development around London was closely allied to the economic decline of substantial areas of the North and of South Wales, and both were part of the much wider problem of industrial location. In the South East, the insured employed population rose by 44 per cent between 1923 and 1934, but in the North East it fell by 5.5 per cent and in Wales by 26 per cent. In 1934, 8.6 per cent of insured workers in Greater London were unemployed, but in Workington the proportion was 36.3 per cent, in Gateshead 44.2 per cent, and in Jarrow 67.8 per cent.

For London, various advisory committees were set up and a series of reports issued on governance and planning.[3] The problems of the North and of Wales extended beyond anything that land use planning might address to a wider need for state intervention through regional policy and economic planning. The awful plight of such areas generated angry polemics, such as Sharp (1935) *A Derelict Area: A Study of the South West Durham Coalfield*. Attention was first concentrated on encouraging migration, on training, and on schemes for establishing the unemployed in smallholdings. Special Commissioners were subsequently appointed for England and Wales, and for Scotland, with powers for 'the initiation, organisation, prosecution and assistance of measures to facilitate the economic development and social improvement' of the special areas, but their main task – the attraction of new industry – proved to be extraordinarily difficult. More successful was the attraction of new light industry to new trading estates, such as the Team Valley in Gateshead.

Nevertheless, there were still 300,000 unemployed in the special areas at the end of 1938, and although 123 factories had been opened between 1937 and 1938 in the special areas, 372 had been opened in the London area. Sir Malcolm Stewart concluded, in his third annual report (1938), that 'the further expansion of industry should be controlled to secure a more evenly distributed production'. Such thinking might have been in harmony with the current increasing recognition of the need for national planning, but it called for political action of a character which would have been sensational. Furthermore, as Neville Chamberlain (then Chancellor of the Exchequer) pointed out, even if new factories were excluded from London, it did not follow that they would necessarily spring up in South Wales or West Cumberland. The immediate answer of the government was to appoint the Royal Commission on the Distribution of the Industrial Population.

The Barlow Report

The Barlow Report is of significance not merely because it is an important historical landmark, but also because some of its major recommendations were for so long accepted as a basis for planning policy.

The terms of reference of the Commission were

> to inquire into the causes which have influenced the present geographical distribution of the industrial population of Great Britain and the probable direction of any change in the distribution in the future; to consider what social, economic or strategic disadvantages arise from the concentration of industries or of the industrial population in large towns or in particular areas of the country; and to

report what remedial measures if any should be taken in the national interest.

These very wide terms of reference represented, as the Commission pointed out, 'an important step forward in contemporary thinking' and, after reviewing the evidence, it concluded that

> the disadvantages in many, if not most of the great industrial concentrations, alike on the strategical, the social and the economic side, do constitute serious handicaps and even in some respects dangers to the nation's life and development, and we are of opinion that definite action should be taken by the government towards remedying them.

The advantages of more urban concentration at that time were clear: proximity to markets, reduction of transport costs and availability of a supply of suitable labour. But these, in the Commission's view, were accompanied by serious disadvantages such as high site values, loss of time through street traffic congestion and the risk of adverse effects on efficiency due to long and fatiguing journeys to work. The Commission maintained that the development of garden cities, satellite towns and trading estates could make a useful contribution to the solution of the problems of urban congestion.

The London area presented the largest problem, not simply because of its huge size, but also because 'the trend of migration to London and the Home Counties is on so large a scale and of so serious a character that it can hardly fail to increase in the future the disadvantages already shown to exist'. The problems of London were thus in part related to the problems of the depressed areas:

> It is not in the national interest, economically, socially or strategically, that a quarter, or even a larger, proportion of the population of Great Britain should be concentrated within twenty to thirty miles or so of Central London. On the other hand, a policy:
>
> (i) of balanced distribution of industry and the industrial population so far as possible

throughout the different areas or regions in Great Britain;

(ii) of appropriate diversification of industries in those areas or regions

would tend to make the best national use of the resources of the country, and at the same time would go far to secure for each region or area, through diversification of industry, and variety of employment, some safeguard against severe and persistent depression, such as attacks an area dependent mainly on one industry when that industry is struck by bad times.

Such policies could not be carried out by the existing administrative machinery: it was no part of statutory planning to check or to encourage a local or regional growth of population. Planning was essentially on a local basis; it did not, and was not intended to, influence the geographical distribution of the population as between one locality and another. The Commission unanimously agreed that the problems were national in character and required 'a central authority' to deal with them. They argued that the activities of this authority ought to be distinct from and extend beyond those of any existing government department. It should be responsible for formulating a plan for dispersal from congested urban areas – determining in which areas dispersal was desirable, and whether and where dispersal could be effected by developing garden cities or garden suburbs, satellite towns, trading estates, or the expansion of existing small towns or regional centres. It should be given the right to inspect town planning schemes and 'to consider, where necessary, in cooperation with the government departments concerned, the modification or correlation of existing or future plans in the national interest'. It should study the location of industry throughout the country with a view to anticipating cases where depression might probably occur in the future and encouraging industrial or public development before a depression actually occurred.

Whatever form this central agency might take (a matter on which the Commission could not agree), it was essential that the government should adopt a much more positive role: control should be exercised over new factory building at least in London and the Home Counties, dispersal from the larger conurbations should be facilitated, and measures should be taken to anticipate regional economic depression.

The impact of war

The Barlow Report was published in January 1940, some four months after the start of the Second World War. The problem which precipitated the decision to set up the Barlow Commission, that of the depressed areas, rapidly disappeared. The unemployed of the depressed areas now became a powerful national asset. A considerable proportion of the new factories built to provide munitions or to replace bombed factories were located in these areas. This emergency wartime policy, paralleled in other fields, such as hospitals, not only provided some 13 million square feet of munitions factory space in the depressed areas which could be adapted for civilian industry after the end of the war, but also provided experience in dispersing industry and in controlling industrial location, which showed the practicability (under wartime conditions at least) of such policies. The Board of Trade became a central clearing house of information on industrial sites. During the debates on the 1945 Distribution of Industry Bill, its spokesman stressed:

We have collected a great deal of information regarding the relative advantage of different sites in different parts of the country, and of the facilities available there with regard to local supply, housing accommodation, transport facilities, electricity, gas, water, drainage and so on . . . We are now able to offer to industrialists a service of information regarding location which has never been available before.

Hence, although the Barlow Report 'lay inanimate in the iron lung of war',[4] it seemed that the conditions for the acceptance of its views on the control of industrial location were becoming very propitious:

there is nothing better than successful experience for demonstrating the practicability of a policy.

The war thus provided a great stimulus to the extension of regional planning into the sphere of industrial location. This was not the only stimulus it provided: the destruction wrought by bombing transformed 'the rebuilding of Britain' from a socially desirable but somewhat visionary and vague ideal into a matter of practical and clear necessity. The need for rebuilding not only provided an unprecedented opportunity for comprehensive planning of the bombed areas but a stimulus for town planning more generally. The war generated a flurry of plans for towns and cities across the country, large and small, industrial and historic, bombed and undamaged, including Abercrombie's famous 1944 (published 1945) plan for Greater London. In the centre of Coventry, one of the most famous postwar redevelopments, complete with pedestrian precincts and inner ring road, the City Architect, Donald Gibson, had been preparing plans for comprehensive development *before* the enemy bombing that destroyed much of the historic centre in November 1940, and it was this destruction that enabled realisation of his plans. In his plan for Exeter, Sharp urged that

> to rebuild the city on the old lines . . . would be a dreadful mistake. It would be an exact repetition of what happened in the rebuilding of London after the Fire – and the results, in regret at lost opportunity, will be the same. While, therefore, the arrangements for rebuilding to the new plan should proceed with all possible speed, some patience and discipline will be necessary if the new-built city is to be a city that is really renewed.
>
> (Sharp 1947: 10)

Lutyens and Abercrombie argued that in Hull,

> there is now both the opportunity and the necessity for an overhaul of the urban structure before undertaking this second refounding of the great Port on the Humber. Due consideration, however urgent the desire to get back to working conditions, must be given to every aspect of town existence.
>
> (1945: 1)

This was the social climate of the war and early postwar years. There was an enthusiasm and a determination to undertake 'social reconstruction' (i.e. public sector intervention) on a scale hitherto considered utopian, and town planning was a central part of this vision for politicians and a significant section of the public alike. This was a period when a book such as Sharp's 1940 Penguin paperback, *Town Planning*, could be a bestseller, achieving sales not enjoyed by a book on planning before or since. The catalyst was, of course, the war itself, which engendered both support for the way of life being fought for and a critical appraisal of the inadequacies of that way of life. Thus, in relation to the second point, the promise of a new and better post-war Britain was important for the morale of troops and citizens alike. The feeling was one of intense optimism and confidence. Not only would the war be won, but also it would be followed by a similar campaign against the forces of want. That there was much that was inadequate, even intolerable, in pre-war Britain had been generally accepted. What was new was the belief that the problems could be tackled in the same way as a military operation. What supreme confidence was evidenced by the setting up in 1941 of committees to consider post-war reconstruction problems: the Uthwatt Committee on Compensation and Betterment, the Scott Committee on Land Utilisation in Rural Areas, and the Beveridge Committee on Social Insurance and Allied Services. Perhaps it was Beveridge who most clearly summed up the spirit of the time, and the philosophy which was to underlie post-war social policy:

> The Plan for Social Security is put forward as part of a general programme of social policy. It is one part only of an attack upon five great evils: upon the physical Want with which it is directly concerned, upon Disease which often causes Want and brings many other troubles in its train, upon Ignorance which no democracy can afford among its citizens, upon Squalor which arises mainly through haphazard distribution of industry and population, and upon Idleness which destroys wealth and corrupts men, whether they are well fed or not, when they are idle.
>
> (Beveridge 1942: 170)

It was within this framework of a newly acquired confidence to tackle long-standing social and economic problems that post-war town and country planning policy was conceived. No longer was this to be restricted to town planning 'schemes' or regulatory measures. There was now to be the same breadth in official thinking as had permeated the Barlow Report. The attack on squalor was conceived as part of a comprehensive series of plans for social amelioration. To quote the 1944 White Paper *The Control of Land Use*, 'provision for the right use of land, in accordance with a considered policy, is an essential requirement of the government's programme of postwar reconstruction'.

The new planning system

The pre-war system of planning was ineffective in several ways. It was optional on local authorities; planning powers were essentially regulatory and restrictive; such planning as was achieved was purely local in character; the central government had no effective powers of initiative, or of coordinating local plans; and the 'compensation bogey', with which local authorities had to cope without any Exchequer assistance, bedeviled the efforts of all who attempted to make the cumbersome planning machinery work.

By 1942, 73 per cent of the land in England and 36 per cent of the land in Wales had become subject to interim development control, but only 5 per cent of England and 1 per cent of Wales was actually subject to operative schemes (Uthwatt Report 1942: 9). There were several important towns and cities as well as some large country districts for which not even the preliminary stages of a planning scheme had been carried out. Administration was highly fragmented and was essentially a matter for the lower-tier authorities: in 1944 there were over 1,400 planning authorities. Some attempts to solve the problems to which this gave rise were made by the (voluntary) grouping of planning authorities in joint committees for formulating schemes over wide areas but, though an improvement, this was not particularly effective.

At the same time, a consensus was beginning to emerge about the purpose town and country planning might have. It was generally (and uncritically) accepted that the growth of the large cities should be restricted. Regional plans for London, Lancashire, the Clyde Valley and South Wales all stressed the necessity of large-scale overspill to new and expanded towns. Large cities were no longer to be allowed to continue their unchecked sprawl over the countryside; the explosive forces generated by the desire for better living and working conditions would be contained and directed. Overspill would be steered into new and expanded towns which could provide the conditions people wanted, without the disadvantages inherent in satellite suburban development. When the problems of reconstructing blitzed areas, redeveloping blighted areas, securing a 'proper distribution' of industry, developing national parks, and so on, were added to the list, there was a clear need for a new and more positive role for central government, a transfer of powers from the smaller to the larger authorities, a considerable extension of these powers and, most difficult of all, a solution to the compensation–betterment problem (discussed further below).

The necessary machinery was provided in the main by the Town and Country Planning Acts, the Distribution of Industry Acts, the National Parks and Access to the Countryside Act, the New Towns Act and, later, the Town Development Acts.

The Town and Country Planning Act 1947 brought almost all development under control by making it subject to planning permission. Planning was to be no longer merely a regulative function. Development plans were to be prepared for every area in the country. These were to outline the way in which each area was to be developed or, where desirable, preserved. In accordance with the wider concepts of planning, powers were transferred from district councils to county councils. The smallest planning units thereby became the counties and the county boroughs. Coordination of local plans was to be effected by the new Ministry of Town and Country Planning. Responsibility for securing a 'proper distribution of industry' was given to the Board of Trade. New industrial projects (above a minimum size) would require the board's certification that the development would be consistent with the proper distribution of industry. More positively, the Board was given powers to attract industries to development areas by loans and grants, and by the erection of factories.

New towns were to be developed by ad hoc development corporations financed by the Treasury. Somewhat later, new powers were provided for the planned expansion of towns by local authorities. The designation of national parks and 'areas of outstanding natural beauty' (AONBs) was entrusted to a new National Parks Commission, and local authorities were given wider powers for securing public access to the countryside. A nature conservancy was set up to provide scientific advice on the conservation and control of natural flora and fauna, and to establish and manage nature reserves. New powers were granted for preserving amenity, trees, historic buildings and ancient monuments. Later controls were introduced over river and air pollution, litter and noise.

The ways in which the various parts of this web of policies operated, and the ways in which both the policies and the machinery have developed since 1947, are summarised in the following chapters. Here a brief overview sets the scene for some of the principal issues that subsequently emerged and the responses that were made in terms of the planning system. However, some important factors, such as local government reforms, are not included and reference should be made to the relevant chapter.

A question of land values

The UK planning system is underpinned by an extraordinary feat of nationalisation incorporated in the 1947 Act, which was passed without the revolution that might have been expected in many other countries. It was the nationalisation of the right to develop land. Surprisingly, the issues were considered to be of a technical rather than a political nature and so the Uthwatt Committee was required to 'make an objective analysis of the subject of compensation and recovery of betterment in respect of public control of the use of land . . . and to advise on possible means of

stabilizing the value of land required for development or retirement'. Effective planning necessarily interferes with the market, sometimes removing a hoped-for increase in value by determining that an area of land will not be developed for a profitable use, at other times boosting the value of land by designating it for development, and the debate on the 1947 Planning Bill was greatly concerned with these issues of losses and gains in land values and how they should be dealt with.

The rights of individual landowners had for some time been restricted in the interest of public health: as noted above, owners had, for example, to ensure that new buildings conformed to certain standards and streets were laid out at a certain width. It had become accepted that these restrictions on individual property owners were in the public interest. However, the development of a more comprehensive planning system moved the law from a role in shaping standards of particular developments to one of selecting the most suitable pieces of land for particular uses. The public control of land use necessarily involves the shifting of land values from certain pieces of land to other pieces: the value of some land is decreased, while that of other land is increased.

In theory, it is logical to balance the compensation paid to owners whose land had decreased in value by collecting a betterment charge from owners who benefit from planning controls (Hagman and Misczynski 1978), but previous experience with the collection of betterment had not been encouraging.[5] The Uthwatt Committee's solution to this problem was the nationalisation of development rights in un-developed land and, essentially, this is what the Town and Country Planning Act 1947 did: development rights and their associated values were nationalised.

To deal with the matter of losses of development value incurred by landowners when the provisions of the new Act became effective, a £300 million fund was established for making payments. Landowners would submit claims to an agency, the Central Land Board, for *loss of development value*, that is, the difference between the *unrestricted value* (the market value without the restrictions introduced by the Act) and the *existing use value* (the value subject to these restrictions). When all the claims had been received and examined, the £300 million would be divided between claimants at whatever proportion of their 1948 value the total would allow.[6] All sales of land to public authorities took place at existing use value.

The question of betterment proved, and has continued to prove, to be even more contentious and intractable. The proposals in the 1947 Bill were opposed in principle by the Conservative Minister for Town and Country Planning in the later war years, William Morrison, on the grounds that they 'offend against two canons of taxation . . . uncertainty and equality . . . the whole bias of the Bill is in favour of the public as against private development' (Hansard, HC Deb, 29 January 1947, Vol. 432, cc. 947–1075). There was also disagreement about whether an increase in value was created by 'the enterprise of the people' or by development being carried out by individual landowners (Hansard, HC Deb, 29 January 1947, Vol. 432, c. 1025). In practice, the operationalisation of the intent of the Act also proved difficult, in part because the ideological differences highlighted in the debate were reflected in changes of government and changes in the commitment to the principle of capturing betterment.

The early years of the new planning system: development and betterment, housing, new towns and green belts

The early years of the new system were years of austerity. This was a truly regulatory era, with controls operating over an even wider range of matters than during the war. It had not been expected that there would be any surge in pressures for private development, but even if there were, it was envisaged that these would be subject to the new controls. Additionally, private building was regulated by a licensing system which was another brake on the private market. Building resources were channelled to local authorities, and (after an initial uncontrolled spurt of private house building) council house building became the major part of the housing programme.

The sluggish economy made it relatively easy to operate regulatory controls (since there was little to regulate), but it certainly was not favourable to 'positive planning'. It had been assumed that most of this positive planning would take the form of public investment, particularly by local authorities and new town development corporations. Housing, town centre renewal and other forms of 'comprehensive development' were seen as essentially public enterprises. This might have been practicable had resources been plentiful, but they were not, and both new building and redevelopment proceeded slowly. Thus, neither the public nor the private sector made much progress in 'rebuilding Britain' (to use one of the slogans which had been popular at the end of the war). The founders of the post-war planning system foresaw a modest economic growth, little population increase (except an anticipated short post-war 'baby boom'), little migration either internally or from abroad, a balance in economic activity among the regions, and a generally manageable administrative task in maintaining controls. Problems of social security and the initiation of a wide range of social services were at the forefront of attention: welfare for all rather than prosperity for a few was the aim. The making of land use plans following the 1947 Act went ahead at a steady pace, essentially physical blueprints, frequently in isolation from wider planning considerations, but even here progress was much slower than expected, and it soon became clear that comprehensive planning would have to be postponed for the sake of immediate development requirements.

For a time, the early economic and social assumptions seemed to be borne out, but during the 1950s dramatic changes took place, some of which were the result of the release of pent-up demand following the return of the Conservative government in 1951. One of the first acts of this government in the planning sphere was a symbolic one: a change in the name of the planning ministry – from 'local government and planning' to 'housing and local government'. This reflected the political primacy of housing and the lack of support for 'planning'. Gone were the utopian dreams of the 1940s – the political priority was now more materially and immediately placed upon housing. Slum clearance proceeded apace as did council housing, with from the mid-1950s the Conservative government setting high targets for new dwellings and giving significant incentives to local government for this to be high-rise construction. Over half a million new council houses were built in 1953–4 alone.

At the same time, the Conservative government wanted to stimulate private development and building. So, the 1948 betterment regulations were seen by the incoming 1951 Conservative government as a barrier to raising levels of private house building and they eventually removed the landowner's obligation to pay the development charge. However, as existing use value remained the price at which land was acquired by the public sector, a dual land market emerged, which in itself was a source of confusion and discontent. At about the same time, all building licensing was scrapped. Private house building boomed. The birth rate (which dropped steadily from 1948 to the mid-1950s) suddenly started a large and continuing rise.

One of the major planning acts of the post-war Labour government was the creation of the new towns programme, following the New Towns Act 1946. The origins of the Act can be traced back to the garden city movement at the beginning of the century, which had developed into a new towns campaign in the inter-war period: new towns were specifically advocated for London by Abercrombie in his 1944 plan. Work began on some eight new towns around London and six elsewhere across the country in north-east England, Northamptonshire, Scotland and Wales. The London region new towns were part of a policy of dispersal of economic activity, but others, such as Newton Aycliffe and Peterlee on the Durham coalfield, were seen has having an important role in regional development. The new towns programme went ahead at a slow pace, accompanied by a constant battle for resources which, so the Treasury argued, were just as urgently needed in the old towns. Only 592 new houses had been built in all the new towns combined by the end of 1950 (Ward 2004). The incoming Conservative government thought about abandoning the programme. However, whilst (with the exception of Cumbernauld in Scotland) further designations were not made during the 1950s, the development of the new towns picked up momentum during the decade, such that some were considered as largely complete by the early 1960s. Most were popular with residents and business, and they were showcases for planning innovation, with pedestrian precincts and organisation into neighbourhood units to Radburn layouts, segregating pedestrian and vehicle circulation systems.

However, the new town programme was not sufficient to stop the pressure for the outward expansion of existing settlements. New policies were forged, foremost of which was control over the urban fringes of the conurbations and other large cities, where an acrimonious war was waged between conservative counties seeking to safeguard undeveloped land and urban areas in great need of more land for their expanding house building programmes. On the side of the counties was the high priority attached to maintaining good-quality land in agricultural production. On the side of the urban areas was a huge backlog of housing need. The war reached epic proportions in the Liverpool and Manchester areas, where Cheshire fought bitterly 'to prevent Cheshire becoming another Lancashire'. The pressures for development grew as households increased even more rapidly than population – a little understood phenomenon at the time (Cullingworth 1960) – and as car ownership spread (the number of cars doubled in the 1950s and doubled again in the following decade). Increased mobility and suburban growth reinforced each other, and new road building began to make its own contribution to the centrifugal forces.

Working in the opposite direction was the implacable opposition of the counties. They received a powerful new weapon when Duncan Sandys, Minister of Housing and Local Government, initiated the green belt circular of 1955. The idea of green belts had been around for some time. Again, a mechanism for creating urban containment in this way can be traced back to the garden city movement and was evident, for example, in Unwin's 'green girdle' as part of his plan for the London region in 1929 and similarly was a feature of Abercrombie's 1944 Greater London plan. The circular provided for green belts to be incorporated in development plans and established ground rules that the purpose of green belts was to check the growth of large built-up areas, prevent coalescence of neighbouring urban areas and/or to preserve the special character of particular towns (a recreational objective was added by the Labour government in the 1960s). A flood of proposals followed and whilst by the early 1960s only London had a fully approved scheme, most of the other major conurbations and a number of the larger free-standing cities had interim schemes.

The affluent society and modernisation

After the harsh austerity in the immediate post-war years, economic growth and disposable incomes began to pick up during the 1950s, which by the end of the decade was fuelling development pressures and consumer demand. For example, the end of rationing and the ending of building licensing restrictions both occurred in 1954. The labour market was restructuring away from traditional heavy industries and towards lighter industries manufacturing consumer durables, including cars. This was combined with demographic and other projections anticipating large population rises, a massive growth in car ownership and further consumer demand. Whilst there were arguments over the details, a broad political consensus developed that was in favour of more planning generally (including, for example, economic planning) and more physical planning specifically, whether in terms of infrastructure (e.g. the creation of the motorway network), the creation of new settlements (more new towns and similar) or interventions in existing settlements (comprehensive redevelopment). However, compared with the late 1940s, a much larger role was now anticipated for the role of private developers and capital in many of these policies.

Whilst the foundations of the 1947 Act remained intact, albeit with some of its radicalism on issues like betterment neutered, it was evident as the 1960s progressed that existing planning mechanisms were struggling to cope. The blueprint development plans were inflexible and it was unclear that they had generally under-allocated development land (Ward 2004). An external advisory group, the Planning Advisory Group, initially created by the Conservative government, presented their report, *The Future of Development Plans*, to the Labour government in 1965 (PAG 1965). The key innovation proposed was a distinction between strategic questions and more detailed issues of land allocation. They emphasised the policy basis of plans and sought to avoid the highly prescriptive detail of the plans that had been produced under the 1947 Act. The essence of their ideas was incorporated into the development planning system made by the Town and Country Planning Act 1968, which established the two-tier system of Structure Plans and Local Plans (which might focus on particular districts, action areas or be thematic).

The other key report to government on planning matters in the 1960s was *Traffic in Towns*, otherwise known as the Buchanan Report, after the consultant Colin Buchanan who led this task group (Buchanan 1963). Britain was experiencing an exponential growth in traffic. The Buchanan Report accepted the need to provide for the car as a given (despite Buchanan himself being considered something of a sceptic on the desirability of massive urban road building). The report outlined options that, whilst providing for traffic, would hopefully both deal with issues of congestion and environmental deterioration, through the use of limited access 'environmental areas' and pedestrian–vehicular separation. One contribution of the report was to link transport and land use planning more closely – a key issue in town centres as commercial pressures grew and public-private partnerships to achieve redevelopment flourished. Ultimately the pace of redevelopment and urban road building caused a negative reaction and was an important stimulus for the developing conservation movement.

Development in this era was not entirely focused on central areas, however. Alarming population projections had appeared which transformed the planning horizon. The population at the end of the century had been projected in 1960 at 64 million; by 1965 the projection had increased to 75 million. At the same time, migration and household formation had added to the pressures for development and the need for an alternative to expanding suburbs and 'peripheral estates'. It seemed abundantly clear that a second generation of new towns was required. Between 1961 and 1971, fourteen additional new towns were designated in two waves. Some, like Skelmersdale and Redditch, were 'traditional' in the sense that their purpose was to house people from the conurbations. Others, such as Livingston and Irvine, had the additional function of being growth points in a comprehensive regional programme for Central Scotland. One of the most striking characteristics of the last new towns to be designated was their much larger size than their predecessors. In comparison with initial targets for the first generation of new towns of 30,000 to 50,000, Central Lancashire's projected population of 500,000 seemed massive. Furthermore, four of them were based on substantial existing towns – Northampton, Peterborough, Warrington and Central Lancashire (Preston-Leyland). In addition, a variety of other new settlements took place, such as Cramlington and Killingworth in the north-east of England, but were not new towns as such.

No sooner had all this been settled than the population projections were drastically revised downwards. It was too late to reverse the new new towns programme, though it was decided not to go ahead with further proposals. In reality, important though the new towns programme was, such was the pace and extent of development, Ward (2004) estimates that only 3.7 per cent of new public and private house building in England and Wales between 1945 and 1969 was in new and expanded towns. As well as redevelopment following slum clearance, large new, private suburban developments were mushrooming around the country.

Economic crisis, challenges to authority, policy shifts and betterment again

The 1970s started with a property boom under the Conservative government of Edward Heath. This juddered to a halt in 1973–74, with scandals about the relations between developers and public officials (which led, for example, to the former leader of Newcastle City Council T. Dan Smith being jailed), a

quadrupling of oil prices imposed by the OPEC cartel of oil-producing nations, and industrial strife. This was really the end of the long post-war boom, as structural weaknesses in the economy became more evident. And, indeed, it was effectively the end of a period of high modernism in planning and architecture, as the physical solutions of the 1950s and 1960s were increasingly seen to be problematic.

At the same time, social challenges to state authority had been developing since the late 1960s. In the sphere of planning, this gave impetus to changes of policy that had begun to take place, also since the latter part of the 1960s, in policy spheres such as housing, conservation and transport, as well as the more systemic area of public participation more generally. Thus issues of participation had been considered in the rather cautious Skeffington Report (Skeffington 1969), provisions for local authority designations of conservation areas were established in the 1967 Civic Amenities Act, and the partial collapse of the Ronan Point tower block in 1968 was an important moment in crystallising the move away from slum clearance and high-rise replacement.

The scale of slum clearance and its insensitivity to community concerns, as well as the inadequate character of some redevelopment schemes, led to an increasing demand for a reappraisal of the policy. Added force was given to this by the growing realisation that demolition alone could not possibly cope with the huge amount of inadequate housing – and the continuing deterioration of basically sound housing. Thus in the 1960s policies began to be introduced to improve, rehabilitate and renovate older housing: changing terminology reflected constant refinements of policy. Increasingly, in the 1970s it was realised that ad hoc improvements to individual houses were of limited impact: area rehabilitation paid far higher dividends, particularly in encouraging individual improvement efforts. Similarly, many proposals for large-scale urban road building were opposed and ultimately abandoned in the 1970s in the face of public opposition ('Get us out of this hell,' cried the families living alongside the elevated M4: Goodman 1972) and worsening economic conditions. Linked to, but distinct, from these movements of grass-roots opposition was the strengthening lobby for conservation, which led to new enhanced legislation and policy. Often these forces came together. Thus at Covent Garden in London, the Greater London Council's modernist plans of comprehensive redevelopment, including substantial road building, were defeated by an alliance of local residents and businesses together with conservationists, when national government was persuaded to list 245 buildings in the area. As part of this general ferment of protest and challenges to the authority was a wave of environmentalism, stimulated by classic texts such as Rachel Carson's 1962 *Silent Spring* and the realisation that fossil fuels were a finite and diminishing resource.

The shift in attitudes towards housing was accompanied by the slow development of a wider perspective on urban poverty. Again, the first stirrings of policy can be seen in the 1960s, but it was in the 1970s that this emerged as a significant problem requiring significant intervention and a recognition that it had multiple dimensions rather than just being a problem of, for example, housing. A major trigger was the rapid rate of deindustrialisation and the massive movement of people and jobs to outer areas and beyond. It was not exclusively an inner-city problem, as peripheral estates often had some of the worst deprivation, nor was it exclusively a northern problem, as London lost three-quarters of a million manufacturing jobs between 1961 and 1984 (Hall 2002a). In response, there was initially little difference between the political parties, but solutions were elusive. The initial culmination was the Inner Urban Areas Act 1978, the main intention of which was to empower inner-city authorities to stimulate economic development. Urban policy was a major feature of the Conservative governments in the 1980s and 1990s, which we return to below.

The turbulent and economically unstable 1970s were always going to be a difficult time to try and crack the problem of betterment. The Labour government elected in 1964 had tried through the Land Commission Act 1967. A Land Commission was set up, and given wide powers to acquire land in advance of requirements, so that it could be available 'at the right time'. A Betterment Levy was imposed at a uniform rate – initially 40 per cent of the development value – when land was sold, leased or realised by development. Its introduction was followed by uncertainty in the land market, which was reflected in the reluctance of landowners to part with their land, feeling that they could wait for a change in government and the abolition of the Betterment Levy. Thus the objects of the Act were not being realised. Land was less, rather than more, readily available, and the proceeds of the levy fell far below what had been expected. Both the Land Commission and the Betterment Levy were abolished by the Conservatives when they came to power in 1970.

The Labour government elected in 1974 revisited the issue of betterment through two pieces of legislation. The first was the Community Land Act 1975, which had objectives along the same lines as its predecessors, 'to enable the community to control the development of land in accordance with its needs and priorities'. The second was the Development Land Tax Act 1976, whose objects were the same as those of the Development Charge and the Betterment Levy, namely 'to restore to the community the increase in value of land arising from its efforts' (quoted in Blundell 1993): the rate of the tax was initially set at 80 per cent of gains. However, the legislation was launched in a difficult economic period in which restrictions on public spending meant that there was pressure on funds for acquisition of land under the Act and an uncertain market for development. The Conservative government elected in 1979 repealed the Community Land Act and reduced the rate of development land tax to 60 per cent before repealing it in 1985.

In addition to the buffeting by conflicting political perspectives of the attempts at capturing betterment, there were also some fundamental problems. The systems were complex: for example, the Development Land Tax Act was accompanied by 94 pages of additional guidance on its operation. Complex systems lead to uncertainty and suspicion and can increase the possibility of avoidance. Generally, the effects of the three attempts were to deter development and the better use of land, to encourage land hoarding by owners and to produce an artificial scarcity of sites. Since the 1976 Act, the issue of betterment has not been confronted in the same way. Using the provisions of section 52 of the 1971 Act and, more recently, section 106 of the 1990 Act, local planning authorities have sought to capture community benefits through local planning agreements. Most recently, the Community Infrastructure Levy has sought to recover a contribution to the cost of community investment from development gains, and this in turn emerged following significant levels of consultation on a 'tariff' on development proposed in the 2001 Planning Green Paper and a 'Planning Gain Supplement' following on from Kate Barker's 2004 report. However, each of these attempts has been the subject of criticism, and issues with the operation of the present approaches are discussed further in Chapter 5.

Planning as a problem, urban regeneration and the environment

In 1979 the country saw the election of a radical Conservative government and the effective end of the post-war consensus over the role of the state in the economy and society – of course political parties always disagreed in this period, but there was broad agreement that the state had an important role to play in a mixed economy and in supporting a welfare state. It was accepted by the new government that there would be a price to pay for this: for example, traditional heavy and manufacturing industries were allowed to decline and unemployment to rise as part of an economic restructuring process. New jobs arose in very different sectors such as banking and finance and service industries such as retailing and tourism. This became visible across the country as working docks were replaced by waterfront housing, new retail parks sprang up on the edge of towns and cities and so on. Much of this was facilitated by public policy, as we will discuss later.

In government rhetoric, planning was part of the problem that had held the British economy back, causing 'jobs to be locked up in filing cabinets'. More generally, local government was perceived as a problem and indeed a layer of government was removed across the conurbations with the abolition of the Greater London Council and the metropolitan county councils. However, in practice, government intervention in the

planning system was less drastic and more nuanced than the rhetoric might have suggested. Thornley (1993) divided the government approach in the 1980s into three:

- The general case, where the market was given greater market freedom and local authority development control powers were weakened. Measures to effect this included strong government guidance for local authorities not to exert control over aesthetic matters and an extension of the degree to which development could proceed without requiring planning permission,
- More extreme liberalization including removing the need for planning permission in some circumstances, sometimes together with other measures such as changed governance, to achieve urban regeneration/transformation e.g. Enterprise Zones, Simplified Planning Zones, Urban Development Corporations,
- Protection of vested interests in defined areas of restraint, such as areas of countryside protection, the green belt and areas of cultural heritage.

Strategic planning certainly came under fire. Any attempt at regional planning was more or less given up and the government intervened to make structure plans more market-orientated and in the conurbations they were replaced by unitary development plans, prepared by the districts and covering both strategic and more local issues. National policy was more laissez-faire and this enabled, for example, the dispersal of retailing from town centres to suburban and edge-city locations. Nevertheless, the essentials of the 1947 planning system – development plans and development control – were left largely intact.

However, to achieve regeneration and urban policy goals, together with its distrust of urban, largely Labour, local authorities, the government was significantly more interventionist. Initiatives included enterprise zones and simplified planning zones, and there was a strong emphasis on property-led regeneration and using public money to leverage private investment. Perhaps the most marked example of this amongst a bewildering flurry of initiatives was the creation of urban development corporations (UDCs) – initially in London Docklands and on the Merseyside waterfront, but subsequently in cities up and down the country. These were modelled on the new town development corporations but with a different private enterprise ethic. The London Docklands Development Corporation (LDDC) initially seemed almost determined to ignore, if not override, the community in which it was located, but this attitude eventually changed, and both the LDDC and later UDCs became more attuned to local needs and feelings.

At the same time some places were insulated from the liberalising agenda. For example, when the possibility of relaxing green belts was mooted, powerful lobbies within and beyond the government resisted these changes, and succeeded in maintaining the status quo (Elson 1986). Indeed, in the sphere of heritage, protection was significantly extended and strengthened (Pendlebury 2000).

After 1990, the Conservative government rhetoric became less strident. The agenda was similar, with the emphasis still firmly on the importance of the market and exerting central control in order for market freedom to operate (Thornley 1998), but in order to achieve policy objectives there was, however, a shift from interfering in day-to-day decisions through the appeal process to an emphasis on a tighter policy framework. Central prescription occurred through the series of *planning policy guidance notes* (PPGs), which in turn formed a key framework for locally produced development plans, given new emphasis with the Planning and Compensation Act 1991 and its reference to the 'plan-led system'. Linked to this was some reining in of the laissez-faire attitudes of the 1980s, so, for example, there was new guidance that sought to concentrate retail development in town centres. Urban policy also continued as a major focus, but again with shifting programmes, including City Challenge and the Single Regeneration Budget. Embedded in here were a softening of attitudes towards local government and the development of regeneration again being seen as a partnership activity.

An issue that re-emerged centre stage across much of the world in the 1990s was the environment, now considered more holistically than in the 1970s, through such landmark documents as the United Nations Brundtland Report, *Our Common Future* (Brundtland 1987), which introduced ideas such as 'sustainable development' and the 'compact city' into the lexicon. After many years of relegating environmental issues to a low level of concern, the British government needed to respond pragmatically when the Green Party did unexpectedly well in the European elections of 1989. A 1990 White Paper *This Common Inheritance* spelled out the government's environmental strategy over a comprehensive range of policy areas (untypically this covered the whole of the UK). Environmental protection legislation was passed, 'integrated pollution control' implemented, 'green ministers' appointed to oversee the environmental implications of their departmental functions, and new environmental regulation agencies established. One example of the shift in approach was a different attitude to transport planning and road building. The year 1989 also saw new forecasts of huge increases in car ownership and use. It was widely considered to be impossible to accommodate the forecast amount of traffic satisfactorily. Traffic calming became part of the vocabulary, road pricing moved onto the agenda for serious discussion (but little action) and road building was slashed.

'Things can only get better': the Labour governments from 1997 to 2010

Whilst the Labour governments of 1997 to 2010 were keen to be perceived as pro-development and there were continuing neo-liberal undercurrents, at the same time it was evident that overall they were more positive towards the public sector, local government and planning than their Conservative predecessors. This new emphasis on planning came to be wrapped up in the 'spatial planning' label. One dimension of this was the re-emergence of regional planning. At the most dramatic level the changes were constitutional, with the creation of both the Scottish Parliament and the National Assembly for Wales, and changes in Northern Ireland following the Good

3   The Royal Commission on the Local Government of Greater London (1921–3); the London and Home Counties Traffic Advisory Committee (1924); the Greater London Regional Planning Committee (1927); the Standing Conference on London Regional Planning (1937); as well as ad hoc committees and inquiries, for example on Greater London Drainage (1935) and a Highway Development Plan (Bressey Plan, 1938).

4   The phrase was coined by Alix Meynell, a senior official at the Board of Trade.

5   The principle had been first established in an Act of 1662 which authorised the levying of a capital sum or an annual rent in respect of the 'melioration' of properties following street widenings in London. There were similar provisions in Acts providing for the rebuilding of London after the Great Fire. The principle was revived and extended in the Planning Acts of 1909 and 1932. These allowed a local authority to claim first 50 per cent, and then (in the later Act) 75 per cent, of the amount by which any property increased in value as the result of the operation of a planning scheme. In fact, these provisions were largely ineffective, since it proved extremely difficult to determine with any certainty which properties had increased in value as a result of a scheme or, where there was a reasonable degree of certainty, how much of the increase in value was directly attributable to the scheme and how much to other factors. The Uthwatt Committee noted that there were only three cases in which betterment had actually been paid under the Planning Acts.

6   Considerable discussion on this issue took place in the debate over the Bill, with many saying the proposed sum was too small, whilst others claimed it was too generous. In 1947 £300 million was a considerable amount of money, particularly for a country seeking to recover from the ravages of war and to launch a welfare state. Whilst it is difficult to produce indisputably equivalent figures for value seventy years on, the basis (beyond the promotion of vested interests) for an impassioned debate can perhaps be illustrated by the fact that £300 million in 1947 is maybe equivalent to £10 billion now. At the time it was a little over 5 per cent of annual public spending, or about 80 per cent of annual spend on education. Today, the equivalent proportion of public spending amounts to about £38 billion. In the event, the estimate proved to be quite accurate, as total claims added up to about £380 million. However, nothing could convince those who argued that no compensation should be paid. Henry McGhee, Labour Member of Parliament for Penistone, put it in these terms in the 1947 debate on the Bill: 'What we are really paying for is to induce these people to cease preventing good planning. The tuberculosis figures, and the disease figures in this country are, in a large measure, due to the fact that our people have been crowded into unhealthy slums, without air and sunlight. Who did that? The Minister says in the White Paper that the owners of development values have done it. But, for that they are to be paid this huge sum. Half the time of our local authorities, and half the time of the great medical profession is being spent in trying to cure the results of private monopoly in land, and the crowding of our people into unhealthy areas.' (Hansard, HC Deb, 29 January 1947, vol. 432, c. 1020).

3   The agencies of planning

I don't care how things are organised. They can have it on the basis of a committee system, on a cabinet basis, on the mayoral system. If they want to introduce it on a choral system with various members of the council singing sea shanties, I don't mind, providing it's accountable, transparent and open. That's all I need to know.

(Eric Pickles MP, interview with *Total Politics*, 23 July 2010)

Introduction

The quote above from Secretary of State Eric Pickles may be a little knockabout in style, but it highlights that we will be as much, if not more, interested in the quality of the decision-making processes than in the details of the structures in which they are made. However, achieving accountability involves an understanding of both the where and the how.

Planning has popularly been seen as a function of local government – planning applications are submitted to and determined by your local authority, according to policies in its adopted local plan. Whilst it is true that the primary interface between the public and planning has been at the local authority level, there has always been a key role for central government. As you will see throughout this book, the shape and nature of the planning system which local government operates is set by central government. This is a matter not only of setting policy guidance – currently exemplified by the National Planning Policy Framework (NPPF) – but also of defining roles and responsibilities of bodies such as the Planning Inspectorate. In some senses and circumstances, it has the role of final arbiter in planning decisions: the judiciary can only arbitrate on the correctness of the process, whilst government

can decide on the merits of an application. More fundamentally, central government can be seen as deciding 'how much' planning we have in that in its decisions on policy and process, it determines how far planning can intervene in property market processes.

However, the range of agencies with a role to play in the planning process has, like much of government, been subject to significant change since the planning system as we know it was established by the Town and Country Planning Act 1947. From the outset, there were bodies, which came to be referred to as 'quangos',[1] with a role to play in planning, with an early example being the New Town Development Corporations, non-elected bodies, which had planning powers in their areas: some of their 'successors in style' have remained with us until recently in the form of Urban Development Corporations such as the Thames Gateway. The Planning Inspectorate was established as a separate 'Executive Agency' in 1992. This effectively put a role which had been carried out within government outside the day-to-day control of ministers. There is a wide range of advisory and executive agencies with a part to play in the planning process and understanding what this role is and how it is carried out is important for a complete understanding of planning in the UK, but this has perhaps become more difficult to achieve with the process of fragmentation and change which has affected government in the last thirty years or so.[2]

Whilst the Treaty of Rome did not give the European Community (as it then was) any competence in town planning, a place for the EU has gradually emerged over recent years. This can perhaps be observed most directly in the area of environmental legislation, such as the requirement to undertake environmental impact assessments. However, it should also be borne in mind that the EU has always had some form of regional policy, which has an effect on patterns of development, and this increased in importance as progress towards the completion of the Single Market accelerated. Closer European integration has meant that the prospects for localities need to be viewed in a wider context, which means there is a 'need for a European perspective on local economic performance, potential and future development' (Roberts et al. 1993: 49) to be included in the process of developing policy. The adoption of the European Spatial Development Perspective (ESDP) in 1999 represented a further element of significance in the emergence of a European dimension to planning. Whilst 'the ESDP does not provide any new responsibilities at Community level, it will serve as a policy framework for the Member States, their regions and local authorities' (CEC 1999a: foreword). So the planning profession in the UK needs to recognise the importance of the European dimension, and also to engage with shaping initiatives that emerge from the European level: the EU needs to be 'recognised essentially as part of the structure of government' (Williams 1996: viii).

At the other end of the spatial spectrum, it is important not to forget that planning is very much a community matter. The direct effects of new developments are felt at the local level and, as experience with regeneration has shown, it is at the community level that the nature of the local context is perceived, experienced and understood. Local and community-level bodies, such as town and parish councils, have long had a formal role in the planning process, but the new enthusiasm for 'localism' has shone a new light on this local level and given new powers to local communities.

The foregoing discussion has been framed largely in terms of administrative structures and, by inference, the formal boundaries they inhabit: local plans and decisions on planning applications are tied to local authority boundaries. However, the 'duty to cooperate', now part of national policy, could be seen as an explicit recognition that the forces we are planning for often do not respect these administrative boundaries. Indeed, the design of some planning initiatives, such as the Thames Gateway mentioned above, do not fit neatly with local authority jurisdictions, and concepts such as 'city-regions' take in several local authorities' areas of responsibility. These 'soft spaces' (Haughton et al. 2010) pose some interesting challenges when we are trying to pin down just who is responsible for a particular development decision.

Finally, in an era of 'spatial planning' it is important to look beyond the 'agencies of planning', in the sense of those bodies which have a clear land use planning responsibility, to a wider constituency, including the range of agencies whose role it is to help – or perhaps hinder – planning policies become reality, and which often do much to shape planning policy and outcomes. The main area in which this makes an impact is at the local level, but in central government much of the direction of recent policy has been influenced by Kate Barker's reports on housing and planning, although much of the driving force behind this work came not from the department with primary responsibility for planning, Communities and Local Government, but from the Treasury, on the basis that planning and its influence over housing supply had a key part to play in shaping the fortunes of the national economy.

This chapter reviews the spatial hierarchy of the 'agencies of planning', from the European to the local, with the objective of identifying who is responsible, and how, for the planning which is delivered. As this introduction has attempted to indicate, there is a considerable range of agencies which have an input in planning. Clearly, in the pages available it would not be possible to give full consideration to them all, but the review below will cover the most important agencies and provide an indication of the range of others involved.

EUROPEAN UNION

The growing influence of Europe

As has been stated, the impact of the European Union has been predominantly in the field of environmental controls but it is now being felt more directly on mainstream planning practice and urban policy. The most striking and perhaps best-known example of EU influence is environmental impact assessment, but other examples in cross-border and transnational spatial planning are emerging. Later chapters identify a range of agricultural, environmental, economic and regional policies of the EU which are having an effect on parts of the British planning system. Chapter 4 includes some discussion of supranational and cross-border planning instruments and policies that have been introduced at the European level.

Britain in the EU

The UK was not an enthusiastic supporter of the post-war moves towards a more closely integrated Europe. Although it favoured intergovernmental cooperation through such bodies as the Organisation for European Economic Cooperation (to become the OECD in 1961, but formed as OEEC in 1948) and the Council of Europe (formed in 1949), it was opposed to the establishment of organisations which would facilitate functional cooperation alongside nation states. It therefore did not join the European Coal and Steel Community (formed in 1952), nor was it a signatory to the 1957 Treaty of Rome which established the European Economic Community (EEC) and the European Atomic Energy Community (EURATOM). However, along with the other members of the Organisation for European Economic Cooperation, it formed the European Free Trade Association (EFTA) in 1960. Britain envisaged that EFTA would form the base for the development of stronger links with Europe. When it became clear that this was not viable, Britain applied for membership of the European Community. This was opposed by France and, since membership requires the unanimous approval of

existing members, negotiations broke down. The opposition continued until a political change took place in France in 1969. Renewed negotiations led finally to membership at the beginning of 1973.

The Treaty of Accession provided for transitional arrangements for the implementation of the Treaty of Rome, which Britain agreed to accept in its entirety. The objectives include the elimination of customs duties between member states and of restrictions on the free movement of goods; the free movement of people, services, and capital between member states; the adoption of common agricultural and transport policies; and the approximation of the laws of member states to the extent required for the proper functioning of a common market. These objectives are often referred to as the 'four freedoms': the free movement of goods, people, services and capital.

Currently (2014) there are twenty-eight member states of the EU, with a combined population of 504 million. For comparison purposes, the USA has a population of 314 million occupying over twice the land area, so the EU is a densely populated area. Perhaps a key feature is diversity in language – in the EU there are twenty-four official languages (and many others that are not used for official purposes). The EU is also easily the world's largest trading bloc, having a share of exports and imports more than three times that of the USA and accounting for 20 per cent of world imports and exports.

The organisational and political structure of the EU can seem complex and, like all such bodies, its actual workings can appear somewhat different from the formal organisation chart. Enlargement in 2004 prompted a major review of the treaties which govern the EU and a new European Constitution was proposed in 2004, but rejected in referenda in a number of countries. The main institutions of the EU and the parts that are of particular interest to planning are shown in Figure 3.1. In brief, there is an elected Parliament which operates as an advisory body and for some matters as a joint legislature with the Council of Ministers. The main legislature is the Council of Ministers, which makes policy largely on the basis of proposals made by the executive, the European Commission. There is also a Court of Justice which adjudicates matters of legal interpretation and alleged violations of Community law. The distribution of competences between Parliament and the executive is very different from that of most national governments.

It is important to separate what actually happens in the EU from the tone of reporting on the EU in much of the British press, which can give the impression that the EU 'imposes' regulations, policies and judgments on the UK. As a member of the EU, the UK plays a full part in the development of policies and, as one of the larger member states, has a significant influence over what happens. Perhaps sometimes it has not always played this role as effectively as it might and it has also at times been reluctant to accept majority opinions: this was particularly apparent during Mrs. Thatcher's administration in the 1980s. In considering the EU as an 'agency' affecting the nature of planning in the UK, it is therefore important to consider not only how developments in the EU might shape approaches here, but also how the UK can act to shape emerging patterns of EU policy and regulation. These 'levers of power' are operated through the institutions reviewed below. For fuller coverage of how these institutions work, a good place to start is the EU's own website, www.europa.eu/index_en.htm.

European Council

A summit of heads of state or government of the member states, together with the President of the European Commission, it provides general political direction for the European Union, considers fundamental questions related to the 'constitution' and construction of the EU, and makes decisions on the most contentious issues (Dinan 1998). It is not the legislature: this is the function of the Council of the European Union. Decisions which require legislation have to go through the normal EU legislation process, but agreements and declarations reached in the European Council are binding on the EU institutions, and have been critical in shaping the evolution of the EU. The Presidency of the Council rotates on a six-monthly cycle.

Council of the European Union (Council of Ministers)

The main decision-making body of the EU is the Council of Ministers. This is the legislature of the Community, a task it shares for some matters with the European Parliament. Unlike most other legislatures it is indirectly elected – being composed of representatives elected in the member states – and it deliberates in private. These characteristics have given rise to the criticism of a 'democratic deficit' in comparison with national legislatures and the European Parliament, which is directly elected and debates in public. But the characteristics reflect the fundamental nature of the EU as a pooling of national sovereignties and legislative powers, rather than a federal structure with a unitary legislature. This requires complex negotiation among the member states. The Council meets in different compositions depending on the topic, with the relevant ministers representing each member state, as for example in meetings (councils) of ministers of transport, environment and agriculture. The nations' representatives must be authorised to commit the member state to the decisions made. There is no formal Council of Ministers responsible for planning but, since 1991, there have been biannual informal meetings of ministers responsible for spatial planning. It consists of officials representing the ministries of member countries with a responsibility for planning. Spatial planning in the EU is discussed in the next chapter.

The Council (in some cases in cooperation with the European Parliament) can make three main types of legislation. *Regulations* have direct effect and are binding throughout the EU. They require no additional implementing legislation in the member states and are used mostly for detailed matters of a financial nature or for the technical aspects of (for example) administering the Common Agricultural Policy (CAP). By contrast, *directives* are equally binding on member states but the method of implementation is left to individual governments. Environmental matters are typically dealt with in this way. The Council can also issue *decisions*, which are binding on the member state, organisation, firm or individual to whom they are addressed. Finally, there are *common*

THE EUROPEAN COUNCIL Meeting of the Heads of State, *c.* 4 times per annum	Gives broad guidance and impetus to action

THE COUNCIL OF THE EU Meetings of ministers (one from each member state). The Council meets in different configurations depending on the issue e.g. The Environment Council The Transport, Telecoms and Energy Council The Presidency of the Council rotates every six months.	Legislature (on some matters, shared with the European Parliament)

	January to June	July to December
2014	Greece	Italy
2015	Latvia	Luxembourg
2016	Netherlands	Slovakia
2017	Malta	United Kingdom
2018	Estonia	Bulgaria
2019	Austria	Romania

INFORMAL MEETING OF MINISTERS OF SPATIAL PLANNING
Generally meets once during each Presidency.
It is not a formal council and has limited powers.

COREPER
Committee of Permanent Representatives – Civil servants from the member states who manage the work of the Council.

SUB-COMMITTEE ON SPATIAL AND URBAN DEVELOPMENT (SUD) of EU Regional Policy Committee (CDCR), previously the Committee on Spatial Development that produced the ESOP.

EUROPEAN COMMISSION 28 Commissioners, one from each country 33 Directorates (departments), including: MOBILITY AND TRANSPORT ENVIRONMENT REGIONAL POLICY (includes spatial planning and Structural Funds)	Applies the Treaties by initiating legislation and implements policy as executive body and works in 20 official languages of the EU.

EUROPEAN PARLIAMENT 766 elected members (73 from UK) 20 standing committees, including: **Agriculture and Rural Development** **Transport and Tourism** **Environment, Public Health and Food Safety** **Regional Development**	Political driving force, supervising and questioning the Council and Commission. Joint power to adopt legislation with Council. Supervises appointment of Commission.

ECONOMIC AND SOCIAL COMMITTEE 353 nominated members from employers, workers and other interests (24 from UK) 6 sections, including: **Agriculture, Rural Development and the Environment** **Transport, Energy, Infrastructure and the Information Society** **Economic and Monetary Union and Economic and Social Cohesion**	A non-political body that is consulted and delivers opinions on proposed legislation.

COMMITTEE OF THE REGIONS 353 members representing regional and local government (24 from UK) 6 Commissions, including: **Commission for Territorial Cohesion** **Commission for Environment, Climate Change and Energy** **Commission for Economic and Social Policy**	Is consulted and delivers opinions where regional interests are involved.

THE COURT OF JUSTICE AND GENERAL COURT 28 judges and 8 advocates general 28 judges, at least one from each country	Interpret the Treaties and apply judgments and penalties in cases of non-compliance.

Figure 3.1 Institutions of the European Union and spatial planning

positions or *actions*, *recommendations* and *opinions*, which have no legal force but offer guidance to governments.

The work of the Council of Ministers is supported by officials in the Committee of Permanent Representatives (COREPER). These are civil servants or permanent ambassadors to the EU of the member states. Indeed, it has been argued that COREPER is where the real decisions are made. It is the officials who conduct often very lengthy negotiations to reach agreement about measures among the member states before proposals are put before the ministers.

Whilst, over the life of the EU, many decisions have been made by unanimous agreement, the Single European Act 1987, which aimed to speed up the process of integration, introduced a system of majority voting on a wide range of policy areas. This system of *qualified majority voting* was designed to ensure that (for example) the larger states could not impose their views on the smaller states. From 2007, following the enlargement of the EU, 255 votes out of the (then) total of 345 were required for a proposal to be adopted (representing 62 per cent of the EU population). From November 2014 and under the provisions of the Treaty of Lisbon, numbers and proportions will be changed to reflect further enlargement and other developments. Under the same Treaty, the range of matters decided by majority voting will be extended. The UK has twenty-nine votes out of 353 in the Council.[3]

European Commission

The main work of the EU is undertaken by the executive of the Community, the European Commission. The Commission's main roles are to: set objectives and priorities for action; propose legislation to Parliament and Council; manage and implement EU policies and the budget; enforce European Law (jointly with the Court of Justice); represent the EU outside Europe (negotiating trade agreements between the EU and other countries, etc.). The Commission is a major driving force within the EU because it has the primary right to initiate legislation. It produces Green Papers, to stimulate debate and consultation, and White Papers, which are concrete proposals for legislation.

The Community's decision-making process is dominated by the search for consensus among the member states and this gives the Commission a crucially important role in mediation and conciliation.

A team of twenty-eight Commissioners – one from each member state – is appointed every five years: the appointment of the President of the Commission (currently – 2014 – José Manuel Barroso from Portugal) and the other twenty-seven Commissioners has to be approved by Council. The term of the current Commission expires at the end of October 2014. The Commissioners lead the work of thirty-four Directorates General, responsible for particular policy areas. Some of these are substantive policy areas, such as Regional Policy or Agriculture and Rural Development, whilst others are 'services' such as translation. Considerable influence over the work of each Directorate is exercised by the personal 'cabinet' of the Commissioner, and in particular by the chair (who is known as the *chef du cabinet*). The departments with an interest in town and country planning or the broader concept of spatial planning are Regional Policy (known as DG Regio) (whose main responsibility is for the Structural Funds, the implements through which EU funds are directed to regions or sectors felt to be in need of assistance), DG Environment, and DG Energy and Transport.

During the 1990s, the influence of the Commission waned under fierce criticism of its perceived greed for power and the acquisition of national competences. The culmination of mounting criticism came at the end of 1998, when the European Parliament, to which the Commission is accountable, threatened to sack all the Commissioners. Although the proposal was defeated, the debate fuelled popular antagonism against the Commission and, in March 1999, the Commissioners resigned en bloc. A new Commission was approved by the European Parliament in September with major reforms to its organisational structure and procedures.

European Parliament

The European Parliament is a directly elected body consisting of 766 members who are elected every five years. Britain has seventy-three representatives. The

Single European Act and Treaty on European Union extended the powers of the Parliament, and the Amsterdam Treaty (which came into force in May 1999) again increased its role in joint decision-making with the Council and its supervisory powers over the Commission. The Parliament is consulted on all major Community decisions, and it has powers in relation to the budget, which it shares with the Council, and in approving the appointment of the Commission. The assent of Parliament is also needed for the accession of new members and international agreements. However, it is important to note that the Parliament was established essentially as an advisory and supervisory body, while the Council of Ministers is the legislature. One reflection of the lack of legislative power is that the Parliament sits in plenary session for only three days each month and bizarrely continues to divide its sittings between two locations – Brussels and Strasbourg.

Parliament is organised along party political (not national) lines. The political groups have their own secretariats and are the 'prime determiners of tactics and voting patterns' (Nugent 1999: 130). Much of their work is carried out by standing committees and through questions to the Commission and Council. The Regional Development, Agriculture and Rural Development, and Transport and Tourism Committees consider matters related to spatial development, including European regional planning policy and the common transport policy.

Committee of the Regions

The Committee of the Regions (COR) is the youngest European Institution, set up following the Maastricht Treaty, and holding its first session in March 1994. It is intended to give a voice to the regions and local authorities in European Union debates and decision-making. It has 317 members representing the regions, including twenty-four from UK local authorities. (The UK representation is made up of fourteen from England, five from Scotland, three from Wales, and two from Northern Ireland.) The COR has taken a particular interest in regional planning and in advocating wider use of the principle of subsidiarity, so as to strengthen the role of regional and local authorities.

The Treaty identifies particular areas on which the COR has to be consulted by the Commission, including trans-European networks, economic and social cohesion, and structural fund regulations. It can also offer opinions in other areas that it thinks appropriate, typically when an issue has a specifically regional dimension. It has issued many opinions on planning, urban and environmental issues. A committee (confusingly known in the COR as a commission) has been established to deal exclusively with regional policy, spatial planning and urban issues, which is known as the Commission for Territorial Cohesion (COTER), and another on Sustainable Development (DEVE).

European courts

The European Court of Justice has one judge per EU country (currently twenty-eight). The court is helped by eight 'advocates-general' whose job is to present opinions on the cases brought before the court. They must do so publicly and impartially. Each judge and advocate-general is appointed for a term of six years, which can be renewed. The governments of EU countries agree on whom they want to appoint. To help the Court of Justice cope with the large number of cases brought before it, and to offer citizens better legal protection, a 'General Court' deals with cases brought by private individuals, companies and some organisations, and cases relating to competition law. The courts have played an important part in extending the competences of the European Union by confirming that actions by the Community are legal under the treaties (Nadin and Shaw 1999), and by promoting harmonisation by ruling that certain actions are illegal (Nugent 1999: 263). These courts are quite separate from the European Court of Human Rights, discussed in the following section.

Council of Europe

The Council of Europe is not to be confused with the EU. It was set up in 1949 with ten member countries to promote awareness of a common European identity, to protect human rights and to standardise legal practices across Europe in order to achieve these aims. It now has forty-seven member countries (including sixteen countries that were formerly part of the communist bloc). It is best known for its European Convention on Human Rights. Anyone who feels that their rights under the Convention have been breached may take a case to the European Court on Human Rights for a decision which will be binding on those states that have signed up to the Convention. The Convention is now incorporated into UK law and its impacts on planning are discussed in Chapter 13.

The Council has been active for many years in the field of regional planning and environment, and perhaps the most notable achievement is the Berne Convention on Conservation of Wildlife and Habitats. It has published conference and other reports on the implications of sustainability for regional planning, the representation of women in urban and regional planning, and many other topics. A conference of ministers of spatial and regional planning (CEMAT) has been meeting since 1970 and its most important contribution has been the European regional/spatial planning charter, known as the *Torremolinos Charter*. This was adopted in 1983 and committed the Council to producing a 'regional planning concept' for the whole of the European territory. It has taken some time but, as noted in the next chapter, CEMAT has published *Guiding Principles for Sustainable Spatial Development of the European Continent* (2000).

The Council was responsible for the European Campaign for Urban Renaissance (1980–2). This led to a programme of ad hoc conferences, various reports and 'resolutions' on such matters as health in towns, the regeneration of industrial towns, and community development. In 1992, the Conference adopted the *European Urban Charter*. This 'draws together into a single composite text, a series of principles on good urban management at local level'. The 'principles' relate to a wide range of issues, including transport and mobility, environment and nature in towns, the physical form of cities, and urban security and crime prevention.

CENTRAL GOVERNMENT

Organisational responsibilities

A large number of governmental departments and agencies are involved in town and country planning. Those having the main responsibility for the Planning Acts are the Scottish Government Directorate for the Built Environment, the Planning Division of the Welsh Government, the Planning and Local Government Group of the Department of the Environment for Northern Ireland (DoENI) and, for England, the Department of Communities and Local Government. There are, of course, many planning functions that fall to departments responsible for agriculture, the countryside, the human heritage, national heritage, nature conservation, and trade and industry. Additionally, an increasing number of functions have been transferred from government departments to agencies and other 'arm's-length' bodies. Figure 3.2 shows the main institutional arrangements, and gives a flavour of their complexity.

Planning responsibilities have evolved over time and, though there have been numerous reorganisations, the machinery inevitably has a patchwork appearance. (As an example of the problems involved: in which department should questions of the rural economy be placed – the one concerned with agriculture, or natural resources, or economic development, or employment? Or should it form a separate department of its own?) The machinery is also unstable: changing perceptions, conditions, problems and objectives demand new policy responses which in turn can lead to organisational changes. For example, increased concern for environmental planning has contributed to the transfer of widespread environmental functions to a number of environment agencies. The agencies establish themselves, they extend their activities and the problems of cooperation and overlapping competences get more attention, which leads to calls to unify and simplify the structure of agencies. Sometimes, different patterns emerge in different parts of the UK. Thus nature conservation and access to the countryside are the responsibility of one agency in Scotland (Scottish Natural Heritage) and in Wales

(Natural Resources Wales), but were divided between two in England (English Nature and the Countryside Agency) until the functions were brought together under Natural England.

Department for Communities and Local Government

At the time of writing, the central government planning department for England is the Department for Communities and Local Government (DCLG), which was created in 2006. The DCLG is shown alongside other central government departments and agencies that have some relationship to the planning system in Figure 3.3. New governments and ministers are prone to reorganise the government machinery, and this has led to several changes to the name and location in the government system of the 'national' department for planning. So, in this book and other sources, many references are made to the DCLG's predecessors when discussing the role of central government. Up to 2006, the department responsible for planning was the Office of the Deputy Prime Minister (ODPM); prior to that the Department of Transport, Local Government and the Regions (DTLR); and prior to that, from 1997 to 2000, it was the Department of Environment, Transport and the Regions (DETR). The big changes were the move of most environment functions to the new Department for Environment, Food and Rural Affairs (DEFRA) in 2000 and the creation of a separate Department for Transport (DfT) in 2002 (which recreated the separate department that existed before 1997).

The changing permutations of competences for planning, local government, environment and transport have a much longer history, as illustrated in Figure 3.3. Since 1942 there has been a separate ministry for land use planning with varying competence for other related policy fields. The Department of the Environment was formed in 1970 with the aim of providing more integration of planning, transport and some environmental policy, but transport responsibilities were moved back to a separate Department of Transport in 1976. Transport was 'reintegrated' with planning by the Labour government in 1997, though

this lasted only until 2002. The old Department of the Environment also had many heritage responsibilities, but these were moved to a separate Department of National Heritage (DNH) in 1992, which was superseded in 1997 by the Department for Culture, Media and Sport (DCMS). Other changes have included the gathering together of the pollution regulation functions within Her Majesty's Inspectorate of Pollution (HMIP), and later the establishment of environmental agencies for England and Wales, and for Scotland; these have taken over the functions of the HMIP, the National Rivers Authority and the waste regulation functions of local government. These have been restless times in Whitehall. Whilst the title of the department – DCLG – was retained following the 2010 election, the style of operation of the department changed with new parties in government and a new team of ministers. The operational detail of this is discussed further below, but a reading of the current departmental priorities and responsibilities (Box 3.1) gives a flavour of the style of the new ministerial team. Whilst it indicates that planning is but one of the responsibilities of the department, it also serves to emphasise that the localism and growth agendas of the Coalition government very much set the tone of the department and inevitably influence approaches to planning. In particular, the issues related to a depressed economy allied to any ideological preferences of the majority Conservative party can be seen reflected in the priority given to 'making it easier for jobs to be created' (in the words of the NPPF), and to the importance accorded to efforts to increase house building rates.

Department for Culture, Media and Sport

The Department for Culture, Media and Sport was established in 1997, superseding the Department of National Heritage. It has a wide range of responsibilities, including the arts, sport and recreation, libraries, museums, broadcasting, film, press freedom and regulation, heritage and tourism. Its overall aim is 'to improve the quality of life for all through cultural and sporting activities, and to strengthen the creative industries'.

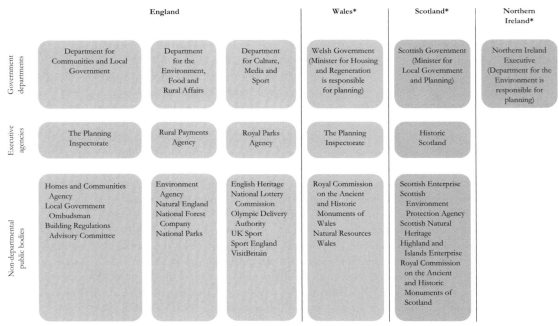

Figure 3.2 **The organisation of central government for planning**

There are now greatly enhanced resources for these worthy objectives by way of the National Lottery. The areas of 'good causes' for which Lottery funds provide support are sport, the arts, heritage, charities, millennium projects, health, education and the environment. The department has important responsibilities for heritage planning, including listed buildings.

The DCMS is responsible for over forty executive and advisory non-departmental public bodies, including the British Library, VisitBritain, the Millennium Commission, the National Heritage Memorial Fund, English Heritage (which is discussed in Chapter 8) and, latterly, the Olympic Delivery Authority.

Department for Environment, Food and Rural Affairs

Other departments of government have special status in respect of town and country planning. From the 1940s, particular status was afforded the Ministry of Agriculture,

Fisheries and Food (MAFF), the functions of which are now part of the much wider DEFRA. An overriding concern of government after the war was the protection of agriculturally productive land. This secured a central place for the MAFF in land use decisions. It had to be consulted on important proposals, and the MAFF classification of agricultural land quality remained a potentially important consideration in development control (see Chapter 10 for further discussion of this role). The influence of the ministry waned somewhat in parallel with the decline of agriculture, but also with the election of the Labour government in 1997, which wanted to shift to a wider rural focus away from a focus perceived as almost exclusively agricultural. DEFRA (like MAFF before it) also has to be consulted on any planning proposal which involves a significant loss of high-quality agricultural land. Such objections have fallen considerably over recent years.[4] At the same time, it has assumed an increasing responsibility for countryside protection functions, such as environmentally sensitive areas (discussed in Chapter 10).

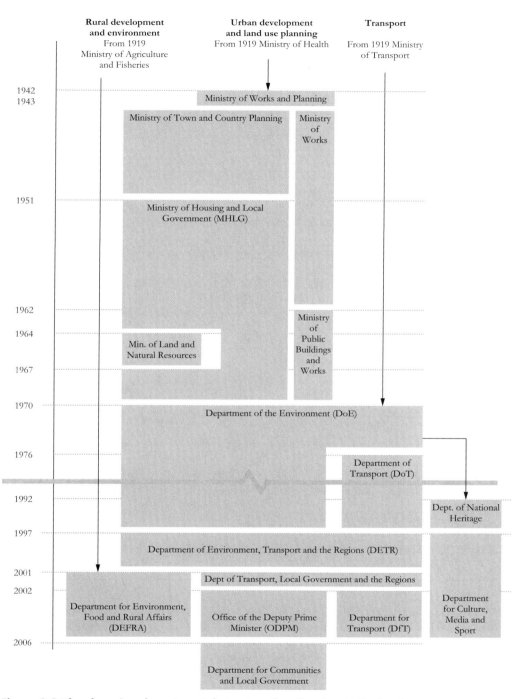

Figure 3.3 The changing departmental structure for planning in England

BOX 3.1 THE ROLE AND RESPONSIBILITIES OF DCLG

What we do

We work to move decision-making power from central government to local councils. This helps put communities in charge of planning, increases accountability and helps citizens to see how their money is being spent.

Responsibilities

We are responsible for:

- supporting local government by giving them the power to act for their community – without interference from central government
- helping communities and neighbourhoods to solve their own problems so neighbourhoods are strong, attractive and thriving
- working with local enterprise partnerships and enterprise zones to help the private sector grow
- making the planning system work more efficiently and effectively
- supporting local fire and rescue authorities so that they're able to respond to emergencies and reduce the number and impact of fires

Priorities

In 2012 to 2013, our priorities will be:

- putting local councils and businesses in charge of economic growth and bringing new business and jobs to their areas
- getting the housing market moving again so there are more homes to buy and to rent at prices people can afford
- ensuring Council Tax payers get value for money and making their local council accountable to them
- turning round the lives of troubled families, giving them the chance of a better life and reducing the cost to the taxpayer
- bringing people together in strong united, communities

Source: DCLG website 2013 (www.gov.uk/government/organisations/department-for-communities-and-local-government)

DEFRA has a role in overseeing sustainable development across central government: the Environment Secretary sits on the key domestic policy Cabinet committees, including the Economic Affairs Committee, to enforce the government's commitment to sustainability across policymaking; and DEFRA takes responsibility for reviewing departments' plans in relation to sustainable development principles. DEFRA plays an important role in the development of international policy on the environment and sustainable development. Box 3.2 shows DEFRA's priorities, and the switch from MAFF's almost exclusive focus on agriculture to a wider rural agenda is readily apparent, although of course it retains an important

BOX 3.2 THE ROLE AND RESPONSIBILITIES OF DEFRA

What we do

We are the UK government department responsible for policy and regulations on environmental, food and rural issues. Our priorities are to grow the rural economy, improve the environment and safeguard animal and plant health.

Responsibilities

We are responsible for policy and regulations on:

* the natural environment, biodiversity, plants and animals
* sustainable development and the green economy
* food, farming and fisheries
* animal health and welfare
* environmental protection and pollution control
* rural communities and issues

Although Defra only works directly in England, it works closely with the devolved administrations in Wales, Scotland and Northern Ireland, and generally leads on negotiations in the EU and internationally.

Priorities

Our priorities are to:

* grow the rural economy
* improve the environment
* safeguard plant health
* safeguard animal health

Our 10 point growth plan underpins our four priorities:

* Growing our sectors and their exports
  * increase exports and competitiveness in the food chain
  * set the conditions to ensure that GM and nanotechnology can play a part in contributing to economic growth
  * improve rural competitiveness and skills, invest in tourism and support micro-enterprises
  * proactively safeguard animal and plant health
  * reduce waste and inefficiency
* Investing in infrastructure
  * improve broadband and mobile phone access in rural areas
  * invest in flood and coastal protection
  * Thames Tideway Tunnel
* Removing regulatory and other barriers to growth
  * unblock growth potential by removing red tape and improving environmental challenge
  * make it simpler and quicker to comply with the Habitats and Wild Birds Directives

Source: DEFRA website (www.gov.uk/government/organisations/department-for-environment-food-rural-affairs)

role in relation to agriculture. DEFRA is responsible for about £2 billion of public spending and employs about 10,000 staff.

DEFRA is responsible for a number of executive agencies, but amongst the most important for planning are Natural England, the Environment Agency, and the Forestry Commission. It is also the lead department for the National Parks. All are discussed in later chapters, but there follows a general explanation of executive agencies and non-departmental bodies.

Executive agencies

The proliferation of new government agencies is confusing. In its review of quangos in 2010, the Public Administration Select Committee noted, 'One reason for this lack of clarity is the complexity of the public bodies' structures; non-departmental public bodies, arm's length bodies, quangos, public bodies, executive agencies, non-ministerial departments, and independent statutory bodies all clutter the landscape.' This difficulty is also reflected in the challenge of counting how many there are: in 2010, the minister responsible 'believed, but could not be certain' that there were 901, but in 1996 the Democratic Audit,[5] by including bodies which operated at the local as well as national level, estimated that there were 6,424. Since Mrs Thatcher's election in 1979, there have been numerous attempts to 'cull' quangos, but with limited success. Flinders and Skelcher (2012) have identified a number of reasons for this lack of success, including both the difficulty of identifying quangos and the fact that they often perform valuable functions. It is also the case that new ones are continually being created as new tasks for government – particularly those of a 'technical' nature – assume enhanced importance.[6] An overriding concern about quangos is their impact on public accountability for services. Generally, their activities are not as open to scrutiny as organisations firmly within the public sector – they may not have open meetings, or be open to the use of ombudsman services and the like – but, whilst they may have some clear, often contractual, performance standards to which they are obliged to adhere (what might be called

managerial accountability), they lack the political accountability associated with public-sector bodies (Day and Klein 1987; Hart *et al.* 1996).

To sidestep deep reflection on taxonomic issues, executive agencies remain part of their department, and their staffs are civil servants, but they have a wide degree of managerial freedom (set out in their individual 'framework' documents). They enjoy delegated responsibilities for financial, pay and personnel matters. They work within a framework of objectives, targets and resources agreed by ministers. They are accountable to ministers, but their chief executives are personally responsible for the day-to-day business of the agency. Ministers remain accountable to Parliament. If this sounds somewhat confusing, that is because it is. However, in principle, the stated intention is to increase accountability. A distinction is drawn between responsibility, which can be delegated, and accountability, which remains a matter for ministers – a contention which is the subject of considerable controversy.[7] The Planning Inspectorate, discussed in the following section, is an example of an executive agency.

As well as executive agencies, there are many other types of 'extra-governmental organisations'. These range enormously in function, size, and importance. They all play a role in the process of national government, but are not government departments or parts of a department. So, within the purview of DCLG, in addition to the Planning Inspectorate, there are executive bodies such as the Homes and Communities Agency and there have been numerous Development Corporations, whilst there are also advisory bodies such as that advising on building regulations. DCMS has bodies such as English Heritage, Sport England and the Olympic Delivery Authority. DEFRA has the Environment Agency and Natural England.

Many of these bodies provide a key input to the planning function, either through undertaking important functions – such as those undertaken by the Planning Inspectorate – providing advice on policy development or planning applications, on which they are often statutory consultees, or undertaking essential implementation functions. The work of some of these bodies is referred to in more detail in other parts of this book but it is important to try to have a picture of the range of organisations out there, how they operate and what their responsibilities are, in spite of the challenges of mapping and enumerating the population noted at the beginning of this section.

Planning Inspectorate

The way in which the aims and objectives of executive agencies are cast is illustrated by the case of the Planning Inspectorate. This is an executive agency of the Department for Communities and Local Government which also does casework for the Welsh Government. It has offices in Bristol and Cardiff and employs about 800 staff. In Scotland, the equivalent is the Directorate of Planning and Environmental Appeals, whilst in Northern Ireland the work is carried out by the Planning Appeals Commission. The major areas of work of the Inspectorate have varied considerably since it was established as an executive agency in 1992. Numbers of planning appeals fluctuate, amongst other things, with the level of economic activity. So, the number determined was 12,619 in 1999–2000, 21,087 in 2008–9 and 14, 253 in 2012–13 in England (524 in Wales in 2012–13). In addition, a sample of other activity by the Inspectorate reveals that 1,723 enforcement appeals were decided in 2012–13, as were fifty-three high hedge appeals: work was done on ten applications called in by the Secretary of State. Between 2004 and 2013, decisions were issued on 377 development plan documents submitted for examination. (The Inspectorate's appeal work is further considered in Chapter 5.)

Other responsibilities of the Planning Inspectorate include listed buildings appeals, rights of way orders, environmental appeals of various sorts, transport appeals, Definitive Map modification orders under the Wildlife and Countryside Act 1981 and casework related to commons. Changed government policy has led to changes in the Inspectorate's workload. In 2012, the Inspectorate took on the work of the Infrastructure Planning Commission, whilst the Growth and Infrastructure Act 2013 added new responsibilities:

from May developers were able to appeal against affordable housing requirements included in section 106 agreements which were felt to render developments unviable; and from October an initial designation of poorly performing local authorities and the subsequent ability of developers to choose to come to the Inspectorate (rather than the local planning authority (LPA)) to determine applications for major developments. The work of the Inspectorate has also been changed by adaptations to procedures, such as an encouragement of the use of written representations rather than hearings, and the introduction of the householder appeals service, which is discussed further in Chapter 5.

Central government planning functions

Relationships between central and local government vary significantly among various policy areas, 'reflecting, in part, the difference in weight and concern which the centre gives to items on its political agenda, and, in part, differences in the sets of actors involved in particular issues' (Goldsmith and Newton 1986: 103).

Under the Town and Country Planning Act 1943 (which preceded the legislation on the scope of the planning system), there was a duty of 'securing consistency and continuity in the framing of a national policy with respect to the use and development of land'. Though this is no longer an explicit statutory duty, the spirit lives on, and the Secretary of State has extensive formal powers. These, in effect, give the department the final say in all policy matters (subject, of course, to parliamentary control – though this is in practice very limited). For many matters, the Secretary of State is required or empowered to make regulations or orders. Though these are subject to varying levels of parliamentary scrutiny, many come into effect automatically. This delegated secondary legislation covers a wide field, including the Use Classes Orders (UCOs) and the General Development Orders (GDOs). These enable the Secretary of State to change the categories of development which require planning permission.

The formal powers over local authorities are wide-ranging. If a local authority fails to produce a 'satisfactory' plan, default powers can be used. The Secretary of State can require a local authority to make 'modifications' to a plan, or 'call in' a plan for 'determination'. Decisions of a local planning authority on applications for planning permission can, on appeal, be modified or revoked. Development proposals which the Secretary of State regards as being sufficiently important can be 'called in' for decision by the minister. And, as was noted above, if a planning authority is deemed to be 'underperforming', then its ability to determine applications can be removed and passed to the Planning Inspectorate.

These powers have been frequently employed in the plan-making process (previously through the now defunct regional offices). In less interventionist times, they were reserved for cases where there was a deadlock between local and central government. This can amount almost to a game of bluff as, for instance, when a local authority wanted to make a political protest, or to demonstrate to its electors that it is being forced by central government to follow a policy which is unpopular. Thus, opposition in Surrey to the M25 was so strong that the county omitted it from its structure plan. The Secretary of State made a direction requiring it to be included. Another battle arose over the Islington unitary development plan, where the Secretary of State took strong objection to the stringent controls which the borough proposed (inter alia) for its thirty-four conservation areas. The Secretary of State issued a direction requiring most of these to be changed. The Borough took the matter to court, which held that it had no power to intervene on the planning aspects of the case, and since the Secretary of State had not acted perversely or in conflict with his own policies, his action was quite legal. Judicial review cannot be used as an oblique appeal. It was therefore the responsibility of the Borough and the Secretary of State to resolve their differences to the satisfaction of the Secretary of State (*Journal of Planning and Environment Law* (*JPL*) 1995: 121–5).

The level of provision for new housing is frequently an area for debate in the development of local planning policy, with a not untypical position being that local opinion and often the draft local plan propose a level of provision which is below the level that policy at a national level sees as necessary. For example, the Berkshire structure plan adopted in 1995 was required to increase the provision for new housing in the county (from 37,500 to 40,000) by the year 2006. Under current plan inquiry procedures, such debates are conducted initially through the inspector operating the 'test of soundness' for policy proposals. So, in September 2013, in a preliminary judgment on the proposals for the Gravesham plan, the inspector concluded that it should 'include a significant addition to the total figure for new housing over the plan period', as 'it is unlikely that I could reach a finding of soundness on the basis of the housing evidence'.[8] Perhaps a classic case of open political conflict was the North Southwark Local Plan, which was formally called in by the Secretary of State. It was only the second plan to be called in and the first (and only plan) to be rejected entirely, because it opposed private investment and was hostile to the London Docklands Development Corporation (Read and Wood 1994: 11). However, cases of high-octane difference are rare, although differences of opinion on policy between the centre and the local level are not uncommon.

The North Cornwall case was something of a cause célèbre, because it was the subject of a documentary on Channel 4 television, 'Cream Teas and Concrete'. Officially, it was handled in a way more consistent with tradition. The local authority was giving planning permissions for development in the open countryside contrary to national policies and the approved county structure plan. Pressure was brought to bear upon the district council by way of a special inquiry carried out by an independent professional planner (Lees 1993). Normally, informal pressures are sufficient: the threat of strong action by the Secretary of State is typically as good as – if not better than – the action itself. With the enhanced position of development plans in the so-called plan-led system, attention now focuses on the provisions of draft plans. The examination and inquiry process should ensure that local plan policies accurately reflect those established at the national level.

In spite of all this, it is not the function of the Secretary of State to decide detailed planning matters. In a ministerial statement, it was explained that

It is the policy of the Secretary of State to be very selective about calling in planning applications. He will, in general, only take this step if planning issues of more than local importance are involved. Such cases may include, for example, those which, in his opinion,

- may conflict with national policies on important matters;
- could have significant effects beyond their immediate locality;
- give rise to substantial regional or national controversy;
- raise significant architectural or urban design issues; or
- may involve the interests of national security or of foreign Governments.

(Hansard, HC Deb, 16 June 1999, c. 138)

This echoes statements made by previous Secretaries of State. Local planning decisions are normally the business of local planning authorities. The Secretary of State's function is to coordinate the work of individual local authorities and to ensure that their development plans and development control procedures are in harmony with broad planning policies and operate to the required standards of performance. That this often involves rather closer relationships than might prima facie be supposed follows from the nature of the governmental processes. The line dividing policy from day-to-day administration is a fine one. Policy has to be translated into decisions on specific issues, and a series of decisions can amount to a change in policy. This is particularly important in the British planning system, where a large measure of administrative discretion is given to central and local government bodies. This is a distinctive feature of the planning system. There is little provision for external judicial review of local planning decisions (Scrase 1999; Keene 1999). Instead, there is the system of appeals to the Secretary of State. The department in effect operates both in a quasi-judicial capacity and as a developer of policy.

The department's quasi-judicial role stems in part from the vagueness of planning policies. Even if policies are precisely worded, their application can raise problems. Since local authorities have such a wide area of discretion, and since the courts have only very limited powers of action, the department has to act as arbiter over what is fair and reasonable. This is not, however, simply a judicial process. A decision is not taken on the basis of legal rules as in a court of law: it involves the exercise of a wide discretion in the balance of public and private interest within the framework of planning policies.

Appeals to the Secretary of State against (for example) the refusal of planning permission are normally decided by the Planning Inspectorate. Inspectors represent or 'stand in the shoes' of the minister. Such decisions are the formal responsibility of the Secretary of State; there is no right of appeal except on a question of law. Inspectors also consider objections made to local development plans, and their binding decisions are conveyed to the local authority.

Planning authorities, inspectors, and others are guided in their decisions and recommendations by government policy. Central government guidance on planning matters has been issued by way of circulars and, since 1988, in policy guidance (as explained in Chapter 4). Since the introduction of planning guidance documents, circulars were concerned mainly with the explanation and elaboration of statutory procedures. Policy guidance has dealt with government policy in substantive areas, ranging from green belts to outdoor advertising. Circulars and guidance were subject to some consultation with local authorities and other organisations prior to final publication, and they have often been supported by research and sometimes prepared in draft by consultants, but the Secretary of State has the final word.

Circulars and the various forms of guidance have been recognised as important sources of government policy and interpretation of the law, although they are not the authoritative interpretation of law (this is the role of the courts), nor are they generally legally binding. Indeed, advice could be conflicting, perhaps as a result of piecemeal revision at different times. Moreover, as is demonstrated repeatedly at public inquiries, differing interests can 'cherry-pick' from elements of the planning policy guidance to show how well their arguments meet central government objectives. The issuing of the revised National Planning Policy Framework was intended to simplify policy guidance, particularly reducing its volume from 'more than a thousand pages' though there were some fears that, in the view of the Communities and Local Government Committee 'the brevity of the NPPF meant it was lacking in important detail, and that this was more likely to lead to greater uncertainty and delay'.

Policy, of course, has to be translated into action. This presents inevitable problems: policy is general; action is specific. In applying policy to particular cases, interpretation is required, and often there has to be a balancing of conflicting considerations – of which many examples are given throughout this book. Policies can never be formulated in terms which allow clear application in all cases, since more than one 'policy' is frequently at issue. Even the most hallowed of policies has to be flouted on occasion: witness developments in the green belts, in protected sites of natural or historic importance, and in national parks. Such developments may be unusual (if only because they attract great opposition – of an increasingly strident nature), but they represent only the most obvious and the most public of the conflicts over land use.

Given the realities of land use controls, policies are usually couched in very general terms, such as 'preserving amenity', 'sustaining the rural economy', 'enhancing the vitality of town centres' or 'restraining urban sprawl'. This is a very different world from that of a zoning ordinance, which is the principal instrument of development regulation in many countries. Such an ordinance may provide (for example) that a building shall be set back at least five metres from the road, have a rear yard of six metres or more, and side yards of at least two and a half metres. Zoning is intended to be clear and precise, and subject to virtually no 'interpretation'. Indeed, it was hoped that it would be virtually self-executing. Though these hopes failed to materialise, it is fundamentally different in approach from the British planning systems. Above all, the British systems embrace discretion and general planning principles rather than certainty for the landowner and developer.

It is important to recognise that discretion means much more than 'making exceptions in particular cases'. The system requires that all cases be considered on their merits within the framework of relevant policies. Local authorities cannot simply follow the letter of the policy: they must consider the character of a particular proposal and decide how policies should apply to it. But they cannot depart from a policy unless there are good and justifiable planning reasons for so doing. The same applies to the Secretary of State, who is equally bound both by formulated policies and the merits of particular cases. The courts will look into this carefully in cases which come before them and, though they will not question the merits of a policy, they will ensure that the Secretary of State abides by it. Thus, in a curious way, discretion is limited. All material considerations must be taken into account and justified. Arbitrary action is unacceptable as it is in the USA, which has written constitutional safeguards (Booth 1996; Purdue 1999). Are the approaches therefore as different as first appears? As stated by Cullingworth (1994: 163):

> American zoning, while appearing to provide certainty, is in fact manipulated with what amounts to great flexibility. On the other hand, while British development control leaves so much room for discretion that there appears to be little certainty, over the years policies have become established to a point that the system displays a great deal of certainty.

Perhaps a final aspect to point out is that a most important central government planning function is deciding the nature of the planning system we have. Many of the changes introduced are detailed in other chapters, but for illustration we can cite the move to two-tier planning – structure and local plans – in 1972, the introduction in 2004 of the Local Development Framework, and the abolition of regional plans (somewhat haltingly, owing to legal challenges) by the incoming government in 2010 as examples of changes in style of planning. Another important area, noted in the introduction, is a decision over 'how much' planning we have, or to put it more accurately how far planning or the market should shape patterns of development. So, the Thatcher government was ideologically committed to reducing the power and reach of the state, and this was reflected in planning by a reduction in importance of planning policy, to the point where it became one, but only one, of the considerations in determining a planning application. In relation to retail development, it was felt that retailers were much better able to judge than planners whether there was a need for new retail development, so tests of 'need' ceased to be a determining factor in deciding applications. However, the impact on town centres of a rash of out-of-centre developments became apparent and tighter controls over the location of retail development were introduced (see Chapter 6). Later, we had the introduction, with the 1990 Act, of section 54A and the requirement that decisions should normally be made in line with development plan policies (the 'plan-led' system, now defined by section 38(6) of the 2004 Act). Such changes can produce much debate and are often a source of tension between LPAs – and the planning profession more generally – and central government.

'Modernising government'

Modernising Government was the title of a White Paper produced by the Blair government in 1999 and it illustrated the style as well as the zeal of its attempt to change the nature of the governmental system. It was not, of course, the first attempt to change the way government works: that might be assigned to the Northcote-Trevelyan Report of 1854, which sought to place the civil service on a 'professional' footing and remove nepotism and corruption in recruitment, principles which still apply today. The quest for 'professionalism' has continued, with perhaps the next major appearance being associated with Harold Wilson's government in the 1960s, which noted that 'reform of the civil service was an essential element in the Government's programme for modernizing our principal institutions'.[9] More recent 'reforms' have tended to emphasise 'businesslike' approaches and a quest for efficiency – maybe 'professionalism' under a new

guise – starting with the 'Efficiency Reviews' under Sir Derrick Rayner promoted by Mrs Thatcher, through the 'Next Steps' programme, which created many separate agencies responsible for areas of government, to the more recent activities of the Coalition government elected in 2010.

There are perhaps three key themes of 'reform' evident in recent attempts to 'modernise' government. The first might be associated with the 1999 Cabinet Office report *Professional Policy Making for the Twenty First Century*, which emphasised the place of evidence-based policymaking in the development of the work of government. This model makes the claim that it 'uses the best available evidence from a wide range of sources' and 'learns from experience of what works and what doesn't' through the evaluation of past policy actions. Despite claims that the Coalition government was going to be radically different, its *Open Public Services* White Paper continues an emphasis on the use of evidence and evaluation results in policymaking and in guiding the work of government (Cabinet Office 2011a). Such an emphasis on evidence is reflected directly in the approaches to planning, where the 'tests of soundness' of development plans are centrally concerned with the adequacy of the evidence base for policies proposed. The 'evidence-based' approach suggests – or maybe hopes – that there can be a rational basis for all policy development, but for that to be possible we would need to take the politics out of policy, so maybe 'evidence-influenced or even evidence-aware is the best we can hope for' (Davies *et al.* 2000: 11) (the use of evidence in planning is further discussed in Chapter 6).

The second key theme is perhaps managerialism, which reflects a belief that private-sector management techniques would increase the economy, efficiency and effectiveness of the public sector. As well as being reflected in the Rayner Efficiency Reviews noted above, these three 'Es', as they became known, were present in a number of initiatives introduced at all levels of government, with emphasis on matters such as performance standards (and the measurement of performance) and 'value for money'. Whilst it is difficult to disagree with an objective of getting best value for money from scarce public resources, whether the use of performance indicators, management by results, 'market testing' and other such approaches are the best way to achieve this has been the subject of much debate: in the view of Rhodes (2013: 493) 'private sector techniques do not fit the context'.

The third theme is an emphasis on delivery and 'choice'. The latter was thought to represent something of a change in culture in the public sector, where members of the public moved from being seen as 'clients', where the professionals knew best and gave them what was needed, to being 'customers' who had a choice and therefore had be treated somewhat differently (Davies 1972). In the language of the *Open Public Services* White Paper, choice only properly exists if service delivery is 'opened up to a range of providers of different sizes and different sectors' (Cabinet Office 2011a: 8–9). This adds another pressure for the fragmentation of government, partly driven by what King and Crewe (2013: 202) identify as the belief that 'private sector firms could almost invariably be counted upon to outperform public sector bodies', further supporting moves towards 'outsourcing' activities. Whilst 'choice' and 'outsourcing' have not featured prominently in the delivery of planning services, aspects of culture change associated with the service delivery agenda have been evident.

Whilst this book is not the place to discuss the success or otherwise of these initiatives – however 'success' may be measured – it is worth noting that the changes that have been introduced have had an impact on how government 'works'. However, in the view of a number of commentators, the impact has not been as great as it might have been hoped by its promoters. Speaking to the House of Commons Select Committee on Public Administration in 2008, Professor Christopher Hood from Oxford University was of the opinion that promise had failed to translate into reality, identifying

> what I call Civil Service reform syndrome. What I am pointing to is a recurrent system which again I do not think is peculiar to one party or another, in which we go through changes in the executive machine in government through a process that involves excessive hype from the centre, selective

filtering at the extremities and what I call attention deficit syndrome at the top, so that we do not get follow-through and we do not get continuity.

DEVOLVED AND REGIONAL GOVERNMENT

Devolution to Scotland and Wales

The campaign for devolution to Scotland and Wales failed at the end of the 1970s, but succeeded twenty years later. The aftermath of the earlier failure proved to be an important factor in the later success. The 1979 collapse of devolution led to the defeat of the Labour government and eighteen years of Conservative governments bent not on devolving power, but on centralising it. During this period, the strength of the movement for devolution increased, particularly in Scotland, where the Thatcher government displayed a marked insensitivity to Scottish feelings. As Vernon Bogdanor has put it:

> The Thatcher Government's policies of competitive individualism were resented in both Scotland and Wales where they were seen as undermining traditional values of community solidarity; and policies such as privatisation and opting out from local authority control had little resonance there. But resented above all was the community charge, the poll tax. Only devolution, so it seemed, could protect Scotland and Wales against future outbursts of Thatcherism.
>
> (Bogdanor 1999: 195–6)

Following the publication of White Papers (*Scotland's Parliament* and *A Voice for Wales*), the Scotland Act and the Government of Wales Act were passed in 1998, with the first direct elections to the two national assemblies taking place in 1999. The very titles of the White Papers point to a major difference between them. Scotland has a Parliament with legislative powers over all matters not reserved to the UK Parliament. Wales has only executive functions, but it does have full powers in relation to subordinate legislation. The latter include environmental, housing, local government and planning functions. Thus Wales can change the provisions relating to the Use Classes Order, the General Development Order, and the General Development Procedure Order, as well as the regulations on planning applications. The Assembly also gives its views to the departments preparing new legislation about special provisions for Wales. While much legislation is shared with England, there are often, and increasingly, special sections devoted to Wales.

The devolution to Scotland and Wales is of importance to England for a variety of reasons. One of these was felt to be its effect on possible pressure for devolution to English regions. This might be fostered if Scotland and Wales were perceived to benefit economically from devolution at the expense of the poorer regions of England. Support for regional government was thought to be stronger in the north but it proved not to be sufficient to vote for regional government. Under the Coalition government there seemed to be an antipathy towards regions, as evidenced by the abolition of the regional tier of planning, so further development of regional voices seems unlikely. Under the Coalition it is also the case that the differences between national systems widened.

Scottish Executive

Scotland has had a special position in the machinery of government since the 1707 Act of Union. It has maintained its independent legal and judicial systems, its Bar, its established Church (Presbyterian) and its heraldic authority (Lord Lyon King-at-Arms). The Scottish Office has a long history and, even before devolution, had a large degree of independence from Whitehall (though note that this refers to the UK government's Scottish department, not the executive which belongs to the Scottish government). Many years of responsibility for Scottish services, the relative geographical remoteness (from London) of Edinburgh (perhaps essentially psychological), the nature of the distribution of people and economic activity, the vast

areas of open land, the close relationship between central and local administrators and politicians – such are the factors which gave Scottish administration a distinctive character. The departments included: Development (SEDD), Enterprise, Transport and Lifelong Learning (ETLLD) and Environment and Rural Affairs (SEERAD). Following devolution, the ministers for these functions are members of the Scottish Parliament.

The Scottish Executive and the Scottish Parliament have taken an active role in the definition of distinctive planning policy for Scotland. A 1999 consultation paper *Land Use Planning under a Scottish Parliament* issued by the Scottish Office set out the potential:

> The form of any national planning policy guidance which emerges from the Scottish Executive could have significant implications for statutory development plans. A national plan would almost certainly be perceived as unduly centralist and excessively rigid. However, guidance produced by the Scottish Parliament and Executive, bringing together the various National Planning Policy Guidelines and incorporating spatial issues more explicitly, might be attractive. This could inform future development in Scotland and provide some degree of consistency in the pursuit of sustainable development. It could be a vehicle for high level coordination of the objectives of the major agencies as they relate to development and land use. It could also prove attractive for those areas where progress with structure plans has been slow.

This gives some idea of early thinking on the way in which the new machinery might work. Later in the document there is a more certain statement: 'There is a clear expectation that all national strategic policy guidance will be subject to scrutiny by the Scottish Parliament.'

Planning in Scotland has developed its own character and characteristics over time. The National Planning Framework (NPF2) published in 2009 set out a national spatial strategy, something which England does not have. There is no formal regional tier of planning in Scotland, but city-region structure plans cover the main conurbations. There is thus a two-tier system of strategic development plans and local development plans in the four main city-regions – Glasgow, Edinburgh, Aberdeen and Dundee – and a single tier of development plans elsewhere. Other distinctive elements of planning guidance reflect the country's distinctive geographical characteristics.

National Assembly for Wales

In Wales, increasing responsibilities over a wide field were gradually transferred from Whitehall to the (former) Welsh Office. This transfer took many years to achieve. Welsh affairs were dealt with by the Home Secretary until 1960, with many services being administered directly by the departments which served England. There has been a minister responsible for Wales since 1951, but it was not until 1964 that the (Labour) government established the Welsh Office and a Secretary of State for Wales (Bogdanor 1999: 157–62).

Following devolution, the National Assembly for Wales (NAW) took over responsibility for a wide range of functions from the Welsh Office and other government departments: these are defined in section 7 of the Government of Wales Act 2006. Relevant to the fields covered in this book are ancient monuments and historic buildings, culture, economic development, environment, highways and transport, housing, local government, tourism, and town and country planning. All these functions are now transferred to the Assembly. Particularly important are the powers of secondary or subordinate legislation. This is in contrast to the Scottish Parliament, which has the wider powers of primary legislation.[10] However, in the field of town and country planning, the effective difference is smaller than might at first sight appear. This is because of the particular character of the British planning legislation. This provides only a very general framework for the substantive measures which are enacted in secondary legislation such as the Use Classes Order, the General Development Order, the General Development Procedure Order and a host of statutory rules and regulations (Bosworth and Shellens 1999). The latter deal

with such matters as advertisements, development plans, environmental impact assessments, inquiries procedures, and planning obligations. Planning in Wales is developing its own character, with planning guidance and the supporting technical advice notes and circulars providing more detail than the NPPF. There is also a Wales Spatial Plan, something missing in England. Additionally, of course, plans are the responsibility of the local authorities, now subject both to Welsh planning guidance and to approval by the Assembly.

The Assembly is both an executive and a deliberative body, and the executive is described as the Welsh Assembly Government (WAG). The Planning Division of the government is responsible for planning matters and falls within the responsibilities of the Minister for Housing and Regeneration. The Wales Spatial Strategy 'represents shared strategic direction, [and] the Wales Spatial Plan provides a canvas against which Welsh Assembly Government investment, both capital and revenue, can be considered and agreed' (WAG 2008: foreword). Heritage matters and the work of CADW, together with a range of environmental matters, fall within the responsibilities of the Minister of Culture and Sport. Matters of transport and economic development are part of the responsibilities of the Housing and Regeneration Ministry. Some of these functions are discussed in later chapters.

Northern Ireland Office

Government in Northern Ireland has had a unique character and structure. National government has performed, either directly or through agencies, virtually all governmental functions; local government has had few responsibilities. Though there are twenty-six elected district councils, their powers have been limited to matters such as building regulations, consumer protection, litter prevention, refuse collection and disposal, and street cleansing. The councils have nominated representatives on the various statutory bodies responsible for regional services such as education, health and personal social services, and the fire service. They also have had a consultative status in relation to a number of services, including planning. All the major services, including countryside policies, heritage, pollution control, urban regeneration, transport, roads, and town and country planning have been administered directly by the Northern Ireland Office. However, April 2015 will see the completion of a long process of local government reform in Northern Ireland. The twenty-six councils will be replaced by eleven new councils which will take over responsibility for a number of functions, including development planning, development management and enforcement, together with the authority in some areas of conservation, housing, regeneration and tourism. Many of these functions have previously been the responsibility of the DoENI. Most housing matters will continue to be administered by the Northern Ireland Housing Executive, which was formed in 1971 to take control of the local authority housing stock. Other departments include Agriculture and Economic Development.

Given the tragic history of Northern Ireland, the Office's priority aims have been significantly different from those of other parts of the UK: 'to create the conditions for a peaceful, stable and prosperous society in which the people of Northern Ireland may have the opportunity of exercising greater control over their own affairs'. Planning has had and will continue to have an important role in this. The Northern Ireland Assembly, created by the Good Friday Agreement of 1998, has a number of 'transferred responsibilities', which are matters under the direct control of the Assembly and the appropriate minister. However, these are matters which are not included amongst 'excepted matters', which Westminster retains indefinitely, and 'reserved matters', which may be transferred to the competence of the Northern Ireland Assembly at a future date. Transferred matters currently include health, education, agriculture and rural development, and policing and justice. Bills passed by the assembly are subject to Royal Assent to complete their process to becoming an Act.

The waxing and waning of English regionalism

England is a little unusual in Europe and the developed world in not having well-developed regional structures.

Whilst, since the Second World War a majority of countries have been developing regional structures (Marks *et al.* 2008), those that had been developed in England have been removed. Meanwhile, the range in inequality between the most and least prosperous regions in the UK is amongst the largest in Europe, and the difference – between London and the rest of the South East and other regions – has been persistent (Tomaney 2013). Attempts to address regional differences were first introduced in England in the late 1920s and early 1930s with the aim of addressing persistent high unemployment in some localities, and support for these localities continued after the war with a mix of incentives to develop in these areas coupled with restraints on development in London and the South East. However, the strength of this regional policy was gradually reduced from 1984.

For the most part, such interventions were an initiative of central government but were not ones which were accompanied by any form of national strategy, spatial or otherwise, and there was no substantial regional structure in place to complement these interventions. In the 1960s there was a form of national strategy supported by regional economic planning councils and boards. According to the minister responsible, George Brown, such bodies would be made up of individuals selected 'not as delegates or representatives of particular interests, but they will be widely representative of different interests within the region'. These interests would include industry, trade unions and other local interests. The economic planning councils' main function would be 'to assist in the formulation of regional plans and to advise on their implementation. They will have no executive powers.' The boards, on the other hand, would 'provide the necessary machinery for co-ordinating the work of government departments concerned with regional planning and development' (Hansard, HC Deb, 10 December 1964, Vol. 703, cc. 1831). These institutions continued to operate until they were abolished by the incomir government in 1979. However, by local authorities in most regions had jo standing conferences to consider re issues. Regional initiatives were bolst

government Green Paper and 1989 White Paper on *The Future of Development Plans*, which proposed the introduction of strong regional guidance within the planning system, and by the government issuing Strategic Guidance at a regional level from 1986 onwards.

According to Armstrong and Taylor (2000: 339), 'The 1980s and 1990s witnessed an extraordinary flowering of regional and local organisations participating in regional policy delivery. This is true not only for the UK but also for every other EU member state.'[11] As noted earlier, in Wales and Scotland this was expressed in the development of devolved administrations, whilst England saw the creation of three distinct regional institutions. First, the government formalised and provided more resources to *regional bodies* (RBs), sometimes known as regional chambers or assemblies. These comprised nominated elected representatives from local government and other community and business interests, which made them, effectively, indirectly elected bodies representing local interests. Second, *regional development agencies* (RDAs) were established with a specific remit to promote economic regeneration, drawing together funding from formerly national sources for a regional agenda. The RDAs worked with local partners to develop their regional agenda but were accountable to national government. Third, central government strengthened and integrated its presence at the regional level through *government offices* (GOs), with the offices also intended to provide some form of two-way conduit between the centre and the regions. This was central government operating at the regional level. While these were three distinct bodies, they shared much of the same agenda for their regions. What got done at the regional level relied on close cooperation and joint working among the three bodies, much of which was informal. While the regional body prepared the regional spatial strategy (for ratification by the Secretary of State), the regional development agency would be centrally involved in trying to ensure that it met its own agenda, and the government office would attempt to supervise the whole process. This sounds quite neat, but in practice things did not work quite so smoothly. Research in

Yorkshire and the Humber by Counsell *et al.* (2007: 400) identified a number of barriers to the emergence of a unified regional approach:

> The different cultures of professional groups involved in regional strategy work, the different requirements of strategies and the different timescales both for projecting policies forward into the future and for preparing and reviewing documents, suggest that it will be difficult if not impossible to achieve a fully and regionally integrated policy process.

The same research noted that steps were to be taken to attempt to address these issues, by requiring regional economic strategies to be prepared within the framework provided by the regional spatial strategy, but it noted that 'a degree of diversity will remain' (ibid).

The 1997 Labour Manifesto made a firm commitment to elected regional government, but only in those regions where there was a popular demand for it. Where regional government might be established, the government said that a unitary system of local government would be expected (that is the complete loss of counties) and thus no increase in tiers of government. Regional government would be given powers over economic development and regeneration (including control of the regional development agency), spatial development, housing, transport, skills and culture. The Secretary of State's powers to effectively approve and publish the regional strategies would have been devolved to regional government, but not call-in powers, which would stay at the centre. Legislative powers were not to be devolved, but would remain with the UK Parliament. The government published the White Paper on regional governance, *Your Region, Your Choice,* on 9 May 2002. It proposed to strengthen the existing regional institutions in England, and take forward the government's manifesto commitment on elected regional government. The Regional Assemblies (Preparations) Bill was introduced to Parliament on 14 November 2002 and the Act received Royal Assent on 8 May 2003. This Act enabled those regions that want to hold a referendum to have that chance.

The first referendum was held in the North East, a region with perhaps the longest and strongest history of regional institutions.[12] The verdict was, in the words of Deputy Prime Minister John Prescott, an 'emphatic defeat': 78 per cent of the 48 per cent of those voting rejected the idea of an elected regional assembly for the North East. There were, no doubt, a wide range of practical and cultural reasons for the rejection, but at the time it was often said that the assemblies would have insufficient power to make a real difference and would therefore be little more than, in the words of one 'no' campaigner, 'expensive talking shops'. Whether this marks the end for the development of regional bodies is hard to say. The effects of the devolved bodies in Scotland and Wales may prove to be a continuing prompt for action, as they were at the start of the 1997 Labour administration. And there does remain the concern that England remains an overly centralised state compared with others: in the view of the Joseph Rowntree Foundation (*Findings*, No. 2155, 2007, p. 1) 'the lack of devolved power in the eight English regions outside London has constrained the pursuit of locally appropriate solutions and weakened the legitimacy of regional and sub-regional bodies'. If such concerns become widely felt, then the pressure for change is unlikely to disappear permanently.

However, the Coalition government elected in 2010 showed a strong antipathy towards regional approaches, preferring any devolution from the centre to come from its 'localism' agenda. Following the announcement of the abolition of Regional Spatial Strategies, Secretary of State at the DCLG Eric Pickles gave a clue to the attitude to regional approaches and structures in several statements, such as: 'We have made it plain that our decision to remove Regional Strategies was based on clear evidence that they did not work.' (23 June 2011). 'The old regional classifications are also misleading . . . They are arbitrary lines on a map that have no resonance . . . Ministers reject the notion of a "Europe of the Regions" . . . Dismantling such arbitrary, unelected regional administrative structures will assist in that goal' (18 September 2012). Such political attitudes, together with the previous experience of the North East referendum, make it

unlikely that proposals for regional institutions will emerge in the near future.

From regional development agencies to local enterprise partnerships

RDAs were created in 1998 and formally launched on 1 April 1999. There was one for each NUTS level 1 region in England and they had five objectives: furthering economic development and regeneration; promoting business efficiency and competitiveness; promoting employment; enhancing skills; and contributing to sustainable development. They also took over responsibility from government offices for administering EU Structural Funds programmes. Each of the nine RDAs was run by a board of fifteen people drawn from business, local government, trade unions and voluntary organisations; the chair was a person from business. RDAs' funds were drawn from funds from six different central government depart-ments and in their final year of full operation (2010–11) their budget was £2.249 billion, which would be enhanced by contributions from other public and private sector bodies. So the National Audit Office (NAO) reported in 2010 that 'for every pound of RDA spend on physical regeneration, an estimated £2.80 is secured from elsewhere, including £1.51 from the private sector' (p. 5).

The Coalition's verdict that the RDAs were among those regional elements that 'did not work' received some limited support from the NAO report cited above. The report noted that 'since 1999, the eight RDAs outside of London have spent £5 billion on physical regeneration programmes. It is questionable, however, whether they could not have achieved even greater benefits from the £5 billion they have commit-ted' (p. 4). However, arguments for abolition were as much ideological as evidence-based and the govern-ment announced the abolition of the nine Regional Development Agencies in England – eight regional agencies through the Public Bodies Bill and the London Development Agency through the Localism Bill – on 22 June 2010. The Coalition Agreement included a commitment that the government would 'support the creation of Local Enterprise Partnerships – joint local authority–business bodies brought forward by local authorities themselves to promote local economic development – to replace Regional Development Agencies (RDAs)' (p. 10).

England is now covered by a network of thirty-nine *local enterprise partnerships* (LEPs) (see Figure 3.4). These were set up after local proposals and are said to be based on economic rather than political boundaries. The LEP board should consist of councillors nominated by participating local authorities, as well as having significant business representation. All LEPs should be business-led. There needs to be a clear system of accountability for the appointment of all board members. According to *Local Growth: Realising Every Place's Potential* (BIS 2010a: 13), 'Local enterprise partnerships will provide the clear vision and strategic leadership to drive sustainable private sector-led growth and job creation in their area.' They were given an extensive list of objectives[13] but initially little resource to underwrite the activity, other than £9 million to aid the start-up process. In the view of Ward and Hardy (2013: 9), 'if LEPs are to function effectively, they need money to do the job'. Michael Heseltine, in his report *No Stone Unturned in Pursuit of Growth* (2012), identified eighty-nine recommendations for promoting local growth, eighty-one of which were accepted by government. A number related directly to the work of LEPs and their resource base, including the creation of a 'single pot' of central government funding against which LEPs could bid for a five-year block of resources, starting in 2015–16; up to that period LEPs should be allocated £250,000 per year to create multi-year strategies; and the management of EU funding should be simplified to help LEPs exercise their functions in relation to the funds.

In addition to a concern about the resourcing of LEPs for their extensive and difficult menu of activities, concern has been expressed about accountability (House of Commons Business, Innovation and Skills Committee 2013), echoing the concerns about accountability of 'extra-governmental organisations' discussed above. Jones (2013: 92) draws parallels with the favoured economic development mechanism of the Thatcher government, training and enterprise councils (TECs):

> TECs, then, like LEPs, were unaccountable by design. Board directors were appointed as individuals and were not representative of any organisation that could remove or replace them. In practice, few TEC boards reflected the composition of the community, with little representation from women, ethnic minorities, and the voluntary sector. This design of governance limited local democracy and, more importantly, sought to secure a market-based hegemony of neoliberalism by preventing exterior interference.

At the time of writing, Labour has promised not to abolish but reform LEPs if they are elected in 2015, so they seem destined to remain part of the local development and regeneration landscape for some time.

The relationship of these bodies – RDAs and LEPs – to the planning process was not clear on their establishment. At the outset, RDAs 'had no formal planning duties', but over the period of their existence gradually 'strayed further into planning territory' (Tewdwr-Jones 2012: 46). This resulted in government setting up a process for the economic and planning strategies to be prepared by the RDA, but the change of government in 2010 meant that this was never put in place. At the time of their inception, LEPs also had no formal planning powers and they are not included in the priority list of those with whom the LPA has a 'duty to co-operate'. However, in some cases areas are preparing joint strategies, although there is no statutory requirement for this. Whether LEPs will follow the route of RDAs in relation to planning or whether the view of the former RTPI president Richard Summers that 'strategic planning and economic development are arguably more disconnected than before' will prevail remains to be seen. However, if the aims of spatial planning are to be realised, it will be important to achieve a level of integration between LEPs and the planning system, as the former have a significant part to play in policy implementation.

Figure 3.4 Network of local enterprise partnerships

1   North Eastern
2   Cumbria
3   Tees Valley
4   York, North Yorkshire and East Riding
5   Lancashire
6   Leeds City Region
7   Liverpool City Region
8   Greater Manchester
9   Humber
10 Sheffield City Region
11 Cheshire and Warrington
12 Derby, Derbyshire, Nottingham and
    Nottinghamshire
13 Greater Lincolnshire

14 Stoke-on-Trent and Staffordshire
15 Leicester and Leicestershire
16 The Marches
17 Black Country
18 Greater Birmingham and Solihull
19 Northamptonshire
20 Greater Cambridge and Greater Peterborough
21 New Anglia
22 Coventry and Warwickshire
23 Worcestershire
24 South East Midlands
25 Gloucestershire
26 Hertfordshire

27 Buckinghamshire, Thames Valley
28 Oxfordshire
29 London
30 Thames Valley, Berkshire
31 West of England
32 Swindon and Wiltshire
33 Enterprise M3
34 South East
35 Coast to Capital
36 Solent
37 Dorset
38 Heart of the South West
39 Cornwall and Isles of Scilly

Greater London Authority

The abolition of the Greater London Council in 1986 left a gap in the machinery of government, which was cumbersome, inefficient and indefensible. London became the only western capital city without an elected city government. Some functions carried out by the GLC were transferred in part to the London boroughs, but many were taken over by a range of joint bodies, committees, ad hoc agencies and such like (including the London Planning Advisory Committee, which prepared strategic planning guidance for the capital). The result was 'a degree of complexity that can be seen not so much as a "streamlining" as a return to the administrative tangle of the 19th century' (Wilson and Game 1998: 54). The election manifesto of the Labour Party promised a referendum to confirm popular demand for a strategic authority and mayor.

The referendum was held in 1998, and though only one-third of Londoners voted, there was an overall 72 per cent majority. The Greater London Authority Act 1999 provides for an elected mayor and an elected Greater London Assembly. The mayor is not a figurehead, but a highly influential leader. In the words of the 1998 White Paper *A Mayor and Assembly for London* (DETR 1998a), 'the Mayor will have a major role in improving the economic, social and environmental well-being of Londoners, and will be expected to do this by integrating key activities'. The main responsibilities include:

- The production of an integrated transport strategy for London (extending to transport issues for which the mayor is not directly responsible) to be implemented by a new executive agency Transport for London (TfL) which will have responsibility for a wide range of services including London's bus and light rail services, the Croydon Tramlink, the Docklands Light Railway, Victoria Coach Station, taxis and minicabs, and river services. It also acquires responsibility for a strategic London road network. Government funding is paid in a single block grant, and capital investment schemes within the budget available do not require central government approval.

- Preparation of strategic planning guidance for London in the form of a new Spatial Development Strategy (SDS). The content of this is for the mayor to decide, but includes transport, economic development and regeneration, housing, retail development and town centres, leisure facilities, heritage, waste management, and guidance for particular parts of London such as the central area and the existing Thames Policy Area (there are also other strategies, including transport). The unitary development plans of the Boroughs are required to be 'in general conformity' with the SDS. Development control remains with the Boroughs, but the mayor is a statutory consultee for planning applications of strategic importance, and has defined powers of intervention, which are already being used for significant applications.

- The setting of an economic development and regeneration strategy for London. A London Development Agency has been appointed by, and is responsible to, the mayor.

- Improvement of London's environment, the development of an air quality management strategic plan, the production of a report every four years on the state of the environment in London.

- Appointment of half the members of a new independent Metropolitan Police Authority, and scrutiny of the policies of the authority.

- Overall responsibility for a new London Fire and Civil Defence Authority, and appointment of the majority of its members.

- Preparation of a strategy for the development of the culture, media and leisure sectors, appointments and nominations to the key cultural organisations.

Clearly this is a highly significant change to the government of London, providing a clear indication of the then government's commitment to a more effective and democratic system of government. The position of mayor is not an easy one, since it involves extensive and intensive negotiation with the London Boroughs and innumerable governmental bodies, as well as many professional and voluntary organisations. The notion of mayors found favour with the incoming Coalition government, and referenda have been held for the election of mayors in a dozen other major cities in England. The current (2014) holder of the post of London mayor, Boris Johnson, has a high profile and has become a significant political figure.

LOCAL GOVERNMENT

The evolution of local government

The first attempt at imposing a coherent structure on local government in Britain was 125 years ago. Prior to that it had evolved in a somewhat piecemeal fashion from feudal patterns of bodies which were not at all well suited to dealing with the impacts of the industrial revolution. In the 1870s it was described as 'a chaos of areas, a chaos of authorities and a chaos of rates' (Byrne 1990: 14). The Local Government Acts 1888 and 1894 provided something of the building blocks for the pattern we see today, including government for London, counties and some unitary authorities (ones which undertake all functions for an area). Since then the fortunes of local government have waxed and waned. The range of functions assumed by the state grew from the 1890s and many of these became the operational responsibility of local government. However, after the Second World War functions such as utilities, health and aspects of social security were lost, and they also lost some of their financial independence as central government grants to local authorities grew in importance. The first radical reform of the structure of local government since 1888 was undertaken in the 1970s. This was not mainly as a result of the trends noted earlier but for a range of other reasons, including the fact that the boundaries of many local government bodies no longer reflected patterns of social and economic life, and some local authorities were too small to operate effectively in a more complex world.

The Royal Commission on Local Government, set up in 1966 and often associated with the name of its chair, Lord Redcliffe-Maude, may not have quite delivered the model for a renewed local government, but the report raised some issues which are still of relevance

today. It identified four principles for local government: the ability to deliver services efficiently; the ability to act as an integral part of the democratic process, through citizens being able to identify with the space and the institution; the ability to have sufficient strength to work in meaningful partnership with central government; and the ability adapt to change in the way people work and live. However, some of these objectives can be seen as conflicting. Generally, larger units are seen as more efficient for service delivery through being able to take advantage of economies of scale, but the larger the local authority becomes, the greater the risk of its becoming more distant from its electors. The recommendation of the Redcliffe-Maude Report was for a minimum size of 250,000 people for a local authority, something not achieved by the current pattern, but this still leaves Britain with the largest units of local government in Europe (Barrington 1991) and if we look at the two extremes, Britain has more than twenty times more people per councillor than France.

The legislation following the Redcliffe-Maude Report was the Local Government Act 1972, which took effect in April 1974. This established a two-tier system of local government in England and Wales consisting of forty-seven 'shire' counties which included 333 district councils; six metropolitan councils in the urbanised areas (such as Greater Manchester) which included thirty-six metropolitan district councils with unitary status; in London there was the Greater London Council with thirty-two unitary district councils, together with the City of London. In Scotland, following the Wheatley Report, the Local Government (Scotland) Act led to the establishment of a similar two-tier system consisting of nine 'regions' linked with fifty-three district councils and three 'island' councils. The two-tier system in London and the metropolitan counties wa'   to drastic change by the Thatcher goverr the banner of 'streamlining the (the Greater London Council county councils of Greater N South Yorkshire, Tyne and V West Yorkshire) was abolis' tier of local government in whilst the argument adva enhanced accountability and removal of waste, the imperatives were to a significant degree political, stemming from the fact that the Labour-controlled metropolitan councils had been particularly resistant to some policies introduced by the Conservative government. The impact of the change was felt to have removed an important element of strategic planning in education, transport and economic development, and left London as one of the few capital cities without a directly elected authority to represent its needs and its citizens.

In the early 1990s, a further review of local government was announced by the Secretary of State for the Environment, together with the Secretaries of State for Scotland and Wales. A presumption of the reviews was that a unitary system was to be favoured. The reviews by the Welsh Office and the Scottish Office proceeded relatively smoothly and resulted in a single-tier system of twenty-one councils in Wales and twenty-eight in Scotland. However, in England a Local Government Commission was established to work within policy and procedural guidance published in a consultation paper *The Structure of Local Government in England*, which stressed the not-unfamiliar criteria of efficiency, accountability, responsiveness and 'localness', alongside the desire to achieve a single-tier system of unitary authorities. This 'guidance' proved to have greater power than the government expected: it had the effect of limiting the changes that could be made to the Commission's proposals. Moreover, because of the consultative way in which the Commission operated, these proposals were significantly influenced by the views of articulate local interests.

During the two years of the Commission's review, district and county authorities sought to justify their existence through an expensive and sometimes bitter propaganda war. In fourteen cases this led to challenges to the Commission's recommendations in the courts. There was also a legal challenge by the Association of County Councils which successfully prevented the Secretary of State from modifying the guidance he had previously given to the Commission in an attempt to strengthen the case for unitary authorities. Despite ___ government's wish to see a unitary structure, the ___ undoubted winners were the counties. The

Commission found little evidence that change would improve service provision. In the main, changes were limited to renewing unitary status for former county boroughs, and abolishing new and contrived counties created in the 1974 reorganisation.

After much debate, the Commission recommended only fifty new unitary authorities. These were mostly former county boroughs (unitary authorities before the 1974 reforms), although a significant number of 'special cases' were included on the basis of 'substantial local support for change'. The Commission explained the modest extent of its recommendations as due to the 'weight of evidence from national organisations pointing to the problems and risks associated with a breaking up of county wide services' – a view that was strongly supported by local opinion. However, these arguments failed to satisfy the many districts which were not proposed for unitary status and which had campaigned for this. More significantly, it did not satisfy the government, which was concerned to further increase the number of unitary authorities. Following these disagreements, the chairman of the Commission resigned, and the new chairman was given the remit to review again the case of twenty-one districts where the government believed there was a strong case for unitary status. Further guidance was issued for this mini-review, stressing the potential benefits of unitary status particularly for areas needing economic regeneration (as in the Thames Gateway). It was argued that the 'single focus' of unitary local government would be more effective in promoting multi-agency programmes in these areas. This final review initially recommended unitary status for ten of the twenty-one districts, but this was reduced to eight after consultation.

Between 1995 and 1998 several more changes were made. These often resulted in the creation of additional unitary authorities. An example might be the City of York, created as a unitary authority (UA) from North Yorkshire County in 1996. Here the tight boundary of the city meant that many elements that were functionally part of the city, including the University of York, were part of other district councils' administrative areas. The city's boundaries were extended to include these 'functional' areas and adjoining rural communities, and the UA was created. In 2009, ten new UAs were created. These involved the county of Bedfordshire being abolished and split into two UAs, and the county of Cheshire being abolished and split into two UAs. Five complete counties were abolished and created as five separate UAs – Cornwall, County Durham, Northumberland, Shropshire and Wiltshire. However, following the election of the Coalition government in May 2010, plans to create two new UAs in Exeter and Suffolk were revoked by Parliament in 2011.

BOX 3.3 LOCAL AUTHORITY TYPES AND FUNCTIONS

Drawing a distinction between the type of local authority and the name it is given can help to reduce the confusion that arises about local government and the local planning authority. For example, the name 'city council' has only ceremonial significance and does not affect the competences of a council. Local authorities also describe themselves as boroughs, but outside London this is just a hangover from structures long abolished. One unitary council in England, Rutland, calls itself a county council, recognising its historical county status.

Scotland and Wales have 'all-purpose councils', known as *unitary authorities*, and they deal with all functions of local government. There are twenty-two unitary authorities in Wales and thirty-two in Scotland. The system in Northern Ireland is, at the time of writing, undergoing change to one of eleven councils with an enhanced range of functions.

England has a hybrid system. In a number of areas there are unitary authorities. There are thirty-six in the former metropolitan council areas, thirty-three London Boroughs (including the City of London) and fifty-six in other parts of the country. Other areas – principally more rural localities – have a two-tier system of twenty-seven *county councils* enveloping 201 *district councils*.

Functions carried out by each type of authority are shown in the table below. In addition, planning powers are exercised by National Parks.
There are also about 12,500 *parish* and *community councils* in England, Scotland and Wales.

Local authority responsibility for major services

	Metropolitan areas		Shire areas				London areas		
	District councils	Single-purpose authorities	Unitaries	County councils	District councils	Single-purpose authorities	London Boroughs	GLA	Single-purpose authorities
Education	✔		✔	✔			✔		
Highways	✔		✔	✔			✔	✔	
Transport planning	✔		✔	✔			✔	✔	

	Metropolitan areas		Shire areas				London areas		
	District councils	Single-purpose authorities	Unitaries	County councils	District councils	Single-purpose authorities	London Boroughs	GLA	Single-purpose authorities
Passenger transport		✔	✔	✔				✔	
Social care	✔		✔	✔			✔		
Housing	✔		✔		✔		✔		
Libraries	✔		✔	✔			✔		
Leisure and recreation	✔		✔		✔		✔		
Environmental health	✔		✔		✔		✔		
Public health	✔		✔	✔			✔		
Waste collection	✔		✔		✔		✔		
Waste disposal		✔	✔	✔			✔		✔
Planning applications	✔		✔		✔		✔		
Strategic planning	✔		✔	✔			✔	✔	
Police		✔				✔		✔	
Fire and rescue		✔	✔	✔		✔	✔	✔	
Local taxation	✔		✔		✔		✔		

Source: Adapted from LGA, *Analysis and Research* (2010)

Parish councils (or community or town councils) can play a role in the democratic process by providing an effective voice for local interests and concerns. Unlike their counterparts in Scotland and Wales, they have statutory functions, though these are limited. Of particular importance (and widely used) is their right to be consulted on planning applications in their areas. They can also play a part in the consultation process for the preparation of development plans, as well as undertake localised planning functions. Under the 'localism' regime, there are a number that have proceeded to play a role in the preparation of neighbourhood plans, though previously several played an active part in the preparation of design guidance for local development (village design statements) and the preparation of parish appraisals.[14]

Further reorganisation into unitary authorities took place in Scotland and Wales in 1996, and a number of unitary authorities were introduced in parts of non-metropolitan England between 1995 and 1997 (although much of the two-tier system remains). In Northern Ireland, a unitary system of local government was set up in 1973 but is due for change in 2015. Thus, while Northern Ireland, Wales and Scotland have a unitary system, England has a varied structure of local government in which some areas have a unitary system and the remainder are two-tier, made up of counties and districts.

Decision-making in local government

Local authorities comprise a representative system of elected councillors and an administrative system of employed staff, a significant proportion of whom will belong to the various professions encompassed by the services within the local authority's area of responsibility. A simplistic description might be that councillors decide on a policy and the officers are responsible for its implementation. However, such a description is indeed simplistic: the officers will have the skills, knowledge, experience and time to provide the basis for the development of a policy, proposals for which can be discussed, if necessary modified and then adopted by councillors; the councillors, on the other hand, will not be uninterested in implementation, as how the policy is experienced on the ground will be of great interest to their electors and to them as individuals. So there is an interdependence between the professional and political elements of local government, and almost all decisions will emerge from a dialogue between these two elements.

Councillors have traditionally had a reputation of being mainly white, middle-class, middle-aged and male (Widdicombe Committee 1986) and there has not been a great deal of change in this picture: the Communities and Local Government Committee in its report *Councillors on the Front Line* (2013: 3) noted that 'the membership of many local authorities does not reflect the demographic make-up of the communities they serve'. Part of the explanation for this lies in other characteristics of the 'job' of a councillor – it is, apart from allowances, unpaid and normally part-time, whilst being time-consuming, taking around 100 hours per month,[15] and all of these are factors which militate against the young employed taking up roles as a councillors. A councillor also has to perform multiple roles

– as a community representative, as a political party member (a majority of elected councillors stand as representatives of a political party), as a representative of particular interests, amongst others (Kitchen 1997) – so it can be a particularly challenging position.

Officers, on the other hand, have traditionally – particularly at a more senior level – had a professional specialism and qualification, marking them out from the established model of civil servants who have – certainly in the past – normally been generalists. However, they share the traditional picture of a civil servant as being politically neutral and thereby a source of 'objective' advice. As with the civil service, there has been some recent concern as to whether this quality has been eroded, for example by the appointment of senior officers who are in sympathy with the objectives of the locally dominant political party, although this might be the basis for establishing a good working synergy between officer and politician. There has also been a concern that changes to the way of working of local government, and in particular the introduction of competition into service delivery, have altered the approach of the service and have eroded the 'public service ethos' of officers (Morphet 2007). Such allwed overlap with questions of ethical behaviour, which is discussed later.

The structures within which councillors and officers work to take decisions were unchanged for many years, being founded on a number of service-based committees of councillors who made decisions on the basis of reports prepared by officers. The first substantial changes last century followed from the Bains Report of 1972 (the Patterson Report of 1973 in Scotland), which attempted to introduce a more corporate – as opposed to departmental/service-based – approach to the management of a local authority, but more fundamental changes were brought about by the Local Government Act 2000. In some ways this could be seen as a continuation of a quest for more corporate/strategic approaches, in that it swapped the committee model for one that to some looked more like Whitehall and to others had an echo of approaches in the USA (Hambleton and Sweeting 2004). Whilst the full council agrees a budget and policy framework, a 'cabinet' of a small number of members takes responsibility for the implementation of these broad policies. The cabinet consists of councillors with responsibility for a portfolio of service areas, with a cabinet leader appointed by the full council. In a small number of cases the cabinet leader may in fact be a directly elected mayor, and an example of a cabinet structure with a directly elected mayor, from Bristol City Council, is shown in Box 3.4.

BOX 3.4 AN EXAMPLE OF LOCAL GOVERNMENT ORGANISATION: BRISTOL CITY COUNCIL

What is the Cabinet?

The Cabinet (or 'Executive') is made up of the Mayor and executive members, whose role is to:

- provide leadership
- propose the budget and policy framework
- implement policy through chief officers.

The Mayor has ultimate responsibility for all major policy decisions. Whilst some of this responsibility may be delegated to/shared with executive members, the Bristol Mayor has decided to retain this function. Key decision taking occurs at Cabinet meetings and the Mayor will take decisions following discussion with his executive members.

Executive members are councillors with special responsibilities over an area of the council's activities, for example environment. Their area of responsibility is known as their portfolio. Executive members may work with council officers and others to develop policy within their portfolio, which then comes to Cabinet for formal approval.

The work and decisions of the Cabinet is scrutinised by the council's scrutiny commissions. Councillors have the right, in certain circumstances, to 'call in' a Cabinet decision for consideration by a scrutiny commission.

The makeup of the Cabinet is determined by the Mayor. The current Mayor's policy has been to invite councillors from all four parties who are represented on the Council into his Cabinet.

Although the Cabinet only includes the Mayor and executive members, the **Full Council** includes all members (ie councillors and the Mayor). The Full Council's role is to:

- approve or reject the budget and policy framework
- adopt the constitution
- appoint committees to be responsible for overview and scrutiny functions and regulatory matters.

Cabinet members and portfolios

The Cabinet has six members in addition to the Mayor. Their portfolios are:

1. Transport, planning, strategic housing and regeneration.
2. Leisure, tourism, licensing and community safety.
3. Finance and corporate services.

4. Neighbourhoods, environment and council housing.
5. Health and social care.
6. Children, young people and education.

The Mayor retains direct responsibility for other policy areas within his own portfolio.

Source: Bristol City Council website (www.bristol.gov.uk)

One of the objectives of the reordering was to remove the grip of a dominant party on decision-making. In some cases, where a political party had an absolute majority on the council, a party meeting was seen as where real decisions were made and this was not an arena open to public scrutiny. In this context, the 'scrutiny' role assumes a significant level of importance. Whether this change has made local government less prone to domination by party groups has been questioned (Ewbank 2011), but it would be more generally agreed that party politics continues to play a role in local government. The nature of party involvement has changed over time, with the emergence from the 1980s of more 'hung' councils – that is ones with no clear party majority, though the term 'coalition' tends not to be used in local government. Paradoxically, when we have had a coalition government in Whitehall, there is evidence of a decline in hung councils in local government (Game 2011).

The Localism Act 2011 has allowed councils to revert to the previous committee system, should they so choose. So far only a limited number of councils have chosen to go down this route. Generally, such a change has not meant the demise of the scrutiny function, so what has happened should not be characterised as 'going back' to a committee system. However, whatever the system of organisation in operation is, decisions in local government are the product of both professional and political inputs. Planning decisions are made within this framework, with decisions on planning applications made in an authority-wide committee or in an area-based committee, depending in part on the size of the authority. However, many decisions of a non-controversial nature are delegated to officers, although there are

guidance and scrutiny processes to regulate this activity. Decisions on planning policies, such as the adoption of a new development plan document, are often matters for the cabinet, though this would frequently be on the basis of advice from a panel of councillors who had given detailed consideration to proposals.

Local government in Scotland

Even before devolution, the cultural history and physical conditions of Scotland dictated that, to a degree, the administration of planning is distinctive. Changes to the law in Scotland require specific legislation, and the Scottish Office had long had administrative discretion within which it could take account of the special circumstances that exist in parts of the country. Nevertheless, the broad thrust and impact of policy have been much the same (Carmichael 1992).

In setting out to reorganise Scottish local government, the government was firmly committed to a single-tier structure, and the 1992 Scottish consultative paper provided options only on the number that were to be established. There were, of course, some political factors involved in this decision: the problem of conflicting interests within the Conservative Party was much less in Scotland, since only a handful of the sixty-five Scottish local authorities were in Conservative control. The consultation document in Scotland was also more forthright about the role of local government reform in direct service provision. While the government confirmed its commitment to 'a strong and effective local authority sector', it also argued that local authorities no longer needed to 'maintain a comprehensive range of expertise within their own organisation', since 'much could be done by outside contractors'.

In reviewing the possible number and size of the proposed unitary authorities, a consultation paper provided four illustrations showing structures ranging from fifteen to fifty-one authorities. The outcome of reorganisation in Scotland was thirty-two unitary councils, each of which had full planning powers for its area. The fragmentation of the strategic planning function across a larger number of authorities threatens a recognised strength of the Scottish system, and the need for special arrangements for strategic planning was acknowledged by the Scottish Office during the review. As noted earlier, subsequent reform of the planning system allowed for the four largest cities and their surrounding areas – Aberdeen, Dundee, Edinburgh and Glasgow – to prepare 'strategic development plans'. These set out long-term visions for the city-regions and deal with region-wide issues such as housing and transport. The plan framework is discussed further in Chapter 4.

Scottish legislation provides for the establishment of community councils where there is a demand for them, under schemes prepared by local authorities. As in England and Wales, their purpose is to represent the local community and 'to take such action in the interests of the community as appears to its members to be desirable and practicable'. Unlike in England and Wales, community councils do not have the right to raise funds by setting a precept on local taxes, and are instead dependent upon local authority funding, which is usually received for running costs only, and other grants for which they can apply. A study of community councils concluded that 'in contemporary moves towards democratic renewal in local government, community councils are seen as having no special status or role by most local authorities, though some do accord them a distinctive role in consultation, and there is a wide variety throughout Scotland in their operations and effectiveness' (Goodlad et al. 1999: 5). Perhaps partly because of their limited powers and resources, a number of community councils have folded, and in 2011 only seven of the thirty-two local authorities could boast a full complement of active community councils. The

1999 McIntosh Report, *Moving Forward: Local Government and the Scottish Parliament*, felt that the community council movement 'stands in need of some renewal', with one of its recommendations being that 'they have adequate resources to do what they have to do'.

The *Local Government in Scotland Act* 2003 made some moves to 'modernise' Scottish local government. In addition to placing an obligation to obtain 'Best Value' on councils, the Act gave local authorities a 'general power to advance well-being' – in effect a power to do many things not prohibited by existing legislation – and a duty to develop a community planning process involving public, private and voluntary sector partners. These partnerships operate in all thirty-two local authority areas. Two further changes were introduced in 2007: a switch to a type of proportional representation for voting in local elections – the single transferrable vote system operating in multi-member wards – and the introduction of salaries for elected councillors to replace the previous system of allowances.

Scottish local government and the Scottish Parliament

The establishment of the Scottish Parliament raises a host of questions concerning local government, some of which have long been of importance (such as public apathy and mistrust: Carole Millar Research 1999), some of which arise because of devolution (particularly relationships between Parliament and local government), while others have arisen on the tide of reform which devolution has created (such as the electoral system). Whatever the reason, there has been a major endeavour to improve the system of governance in Scotland.

The Commission on Local Government and the Scottish Parliament (the McIntosh Report) recommended a number of wide-ranging proposals for reforming local government in the context of devolution. Its starting point is a declaration that 'relations between local government and the Parliament ought to be conducted on the basis of mutual respect and parity of esteem . . . Councils, like Parliament, are democratically elected and consequently have their own legitimacy as part of the whole system of governance.' It is on the basis of this type of thinking (and a long list of specific recommendations) that the Commission stressed that major changes lay ahead for Scottish local government. Among its recommendations are the ratification of a covenant between the Parliament and the thirty-two councils setting out their working relationship. This concept of a direct working relationship between local and central government is 'without parallel or precedent at Westminster', though it is in harmony with the *European Charter of Local Self-Government*, as well as the Hunt Report.[16]

These recommendations were translated into reality by the 'concordat' of 2007. This was described as 'a fundamental shift in the relationship between the Scottish government and local government, based on mutual respect' (Scottish Government and COSLA 2007: 1), and as such would give 'local authorities substantially greater flexibility and also greater responsibility. In future, the onus will be increasingly on authorities to reach decisions on where money should be spent to achieve agreed outcomes' (p. 7). In practice, this meant a reduction in 'ring fencing' central funding, freezing council tax and the introduction of 'Single Outcome Agreements'[17] between the community planning partnerships and the government. However, public expenditure cuts from 2010 placed some strain on the relationship underpinning these agreements.

This selection of recommendations gives some flavour of the extent to which Scottish local government was under fundamental review. In addition, a new ethical framework for local government in Scotland was established.[18] This included a review of aspects of the planning process, such as the training of members for the work of a planning committee, and the introduction of 'best practice' (see the discussion below). As with the somewhat more modest English ideas for modernising local government, these were big aspirations which could be met only by major changes in the culture of local government (Brooks 1999: 43).

Local government in Wales

The reorganisation of local government in Wales proceeded more quickly than in England. The review was carried out by the Welsh Office (rather than by an independent commission) and the country was considered as a whole (rather than by separate areas). After a two-year period of consultation, in 1993 a White Paper *Local Government in Wales: A Charter for the Future* was published, setting out detailed proposals. There was widespread agreement that the new structure should be unitary in character, and the debate was focused on the number and boundaries of the new local authorities.

The underlying thinking included a restoration of authorities which had been swept up in an earlier reorganisation (Cardiff, Swansea and Newport) and some of the traditional counties, such as Pembrokeshire and Anglesey. However, to fit into a unitary structure, the boundaries had to be stretched somewhat, and a number of counties had to be amalgamated. After consideration of proposals for thirteen, twenty and twenty-four unitary councils, the final outcome was twenty-two authorities. (In a reflection of their history, these have varied formal names such as county borough, city and county council, but they all have the same functions.) They range in population from 66,000 in Cardiganshire to 318,000 in Cardiff.

In the White Paper, the unitary system was commended for its administrative simplicity, its roots in history, its familiarity and the relative ease with which residents could identify 'with their own communities and localities'. The intention was to create 'good local government which is close to the communities it serves'. The White Paper continued:

> Its aims are to establish authorities which, so far as possible, are based on that strong sense of community identity that is such an important feature of Welsh life; which are clearly accessible to local people; which can, by taking full advantage of the 'enabling' role of local government, operate in an efficient and responsive way; and which will work with each other, and with other agencies, to promote the well-being of those they serve.
>
> (Welsh Office 1993: 2)

These desirable objectives do not all work in the same direction, of course, and some compromise was inevitable. Some of the areas are very large: Powys, for example, is over 500,000 hectares.

A network of 730 community councils exists, much like parish councils in England, and has been described by the Welsh government as 'the grassroots level of local governance in Wales' (WAG 2014). Also like parish councils in England, they vary in size, from 200 to 45,000, and in what they achieve.

Welsh local government and the Welsh Assembly

Preparations for devolution in Wales were far less advanced than they were in Scotland, where there was a much firmer expectation that devolution would in fact take place. The first steps included the mounting of a consultation exercise on the establishment of a Partnership Council with local government which was mandated by the Government of Wales Act 1998. It now consists of twenty-five members together with four observers: twelve from the Welsh government, eight from the local authorities, one from the community councils, and one each from the police authorities, the fire authorities, health and the national park authorities. The Government of Wales Act 2006 effected a formal separation between the Welsh Government and the National Assembly for Wales, and, as a consequence, the Welsh Ministers are responsible for discharging the various duties previously laid on the National Assembly for Wales to establish statutory partnership arrangements. The Local Government Partnership Scheme of 2008, which deals with matters such as the organisation and improvement of services and a budgetary framework, states that 'The Welsh Government and local government in Wales are committed to working together in partnership, within an atmosphere of mutual trust and respect, recognising the value and legitimacy of the roles both have to play in the governance of Wales.' However, a report by Martin *et al.* (2013: 7) found that 'only four in ten [local government officers surveyed] believed that the Government is willing to collaborate. Most described

their interactions with government in terms of command and competition, with partnership coming in a distant third.' The same report suggested that, in order to make best use of resources in Wales, 'Government needs to continue to encourage partnership working between authorities, or to reorganise local government in order to create larger councils and, perhaps, consider whether the current division of responsibilities between local and central government is right' (p. 9).

Planning in Wales is based on the English system but it has developed some distinctive features since planning has been devolved to the Welsh Government. In particular, Wales now has a Spatial Plan, last updated in 2012, and Welsh Ministers have a duty under the Government of Wales Act 2006 to promote sustainable development: it is a principle of the Wales Spatial Plan that development should be sustainable. The context for planning policy in Wales is contained in two main documents, *Planning Policy Wales*, which provides guidance on the preparation and content of development plans and advice on development control decisions and appeals; and *Minerals Planning Policy Wales*, which gives guidance for the extraction of all minerals and other substances in, on or under land. They are supplemented by a series of twenty-one topic-based *technical advice notes* (TANs) and two *minerals technical advice notes* (MTANs), together, with circulars. This means that there is more detailed planning guidance in Wales than is now offered by the NPPF in England. Further details of planning policy in Wales are given in subsequent chapters.

Local government in Northern Ireland

Local government in Northern Ireland was reorganised in 1973, when thirty-eight authorities, made up of counties, county and municipal boroughs and urban districts, were replaced by a single tier of twenty-six district councils. Although this reduced the enormous variation in the size of districts (previously ranging from 2,000 to over 400,000), a wide variation remained, from Moyle with a population of some

15,000 to Belfast City with a population approaching 300,000. Planning powers were centralised under the then Northern Ireland Ministry of Development. Following the demise of the power-sharing Northern Ireland Assembly in 1974, planning, like all public services, was subject to 'direct rule' under the supervision of the Secretary of State for Northern Ireland. The preparation of plans and the control of development are functions of the Department of the Environment for Northern Ireland, which it exercises through the Planning Service (an executive agency). Local government is consulted only on the preparation of plans and development control matters.

The lack of accountability through local government (described as the 'democratic deficit') obviously needed to be seen in the light of the very special circumstances, though it has been judged to have operated with a 'considerable measure of success' (Hendry 1992: 84). Nevertheless, local councillors were able to attack planning and to 'represent themselves as the champions of the local electorate against the imposed rule of central government' (Hendry 1989: 121). Even when the central bureaucracy made determined efforts to open decision-making and involve local people, it was accused of having ulterior motives (Blackman 1991).

The promise of a 'lasting peace' in Northern Ireland brought with it ideas for reform which will come into effect in 2015. The twenty-six existing councils will be replaced by eleven new councils which will assume a range of functions, including planning, roads, urban regeneration, community development, housing, local economic development and local tourism. The transfer of functions will also include some elements of the delivery of the EU Rural Development Programme, spot listing of buildings and greater involvement of local government in local sports decisions. It is also proposed that councils will have a new statutory duty of community planning and a general power of competence. Community planning will provide a framework within which Councils, departments, statutory bodies and other relevant agencies and sectors can work together to develop and implement a shared vision for promoting the economic, social and environmental well-being of their area based on effective engagement with the community. The General Power

of Competence will enable a council, in broad terms, to act with similar freedom to an individual, unless there is a law to prevent it from doing so.

'Joining up' at the local level

The shift from local government as 'provider' to 'enabler' and the privatisation of much service provision since the 1980s have been accompanied by the dispersal of competences (responsibilities and powers) to many other agencies of government, voluntary bodies and the private sector. This has been described as a shift from government to governance. With many actors involved, getting sensible coordination of policy and action is a considerable challenge. In many ways it has exacerbated a longer-standing problem sometimes referred to as 'silo government', where the determining factor of what gets done is functional – professional, policy or spending groups – as opposed to an approach that looks at the needs of an area or a group and deploys interventions from all functional professional/policy/spending silos to meet these needs. In the 1990s, terms such as 'joined-up government', 'wicked issues', 'area-based initiatives' and 'partnership working' were used to refer to attempts to address this problem. HM Treasury's *Autumn Statement 2012* pointed to the genesis of the preferred solution to these problems lying in increased devolution of decision-making to local partnerships (pp. 40–1). The logic for this was explained by Willis and Jeffares (2012: 542) in the following terms:

> The doctrine that the solution to such 'wicked issues' was to encourage or require partnership working was endorsed in both academic and political fora with widespread enthusiasm. It was variously argued that partnership working would result in the more efficient and effective use of public resources and community empowerment.

As highlighted in Chapter 11, 'partnership' or at least inter-organisation working has been the hallmark of much policymaking and implementation, especially in the field of regeneration. This has benefits, but the complexity of partnership working, with large numbers of overlapping and often ad hoc arrangements, is not without its problems, not least in the areas of accountability and transparency.

The 2000s saw the development of a number of initiatives intended to address problems of 'joining up'. The ODPM promoted the notion of *local strategic partnerships* (LSPs) 'to provide a single overarching local coordination framework within which other, more specific local partnerships can operate'. LSPs included representatives from public, private, community and voluntary sectors and took a lead role in the preparation of the community strategy (there is some discussion on community strategies in Chapter 4). *Local area agreements*[19] (LAAs) were three-year agreements between central government and a local area, working through its Local Strategic Partnership. They contained a set of improvement targets which local organisations are committed to achieving and a delivery plan setting out what each partner is intending to do to achieve those targets. They bear significant similarities to the arrangements in the devolved administrations. In October 2010, Eric Pickles announced that he was abolishing LAAs, ostensibly on the grounds that they were 'too bureaucratic', adding that he did not want to be an 'overbearing parent, handing out pocket money and telling you how it should be spent' (Pickles 2010). The Total Place initiative had similar objectives. It was piloted in 2009, with a report produced by HM Treasury in March 2010 with the title *Total Place: A Whole Area Approach to Public Services*. Although progress was interrupted by the change of government in May 2010, there were similarities between its approach and that of the Coalition government. The Total Place paper stated:

> We will work with consistently high performing places to develop a 'single offer' for those places. This offer will give places a range of freedoms (freedoms from central performance and financial control as well as freedoms and incentives for local collaboration) for working in partnership with central government to co-design services and arrangements to deliver greater transparency, efficiency and value for the citizen and the public purse.
>
> (HM Treasury 2010a: 10)

Pressure on public sector spending gave an additional impetus to efforts to integrate public-sector programmes, as integrated approaches are seen as a way to achieve more cost-effective delivery and eliminate duplication. However, a report by the National Audit Office (2011: 6) commented that past efforts 'did not lead to widespread or fundamental changes in local public services or in the relationship between central and local government'. Nonetheless, a programme of 'whole place community budgets' was launched in 2011. There are indeed significant similarities between the objectives of this programme and Total Place. However, there are also significant similarities between the initial verdict of the House of Commons Communities and Local Government Committee on whole place community budgets and earlier research into other 'joining-up' initiatives.[20] A witness to the Committee commented:

> If you look back at the previous year, not one pound from other Departments [than DCLG] was committed to the budgets, so commitment on a systemic basis is just not there. Budgets are not the whole answer – absolutely not, but they are a sine qua non, in our view, of getting a formal commitment to the need to join up services.

'City Deals' announced in 2011 and commenced in 2012 is a similar initiative to that of community budgets, but they were developed by the Treasury working together with the Deputy Prime Minister. They consist of central government offering the devolution of budgets and decision-making power on a bespoke basis to individual local authorities or groups of authorities. Their main focus is on economic investment, transport, employment, skills and education. There are overlaps between city deals and the establishment of local enterprise partnerships. The report *No Stone Unturned*, produced by Lord Heseltine in late 2012, proposed the amalgamation of a large number of budgets into single economic development-related pots for local areas, which would then be devolved to local enterprise partnerships. This proposal was in part accepted by government.

However, moving in the direction of more joined up approaches will require fundamental changes in government and particularly central–local relations. Early findings from the national evaluation of LSPs highlighted the fundamental problem. Lambert concluded that 'the long standing silo culture of the UK government system is confirmed, driven by sector specific objectives and performance measures, and embedded in established policy and professional communities, legal frameworks and funding regimes' (Lambert 2004: 5). Even though Eric Pickles has espoused the idea of fewer centrally determined targets for such initiatives, the impact on their success has yet to be convincingly demonstrated.

Managing planning at the local level

For town and country planning, the apparent and seemingly paradoxical outcome of change in the 1980s and 1990s was a larger and stronger body of planners with strengthened statutory functions. However, financial austerity and its application to local government some years later acted to reduce the resource available. The indirect effect of market deregulation, the increasing complexity of development issues and the growing emphasis on environmental protection were bound to lead to a greater demand for planning skills (Healey 1989); however, a corollary of the fall in volume of development after 2008 was a fall in numbers of applications handled and a pressure for reductions in staff.

The emergence of the concept of an enabling local government also increased the need for strategic thinking and focused attention on the corporate (rather than land use) planning function (Carter *et al.* 1991) within local government. The direct impact on the way in which a spatial planning service is delivered has taken longer to come through, perhaps because it represents, for some, a cultural change. During the 1990s, planning was subject to only minimal change in comparison with other local authority services, and

partly because of its statutory and regulatory functions was somewhat protected from the pressure for change. Compulsory competitive tendering (CCT) introduced by the Conservative government in the 1980s was not applied to planning. However, in the new century there was some development of outsourcing of planning services, with examples being found in Salford, North East Lincolnshire, North Tyneside and Barnet, with the argument in favour being one of enhancing 'efficiency'. These decisions were sometimes a source of controversy.[21] More usual has been the outsourcing of specialist aspects of planning, such as analyses in the areas of housing, employment and retail, where skills might not be available 'in-house', especially in the case of small authorities.

Over the past thirty years, there has been strong pressure for change involving both sticks and carrots. The spread of auditing and value for money (VFM) was given a new gloss by the Labour government with the concept of 'Best Value'. These are not easy concepts to define for planning because of the difficulty of assessing quality in plans and planning decisions. The Audit Commission provided guidance for local authorities and district auditors on performance indicators for all services, including planning, but these were criticised for the reliance on quantitative measures, the classic example of which was that of the proportion of applications decided within eight weeks.

Best Value required that performance reviews were expected to look ahead over a five-year period, starting with areas of work in which there were problems. The reviews had to challenge why and how the service was being provided; invite comparison with others' performance across a range of relevant indicators; involve consultation with local taxpayers, service users and the wider business community in the setting of new performance targets; and embrace fair competition as a means of securing efficient and effective services. The reviews produced new performance targets to be published in an annual local performance plan together with comparisons with other authorities; identification of forward targets for all services annually and in the longer term (at least five years); and commentary on how the targets will be achieved, including proposed changes to procedures. Local performance plans are audited.

The Audit Commission's 1992 report *Building in Quality* addressed criticisms of the accent on efficiency rather than effectiveness of the planning system in performance review. It made a real attempt to introduce a wider assessment, recognising that there were many ancillary tasks in providing advice and negotiating with applicants and making 'complex professional and political judgments'. After consultation, the Audit Commission settled on six key 'best value performance indicators' (or BVPIs as they are inevitably called). As with the earlier version, these concentrated on matters of efficiency rather than on the effectiveness of the system, though the added breadth of performance review was felt to be a significant improvement on previous practice. Because it is easier to measure the throughput of applications rather than the achievement of strategic objectives, the indicators were mostly concerned with the development management function and are discussed further in Chapter 5.

The 1998 White Paper *Modern Local Government: In Touch with the People* (DETR 1998c) succinctly described the duty of local authorities 'to deliver services to clear standards – covering both cost and quality – by the most effective, economic and efficient means available' (para. 7.2). This is essentially a positive recasting of the enabling concept. (It is the norm in western Europe, with local government implementing its functions through a diverse range of agencies, often in partnership, but essentially seeing its role as prioritising community needs and acting as the focus for local political activity.) Best Value was also seen as an aid to local government 'to address the cross cutting issues facing their citizens and communities, such as community safety or sustainable development, which are beyond the reach of a single service or service provider' (para 7.3). The very best-performing councils were eligible for 'beacon' status normally for particular services. Applicant councils were chosen by an independent advisory panel, and rewarded by being given wider discretion in the operation of the beacon service.

The Coalition government continued the emphasis on achieving speedy development management decisions and a rapid coverage of development plans. The emphasis here seemed to be somewhat more on sticks than carrots, in that the Growth and Infrastructure Act 2013 set out provisions which enabled the Secretary of State to designate an LPA as poorly performing on the basis of its performance on major applications. Designation will mean that applicants may choose for their major application to be handled by either the LPA or the Planning Inspectorate. This provision chimed with the general rhetoric of the time, characterising planning as a barrier to growth and seeking to remove 'bureaucratic' barriers to development.

The ethical local authority

Local government has long had a reputation for probity, particularly in planning, where foreign observers are quick to point out the obvious opportunities (some may say temptations) for corruption. That the temptations have not always been resisted is now well known. It was in the 1970s that the Poulson Scandal blew up. Several local authority politicians and officials were found guilty of securing contracts for the architectural business of John Poulson. A number of well-known figures went to jail. It was an extreme case which shocked the local government world. It led to the setting up of a Royal Commission on Standards of Conduct in Public Life and to the introduction of a National Code of Local Government Conduct. There have been other cases of local government impropriety particularly in the planning arena, such as those in Warwick, Bassetlaw (Nottinghamshire) and Newark.[22] Perhaps the most high-profile case of recent years was in Doncaster, which came to be referred to as 'Donnygate'[23] and which led to lengthy jail terms for a councillor and a developer.

Though only a small number of such cases have arisen, they are clearly unacceptable. In fact, the numbers involved were certainly lower than the number of cases of Westminster 'sleaze' (a conveniently vague and all-embracing term) in the later years of the Conservative government.[24] This led to the appointment of a Committee on Standards in Public Life, under the chairmanship of Lord Nolan, which embarked upon a series of inquiries into various areas of public life and identified a number of 'principles of public service'.[25] Its third report, devoted to local government, was published in 1997.[26]

The Nolan inquiry was concerned not with putting local government on trial but with providing guidance on what standards of conduct should apply and how they could be maintained. The Local Government Act 2000 went beyond what had been proposed. It created a statutory code of conduct for local authority members, independently chaired statutory standards committees for each principal local authority, an independent regulator of local authority standards (the Standards Board for England) and a separate independent body (the Adjudication Panel for England) to which the most serious cases could be referred. Adjustments were made in 2007, following recommendations from the Committee in the light of criticism of the system as bureaucratic and bogged down with trivial complaints. The Standards Board (which became Standards for England) was made more strategic. Local standards committees and monitoring officers were given responsibility for filtering complaints, and 2007 also saw the issue of guidance on behaviour for councillors in the Local Authorities (Model Code of Conduct) Order 2007 (SI 2007/1159).

The Localism Act 2011 abolished Standards for England, removed the ability of local authorities to suspend members as a sanction for poor behaviour and disbanded local standards committees. It introduced a new offence of failing to declare or register a pecuniary interest. It also required local authorities to develop their own code of conduct based on the seven principles of public life and to appoint an Independent Person to be consulted during the investigation of any complaint. However, in its 2013 report *Standards Matter*, the Committee expressed some concerns about these changes:

> The new, slimmed down arrangements have yet to prove themselves sufficient for their purpose. We have considerable doubt that they will succeed in doing so and intend to monitor the situation closely. The arrangements place a particular onus on the Local Government Association to provide leadership for the sector and to ensure that they work in practice.
>
> (Committee on Standards in Public Life 2013: 22)

It also drew attention to the additional challenge posed by the changes in patterns of service delivery becoming increasingly common in local government, where more services are likely to be delivered by contractors. It stated, 'We strongly believe that the ethical standards captured by the seven principles should also apply to such individuals and their organisations' (p. 24) and recommended that the seven principles should be enshrined in contracts.

Planning has sometimes been seen as a particularly difficult are, in which additional problems may be faced and additional guidance can be needed. One of the reasons for this is that decisions can be particularly controversial. A guide issued in 2013 by the Planning Advisory Service noted that this was related to the fact that 'planning decisions are based on balancing competing interests and making an informed judgement against a local and national policy framework', 'the openness of a system which invites public opinion before taking decisions and the legal nature of the development plan and decision notices' (PAS 2013a: 4). This complexity has been seen as a reason for ensuring that councillors receive training to prepare them for the challenges.

Further reading

European government

The institutions and policies of the EU are well explained on the EU website (http://europa.eu/index_en.htm), and this has the virtue of being up to date. For a summary of the history of the EU, see Borchardt (1995) *European Integration: The Origins and Growth of the European Community*.

There are a great many general accounts of the making of the European Union, including the very comprehensive *Encyclopaedia of the European Union* edited by Dinan (1998). For a more critical account, see Chisholm (1995) *Britain on the Edge of Europe*; for a more theoretical view of European integration, see Nelsen and Stubb (2003) *The European Union* and Emerson (1998) *Redrawing the Map of Europe*. Williams (1996) *European Union Spatial*

Policy and Planning gives an account of the European institutions from a planning perspective.

A chronological review of how Europe has influenced planning is given in Nadin (1999) 'British planning in its European context'. Two DETR research reports address the consequences of the development of European policies for planning in the UK: Nadin and Shaw (1999) *Subsidiarity and Proportionality in Spatial Planning Activities in the European Union* and Wilkinson *et al.* (1998) *The Impact of the EU on the UK Planning System*. See also Bishop *et al.* (2000) 'From spatial to local: the impact of the European Union on local authority planning in the UK' and Shaw *et al.* (2000) *Regional Planning and Development in Europe*. Further references on European spatial planning are given at the end of Chapter 4.

On European comparative planning systems, including the organisation of government, see the *EU Compendium of Spatial Planning Systems and Policies*. Whilst this was compiled in 1997 and covers the 'Europe of the 15' as opposed to the current group, it is a good place to start (there are volumes describing the systems and policies of spatial planning in each member state, but also a summary volume available at http://commin.org/upload/ Glossaries/European_Glossary/EU_compendium_ No_28_of_1997.pdf (accessed 13 June 2014)).

Central government

A good, up-to-date introduction can be found in *Politics UK* by Jones and Norton (2013). A more reflective review of how government works can be found in Rhodes (2011) *Everyday Life in British Government*.

A summary description of government departments and their functions has been published as *Britain: An Official Handbook*, prepared by the Office for National Statistics, but this has been superseded by web-based resources such as can be found on the www.gov.uk and www.parliament. uk sites. Legislation (dating back to Acts passed in 1267 for those who are interested) is accessible via the government's website (www.legislation.gov.uk). The gov. uk site is intended to provide ready access to current policies and debates but offers less easy access to historical

material than was found on the departmental websites it replaced. A principal source of information on the work of government departments is the departmental annual reports. These are the government's expenditure plans for the forthcoming three years and are sometimes referenced in this way. The devolved administrations also offer useful web-based resources on functions and policies.

The relationship between central and local government has been reviewed by the House of Commons Political and Constitutional Reform Committee, and their report *Prospects for Codifying the Relationship between Central and Local Government* (2013) provides an interesting overview. Walker (2000) *Living with Ambiguity: The Relationship Between Central and Local Government* offers further interesting insights.

Devolved and regional government

On devolution an essential book is Bogdanor (1999) *Devolution in the United Kingdom*, which has an extensive bibliography. Prior to devolution there were several texts produced on the likely consequences, which now make interesting reading. Amongst these are Hazell (1999) *Constitutional Futures: A History of the Next Ten Years*; Connal and Scott (1999) 'The new Scottish Parliament: what will its impact be?'; McCarthy and Newlands (1999) *Governing Scotland: Problems and Prospects – The Economic Impact of the Scottish Parliament*; and Bosworth and Shellens (1999) 'How the Welsh Assembly will affect planning'. The House of Commons Library produced a useful research paper in 2003, *An Introduction to Devolution in the UK* (Leeke *et al.* 2003), whilst a current view of the position, 'Devolution: how it affects policies and services', is available at www. gov.uk/browse/citizenship/government/devolution-how- it-affects-policies-and-services (accessed 13 June 2014).

Mawson (1996) reviews 'The re-emergence of the regional agenda in the English regions: new patterns of urban and regional governance', while a review of the history of the two strands of regional planning (inter-regional economic and intra-regional land use) and the move towards a more integrated and comprehensive approach is well analysed in Roberts and Lloyd (1999) 'Institutional aspects of regional planning, management, and development:

models and lessons from the English experience'. Bradbury (2008) provides a more recent review in *Devolution, Regionalism and Regional Development: the UK Experience*.

Hall's essay on 'The regional dimension' (1999) gives an overview of post-war regional economic policy, with a short comment on regional land use planning. Wannop (1995) *The Regional Imperative: Regional Planning and Governance in Britain, Europe and the United States* is an excellent account by a knowledgeable practitioner of the endeavours to plan on a regional scale. Swain *et al.* (2012) gives a review of the experience of regional planning in the Blair/Brown years in *English Regional Planning 2000–2010: Lessons for the Future*.

Local government

The principal textbooks giving a general introduction to local government structure and organisation are Chandler (2007) *Explaining Local Government: Local Government in Britain since 1800*, and the fourth edition of Wilson and Game (2006) *Local Government in the United Kingdom*. Although a little dated, Byrne's (2000) *Local Government in Britain: Everyone's Guide to How it All Works* is very readable. For an overview of the politics of local government, including the roles and relationships between councillors, officers and political parties, see Stoker (1991) *The Politics of Local Government*; for a planning 'take' on local government, Kitchen's (1997) *People, Politics, Policies and Plans* is a very informative read. On changing management approaches, see Stoker's volume of essays on *The New Management of British Local Governance* (1999); Stoker and Wilson (eds) (2004) *British Local Government into the 21st Century*; and Stewart (2003) *Modernising Local Government*. Morphet (2007) in *Modern Local Government* provides an insight with the benefit of experience of managing in local government. For an examination of the relationship between planning and changing government, see Vigar *et al.* (2000) *Planning, Governance and Spatial Strategy in Britain*. European comparisons are given in Hirsch (1994) *A Positive Role for Local Government: Lessons for Britain from Other Countries*.

Parish and town councils could be seen as a somewhat underresearched and underdocumented area, but a good introduction to the range of their activities can be found on the National Association of Local Councils website (www.nalc.gov.uk/). Insights into specific roles are given by Owen (2002) in relation to planning in 'From village design statements to parish plans: some pointers towards community decision making in the planning system in England', whilst Newman (2005) reviews the part they can play in regeneration in a study for Joseph Rowntree, *Parish and Town Councils and Neighbourhood Governance*. The Labour government placed an emphasis on increasing the part played by parish and town councils with the launch of a 'quality' scheme. An evaluation of this can be found in a report for DEFRA by Woods, Gardner and Gannon (2006), *Research Study of the Quality Parish and Town Council Scheme*.

The rate of change in local government under the Blair government resulted in a large number of official publications discussing and proposing changes in the operation of local government. Of particular importance are *Modern Local Government: In Touch with the People* (DETR 1998c), *Local Leadership, Local Choice* (DETR 1999) and *A Mayor and Assembly for London* (DETR 1998a). The changes were much discussed at their inception and subsequently. A review of some of the research is provided in Martin and Bovaird (2005) *Meta-evaluation of the Local Government Modernisation Agenda: Progress Report on Service Improvement in Local Government*. For a reflection on how the modernisation process interacted with planning, see Gunn and Vigar (2012) 'Reform processes and discretionary acting space in English planning practice, 1997–2010'.

Notes

1   QUANGO, or quasi-autonomous non-governmental organisation, was a term coined by Anthony Barker in 1969–70 and incorporated in his book *Quangos in Britain* (1982) to describe bodies which were not officially part of government but which were used by government to deliver its policies.

2   There have been many attempts to change the management of the civil service. The creation of the Prime Minister's Efficiency Unit in 1979 and the

Financial Management Initiative in 1982 involved industrialists – initially Derek Rayner of Marks & Spencer and Robin Ibbs of ICI – and sought to introduce new cultures of management. A subsequent report by the Unit, *Improving Management in Government: The Next Steps* (1988), led to the creation of what were known as 'Next Steps Agencies', which were 'spun out' of government departments to focus solely on service delivery (as opposed to policy development), on the assumption that this would make improvements in efficiency easier to capture. If the number of these agencies is added to the significant number of other forms of executive and advisory bodies which have been established, then it has been estimated that there is one such body for every 10,000 people and they are responsible for a third of all public spending. For some, this has raised the question of 'democratic accountability, especially as they have become a means of shifting responsibility for essential public services from elected government to unelected bodies' (Weir and Hall 1994: 4). This concern has been the subject of a number of reports by the Select Committee on Public Administration.

3   Recent research has shown that the UK has either voted against or abstained (which has a similar effect) in most votes in the Council (Miller 2013).

4   Curiously, in 1999, the ministry made the first-ever use of its powers under section 43(6) of the Planning Act to veto the allocation of high-grade farmland in East Yorkshire for development. This was overruled by the Secretary of State for the Environment. See *Planning*, 26 February 1999, p. 1.

5   Democratic Audit is a project funded by the Joseph Rowntree Charitable Foundation to 'monitor democracy throughout Britain'. It has published a number of reports from the early 1990s on various aspects of democracy and freedom, including the work on quangos referred to here. One of the team producing the work was Tony Wright MP, who later became chair of the HoC Public Administration Select Committee and carried out further work on the audit of quangos. At the time of the publication of this work Democratic Audit was based at the University of Essex; from 2013 it was based at the London School of Economics.

6   Whilst the case against quangos is well rehearsed, it is not difficult to find examples of new quangos, with perhaps one of the most high-profile created by the Coalition government being the Office of Budget Responsibility. In spite of the rhetoric, they are useful because they can provide specialist expertise and have a longer-term focus than is afforded in a highly politicised environment. They can also benefit from the heightened authority resulting from their relative freedom from political considerations. More cynically, they can serve to distance ministers from difficult decisions and direct responsibility for their consequences.

7   A prime example of these tensions was provided early in the life of 'Next Steps' agencies by an escape of prisoners from Parkhurst prison, then the responsibility of the Prisons Service, an agency. Michael Howard, then Home Secretary, responded to a critical inquiry (the Learmont Report) by sacking the Director General of the Service, Derek Lewis, on the grounds that he had responsibility for how the service operated. Lewis had refused to resign, claiming that his job had been made impossible by excessive Home Office interference. In a debate on this matter in the House of Lords, Lord Rogers noted that 'there is every reason to believe that he has not had the support of the Home Secretary in those areas which should properly be his if the next steps agency is to function in the way intended' (Hansard, HC Deb, 16 October 1994, Vol. 566, c. 606), highlighting the fact that there is not always an easy division between policy and operational matters, together with a concern about diminished political accountability.

8   On the basis of evidence received at the inquiry, the inspector found that the trajectory of housing development required revision and amendments needed to be made to the plan to ensure a five-year supply of housing land. Revising the plan was one of a number of options offered by the inspector, including proceeding with the inquiry and finding the plan unsound and the LPA withdrawing the plan.

9   On 8 February 1966, Prime Minister Harold Wilson announced in the House of Commons the appointment of a Committee on the Civil Service (which became known as the Fulton Committee) 'to examine the

structure, recruitment and management, including training, of the Home Civil Service, and to make recommendations' (Hansard, HC Deb, 8 February 1966, Vol. 724, c. 209). The Fulton Report found, among twenty-two recommendations, that the service continued to be based on the 'generalists' who were felt suitable in the 1800s but not for the present, when more specialist technical and managerial skills were needed. Many of the changes which occurred in the Civil Service in the period after 1967 arose out of the Fulton recommendations.

10 The decision to grant legislative powers to Scotland, but only executive to Wales 'had as much to do with political compromise and accident as with any rational argument' (Osmond 1977: 149).

11 Whilst the development of regional institutions may be relatively stunted in Britain, and the strength of national regional policy waned from 1984, the development of the EU Structural Funds and programmes of regional assistance (a 'programme' rather than a project-focused approach developed from 1984) led to a flowering of inter-agency cooperation and the emergence of regional or sub-regional regeneration strategies. However, the lack of a developed regional capacity was sometimes felt to act as a brake on the success of such activity (Roberts and Hart 1997).

12 Tomaney (2013) notes that the North East has a history of regional institutions dating back to 1935, when the North East Development Board was established.

13 Objectives of LEPs, from *Local Growth* White Paper:
'We envisage that local enterprise partnerships could take on a diverse range of roles, such as:
- working with Government to set out key invest-ment priorities, including transport infrastruc-ture and supporting or coordinating project delivery;
- coordinating proposals or bidding directly for the Regional Growth Fund;
- supporting high growth businesses, for example through involvement in bringing together and supporting consortia to run new growth hubs (see Annex B);

- making representation on the development of national planning policy and ensuring business is involved in the development and consideration of strategic planning applications;
- lead changes in how businesses are regulated locally;
- strategic housing delivery, including pooling and aligning funding streams to support this;
- working with local employers, Jobcentre Plus and learning providers to help local workless people into jobs;
- coordinating approaches to leveraging funding from the private sector;
- exploring opportunities for developing financial and non-financial incentives on renewable energy projects and Green Deal; and
- becoming involved in delivery of other national priorities such as digital infrastructure.'

14 Community-based planning initiatives (in the broad-est sense) have been common in rural communities for a number of years. Village design statements (VDS) are community-prepared documents which give guidance to developers and individuals to encourage good design of the type that will enhance and protect the individual character of the locality. VDS can be adopted as a supplementary planning document and become part of the local development framework if they are prepared in the right manner. A parish plan is a comprehensive plan which addresses issues of concern to the community in one document. It is prepared by a 'bottom-up' approach and based on information provided through research, survey and consultation. The parish plan can form a blue-print for use by the parish council and evidence for bidding for funds, as well as be an indicator for delivery of services from statutory service providers.

15 *Councillors, Officers and Stakeholders in the New Council Constitutions – Findings from the 2005 ELG Sample Survey* (DCLG, October 2006) found that backbench councillors spend on average 81 hours monthly on council business, while executive councillors spend 112 hours. The most significant items are engaging with constituents and reading reports.

16 The Hunt Report (1969) is the *Report of the Lords Select Committee on Relations between Central and Local*

Government. The *European Charter of Self-Government* is reproduced in an appendix to the McIntosh Report (1999).

17 Single Outcome Agreements are agreements between the Scottish Government and community planning partnerships which set out how each will work towards improving outcomes for local people in a way that reflects local circumstances and priorities, within the context of the government's National Outcomes and Purpose.

18 SO (1998) *A New Ethical Framework for Local Government in Scotland: Consultation Paper*. This stems from the Nolan Report, *Standards of Conduct in Local Government* (Cm 3702, 1997).

19 A multi-area agreement was an English political framework that aimed to encourage cross boundary partnership working at the regional and subregional levels. It was defined by the DCLG as a voluntary agreement between two or more top-tier (county councils or metropolitan district councils) or unitary local authorities, their partners and the government to work collectively to improve local economic prosperity.

20 Attempts to 'join up' action are familiar from experience with urban regeneration but a number of barriers to success have been found. Whilst local partnerships can be seen as a good vehicle to achieve focus, barriers have been identified which stem from a lack of 'joining up' at the national level, resulting from central government departments continuing to work according to nationally determined departmental priorities (Carley *et al*. 2000; CCRS 2005).

21 In some instances, fundamental doubts have been expressed about the extent to which a local authority can delegate decisions for which it has a regulatory function. Legal judgments have indicated that they should not delegate so much of the process that they are merely acting as 'rubber stamps'. The North Tyneside example quoted raised another interesting issue. The council had a Conservative elected mayor but a Labour council; the Labour council wanted to prevent the outsourcing but, whilst it could set the mayoral budget, it could not dictate how it was spent, so the mayor was able to go ahead with outsourced contracts to deliver planning and other property-related services.

22 See *External Enquiry into Issues of Concern about the Administration of the Planning System in Warwick District Council* (1994) and *Report of an Independent Inquiry into Certain Planning Issues in Bassetlaw* (1996). On the Newark case, see *Planning*, 29 October 1999, p. 2, and 5 November 1999, p. 15. Note also the North Cornwall case referred to earlier in this chapter.

23 The *Guardian* described the 'Donnygate' affair as the 'worst local government corruption case since Poulson' (13 March 2002). The unravelling of Donnygate, which involved seventy-four arrests and 2,000 interviews, triggered much tougher Labour Party procedures for selecting local councillors, including tests which saw a number of long-standing councillors failing to win reselection. Police put in place confiscation procedures for a farmhouse given to the councillor by the developer in exchange for planning permission for houses on an area of farmland owned by the developer which he later sold for £2.25 million.

24 Allen has commented that 'central agencies are often at least as incompetent, inefficient or corrupt as local bodies; local authorities are perennially in the news for alleged corruption and graft; one or two notorious cases can suffice to keep the whole concept of local government in disrepute' (1990: 12). For a catalogue of cases of corruption in public administration, see Doig (1984).

25 These seven principles are selflessness, integrity, objectivity, accountability, openness, honesty, and leadership. The Commission has stated that these should apply to all involved in public services, both those directly employed and contractors.

26 The first report was on members of parliament, ministers and civil servants, and executive non-departmental public bodies (Cm 2850, 1995). This was followed by a report on further and higher education bodies, grant-maintained schools, training and enterprise councils, and housing associations (Cm 2170, 1996). The report on local government was published in 1997 (Cm 3702).

4 The framework of plans

There is a paradox at the heart of planning: everybody wants it but cannot resist criticising it!

(UN-Habitat Global Report 2009)

Introduction

What constitutes planning and its role has been the focus of intense professional and academic debate, and the subject of political struggle. In Britain, the constant flow of reforms of the planning systems which gathered an unprecedented pace in the 2000s is to some extent a reflection of the contested nature of planning. However, one aspect of the British planning system which has remained constant since its formal inception in 1947 is the existence of its two principal components: development plans and development management. Together they are meant to achieve a balance between certainty and flexibility in the governance of places. Such a combination is almost a unique feature of the planning system in Britain. The focus of this chapter is on development plans and the subsequent chapter discusses the development management process.

Formulating and implementing 'plans' are often an integral part of spatial planning. The term 'plan' in this context can refer to a multitude of tools such as statements, diagrams, written policies and perspectives, or other documents expressing intentions for the future development of an area. Their form and contents are often shaped by other plans, policies and strategies. As Hopkins (2001) observes, plans are often expected to perform as one or more of the following:

- an agenda: a list of actions to be undertaken;
- a policy statement: principles or rules to guide subsequent actions;
- a vision: an image of what could come about;
- a design: a fully worked-out development scheme;
- a strategy: guidance on sets of interrelated decisions about action now, linked to specific contingencies anticipated in the future.

The power of a plan depends on the authority given to it in formal law, through national policy or sometimes customary practices. In planning systems where the right to develop is enshrined in a zoning ordinance (such as those in parts of the USA), the plans which define such zones carry a lot of weight in deciding what can and cannot be developed on a specific site. In more discretionary systems (such as that in the UK), a plan is more of a guidance by which local governments state what they would like to see happen in their area. Here, plans act as an important point of reference and shape the decisions of those involved in development (UNHGR 2009).

Plans can be formulated at one or more spatial levels, such as neighbourhoods, cities, city-regions, regions, national, transnational and supranational levels. While acknowledging that these spatial scales are not fixed and their boundaries are 'fuzzy' (Davoudi 2012a; Allmendinger and Haughton 2009a), we use these administrative demarcations to structure this chapter and describe the framework of plans at each level, starting from the larger scale. Our concerns are mainly with the form and scope of policy instruments and the procedures by which plans are created rather than their policy content, which is explained in later chapters. Again we acknowledge that there is a large degree of interdependence between the contents (or substance) of plans and the processes (or procedures) through which they are produced.

The framework of plans is established by a huge library of statutes, rules, regulations, directions, policy statements, circulars, guidance and other official documents. However, it is important to appreciate at the outset that the formal system is one thing; the way in which matters work in practice may be another. Although the informal planning system operates largely within the formal structure, it may continue with little modification even when major legislative changes are made. Alternatively, there may be significant changes in practice within a stable formal system. Political ideas, professional attitudes, institutional cultures and management styles all affect the ways in which the system operates in practice.

It is also necessary to note that much development (in the everyday, rather than the legal, sense of that word) takes place without any help or hindrance from the planning system. Even where the development is clearly related to some action within the statutory framework for planning, the actual outcome is affected by 'extraneous' factors, and it may not be at all clear what effect planning has had on outcomes.

It is government policy to ensure that planning decisions are made with reference to an explicit and widely agreed framework of policies at national, regional and local levels, set out in plans and other policy instruments. This is described as plan-led development management. Reference to policy in plans reduces the amount of ad hoc decision-making and the need for resolving conflicts over individual development proposals. Plan-led decision-making may be more efficient and consistent than decision-making on a project-by-project basis. Explicit policy statements can help to ensure accountability, as the decision-makers are making their 'decision rules' or criteria explicit. Policy also provides a measure of certainty and coordination for the promotion of investment (Healey 1990).

A plan-led system requires a comprehensive and up-to-date set of national policy, regional strategies and local development plans. Where this is available for particular places and subjects, it is an important factor in decision-making, but it is always going to be difficult to establish and maintain a comprehensive policy framework for a number of reasons, such as rapid social, environmental, economic and technological changes, changes in political ideas and aspirations of the government, and the considerable variation in the capacity of local planning authorities for preparing development plans and keeping them up to date. Furthermore, it is not always desirable to have a set of long-term fixed policies which are not capable of adapting to fluid and highly dynamic societal changes. The discretionary nature of the planning system in Britain is designed to respond to the interplay of a desire for certainty and the state of flux. Therefore, while decisions are guided by the policies in the plan, they are also influenced by other material considerations, which may be specific to a given time and place. As we shall see, there has been a constant flow of adjustments to the system of plan and policy-making since the mid-1960s, in order, first, to reflect the political preferences of the ruling government and, second, to make the system more relevant and responsive to demands of the time. There have been periods of systematic reforms too, probing deeper into the operation of the system and leading to more extensive change. While the pace of reviews has been accelerated significantly in recent years, the fundamental characteristics of the system have remained much the same. Each review of the development plans has tried to tackle questions such as:

- What framework of plans will ensure the accountability of decision-makers and safeguard the interests of those affected by planning, yet be expeditious and efficient in operation?
- How can the framework provide a measure of certainty and commitment, yet allow for flexibility to

cope with changing circumstances, local conditions and new opportunities?

- What objectives should plans pursue, and how will these shape their form and content?
- What should be the scope of plans and what social, economic and environmental issues should they address?
- Who should have influence in the planning process, and what should be the respective roles of central and local government and of local communities?

These perplexing questions have no easy answers and that is why the system is under almost constant appraisal and review. Yet, 'strikingly similar assertions have been used by governments to describe the perceived and actual deficiencies of the planning

system and to justify its reform. In the 2000s, the pace of reforms reached a record breaking height as planning experienced three "radical" changes in 2004, 2008 and 2011' (Davoudi 2011a: 92; Gunn and Hillier 2012). Acceptable answers rarely have stability, since conditions and attitudes change over time. The biggest changes to the framework of plans in recent years have been the dismantling of the regional level of planning and the revocation of regional spatial strategies on the one hand; and the introduction of a sub-local level of planning and neighbourhood plans on the other hand, while maintaining the 'spatial planning approach'. Figure 4.1 gives an overview of the main instruments used to express planning policy in the UK. The discussion of this framework begins with the system at the EU level and works down through the national, regional, local and neighborhood levels.

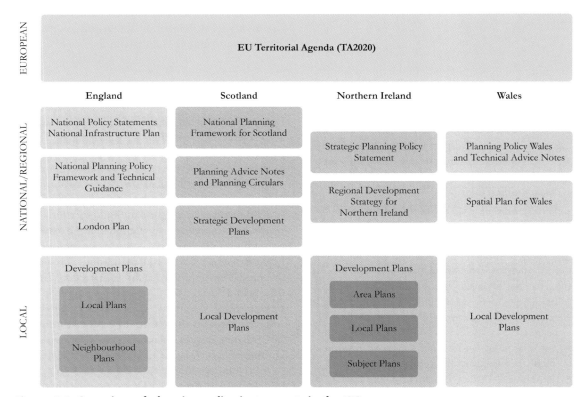

Figure 4.1 Overview of planning policy instruments in the UK

SUPRANATIONAL PLANNING

All in all, in the past decade, the new institutional context posed by the EU has fundamentally changed the relationship between Member States and their territory, despite the lack of a formal European competency to engage in spatial planning. Although it certainly remains necessary to conduct spatial policy at the national level – if for no other reason than to coordinate EU sectoral policies and integrate them into the planning systems – doing so without regard to the growing influence of Brussels will doom it to failure.

(van Ravesteyn and Evers 2004: 9)

The rationale for planning at the European scale

In an increasingly globalised world, it is hard to think how any country can plan in isolation from international influences. This is particularly the case when geographic proximity adds to the wider relational networks. In Europe, development in one country may have significant impacts in other countries. For example flooding in the Netherlands may be affected by development in the same river catchment in Germany. Prospective house buyers and job seekers in one country may seek housing and employment in another because of availability, affordability and salary differentiation.

However, until the 1990s, the idea of a European dimension to planning was considered incomprehensible to many British planners. This is despite the fact that a large number of the EU investments, policies, programmes and initiatives have profound implications for spatial development patterns in the member states. The influence of the EU has no single locus but is exerted through an array of activities such as:

- regional policy and financial supports for development, regeneration and infrastructure;
- legislation on environmental protection and improvement;

- facilitation of innovation and transfer of experience between member states;
- development of spatial development policy options for the EU as a whole.

Many of the EU sectoral policies, especially in the fields of regional policy, transport, environment and agriculture (as explained in relevant chapters of this book), have major implications for spatial development, but these are not always explicitly considered in the policymaking process. For example, in transport the Community has pursued harmonisation of national transport policies and the development of Trans-European Transport Networks (TEN-T). In the environment field, the Community has produced directives on Environmental Assessment, Air Quality, Waste, Birds, Habitats, Water and Soil, all with significant implications for planning systems. The Common Agricultural Policy and other specific measures, such as designation of environmentally and culturally sensitive areas, have been instrumental in changes in spatial development in urban and rural areas.

EU competences in spatial planning

Despite the above de facto spatial planning, there has been no explicit or coordinated spatial strategy for the entire EU territory. A major study undertaken by Robert *et al.* (2001: 158) to examine the cost of non-coordination confirms that 'the degree of horizontal co-ordination between the various Community Institutions is relatively low and no procedure exists which aims at creating spatial coherence between all Community policies'. This is partly due to the lack of EU competence in spatial planning – i.e. spatial planning is not a specified EU objective in the Treaties (Nadin and Shaw 1999). Nevertheless, over the years some form of a strategic spatial policy agenda has emerged through the efforts of both the European Commission and the member states working 'inter-governmentally' (Davies 1989; Williams 1996).

All the European institutions have now recognised the importance of spatial development that cuts across national borders. Member states are encouraged to work cooperatively on spatial planning in order to coordinate the spatial impacts of sectoral policies and promote more sustainable forms of development and economic competitiveness. It is argued that there are important cross-border and transnational dimensions to spatial planning which need to be taken up through appropriate institutions and instruments at jurisdictional levels above the nation state. But there is a question about the legitimacy of such action, because, until recently, competence over spatial planning (in its various forms) has been considered to rest solely with the member state governments (or in some countries subnational governments). To what extent should there be a sharing of powers on spatial planning between the member states and the Community? Box 4.1 summarises key arguments for and against a European dimension to spatial planning.

BOX 4.1 ARGUMENTS FOR AND AGAINST A EUROPEAN DIMENSION TO SPATIAL PLANNING

Issue	For	Against
Spatial policy	Policies and actions of the EU have spatial dimensions and impacts that constitute de facto spatial planning and should be organised and coordinated to achieve fundamental Community goals.	The spatial dimension of sectoral policies can be managed within sectors and by using other non-planning coordinating mechanisms. Policy coordination can be done at national and lower levels.
Spatial equity	Spatial concentrations of economic activity reinforce great disparities in social and economic opportunity and environmental problems. Spatial planning is needed to develop and implement a policy of balanced spatial development.	Concentrations of economic activity in global cities and agglomerations are a competitive advantage, the benefits of which are transferred to other regions. Spatial planning is a blunt tool for tackling spatial equity and redistribution and its impact on the location of economic activity and population is limited.
Transnationality	Globalisation and the network society spell the end of the nation state. New forms of territorial governance are needed that deal more effectively with functional regions and the 'boundary problem'.	Nation states will continue to be the main focus for providing legitimate and accountable interventions in spatial development.
Subsidiarity	Some aspects of spatial development can only be addressed at the European and transnational levels so competences for spatial planning must be ceded to a higher level.	Member states can adequately deal with transnational impacts through intergovernmental and bilateral cooperation. Management of a territory is a fundamental competence of nation states.
Accountability	A European dimension to spatial planning can help to address the democratic deficit by supplementing normal democratic processes and supporting weaker interests.	European spatial planning initiatives have so far only involved an elite group of experts and very little wider stakeholder involvement. Powerful interests pursue their objectives through spatial planning.

officers in the Province made use of policy statements in other parts of the UK to keep in touch with policy developments (Dodd and Pritchard 1993). Since 1995, national planning policy statements have been published by the Planning Service for Northern Ireland. The statements are similar in form to those in the rest of the UK but reflect the special planning and political circumstances in Northern Ireland – not least the centralisation of planning activity in the Planning Service. *Planning policy statements* (PPSs) set out the Northern Ireland Department of the Environment (DoE) policy on particular aspects of land use planning. Produced, currently, by the DoE, they set out the main planning considerations that the DoE takes into account in assessing proposals for various forms of development and are relevant to the preparation of development plans. The PPSs must conform to the Regional Development Strategy, discussed below. They are complemented with *development control advice notes* (DCANs), which provide more detailed guidance on good practice. In 2014, it was announced that the current twenty PPSs will be consolidated into a single *strategic planning policy statement* (SPPS) by the end of the year, ahead of a transfer of planning powers to councils in April 2015. Like the *National Planning Policy Framework* in England (discussed later in this chapter), the SPPS will set out the core principles that will underpin the planning system and promote a plan-led system, sustainable development and support for good design.

Regional Development Strategy for Northern Ireland

The first regional plan in Northern Ireland was the Belfast Regional Plan published in 1964 (the Matthew Plan). This proposed the stopline, a system of radial motorways, and a major new town, Craigavon, modelled on the English experience (Hendry 1989). Like its counterparts in England, the plan was overtaken by the effects of dramatic economic recession. The subsequent *Regional Physical Development Strategy 1975–85* sought to concentrate growth in the province to twenty-six key centres, but the depressing effects on other areas were widely challenged (Blackman

1985). A new rural planning policy published in 1978 took a much more relaxed approach to development in three-quarters of the rural territory, which led to extensive development of single houses in the countryside and ribbon development. This led to a reappraisal of the need for regional planning and the publication of *A Planning Strategy for Rural Northern Ireland* in 1993. This included both strategic objectives for the overall development of the territory and detailed development control policies, and could only have been the product of a system where central government sets the strategy, makes local plans and undertakes development control. The strategy introduced new restrictions on development in the countryside while introducing the novel designation of 'dispersed rural communities'.

In September 2001, the Northern Ireland Assembly approved the *Regional Development Strategy for Northern Ireland 2025* (RDS) (commonly known as *Shaping Our Future*). This was the culmination of a strategic planning process that had commenced in 1997. The RDS is to some extent in the mould of EU developments in spatial planning (reflected in the use of terms such as 'hubs', 'corridors' and 'gateways'), placing Northern Ireland in its European and global context and seeking to integrate concerns about the physical development of its territory with social, economic and environmental objectives. It describes itself as 'not a fixed blueprint or master plan. Rather, it is a framework, prepared in close consultation with the community, which defines a Vision for the Region and frames an agenda which will lead to its achievement' (p. 2). To that extent, the RDS is also consistent with the *National Spatial Strategy for Ireland 2002–20* published in 2002 (for a full account of its process and content, see Walsh 2009).

At the time when the RDS was being prepared, Northern Ireland was experiencing major political changes following the paramilitary ceasefires in 1994 and area-based peace and reconciliation partnerships from 1995, and most importantly the signing of the Belfast Good Friday Agreement in 1998. Murray (2009: 125) argues that:

> Strategic planning in this context . . . was charged with contributing to a negotiated consensus on

emergent spatial relationships within Northern Ireland, on the island of Ireland, across the Irish Sea to Great Britain and beyond. Not surprisingly, these axes of endeavour reflect the political complexities of a deeply divided society with longstanding contested regional allegiances and identities.

In 2012, a revised *Regional Development Strategy – RDS 2035: Building a Better Future* was agreed by the Northern Ireland Executive. This followed a 2008 amendment to the 2001 RDS. The new revised Strategy has four key elements (p. 14):

* a *Spatial Framework* which divides the region into 5 components based on functions and geography;
* *Guidance* at two levels: a) Regional level and b) Specific guidance for each element of the Spatial Framework;
* a *Regionally Significant Economic Infrastructure* section identifying the need to consider strategic infrastructure projects;
* *Implementation,* setting out how the strategy will be implemented.

The making of the 2012 RDS paralleled that of an all-Ireland consultation paper on strategic spatial planning which once again indicated 'a new strategic understanding of the intentions and inter-relations between the spatial planning agendas of the two jurisdictions' (Lloyd and Peel 2012: 181).

England

National Planning Policy Framework for England

Prior to 2004, national planning policy in England was expressed through *planning policy guidance notes* (PPGs) and *minerals planning guidance notes* (MPGs). While national planning policy in England has clarified the national criteria for decision-making, it has tended to be more general than its equivalent in, for example, Scotland, broader in scope, and not at all

location-specific. An evaluation of the effectiveness of the PPGs concluded that they had 'assisted greatly in ensuring a more consistent approach to the formulation of development plan policies and the determination of planning applications and appeals' (Land Use Consultants 1995: 47). This is because they are important material considerations in development control and have a determining influence on the content of development plans. The study found that most professional planners had a high regard for PPGs and welcomed the order and consistency that they brought. Councillors were generally more sceptical, because they constrained their discretion to respond according to their interpretation of local needs.

Following the 2004 reform of the planning system in England, the PPGs and MPGs began to be gradually replaced by a growing number of PPSs and *minerals policy statements* (MPSs) respectively, accompanied with lengthy and detailed 'good practice guides'. In introducing the reform in general and PPSs in particular, the Green Paper stated that 'national planning guidance is long and often unfocused. It mixes key planning policy principles which must be followed, with good practice advice' (DTLR, 2001: 3). Six years after the introduction of PPSs, they had grown into over 1,000 pages of national policy and guidance. However, calls for more statements on new topics continued, while concerns were voiced over perceived contradictions between one statement and another, and between the series and other government policy statements. One example (from the Land Use Consultants 1995 study) was concern over the different explanations of the term 'sustainable development' in government policy.

The 2011 reform of the planning system took a more radical approach and replaced almost the entire series of PPSs (some twenty-five statements) with a single, fifty-two-page document: the *National Planning Policy Framework.* The publication of its draft version in 2011 attracted unprecedented media attention, with extensive coverage in national newspapers and prime-time television and radio. Indeed, it 'propelled planning into the limelight the like of which it had not seen for decades' (Davoudi 2011a: 93). At the centre of what soon turned into a polarised debate were the NPPF's promotion of a 'presumption in favour of sustainable development' and its suggestion that 'the default answer to development' should be 'yes' (DCLG 2011a). This sparked a national debate between two opposing camps, fuelled by the *Daily Telegraph*. On the one side was the powerful pro-countryside lobby consisting of the National Trust and the Campaign to Protect Rural England. On the other side was a powerful pro-development lobby, including eleven bodies representing the housing, construction, business and energy sectors, with the RTPI trying to hold the middle ground and calling for calm and constructive debate. The proponents endorsed the streamlined NPPF's principles and argued that, 'for too long developers, communities and local authorities have found the planning systems negative and confrontational' (*Planning* 2011: 5). The opponents suggested that the presumption would lead to 'unchecked and damaging development in the undesignated countryside on a scale not seen since the 1930s' and that the Government had declared 'open season' on countryside (ibid.).

The final version of the NPPF was published in 2012, setting out planning policies for England and how they are expected to be applied. It provides guidance for local planning authorities in both drawing up plans and making decisions about planning applications. However, the NPPF does not have specific waste policies (because national waste planning policy will be published as part of the National Waste Management Plan for England, as discussed in Chapter 7). Neither does the NPPF have planning policy for travellers' sites. Additional guidance to local authorities, especially on development in flood risk areas and in relation to mineral extraction, is provided through the *Technical Guidance to the National Planning Policy Framework*, which was also published in 2012. In addition to its 'headline' provision to promote sustainable development, the NPPF includes a number of provisions that are not very different from previous guidance. For example, it seeks to:

- plan proactively to help achieve economic growth;
- ensure the vitality of town centres, partly through the continued application of the sequential test;

- support a prosperous rural economy;
- promote sustainable transport through local planning strategies and travel plans;
- support the expansion of electronic communications networks;
- boost significantly the supply and choice of housing;
- plan positively for high-quality design;
- promote healthy communities through enhancing physical and social capital;
- protect the green belt (echoing previous policies);
- meet the challenges of climate change and flooding;
- contribute to and enhance the natural and local environment;
- set out in their local plan a positive strategy for the conservation and enjoyment of the historic environment;
- facilitate the sustainable use of minerals.

Another significant step towards the streamlining of the planning system has been the review of national planning practice guidance (which was not cancelled by the NPPF) led by Lord Taylor of Goss Moor in 2012. The guidance, seen as complex and repetitive, was catalogued in 230 separate documents and a total of 7,000 pages, and was considered by the Taylor Review (2012: 6) as 'almost impossible for residents and businesses to use effectively'. The Taylor Review recommended that national planning practice guidance should be clear, up to date, coherent and easily accessible, provide essential information and exclude best-practice and case-study material, and be made freely available through a web-based, live resource, hosted on a single site as a coherent up-to-date guidance suite. It should also be actively managed to keep it current and reviewed annually, using open-source methods. It also recommended that the DCLG Chief Planner should act as gatekeeper on material to go on the new guidance website. Planning Inspectorate guidance should also be incorporated into the new guidance set and the Planning Inspectorate should advise on future material. This guidance was launched in a final form in March 2014 and is available on the *Planning Portal*.

Furthermore, policies for nationally significant infrastructure projects are not included in the NPPF, because, under the provision of the Planning Act 2008, twelve national policy statements are to be produced by the relevant sectoral ministries for energy, transport, and environment. These would then be the main consideration for decision-making by the Infrastructure Planning Commission (IPC). As Marshall (2013a: 131) explains, a 'critical feature of the NPS process as it is evolving is how unspatialized the documents are. The only ones that identify sites are those for nuclear power, and, in much lower key mode, for wastewater.' It appears that the lack of a spatial dimension reflects the government's reluctance to engage with difficult spatial/locational issues, at least in England. In 2010, the first National Infrastructure Plan for England was published and was then updated in 2011.

A lack of national spatial plan

As the Heatherington Report, commissioned by the Town and Country Planning Association (TCPA), suggests, England has remained 'not only the sole country in British Isles not to have a national spatial strategy; it was the only country in Western Europe without some sort of national spatial strategy that looks to the long tomorrow' (TCPA 2011a: 6). This is despite the ongoing campaign by professional bodies (Wong *et al.* 2012a, 2012b; TCPA 2012).

In 2005, some related work was undertaken by a consortium comprising the English Regions Network, regional development agencies (RDAs), the ODPM and the DfT, and led by Ove Arup and Partners on *Regional Futures: England's Regions in 2030*. This, however, was not meant to be the first stage in the creation of a spatial plan for England, but rather 'to develop a "national perspective" on how England's regions (including London) relate to each other and to underlying forces in the economy, and how these relationships have been changing and will change in the future' (p. 1). The study summarised the prosperity gap between the South East and regions in the North and Midlands and showed how the gap would widen in the coming years. Earlier lobbying after a wide-ranging study found that, in principle, it was both desirable and feasible to have a UK spatial planning framework (Wong *et al.* 2000). It was argued that the main purpose of such a framework should be issued by the Cabinet Office and would seek to join up and fill in the gaps between existing strategies and monitor spatial policy targets. Its main task would be to cut across the compartmentalised sectoral thinking of government.

What came closest to a spatial planning framework for England was *Sustainable Communities: Building for the Future* (ODPM 2003a). However, it focused entirely on housing. It promoted housing growth in the South and the tackling of low demand in the North and Midlands. Therefore, in the South it identified new housing growth areas in Thames Gateway (which was severely flooded in 2014), London, Stansted and the Cambridge Corridor and Milton Keynes-South Midlands Corridor. In the North, it identified nine 'pathfinder' areas for which a renewal fund was created to tackle the problem of low demand, partly through a controversial programme of demolition. As with the Ove Arup study mentioned above, *Sustainable Communities* and also the *Eddington Transport Study* (Eddington 2006) followed the logic of concentrating development in places with greatest potential for growth rather than greatest need for investment – a logic which contrasts sharply with the ideas of balanced development and territorial cohesion which were promoted in the European spatial planning fora as mentioned above.

Summary

In whatever form, national planning policy and plans issued by national government and the devolved administrations carry considerable weight. But though local planning authorities are required to have regard to national policy, they are not bound by it. Indeed, other material considerations may be of greater importance in particular cases, and planning authorities may wish to take a different line, so long as they can give adequate reasons. Moreover, the advice in one statement may contradict another, perhaps as a result of piecemeal revisions at different times. Nevertheless, national policy commands a great deal of respect and is closely followed in development management and development planning, and is quoted profusely in the decision-making process, especially at inquiries. Some national policy is still to be found in circulars and also, from time to time, in ministerial statements. Major changes in policy are often published in White Papers. All of these documents can be regarded as material considerations in planning, and thus central government has an array of instruments in which national policy can be expressed. Indeed the result can be confusing, even for professionals, including planning inspectors. However, in the desire to streamline national policy guidance, there is a risk of ambiguity and inconsistency in decision-making across the country. A critical task is engaging other sectoral policymakers and operators such that there is wider ownership of national spatial policy among government departments, agencies and service providers.

REGIONAL PLANNING IN ENGLAND

In parallel with developments at the national level, regional planning policy was gradually strengthened in England from the mid-1980s onwards, and this involved three major steps. The first began with the 1986 consultation paper on *The Future of Development Plans*, which recognised the value of voluntary cooperation among local authorities in some regions, such as East Anglia and the West Midlands, in producing a general regional overview to guide the production of development plans. Official encouragement was given in this consultation paper to the formation of other regional groupings, though no precise procedures were suggested. Regional planning conferences of local authorities were invited to prepare draft regional planning guidance looking ahead twenty years or more. The Secretary of State would then consider and publish final guidance. The 1989 White Paper recommended the involvement of business organisations and other bodies as well as local authorities in the preparation of guidance. Conservation and agricultural interests were

later added to this list. Strategic guidance was also produced in the metropolitan counties, but this was gradually merged with the regional guidance. The full set of regional planning guidance took some time to prepare but was welcomed by all sides. Early efforts came in for considerable criticism, being described as little more than a detailing of national guidance and restatement of current policies (Minay 1992), though Roberts (1996) gave a more positive appraisal. A later review by Baker (1998) noted the increasing specificity of regional guidance through the use of subregional divisions, the first signs of attempts to integrate a wider range of sectoral policy interests (particularly transport and economic development) and a growing institutional capacity for planning at the regional level. The most contentious task was in defining and allocating regional housing targets, which led to considerable central government amendments to regional guidance, especially in the South.

From 1997, the new Labour government's commitment to regionalisation provided the context for a second step to be taken on strengthening regional planning. In February 1998, a ministerial statement was published on *Modernising Planning*, together with a consultation paper on *The Future of Regional Planning Guidance*. One of the main themes of the ministerial statement was the need to strengthen strategic planning capabilities at the regional and subregional levels. The consultation paper accepted the validity of criticisms of regional planning guidance in not providing a real strategic direction for the regions and not having the confidence of regional stakeholders. Proposals were made for extending the scope and specificity beyond land use, by making the process for its production more inclusive and transparent, and by giving the regional bodies the main competence for its production. The proposals stopped short of recommending that regional guidance should become a statutory document, on the basis that this would require primary legislation. Local authorities were encouraged to take the principles of the consultation document forward prior to the publication of formal guidance. The renewed emphasis on the regional level was reflected in the publication of PPG 11 *Regional Planning*, published in 2000.

European spatial planning, as explained above, had a considerable influence on the new approach to regional planning. The Guidance recommended that the RPG should act as a spatial strategy for the region, with a planning horizon of fifteen to twenty years and covering housing, the environment, transport, infrastructure, economic development, agriculture, minerals and waste management. It was also expected to provide a strategic context for the preparation of local transport plans and regional economic strategies (PPG 11, para. 1.03). There was a strong emphasis in the new arrangements on the need for more concise and regionally specific guidance, not repeating national policies and with more attention to how planning could help to deliver regional policies, including those of the regional economic strategies. Involvement of the stakeholders was another important concern. A study examining this across eight English regions concluded that, 'for RPG to have sufficient 'teeth', it has to balance 'the different views of stakeholders with the strategic visions for the region' (Baker *et al.* 2003: 35).

The third step followed soon afterwards. The 2001 *Planning Green Paper* (discussed in more detail later) repeated many earlier criticisms: regional planning guidance was still too long and too detailed; it was poorly integrated with other regional strategies (especially the regional economic strategies); there was duplication of effort at the regional and county level; and difficult decisions were being avoided, not least the provision of sufficient housing land in the South East. All this needs to be seen in terms of the weak institutional arrangements at the regional level and the very limited resources available for the preparation of regional guidance, After decades of neglect, the regional planning function was weak and 'dependent upon local government officers coming together and carrying out the necessary studies on the back of their mainstream jobs' (Kitchen 1999: 12). The Green Paper addressed these issues with a proposal, later implemented in the Planning and Compulsory Purchase Act 2004, to replace the RPG with *regional spatial strategies* (RSS). These would be more focused on regional-level strategic issues and used as a tool to integrate strategies at the regional level. There would also be subregional strategies that would replace

structure plans (see p. 109 of the Green Paper), which were given a life of only three years (to 2007).

The earlier changes had stopped short of giving the regional guidance statutory status, but the 2004 Act, following proposals in the Green Paper, gave the RSS statutory weight. Furthermore, at a stroke it recast all existing RPG notes as RSSs so as to give them this status. This means that the RSS became part of the statutory development plan and hence local development documents (local plans) had to be in conformity with them, while development management decisions had to be made in accordance with them (see Chapter 5). Although fierce arguments about meeting regional targets for housing development continued, it was hoped that the RSS could provide solid justification for targets and resolve the perennial disputes, especially in the South East. Some regions were certainly making use of the broader and stronger remit to recast regional strategy so as to address more vigorously the objectives of economic competitiveness and sustainable development.

In 2004, PPS 11 *Regional Spatial Strategies* set out the form and content of the RSS and the procedure for its adoption. This showed that the *regional transport strategy* had to be an integral component of the RSS, though separate advice was given on its preparation. Thus local transport plans had to be in conformity with the RSS. The RSS had to contain a vision statement, a spatial strategy with a key diagram, and an implementation plan. Subregional strategies could be prepared for parts of the region and these would throw up some of the most difficult issues cutting across local authority boundaries, especially where city and rural authorities meet. Indeed the rationale given for replacing structure plans with the RSS was the need for policy that cuts across county boundaries in addressing functional subregions such as travel-to-work or housing market areas. Exceptionally, government policy accepted that additional non-statutory strategic planning policy could be prepared for functional regions that cross over regional boundaries. The Thames Gateway was one such case, involving three regions. The responsibilities for the preparation of the RSS were in some ways clearer than for the RPG, and also to some extent stronger. The former regional planning bodies were charged with the preparation of the RSS, but they lacked the authority of directly elected regional governments. As a result, the former government offices were heavily involved in the process, along with the former RDAs, which had considerable influence and interest in implementation. County planning authorities were involved in partnership with the regional body in preparing the subregional elements of the RSS. Along with the national parks and unitary districts, they could help in preparation of subregional strategies and in other work such as monitoring and analysis. Arrangements for London were made separately and earlier, so that the *London Plan: Spatial Development Strategy for Greater London* was the first 'regional' strategy to be prepared and adopted. In this case, the Greater London Authority is and remains the responsible body for both preparation and 'approval' of the London Plan (Figure 4.2).

The increased status of regional planning policy together with demands to improve community involvement required the introduction of even more testing procedures for its preparation and approval. In summary, the former regional planning body prepared a draft strategy in close collaboration with the former government office and RDAs. This was submitted to the Secretary of State, who then appointed a panel to hold an examination in public. The panel reported to the Secretary of State, who then amended the strategy and consulted further before issuing the final strategy.

Two other aspects carried forward into the 2004 reforms were the requirement for sustainability appraisal (considered further in Chapter 7) and the identification of clear targets and performance indicators. The DETR commissioned research on both matters, which provided the basis for good practice guidance for the regional planning bodies (ECOTEC 1999; Baker Associates 1999). The ECOTEC study on targets and indicators highlighted the haphazard proliferation of targets and indicators, and a lack of their systematic consideration in policy objectives. They argued that targets mostly represented general aspirations for the region. Lack of data and a weak evidence base for regional strategies (and not just the RSS) were seen as critical factors, even though the newly formed regional observatories were making some progress on providing more regionally

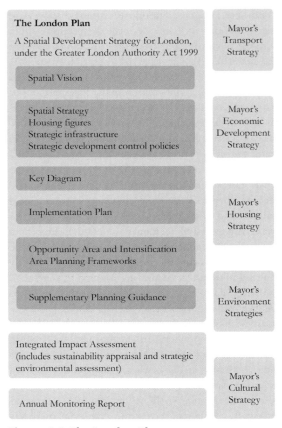

Figure 4.2 The London Plan

relevant data. The requirement for an annual monitoring report on the RSS also demanded better data and information.

The RSS approach lasted five years, from 2004 to 2009, before being replaced with *regional strategies* as discussed below. Morphet (2011: 209) identifies a number of reasons for their short life and for their being 'beset with many of the problems that had faced the previous approach to RPGs'. First was their lack of authority to be a single overarching strategy for the region. This meant that, 'if you talk to the politicians . . . they dismiss the regional assembly as being a toothless tiger' (Davoudi (2005a: 505, quoting from an interview with a planning officer in 2002). Members of the Assemblies knew that the locus of power was elsewhere, at the local level, where final decisions were made. Second, their preoccupation with the single issue of housing was at the expense of other policy concerns, such as the management of inter-regional flows of waste (Davoudi 2000a; Davoudi and Evans 2005). Even worse was a lack of delivery on the supply of affordable housing (Barker 2004). The third weakness was their limited ownership by a wide range of stakeholders (Baker *et al.* 2010; Beebeejaun 2012). The RSS approach recognised the need to involve regional stakeholders more fully in the process and its statutory status gave more incentive to them to take part. Nevertheless, the expectations that an enhanced regional planning policy could command the commitment of a whole range of regional and national actors seemed ambitious in the face of the dominance of a few corporate stakeholders (Vigar and Healey 1999). However, practice varied across the country, and where there was a common enemy or crisis (as in the North), there was likely to be more cooperative working. For much of the country, especially on the fringes of the metropolitan areas and big cities, conflict, not cooperation was the norm. In these places the more difficult decisions were sometimes swept under the carpet, and it was harder to come up with radical solutions. This should be surprising, because the rescaling of the planning decisions to the regional level does not necessarily neutralise them: it just shifts the conflict over a set of inherently political decisions to another arena (Cowell and Murdoch 1999; Davoudi and Evans 2005).

The relationship between the RSS and RDA strategies was critical. The idea was that regional economic strategies would operate within the spatial strategy (i.e. conform to the RSS) and, in turn, the spatial strategy would reflect the economic strategy's analysis of the regional economy so as to devise a strategy that could support objectives such as the promotion of the knowledge economy and improving productivity in the regions. Kitchen asks:

> What will happen when push comes to shove, as at some time in most regions it will? Will the RPG (or indeed RSS) with its environmental and sustainability appraisals have sufficient teeth to make a real difference to what RDAs actually do?
>
> (Kitchen 1999: 130)

The simple answer is that eventually the RDA took over the preparation of regional strategy. In 2009, the RSS was replaced with *regional strategies* (RS). These became the responsibility of not the regional assemblies but the RDA jointly with a leaders' board (consisting of the council leaders of the constituent local authorities). Given the leading role of the RDA in the process, more emphasis was put on economic growth. To strengthen the process, four other institutional components were established, including a regional ministerial system to bring the partners around the table; Parliamentary Select Committees to scrutinise the inter-regional concerns; a Cabinet Committee to address cross-governmental decisions related to investments in the regions; and the Homes and Communities Agency, which brought together a range of funding streams at the local level (Morphet 2011: 211). By 2011, commentators could see cracks in the relationship between the RDAs and leaders' boards, which raised concerns about the limited role of spatial planning in the RS approach (Morphet 2011; Baker and Wong 2013). However, none really anticipated what happened next, the almost overnight dismantling of the entire regional institutions (including RDAs) and the abolition of almost all forms of regional strategies (including RS). This not only dug a major hole in strategic planning in England, but also led to the disappearance of valuable documented knowledge which was accumulated as the evidence base of the regional strategies, as well as the loss of the intellectual capital of those planners and other professionals who were involved in the process of regional strategy development.

The Localism Act 2011 provided for the abolition of regional strategies in a two-stage process. The first stage was to remove the regional planning bodies and prevent further strategies from being created. The second stage was to abolish each existing regional strategy by secondary legislation. Legal challenges had delayed the revocation and the abolition of regional spatial strategies, but the process was eventually completed in 2013, when the final Order abolishing the last regional spatial strategy took effect.

However, under sections 103–13 of the Local Democracy, Economic Development and Construction Act 2009, 'combined authorities' may be set up, by the Secretary of State, at the request of local authorities in a specified area in order to undertake joint functions under the aegis of a public body with its own legal personality. The Greater Manchester Combined Authority (GMCA) was the first such combined authority, set up in 2011 and consisting of ten indirectly elected members and a directly elected councillor from each of the ten metropolitan boroughs that comprise Greater Manchester. A number of other combined authorities are proposed.

Furthermore, the Coalition government's introduction of a 'duty to cooperate' (see Box 4.2) through the Localism Act 2011 has also encouraged collaboration across the administrative boundaries, particularly those

BOX 4.2 DUTY TO COOPERATE IN STRATEGIC PLANNING

The duty to cooperate, introduced by the Localism Act 2011, requires cooperation between local planning authorities and other public bodies to maximise the effectiveness of policies for strategic matters in local plans. The Act defines cooperation as requiring all bodies and persons concerned to engage constructively, actively and on an ongoing basis in:

(a) preparation of development plan documents;
(b) preparation of other local development documents;
(c) preparation of marine plans under the Marine and Coastal Access Act 2009 for the English inshore region, the English offshore region or any part of either of those regions;
(d) activities that can reasonably be considered to prepare or strategically support the above activities.

In this context, 'strategic matters' are defined as:

(a) sustainable development or use of land that has or would have a significant impact on at least two planning areas, including (in particular) sustainable development or use of land for or in connection with infrastructure that is strategic and has or would have a significant impact on at least two planning areas; and
(b) sustainable development or use of land in a two-tier area if the development or use is a county matter, or has or would have a significant impact on a county matter.

The Town and Country Planning (Local Planning) (England) Regulations 2012 (SI 2012/767), Part 2 lists relevant public bodies as:

- Environment Agency
- English Heritage
- Natural England
- Mayor of London
- Civil Aviation Authority
- Homes and Communities Agency
- primary care trusts
- Office of Rail Regulation
- Transport for London
- integrated transport authorities
- highway authorities
- Marine Management Organisation
- local enterprise partnerships.

which relate to the strategic priorities. However, duty to cooperate does not necessarily mean duty to agree, so it is not the same as producing a jointly agreed regional strategy. The NPPF also encourages local authorities to think beyond their administrative areas and take account of the 'functional urban areas' (FUA) or 'city-regions' (Morphet 2011; Baker and Wong 2013). These are defined in different ways using various methodologies (see Davoudi 2008 for a critical review; Harrison 2012) but the most common delineation of FUAs is based on travel-to-work areas and/or housing market areas. The NPPF (para. 181), therefore, requires that

> Local planning authorities will be expected to demonstrate evidence of having effectively cooperated to plan for issues with cross-boundary impacts when their Local Plans are submitted for examination. This could be by way of plans or policies prepared as part of a joint committee, a memorandum of understanding or a jointly prepared strategy which is presented as evidence of an agreed position.

DEVELOPMENT PLANS

Formulating and implementing plans is an integral part of the planning system in the UK. Plans and plan-making processes often represent the long-term, forward-looking and proactive face of planning (Davoudi 2000b; Poxon 2000). However, the more recent history of the planning systems is of a period of the rediscovery of the importance of plans. In theory, they play a strategic, integrating and coordinating role and provide a degree of certainty and consistency in decision-making. In practice, their role has waxed and waned over the years. Nevertheless, they have remained a crucial part of the system. Both the contents of plans and the processes by which they are made have been subject to numerous changes. 'While the decision making procedure is tightly defined by legislation, there is little legal specification as to substance' (Davoudi, *et al.* 1996: 422). In broad terms, development plans tend to deal with, explicitly or implicitly, (and are often judged by) some key aspects of substance and process (Healey 2004a). These are summarised in Box 4.3.

In the following subsections, we provide a historical overview of the evolution of development plans, using major legislative changes as milestones in the journey so far.

Development plans post 1947

The main instrument of land use control in Britain until 1947 was the *planning scheme*. This was, in effect,

BOX 4.3 KEY DIMENSIONS OF PLANS' CONTENTS AND PROCESSES

Contents

- Interpretation of spatiality
- Conceptions of place
- Spatial organising principles
- Understanding of scale
- Concepts of future
- Treatment of time
- Visualisation and representation.

Processes

- Perceived role of planners
- Knowledge and skills employed
- Methods of engagement
- Institutional/governance structures and power relations
- Modes of implementation.

Source: adapted from Davoudi and Strange 2009: 11, Table 1.1

development control by zoning. As discussed in Chapter 2, zoning was replaced in 1947 by a markedly different system which attempts to strike a distinctive balance between flexibility and commitment. The approach is, in many important ways, the same in 2014 as it was in the 1950s. It is, as mentioned earlier in this chapter, fundamentally a discretionary system in which decisions on particular development proposals are made as they arise, against the policy background of a generalised plan. The Town and Country Planning Act 1947 defined a development plan as 'a plan indicating the manner in which a local planning authority proposes that land in their area should be used'.

Unlike the pre-war planning schemes, the development plan did not of itself imply that permission would be granted for particular developments simply because they appeared to be in conformity with the plan. Though developers were able to find out from the plan where particular uses were likely to be permitted, their specific proposals had to be considered by the local planning authority. When considering applications, the authority was expressly directed to 'have regard to the provisions of the development plan', but the plan was not binding and, indeed, authorities were instructed to have regard not only to the development plan but also to 'any other material considerations'. Furthermore, in granting permission to develop, local authorities could impose 'such conditions as they think fit'.

However, though the local planning authorities had considerable latitude in deciding whether to approve applications, it was intended that the planning objectives for their areas should be clearly set out in development plans. The development plan consisted of a report of a survey, providing background to the plan but having no statutory effect, a written statement, providing a short summary of the main proposals but no explanation or argument to support them, detailed maps at various scales, and a programme map, showing the phasing of development. The maps indicated development proposals for a twenty-year period and the intended pattern of land use, together with a programme of the stages by which the proposed development would be realised. The plans were approved by the minister (with or without modifications)

following a local public inquiry. Initially, a three-year target was set for submission of the plans, but only twenty-two authorities met this and it was not until the early 1960s that they were all approved.

By this time, the requirement to review plans on a five-yearly cycle had brought forward amendments, many taking the form of more detailed plans for particular areas. These had to follow the same process of inquiry and ministerial approval as the original plans, and many authorities were still engaged on the first review in the mid 1960s. Furthermore, although the system of development control guided by development plans operated fairly well without significant change for two decades, 1947-style plans did not prove flexible in the face of the very different conditions of the 1960s. This was partly due to the then understanding of the process of plan making which was largely inspired by Patrick Geddes' idea of 'survey–analysis–plan'. This linear view of the planning process led to the creation of blueprints specifying an end state for the use of land.

The statutory requirement for determining and mapping land use led inexorably to greater detail and precision in the plans and more cumbersome procedures. The quality of planning suffered, and delays were beginning to bring the system into disrepute. As a result, public acceptability, which is the basic foundation of any planning system, was jeopardised. A major shortcoming of 1947 development planning was a complete lack of public consultation or participation.

Development plans post 1968

In response to the shortcomings of the 1947 style, the Planning Advisory Group (PAG) was set up in May 1964 to review the broad structure of the planning system and, in particular, development plans. In its report, published in 1965, PAG proposed a fundamental change, which would distinguish between strategic planning issues and detailed tactical ones. Only plans dealing with the former would be submitted for ministerial approval; the latter would be for local decisions within the framework of the approved policy. Legislative effect to the PAG proposals was given in the Town and Country Planning Act 1968 (for England and Wales) and 1969 (for Scotland). This created a two-tier system of structure plans and local plans, which we elaborate in turn, as well as opportunities for public participation in planning.

Structure plans provided a strategic tier of development plan and, until 1985, were prepared for the whole of England by county councils (and the two former national park boards). They were originally subject to the Secretary of State's approval, but between 1992 and their demise in 2004 they were adopted by the planning authority itself. They consisted of a written statement of strategic policies and proposals (but not detailed land allocations) for the area, a key diagram (not a map, to avoid the identification of particular parcels of land) and an explanatory memorandum in which the authority summarised its justification for the policies. The detailed arrangements for structure planning were amended on numerous occasions and central government vacillated on their legitimate scope and content. The initial conception was that they should be wide-ranging, but the government narrowed the range of competence of structure plans over the years, only to widen it again in 1999 and then replace it with the regional spatial strategy in 2004, as discussed above. The idea was that general spatial policies could be determined before detailed land use allocations were made, albeit not always to the liking of those affected by later more detailed plans. In practice, over much of the history of structure planning, counties formulated their 'policy and general proposals' in greater detail than anticipated by government, including quite detailed land allocations and development control policies in some cases. An argument in favour of more detail was that few local plans were being produced, but more detail also gave the county council more control over the implementation of policy.

Local plans provided detailed guidance on land use. They too were replaced under the system introduced in 2004. They consisted of a written statement, a proposals map and other appropriate illustrations. The written statement set out the policies for the control of development, including the allocation of land for specific purposes. The proposals map had to be on an ordnance survey base, which thus showed the effects of the plan on precise and identifiable boundaries. Under the 1968 system, there were three types of local plan: general plans (referred to as 'district plans' before 1982), action area plans and subject plans. *General local plans* were prepared 'where the strategic policies in the structure plan need to be developed in more detail'. *Action area local plans* dealt with areas intended for comprehensive development and *subject plans* dealt with specific planning issues over an extensive area, typically minerals and green belt, but many others such as caravans and pig farming.

Local plans were not subject to approval by the Secretary of State, but were adopted by the local planning authority (although the Secretary of State rarely used powers to call in plans and to require modifications). The original rationale for this was that a local plan would be prepared within the framework of a structure plan, and since structure plans would be approved by the Secretary of State, local authorities could safely be left to the detailed elaboration of local plans. This went to the very kernel of the philosophy underlying the 1968 legislation, namely that central government should be concerned only with strategic issues, and that local matters should be the clear responsibility of local authorities.

This division of plan-making functions was predicated on the creation of unitary planning authorities responsible for preparing both the structure and the local plan. But the Local Government Act 1972 established two main types of local authority in England and Wales and divided planning functions between them. The two levels of local government did not share the same views about planning policy across much of Britain, which exacerbated conflict in the system. For some this was a fundamental weakness of the system, which led to calls for the abolition of structure plans, but for others it was a useful separation of powers, with the conflict usefully exposing critical issues in planning. Two mechanisms were introduced to promote effective cooperation in the planning field and to minimise delay, dispute and duplication – the development plan scheme (later the local plan scheme) and the certificate of conformity given by the structure planning authority.

A decade after the start of the 1968 system, Bruton (1980: 135) summarised the problems as 'delay and lack of flexibility; an over-concentration on detail; [and] ambiguity in regard to wider policy issues'. The same is probably true nowadays. Plans were very slow in coming forward to statutory approval and adoption. The first structure plan cycle took fourteen years to complete and many took more than two years to get approval once prepared. They were long, complex and contained policies thought to be not relevant to planning, for example costs of waste collection, the development of cooperatives, standards of road maintenance and even 'nuclear-free zones'. These delays held back the adoption of local plans; indeed, the first local plan was not adopted until 1975. However, the rate of deposit and adoption increased sharply in the 1980s and, by 1987, 495 local plans had been adopted in England and Wales (Coon 1988). Unfortunately, many of the plans were out of date by the time their processing, which took an average of five years, was complete. (Again, this problem is still as relevant in 2014.) One reason in the 1970s and 1980s was the proliferation of non-statutory planning documents ('informal policy'), which outnumbered statutory plans by about ten to one (Bruton and Nicholson 1985). They took many forms, from single-issue or area policy notes to comprehensive but informal plans, but much of it was correctly described as 'bottom drawer policy' which had been subject to little consultation. While there is always a need for some policy or guidance to be supplementary to the statutory plan and not subject to the same statutory procedures, much informal policy in the 1970s and 1980s was prepared in this way to avoid public scrutiny and/or the formal procedures.

Structure planning was not coming up to the expectations of the PAG report. Though it undoubtedly provided a forum for debate about strategy, it did not provide the firm lead that was promised. The uncertainties and complications of structure planning in practice carried over to local planning and contributed, in some areas, to a professional culture that was at best indifferent to statutory plans (Shelton 1991). There were more positive attitudes in other areas. Where the stakes involved in development applications were high, as in London and counties such as Hertfordshire (where full statutory plan cover was completed during the 1980s), statutory plan making was vigorously pursued. Also, despite turbulent economic conditions, the plans proved to be reasonably robust and effective in implementing policy and defending council decisions at appeal. This variation in practice was identified in research at the time: Healey et al. (1988) concluded that plans had proved to be effective in guiding and supporting decisions, and in providing more certainty, clarity, and consistency to private-sector investors and a framework for the protection of land. They were particularly useful in shaping private-sector decisions, especially in the urban fringe. Conversely, the difficulty encountered in controlling public-sector investment in housing, economic development, inner-city policy and infrastructure provision was shown to be an impediment to effective implementation of strategy (Carter et al. 1991). Davies et al. (1986a, 1986b, 1986c) concluded that plans played only a small part in guiding development control decisions overall, but were much more important when a case went to appeal – what they termed the 'pinch points' of the system. They suggested that this reflected the system's chief virtue: its ability to enable a sensitive response to local conditions. It was recommended that the government should encourage local authorities to provide better written policy cover, reduce its complexity, use statutory plans, and facilitate speedier adoption (Rydin et al. 1990).

Beyond these procedural details, the 1970s style of planning marked a 'paradigm shift' in planning thoughts and practices (Hall 2002a). First, Geddes' linear process of survey–analysis–plan was replaced with 'procedural planning theory'. This considered planning as a technical rational process (and not a blueprint) that follows a cycle of logical steps, including definition of problems and/or policy goals, identification of alternatives, evaluation of alternatives, implementation of the preferred option, and monitoring and review. Second, under the influence of systems theory, a city was considered as a complex system and, as such, subject to systems analysis and control. Both of these assumptions have since been criticised by planning theorists and other commentators.

Development plans post 1985

During the Thatcher government, planning underwent a major scrutiny and attempts were made to modify, bypass and simplify it (Thornley 1993), with particular emphasis on restricting the scope of plans to land-use policies only. For example, the Secretary of State removed 40 per cent of the policies of the Manchester Structure Plan and substantially modified 25 per cent of them in 1980. Attempts were also made to reduce the role of plans in development control decisions, as shown in the following extract from a 1985 circular by the then Department of the Environment: 'The development plans are one, but only one, of the material considerations that must be taken into account in dealing with planning applications' (DoE Circular 14/85 *Development and Employment*).

The formal planning procedures were bypassed in central government's designated areas, notably the urban development areas and the enterprise zone areas. Furthermore, Mrs Thatcher's precipitate decision to abolish the Greater London Council (GLC) and the six metropolitan county councils (MCCs) in 1985 forced hasty changes to the planning system in these areas, introducing, for the first time, two different systems of development planning in England: a unitary system in London and six other metropolitan areas, and a two-tier system for the rest of England. We discuss these in turn.

Development plans in London and metropolitan areas

In London boroughs and metropolitan districts, which became 'unitary' planning authorities after the abolition of the GLC and the MCCs, a new *unitary development plan* (UDP) replaced structure and local plans. In precisely those parts of the country where there is a particular need for a strategic approach to planning, the structure planning at supra-borough/district level was lost. It was replaced by a tier of *strategic guidance* produced on a cooperative basis by the constituent districts but issued by the Secretary of State. To fill the gap, in Greater London, a joint planning committee was established – the London

Planning Advisory Committee. Subject and action plans also disappeared in favour of the new 'unitary' approach to plan making.

UDPs were in two parts: Part I was the structure plan element and had the characteristics of the structure plan described above. Part II was the local plan element, with a written statement of the authority's policies and proposals, a map showing these proposals on an Ordnance Survey base and a reasoned justification of the policies. The two parts of the UDP were presented in one document and had a ten-year time horizon. The UDP was adopted by the district council and was not subject to the approval of the Secretary of State (although reserve powers of central intervention were maintained).

Among the critics, concerns were raised about the future of strategic thinking in the metropolitan areas, difficulties of cooperation between districts, and problems of participation and coping with the statutory right to objection in plans which embrace such large areas. For the districts themselves, many of these worries had proved unfounded. It had been possible to accommodate policy and political differences among districts, but this had been very much on a lowest common denominator level (Hill 1991; Williams *et al.* 1992). Serious concerns were voiced about the extent to which the public, interest groups and even some professionals could engage effectively in the process. There were considerable delays in some metropolitan districts, where very detailed UDPs were produced in particularly contentious circumstances, generating great conflict and many thousands of objections having to be dealt with in public inquiries.

Development plans outside London and metropolitan areas

In 1986, the Green Paper *The Future of Development Plan* was published for consultation. It proposed the abolition of structure plans in England and Wales (but not in Scotland) and their replacement by statements of county planning policies on a limited range of issues (to be specified by the Secretary of State), more policy at regional and subregional level, and the introduction of single-tier district development plans covering the whole of each district. The Green Paper was prompted by the publications of two White Papers, *Lifting the Burden* (1985) and *Building Businesses, Not Barriers* (1986), and their insistence on 'freeing' enterprise from unnecessary restraints. These denigrated both structure and local plans, and criticised the procedures for preparing plans as 'too slow and cumbersome'.

To some extent, the proposal to replace structure plans was a response to growing dissatisfaction about the making of many ad hoc and apparently inconsistent decisions by both the Secretary of State and local authorities. The campaign for change included both the development and the conservation lobbies, who had a common demand for more certainty in the system and a reduction in the growing number of speculative applications. There was also some dissatisfaction among government supporters about decisions taken centrally which went against local (often Conservative) opinion. Local authorities were concerned that more of their decisions were being overruled, and complained at the lack of clarity in central policy. By comparison, matters looked better in Scotland and in the emerging system in the metropolitan counties.

Many of the 500 responses to the Green Paper argued strongly against the proposed abolition of structure plans, and (for the time being) they were saved. Meanwhile PPG 12 *Development Plans* was published in 1988 urging local authorities to extend statutory plan coverage, normally by district-wide plans, and to replace non-statutory policy, which it described as 'insufficient and weak'. Strategic green belt boundaries were singled out as requiring further specification in detailed local plans. In return, the government offered an enhanced status for plans in the 1989 White Paper *The Future of Development Plans*, which echoed the PPG 12 proposals and added a mandatory provision for all counties to prepare minerals development plans. County councils were urged to press ahead with the revision and updating of structure plans and to cooperate on the elaboration of regional guidance. The counties, for their part, were to ensure that plans were less bulky and concentrated on strategic issues. Shortly afterwards, the government announced

proposals to end the requirement for the Secretary of State to approve all structure plans and alterations in favour of adoption by the local authority. During debate on the Planning and Compensation Bill that followed, provisions were added to further increase the status of the statutory plans in development control.

Development plans post 1991

The Planning and Compensation Act 1991 made a number of major changes to the planning framework. The most important one was that it made the plan the primary consideration in development control decisions. In commending the amendment (formerly section 54A of the 1990 Act), Sir George Young coined a phrase in saying that 'the approach shall leave no doubt about the importance of *the plan-led system*'. In sharp contrast to the DoE's Circular 14/85 mentioned above, the Act stated:

> Where, in making any determination under the planning acts, regard is to be had to the development plan, the determination shall be made in accordance with the plan unless material considerations indicate otherwise.

The Act also made the adoption of *district-wide local plans* mandatory, and abolished the need for central approval of structure plans (although central government retained its powers of intervention). Small area local plans and subject plans were abandoned except for minerals and waste, which henceforth would be prepared for the whole of the authority's area.

In the first part of the 1990s, it seemed that the framework of local planning policy in England and Wales was to become more coherent, albeit differentiated in metropolitan and non-metropolitan areas. At that time, most local authorities had little coverage of statutory plans, but a mix of interlinked subject and small area-based policy documents and informal plans. The 1991 Act offered a much clearer system. Those needing to know about planning policy in non-metropolitan areas would make reference to the structure plan, the district-wide local plan and the minerals and waste plans, with more chance that they would exist. In the event, the prospect of an orderly framework of development plans in England was dashed by local government reorganisation in the mid-1990s.

Reorganisation of local government in England between 1994 and 1997 is explained in Chapter 3. It affected the two-tier system of counties and districts beyond the metropolitan areas by introducing unitary district councils. Some counties were abolished completely to be replaced by unitary councils. Other unitary councils were established in counties, mostly for the provincial cities, which created an island unitary authority within the remaining two-tier structure. Where the two-tier system remained, the planning framework was not affected: counties prepared the structure plan and waste and minerals plans (or one plan for both topics) and districts prepared the district-wide local plan. Where new unitary authorities were created, they took on the county as well as district functions. Most prepared joint structure plans with the neighbouring county councils. The exceptions were, at the time, Halton, Warrington, Herefordshire, the Isle of Wight and Thurrock, which prepared unitary development plans (as in the metropolitan districts). All the other unitary authorities prepared their own district-wide local plans. The metropolitan districts were unaffected by local government reorganisation and continued with their unitary development plans (see Figure 4.3a). The content of structure plans, unitary development plans and local plans remained in currency until it was replaced or reviewed by development plan documents prepared under the arrangements established in 2004.

Overall, planning seemed to emerge 'leaner but fitter' from the scrutiny of the 1980s. One of the reasons for the survival of the planning system was that it could easily accommodate the EU requirements for environmental impact assessment of major development proposals. The Planning and Compensation Act 1991 restored the status of plans but the plan-led system also led to the shifting of conflict mediations from specific sites to the plan-making arenas, turning them into 'battle grounds'. It seemed that 'the planning-by-appeal' approach of the 1980s [was] giving way to the 'planning-by-public inquiry' of the 1990s (Davoudi 2000b: 126).

Development plans post 2004

The Planning and Compulsory Purchase Act 2004 introduced further substantial changes to the planning system. The New Labour agenda for modernising and 'joining up' government in pursuit of priority outcomes was central to the reform process, alongside oft-repeated criticisms that planning is a brake on economic growth and does not do enough to protect the environment or promote social cohesion. Wider discourse on 'spatial planning' at the EU level also provided inspiration for the direction of change (Tewdwr-Jones *et al.* 2010). Recommendations on reform varied, but there was a measure of agreement that planning had become marginalised in government decision-making at a time when there was an urgent need for more effective coordination of the impacts of disparate strategies, policies and actions for particular places. Concerns were raised with regard to inconsistencies in policies at different levels, the complexity of national policy guidance and delays in the adoption of plans. By 2002, 13 per cent of relevant authorities in England had not adopted a local or UDP and 214 plans were 'out of date'. It took twenty-one county authorities more than five years to revise structure plans after the approval of new regional guidance. Leeds UDP took nearly ten years to be adopted owing to receiving over 20,000 objections to a small number of its policies. There were also delays in decisions on major development applications, notably major infrastructure projects. The latter was epitomised in

> the Heathrow Terminal 5 'saga' which took nearly nine years (February 1993 to November 2001), cost about £60 million in fees, generated a 600-page report, and despite nearly four years of public inquiry left many people disenchanted with the planning system.
>
> (Davoudi 2011b: 34)

Highlighting this paradox, the *Economist* wrote: 'Few countries have ended up with a planning system which manages both to hold projects up for decades, and to give people the feeling that they don't have any say at all' (*Economist* 2001: 38). In broad terms, there were two parallel but fundamentally different calls for change. One was based on a desire to improve the *quality* of planning decisions that affect *places*; the other was based on a concern to *speed* up and bypass planning decisions that affect individual *clients* (Vigar *et al.* 2000).

In response to these concerns, the government published the Green Paper *Planning: Delivering a Fundamental Change* in 2001, which, following devolution, related only to England. Four 'daughter documents' were published at the same time, alongside the Green Paper, with proposals for change on procedures for dealing with major infrastructure projects, planning obligations (planning gain), compulsory purchase and the Use Classes Order.

In the foreword to the Green Paper, the then minister echoed the above-mentioned criticisms of the planning system, suggesting that:

> Good planning can have a huge beneficial effect on the way we live our lives. . . . But some fifty years after it was first put in place, the planning system is showing its age. What was once an innovative emphasis on consultation has now become a set of inflexible, legalistic and bureaucratic procedures. A system that was intended to promote development now blocks it. Business complains that the speed of decision is undermining productivity and competitiveness. People feel they are not sufficiently involved in decisions that affect their lives.

The Green Paper's analyses of the shortcomings of the planning system were better on symptoms than causes. The main problems were seen as being the failure to tackle issues lying at the boundaries between authorities and between policy sectors; the 'weight' of local planning policy arising from too much attention to comprehensive coverage and not enough concentration of effort where change was anticipated or needed; an unwillingness on the part of politicians and communities to accept new development and make difficult decisions; the questionable qualities of the eventual outcomes of new development from the planning process; and poor management of the system. The main proposals for the plan framework (which by and large were carried through) were the abolition of structure and local plans and their replacement with RSS (mentioned earlier in this chapter) and local development frameworks, along with a better evidence base and increased engagement with the community.

The Green Paper attracted over 15,000 responses from a wide range of groups, including the development industry, community, environment, government and business interests. While there was general support for the reform (Smith & Williamson Management Consultants 2002), 88 per cent of respondents did not want to see local plans replaced by local development frameworks (96 per cent of members of the public) and 90 per cent disagreed with the proposal to abolish structure plans and replace them with regional strategies. Only local government and the professional planning institutions showed some enthusiasm for the change. In evidence given to the 2002 Transport, Local Government and the Regions inquiry into the reform of planning, David Lock, representing the Town and Country Planning Association suggested a number of minor adjustments instead of radical changes to the system, arguing that adjustments could 'get us closer to a faster, fairer and speedier system than chucking the whole lot out and spending several years constructing a brand-new one which is not yet designed' (HC 476 III 24/04: para. 529).

Despite such reasoning, the government went for the more radical solution, while presenting it as a means to engage the community more effectively (which was supported by 80 per cent of respondents). In 2002, the Planning and Compulsory Purchase Bill was produced and became the longest running Bill in Parliament until it was enacted in 2004. For the first time, the Planning and Compulsory Purchase Act 2004 defined a statutory purpose for planning, stating:

> It is a statutory duty for plans to contribute to sustainable development. This means high and stable levels of economic growth and employment, social progress, effective protection of the environment and prudent use of resources.

It is, however, important to note that the emphasis was predominantly on economic growth (Baker and Wong

2013). This continues to be the case, as demonstrated by the Coalition government's underlying rationale for the planning system and the requirement that it should be 'light-touch, fast and responsive' (HM Government 2010a: para. 3.14) and the system should 'transform . . . from an impediment to economic development into a means of economic growth' (para. 3.6). The main aims of the 2004 reform of the planning system were:

- *flexibility* to respond to change and review quickly
- *community involvement* from the outset and throughout the preparation;
- *front loading* to take key decisions early and avoid late changes;
- *sustainability appraisal* at various stages;
- *programme management* in accordance with a defined scheme;
- *integration* with other strategies and plans;
- *soundness* in terms of content, process and evidence base.

Figure 4.3b provides an overview of the framework for planning in England after 2004. This shows that all levels of planning policy were affected, though the biggest changes were at the county and district levels. The implications for national and regional policy were explained earlier in this chapter. At the local level, all plans (structure plans, local plans and unitary development plans) had to be replaced with mandatory local development frameworks (LDFs) prepared by a single tier of planning authority. This meant the abolition of structure planning at county level. However, counties continued to prepare minerals and waste LDFs and helped with subregional strategies.

The concerns of the majority who opposed the abolition of structure and local plans were largely borne out. The outcome of the reform was a very complex system with a battery of new acronyms and terms which made it incomprehensible even to the professional planners, let alone members of the public. The government's attempt to clarify the system through a growing catalogue of 'good-practice guidance' (which amounted to thousands of pages as mentioned above) did not help and probably made the situation worse.

The puzzle was that many local authorities wanted yet more guidance, while the new system was supposed to give more flexibility, and enable innovation and the creation of more locally relevant solutions. One example is PPS 12 *Local Development Frameworks* and its sister document *Creating Local Development Frameworks: A Companion Guide to PPS12*, whose attempt to clarify matters created further confusion through statements such as:

> The local development framework will be comprised of local development documents which include development plan documents, that are part of the statutory development plan and supplementary planning documents which expand policies set out in a development plan document or provide additional detail.
>
> (para. 1.4)

The LDF itself was not a plan: it was a folder or portfolio that contained all the local development documents of the local authority, together with other related information. It was a non-statutory term not defined in the Act and was only used to describe the full portfolio of local development documents. The LDF had six main components, as shown in Box 4.4.

Development plan documents are subject to a *sustainability appraisal report*, which is required by UK law and considers economic and social as well as environmental effects. A SAR incorporates the mandatory *strategic environmental assessment* (SEA) report which is required by EU Directive 2001/42/EC for plans and relates to all stages of plan preparation. The NPPF (paras 165–6) states:

> A sustainability appraisal which meets the requirements of the European Directive on strategic environmental assessment should be an integral part of the plan reparation process, and should consider all the likely significant effects on the environment, economic and social factors.

Local Plans may require a variety of other environmental assessments, including under the Habitats Regulations where there is a likely significant effect on a European wildlife site (which

decentralisation. The Government's approach in practice, however, has thus far been marked by inconsistency and incoherence, not helped by a definition of localism that is extremely elastic. . . . Some policy areas remain notably more centralised than others.

The Government has not produced a compelling vision of what its imagined localist future will look like and the functions and responsibilities of the players within it. Greater clarity and certainty is needed.

The main provisions of the Localism Act 2011 fall under four headings (DCLG 2011b: 6):

- new freedoms and flexibilities for local government;
- new rights and powers for communities and individuals;
- reform to make the planning system more democratic and more effective;
- reform to ensure that decisions about housing are taken locally.

Reform of the planning system is a key element of the 2011 Act, which abolished RSSs and imposed a 'duty to co-operate', limited the role of the Inspectorate in the local planning process and in the Community Infrastructure Levy, and made other alterations with regard to pre-application consultations and enforcement (discussed in Chapter 5). It also abolished the Infrastructure Planning Commission and transferred its functions to the National Infrastructure Directorate (NID) within the Planning Inspectorate. This implied the final decisions are made not by an independent body but by ministers, whose decisions are informed by NID's recommendations. Some commentators argue that this is 'good news for democracy' but question whether it is 'good news for speed and efficiency too' (Davoudi 2011b: 34) and that 'this evidently restores legitimacy, but will bring back political difficulties on matters such as nuclear power in exactly the way the inventors of the regime were trying to avoid' (Marshall (2013a: 132).

The most significant change to the planning system was the introduction of a new neighbourhood planning regime. The Localism Act 2011 allows parish councils and groups of people from the community, called neighbourhood forums, to formulate Neighbourhood Development Plans (NDPs), Neighbourhood Development Orders (NDOs) and Community Right to Build Orders (CRBOs) (the Orders give permission for small-scale, site-specific developments by a community group), all of which must have regard to national policies and conform to local strategic policies. They should also pass an independent check and subsequently be put to a local referendum (see below). The local planning authority has a statutory role to support neighbourhood planning by, for example, providing advice and assistance, organising the independent examination of the NDPs, NDOs or CRBOs and making arrangements for a referendum. In order to incentivise neighbourhood planning, in 2013 the government introduced a package of financial support for both local planning authorities (LPAs) and the communities interested in neighbourhood planning. We describe these in turn.

LPAs could claim up to a maximum of £50,000 for area designations in the 2012 financial year up to a maximum of ten (with an overall limit of £1.5 million on designation payments). From April 2013, they can claim for up to twenty designations (£100,000) in the financial year 2013–14. The overall limit for designation payments in these years has risen to £5 million. In recognition of the LPAs' duties to support neighbourhood planning, they can claim up to £30,000 for each neighbourhood plan, to be paid in three instalments, with the last one (£20,000) being made on successful completion of the neighbourhood planning examination.

As regards the communities, those who want to prepare a neighbourhood plan can bid for grants up to £7,000 each to contribute to the costs of preparing their proposal. Detailed guidance on procedures was then set out in the government's publication of 2011, *Supporting Communities in Neighbourhood Planning 2013–15*. This was followed by further guidance, *Community Led Project Support Funding – Planning Application Route: Application Guidance*, published in August 2013. The latter stated that the rules for accessing the fund would be widened to enable communities to apply who wished to take the more traditional route of applying for planning consent, rather than using a CRBO. The aim is clearly to encourage more take-up from the self-build housing sector (Smith 2013). Another incentive which came into force in April 2013 through the Community Infrastructure Levy (Amendment) Regulations 2013 (SI 2013/982) enables neighbourhoods to receive 25 per cent of the revenues from the Community Infrastructure Levy arising from the development that they have chosen to accept in areas where there is a neighbourhood development plan in place. The money is paid directly to parish and town councils, and could be used for community projects such as re-roofing a village hall, refurbishing a municipal pool, or taking over a community pub.

Neighbourhood plans

As mentioned above, neighbourhood forums and parish councils have been given the power to produce neighbourhood plans, while the local planning authorities continue to produce local plans (previously called LDFs). Local plans provide the strategic context within which neighbourhood plans are positioned. Government guidance has made it clear that neighbourhood plans have to be positive towards development proposals that are already part of the wider local plan. However, they can shape and influence where that development should go and how it should look. Neighbourhood plans have to meet the following conditions: they must have regard to national planning policy; be in general conformity with strategic policies in the development plan for the local area (i.e. such as in a core strategy or a local plan); and be compatible with EU obligations and human rights requirements. These and other considerations are checked by an independent qualified person, normally a planning inspector, before the neighbourhood plan is voted on in a local referendum. If the majority (50 per cent plus of participants, not population) vote in favour, the local planning authority has a legal duty to adopt the plan, unless it conflicts with the European Convention on Human Rights or EU policy. Once adopted, the neighbourhood plan becomes part of the legal framework of the development plan, and planning decisions for the area must take account of the policies in both the local plan (see Figure 4.3c)

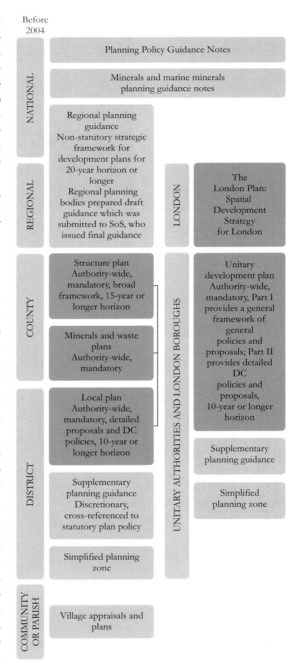

Figure 4.3 The planning policy framework before 2004, 2004–11, and after 2011

(a) Before 2004

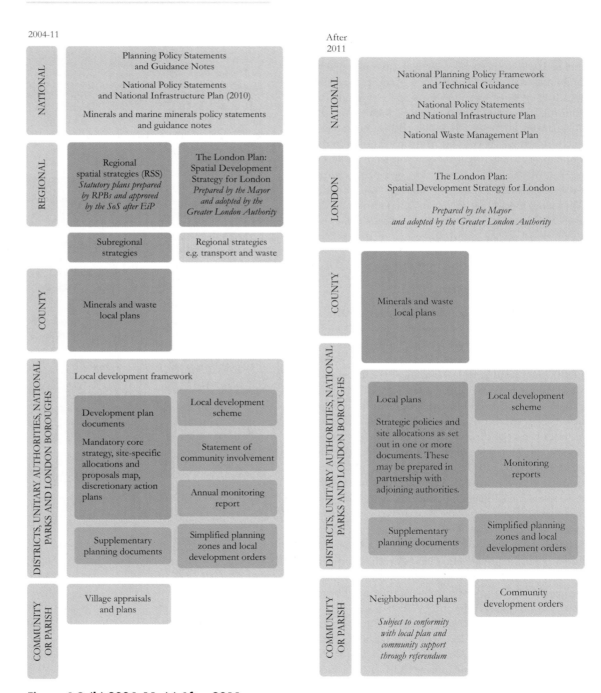

Figure 4.3 (b) 2004–11, (c) After 2011

and the neighbourhood plan (see Figure 4.4). There is no prescribed content for neighbourhood plans, so they can be comprehensive or focused on one or two issues only; they can include detailed proposals or general principles to guide new development. It all depends on what the local community wants, as long as they do not 'block' new development and remain within the scope of existing policy in the local plan. To map the progress in neighbourhood planning in practice, the DCLG developed a dedicated website and a series of newsletters. The British Property Federation (BPF) also launched a website in November 2013 to support business involvement in neighbourhood planning.

In March 2013, Upper Eden in Cumbria became the first area to officially adopt a neighbourhood development plan. The Exeter St James (Devon) Neighbourhood Plan was the first to be drawn up by a neighbourhood forum (as opposed to a parish council) in an urban area. By December 2013, 800 applications had been made for neighbourhood area designation in over half of all the local planning authorities in England. DCLG data indicated that over 630 neighbourhood areas were designated, fifty-four draft plans were published, twenty-five plans submitted to examination, nine passed examination, six plans passed referendum and four plans were in force.

While the potential of neighbourhood planning for progressive localism and for leading production of 'self-build neighbourhoods and cities' has been recognised (Farnsworth 2013: 481), concerns have been raised about the varying capacity of communities to engage with neighbourhood planning and the potential for asymmetric coverage of neighbourhood planning across England, because 'not all communities will be able to find the £100,000+ that the Thame Neighbourhood Plan (a Frontrunner) cost, and the level of support made available to communities from proactive councils' (Brownill and Downing 2013: 374). A 2013 survey, undertaken by *Planning* journal, showed that 'town halls in England's most deprived areas are least likely to have received applications from local groups to take up neighbourhood planning powers'. Only a tenth of the 433 applications received by early 2013 were from the 20 per cent most deprived local authorities (*Planning* 2013b: 4), of which less

Figure 4.4 The process of neighbourhood planning

than half were designated (twenty out of forty-five). By contrast, from the top 20 per cent least deprived authorities, ninety-two applications were received, of which sixty were designated. Drawing on these, Davoudi and Cowie (2013: 565) argue that neighbourhood forums 'have, until early 2013, failed to expand the diversity of political representation in the planning processes by engaging with those who do not have the time, the capacity, the knowhow, or the political resources to participate'. A related concern has been about the low turnout in referenda. For example, the turnout in the referendum for Exeter

St James (Devon) NP, mentioned above, was 20.8 per cent, and Thame in Oxfordshire (drawn up by parish councils) was 39.8 per cent. However, of those who did turn out, a decisive majority voted in favour of the plan, but that is beside the point (*Planning* 2013a: 8).

Overall, the 2010 reform signalled yet another period of transformation of the planning system in England, which in many respects was a move away from the 2004 focus on 'spatial' planning and strategic thinking (Baker and Wong 2013; Haughton and Allmendinger 2013).

Development plans in Northern Ireland

The formal change to a discretionary system of development plans and control did not come to Northern Ireland until 1972. Prior to this, the system was much the same as for the rest of Britain before 1947, with local authorities able to prepare planning schemes. Practice was similar also in that very little progress was made on the preparation and approval of such schemes, and a system of interim development control operated. The Planning (Northern Ireland) Order 1972 (SI 1972/1634) introduced the development plan, with similar status to those in the rest of the UK. There were three types of development plan (area, local and subject plans) which were produced and adopted by the DoENI, with the twenty-six districts acting as statutory consultees. Area plans which could cover the whole or a substantial part of one or more district council areas are the main reference for development control, and include both strategic and detailed policies.

The provisions of the 1991 Act amended by the 2004 Act, which make the development plan the first and primary consideration in development control, were not implemented in Northern Ireland, although in 1999 the NI Office consulted on proposals to make the plan the primary consideration. This was in response to the House of Commons Northern Ireland Affairs Committee's 1996 report, *The Planning System in Northern Ireland*. The Committee expressed serious concerns about the lack of a clear strategy for the Province as a whole (which were responded to by the creation of the *Regional Development Strategy*) and the inadequacy of the development plans system.

Until recently, area and local plans continued to be prepared by the six divisional offices of the DoENI. In 2005, consultation began on fundamental reforms to the planning system in Northern Ireland, which follow change made in England and Wales by the 2004 Act. Meanwhile, a noteworthy example of plans which were prepared in the context of the RDS is the Belfast Metropolitan Area Plan 2015. In 2006, the Planning Reform Order (Northern Ireland) 2006 (SI 2006/1252) represented the formal initiation of the current planning reform in Northern Ireland, which has also been influenced by the Semple Review (2007). Whilst the review was principally concerned with housing sector, its recommendations were of close relevance to planning. In 2008, the Emerging Proposals paper was published, which confirmed the main objectives of the reform agenda. Following the publication of a subsequent consultation paper in 2009, the Planning Bill was published on 6 December 2010 which stated that 'the aim is to create a planning system which is quicker, clearer, more accessible and with resources better matched to priorities' (DoE 2010: 2). The Bill received Royal Assent in May 2011. The most significant change introduced by the Planning Act (Northern Ireland) 2011 is the decentralisation of plan making (Lloyd and Peel 2012) by replacing existing plans drawn up centrally by the DoE with development plans to be drawn up by councils in conformity with the Regional Development Strategy, mentioned above. These development plans will comprise a plan strategy and a local policies plan. There will also be an emphasis on 'soundness', as in the English system, encouragement of joint working between councils and a statement of community involvement. The change has been accompanied by a proposed restructuring of the institutional landscape, as discussed in Chapter 3. The Planning Act (Northern Ireland) 2011, when fully in force, will provide a two-tier planning system, with the majority of planning decisions made by councils, with the DoE dealing with regionally significant development.

In Northern Ireland, neighbourhood/community plans are not currently proposed. However, under the Local Government Bill, introduced to the Assembly in 2013, local authorities will be required to produce and publish a community plan for their area. This will take place when functions transfer from central government to local authorities, which is due in 2015 (Cave *et al.* 2013). Figure 4.5 presents the planning policy framework in Northern Ireland.

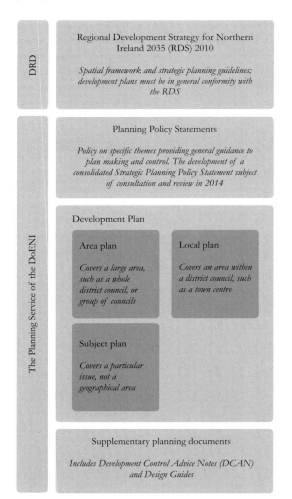

Figure 4.5 The planning policy framework in Northern Ireland

Development plans in Scotland

The Scottish system differs in several significant ways from that in England and Wales, but the two-tier system of development plans and the procedures for the adoption and approval were broadly similar until 1996. Some differences could be attributed to the particular geographical characteristics of Scotland; others may legitimately be attributed to a desire to avoid some of the difficulties of the English system. Because of the different administrative structure and larger planning areas in Scotland, there was a slightly different emphasis in the functions of structure plans, which were to indicate policies and proposals concerning the scale and general location of new development, and to provide a regional policy framework for accommodating development (PAN 37, p. 7). Progress on the approval of structure plans was a significant problem, with an average of seventeen months needed for Secretary of State's approval. It was not until 1989 that full structure plan cover was achieved. However, progress on local plans was better overall than in England and Wales, mainly because of the earlier introduction of a mandatory requirement for full cover.

The 1991 Act brought some of the same changes made in England and Wales to Scotland, notably the enhanced status of development plans in development control and the insertion of section 18A into the 1972 Act, with the same effect as section 54A (now section 38(6) of the 1990 Act) in England and Wales.

Local government reorganisation created unitary authorities in Scotland in 1996, and although the two-tier system of structure and local plans was retained, joint working became necessary for the production of some structure plans. The Scottish Office designated seventeen structure plan areas, six of which covered more than one unitary authority, while local planning continued unchanged in the new unitary districts.

In 2001, the Scottish Office launched a review of strategic planning in Scotland, which suggested that there was no real need for a second higher tier of development plans for much of Scotland, especially with the proposal for a national planning framework (as discussed earlier in this chapter). This met with some agreement, and legislation was amended to require structure plans only for the four major city-regions (Glasgow, Edinburgh, Aberdeen and Dundee) and unitary plans for the rest of Scotland. A 2004 consultation paper *Making Development Plans Deliver* followed a 2003 paper *Options for Change* and explained that the new *city-region plans* (CRPs) would be more selective and strategic but continue to be part of the statutory development plan and require ministerial approval. The same paper made proposals for changes to local plans, and introduced a new local development plan reflecting a specific Scottish agenda, but echoing the thrust of change in England and calling for 'more urgency and confidence in the process with a greater focus on content and outcomes' (para. 1). The review of progress illustrated the need for change, with seven out of ten local plans being more than five years without review and 20 per cent adopted more than twenty years before. The experience of delivering development plans in Scotland was similar to that in England, with local government reorganisation delaying the production of plans (Hillier Parker *et al.* 1998).

The framework for the planning system in Scotland underwent a major change through the Planning etc. (Scotland) Act 2006 and its accompanying secondary legislation, which came into force in 2009. The Act replaced structure plans and local plans and introduced a differentiated system for the four city-regions and the rest of Scotland. In addition to granting statutory status to the national planning framework (as discussed earlier in this chapter), it required the four city-regions (Figures 4.6 and 4.7) to produce strategic development plans which address land use issues that cross local authority boundaries or involve strategic infrastructure. Local development plans are required to be produced by each of the thirty-two local authorities and two national parks, and to be supported by supplementary guidance. Neighbourhood and community plans are not a formal feature of the Scottish planning system. However, a *community planning* system is in place, with the aim of bringing together public bodies and local communities to improve service delivery (Cave *et al.* 2013). In providing guidance for development plans, the *Scottish Planning Policy 2010* (p. 54) states:

The planning system should be judged by the extent to which it maintains and creates places where people want to live, work and spend time. This is a major challenge which will require permission for inappropriate development to be refused, conditions imposed to regulate development and agreements reached on actions to mitigate impacts on amenity, natural heritage, historic environment and communities.

In contrast to the Coalition government's portrayal of planners as 'enemies of enterprise', the above statement indicates a greater confidence in the role of planning to deliver sustainable development and 'exhibits a clearer belief in the value of planning as a positive means of steering spatial development' (Tomaney and Colomb 2013: 379). Figure 4.6 presents the planning policy framework in Scotland and Figure 4.7 shows the Strategic Development Plan Areas in Scotland.

Figure 4.6 The planning policy framework in Scotland

Figure 4.7 Strategic development plan areas in Scotland

Development plans in Wales

In Wales, the system of development plans was virtually the same as that for England until 1996. One important variation was that the responsibility for waste rested with the districts (rather than counties) and thus waste policies were included in local plans rather than separate county-wide subject plans.

Local government reorganisation created unitary councils in 1996, and the plan framework was amended to require each authority (including the national parks) to prepare a unitary development plan. The form of the Welsh UDP was similar to that of the UDPs in England with a Part I and Part II (as discussed above). Provisions were made for joint UDP preparation (though this was always unlikely, given the very large area of Welsh local authorities), while Part II of the plan could be organised with smaller areas. Arrangements for the transition to the new framework allowed planning authorities to seek approval from the Secretary of State to continue through to the adoption of plans already in preparation. There was barely time to establish this system

before the 2004 Act brought further changes. Section 6 of the Act made separate provisions for development plans in Wales (their implementation in detail in regulations and orders made by the Assembly) and made the Wales Spatial Plan (discussed earlier in this chapter) a statutory requirement. The Assembly had previously canvassed views on a Welsh approach to reform of development plans through *Planning: Delivering for Wales* in 2002. They were later to consult on more detailed proposals in *Delivering Better Plans for Wales* in 2004.

The 2004 Act retained a unitary structure of development plans in Wales, but the UDP was replaced by the *local development plan* (LDP) and the two-part structure was abandoned (Figure 4.8). The same broad changes were made in Wales as in England, with the intention of creating simpler and more focused general policies for the whole of the authority's area linked to an overall vision for the area, of providing detailed planning policy only where it was needed, of addressing interdependencies with other plans and programmes

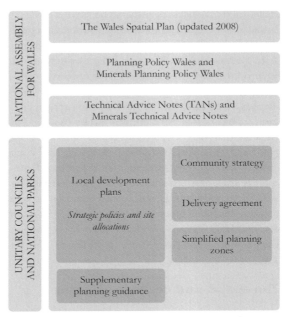

Figure 4.8 The planning policy framework in Wales

and across administrative boundaries and of introducing apparently simpler procedures for the adoption of plans (as discussed below). However, the detailed arrangements were neater than in England, with the LDP as a single document. Where there were UDPs in place (only a few), they remained in force until they were replaced by local development plans. Those authorities that had not adopted a UDP continued to use structure and local plans as the extant development plan policy, while they adjusted their UDP preparation to the revised approach. A number of options were open to authorities in the path they took from UDP to local development plan preparation depending on their progress, but all were required to move immediately to 'LDP principles', which meant, in short, cutting the length and complexity of draft plans, demonstrating effective community engagement and ensuring that plans were subject to the sustainability appraisal process.

Following the 2004 Act (and the supplementary 2008 Act) every local planning authority is now required to prepare an LDP for its area. The LDP will be the statutory development plan for the local planning authority area (i.e. of the county or county borough council and national park authority). Detailed guidance on the content of, and procedure for, development plans is provided in Planning Policy Wales and its updates.

In Wales, there is no power given to communities to produce Neighbourhood Development Plans, although a direct relationship between local development plans and local authority-led community strategies is emphasised in Planning Policy Wales. The Independent Advisory Group on the future of the planning system in Wales has suggested that community and town councils and other broad-based community organisations should be empowered to work with their local planning authority to identify and take forward supplementary planning guidance for their communities (Cave et al. 2013). Figure 4.8 presents the planning policy framework in Wales.

The scope and content of plans

The scope of plans has fluctuated over time: expanding after the 1968 Act, contracting in the 1980s and to lesser degree in the 1990s, only to expand again after the 2004 Act. The key question has been whether the scope of plans should be limited to land use matters only, or expanded to cover a wider range of policy concerns; and if plans are to cover the wider issues, what should the right balance be between them? With regard to this latter question, the post-war history of the planning system in the UK can be interpreted 'as an account of the interplay of economic priorities, social values and environmental concerns in relation to the regulation of land use and development' (Davoudi et al. 1996: 421). However, as will be discussed in Chapter 7, often environmental and social concerns have been crowded out by economic imperatives.

The 1947 legislation was largely concerned with land use: 'a development plan means a plan indicating the manner in which a local planning authority propose that land in their area should be used'. The 1968 Act signalled a major shift in focus: emphasis was laid on major economic and social forces and on broad policies or strategies for large areas. It was held that land use planning could not be undertaken satisfactorily in isolation from the social and economic objectives which it served. Thus the plans were to encompass such matters as the distribution of population and employment, housing, education and leisure.

This broader concept of planning did not survive, and by 1980 central government had moved back to a predominantly land use approach. This radical departure from the ideas of 1968 and the contraction of the scope of structure plans have been widely documented (Cross and Bristow 1983; Healey 1986). Central government also intervened to restrict plan content significantly. Thornley (1993) provides a useful summary of what he describes as the 'attack on structure plans'. During the 1990s departmental advice about plan content became increasingly specific and restrictive. The impact was that, while local plans embraced wide-ranging social and economic objectives, their proposals nevertheless were 'primarily about land allocation' (Healey 1983: 189). Moreover, while local plans varied substantially in form, and appeared 'local in orientation and specific to particular areas and issues', there was considerable (and one might say unhelpful) similarity in scope and content. This arose from the need for conformity with central government policy, which was enforced by planning inspectors and former government offices which tended to strip out policies thought not to be relevant to land use matters and if necessary, exert powers of direction or intervention in the adoption process. Another reason was the professional training and reproduction of a particular culture in the planning profession, which was firmly rooted in land use and physical concerns and the regulation of development.

Criticisms of the system led to a change of attitudes and approaches to planning during the late 1990s, and government began to promote a wider scope for development plans, which was reflected in the 1999 revision of PPG 12. While there appeared to be no succinct list of the matters to be considered as within the scope of plans, there was much reference to having regard to the plans and programmes prepared in other sectors. The 2004 Act and associated policy documents had three core and related messages about the scope of plans – to ensure that development is sustainable, to deliver for the economy and to adopt the spatial planning approach. It was, however, not clear how the potential conflicts arising from the implementation of these seemingly harmonious objectives could be resolved and how a balance could in practice be achieved. Following the 2011 Act, instead of a contents list – which would inevitably have a short shelf life – the NPPF defined a set of '12 core land-use planning principles in England which should underpin both plan-making and decision-taking'. These are presented in Box 4.5.

As regards sustainable development, the NPPF provides a new and even more contested view than PPG 12, which is summarised in Box 4.6. While some of the important dimensions of sustainable development are still outside the planning system, the solution is not the ever-expanding scope of plans. Simply adding new layers to the content of a plan does not necessarily help the delivery of sustainable development. What is needed is a deeper understanding of how these various concerns are played out in a particular locality and at a particular time, and what value a spatial approach (i.e. a focus on space and place) can add to meeting the sustainable development goals.

The 2004 spatial planning approach suggested a considerable widening of the scope of plans in seeking not only to 'have regard to' but also to influence strategies and investment in other sectors. Success was seen as depending very much on the planning authority's skills in building networks, establishing collaborative relationships and planning processes with other sectors, rather than any formal powers of control and influence. And this in turn depended crucially on the objectives that were set in those sectors (at national level), which were often narrow and driven by targets. Building capacity to engage with the wider agenda for spatial planning was also critical. But the change of attitude on the scope of planning and the wider government 'joining-up process' across all sectors provided opportunities, although in the context of the 'congested state', with its proliferation of agencies, special initiatives and collaborative structures (Sullivan and Skelcher 2002). In the NPPF, this has been replaced by an emphasis on community participation, the duty to cooperate and local enterprise partnerships (see para. 160).

The Local Government Act 2000 required local authorities to produce 'community strategies' for the purpose of 'promoting or improving the economic, social and environmental well-being of their area and contributing to the achievement of sustainable development in the United Kingdom'. Non-statutory local strategic partnerships (LSPs) were established to co-ordinate and implement the community strategy process. A study on the relationship between local development frameworks and community strategies, (Entec 2003) concluded that a better relationship could improve understanding of community needs and aspirations in the development plan process, give a more integrated approach to future development and help to join up the approach to community planning. But it also acknowledged the variability in the quality of community strategies and 'the great difference in purpose and procedure of community strategies and development plans'. The community strategy is prepared through voluntary cooperation and although there is a statutory requirement for their production in each authority in England and Wales, there is no statutory procedure to be followed. The community strategy is intended to

BOX 4.5 CORE PRINCIPLES UNDERPINNING PLANNING IN ENGLAND

Planning should:

- be genuinely plan-led, empowering local people to shape their surroundings, with succinct local and neighbourhood plans setting out a positive vision for the future of the area. Plans should be kept up to date, and be based on joint working and cooperation to address larger than local issues. They should provide a practical framework within which decisions on planning applications can be made with a high degree of predictability and efficiency;
- not simply be about scrutiny, but instead be a creative exercise in finding ways to enhance and improve the places in which people live their lives;
- proactively drive and support sustainable economic development to deliver the homes, business and industrial units, infrastructure and thriving local places that the country needs. Every effort should be made objectively to identify and then meet the housing, business and other development needs of an area, and respond positively to wider opportunities for growth. Plans should take account of market signals, such as land prices and housing affordability, and set out a clear strategy for allocating sufficient land which is suitable for development in their area, taking account of the needs of the residential and business communities;
- always seek to secure high quality design and a good standard of amenity for all existing and future occupants of land and buildings;
- take account of the different roles and character of different areas, promoting the vitality of our main urban areas, protecting the green belts around them, recognising the intrinsic character and beauty of the countryside and supporting thriving rural communities within it;
- support the transition to a low-carbon future in a changing climate, taking full account of flood risk and coastal change, and encourage the reuse of existing resources, including conversion of existing buildings, and encourage the use of renewable resources (for example, by the development of renewable energy);
- contribute to conserving and enhancing the natural environment and reducing pollution. Allocations of land for development should prefer land of lesser environmental value, where consistent with other policies in this Framework;
- encourage the effective use of land by reusing land that has been previously developed (brownfield land), provided that it is not of high environmental value;
- promote mixed-use developments, and encourage multiple benefits from the use of land in urban and rural areas, recognising that some open land can perform many functions (such as for wildlife, recreation, flood risk mitigation, carbon storage, or food production);
- conserve heritage assets in a manner appropriate to their significance, so that they can be enjoyed for their contribution to the quality of life of this and future generations;
- actively manage patterns of growth to make the fullest possible use of public transport, walking and cycling, and focus significant development in locations which are or can be made sustainable; and
- take account of and support local strategies to improve health, social and cultural well-being for all, and deliver sufficient community and cultural facilities and services to meet local needs.

Source: Adapted from NPPF 2012, para. 17

BOX 4.6 PRESUMPTION IN FAVOUR OF SUSTAINABLE DEVELOPMENT

The NPPF states that:

> At the heart of the National Planning Policy Framework is a **presumption in favour of sustainable development**, which should be seen as a golden thread running through both plan-making and decision-taking.

For **plan-making** this means that:

- local planning authorities should positively seek opportunities to meet the development needs of their area;
- Local Plans should meet objectively assessed needs, with sufficient flexibility to adapt to rapid change, unless:
  - any adverse impacts of doing so would significantly and demonstrably outweigh the benefits, when assessed against the policies in this Framework taken as a whole; or
  - specific policies in this Framework indicate development should be restricted.

For **decision-taking** this means:

- approving development proposals that accord with the development plan without delay; and
- where the development plan is absent, silent or relevant policies are out-of-date, granting permission unless:
  - any adverse impacts of doing so would significantly and demonstrably outweigh the benefits, when assessed against the policies in this Framework taken as a whole; or
  - specific policies in this Framework indicate development should be restricted.

Source: NPPF 2012, para. 14, emphasis in original

'allow communities to articulate their aspirations, needs and priorities; coordinate the actions of the local authority, and the public, private, voluntary and community organisations that operate locally; and focus and shape existing and future activity' (Entec 2003: 4). The report acknowledges that 'some planners tend to regard LSPs as ephemeral, highly self-selecting and unrepresentative partnerships of the powerful' (p. 19). Lambert (2004: 4) makes a similar point:

> Where LSPs were newly established strategy preparation tended to be a tentative and drawn out

process. The product is seen by some partners as too broad brush and aspirational, or as a regurgitation of existing plans and strategies, and the key objectives can have something of 'motherhood and apple pie' flavour.

The statutory duty to prepare a community strategy will be repealed by the Deregulation Bill 2014. However, experiences of best practice in integrated partnership approaches to strategic coordination remain centrally relevant to the requirements for sound local plans.

Statutory procedures and management of the plan process

The general procedure for the preparation and adoption of development plan documents in England is illustrated in Figure 4.9. While practice is similar across the UK, devolution is resulting in a more varied picture. The procedure for the adoption and approval of plans provides 'safeguards' to ensure the accountability of government and the consideration of many interests in the planning process. It also upholds the rights of private property interests to have their say when proposals affect their interests. This is particularly important in the UK, where there is no constitutional safeguard of private property or other rights (other than that provided by the European Convention on Human Rights) and where there is wide administrative discretion in decision-making. There is no appeal to the courts on the policy content of plans, although they may be used to ensure that statutory procedures are followed. The Secretary of State is the final arbiter on the content of plans through powers of direction in the process. The process of open discussion and formal adoption lends authority and standing to plans. Box 4.7 provides a list of consultation bodies in the development planning process.

In the following discussion, the focus is on the key safeguards, the main criticisms of the procedure and recent amendments. The knotty questions about the extent to which the public and other objectors are effectively able to make use of the safeguards and how this influences plan content are discussed in Chapter 13. The main safeguards in plan preparation and adoption are:

- the opportunity for all interests to be consulted in the formative stages of plan preparation;
- the need for authorities to consider conformity between plans and strategic and national guidance;
- the right to make representations to both strategic and local development plan documents (which may be objections or indications of support);
- the right to have representations to local development plans considered and, if desired, heard before an independent inspector;

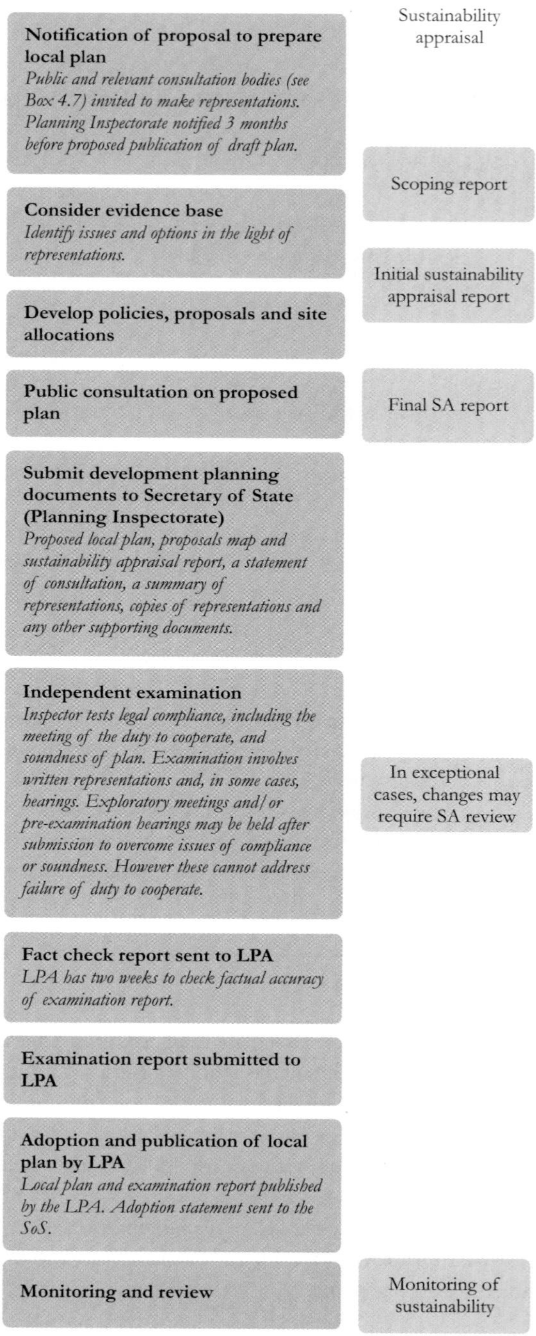

Notification of proposal to prepare local plan
Public and relevant consultation bodies (see Box 4.7) invited to make representations. Planning Inspectorate notified 3 months before proposed publication of draft plan.

Consider evidence base
Identify issues and options in the light of representations.

Develop policies, proposals and site allocations

Public consultation on proposed plan

Submit development planning documents to Secretary of State (Planning Inspectorate)
Proposed local plan, proposals map and sustainability appraisal report, a statement of consultation, a summary of representations, copies of representations and any other supporting documents.

Independent examination
Inspector tests legal compliance, including the meeting of the duty to cooperate, and soundness of plan. Examination involves written representations and, in some cases, hearings. Exploratory meetings and/or pre-examination hearings may be held after submission to overcome issues of compliance or soundness. However these cannot address failure of duty to cooperate.

Fact check report sent to LPA
LPA has two weeks to check factual accuracy of examination report.

Examination report submitted to LPA

Adoption and publication of local plan by LPA
Local plan and examination report published by the LPA. Adoption statement sent to the SoS.

Monitoring and review

Sustainability appraisal

Scoping report

Initial sustainability appraisal report

Final SA report

In exceptional cases, changes may require SA review

Monitoring of sustainability

Figure 4.9 The procedure for the adoption of local plans in England

BOX 4.7 DEVELOPMENT PLAN CONSULTATION BODIES

Two sets of bodies must be consulted by the LPA, as they consider relevant or appropriate, in the preparation of local plan documents:

1 General consultation bodies:
   • voluntary bodies whose activities benefit any part of the LPA area
   • bodies which represent the interests of different racial, ethnic or national groups in the LPA area
   • bodies which represent the interests of different religious groups in the LPA area
   • bodies which represent the interests of disabled persons in the LPA area
   • bodies which represent the interest of business in the LPA area.

2 Specific consultation bodies:
   • Coal Authority
   • Environment Agency
   • English Heritage
   • Marine Management Authority
   • Natural England
   • Network Rail Infrastructure Limited
   • Highways Agency
   • relevant authorities whose area is in or adjoins the area of the planning authority
   • electronic communications code network operators and owners of electronic communications apparatus
   • primary care trusts
   • electricity companies
   • gas suppliers
   • sewerage undertakers
   • water undertakers
   • Homes and Communities Agency
   • Mayor of London (for London Boroughs)

Source: Town and Country Planning (Local Planning) Regulations 2012 (SI 2012/767), Part 2

• the overarching right of the Secretary of State to intervene and to direct changes; and
• a limited right to challenge the plan in the courts on procedural matters.

The central focus of the formal adoption procedure is the examination of local plans (and the Mayor's London Plan). For many years the examination of local plans at the local level was known as a public local inquiry (PLI) and structure plans were heard at an examination in

public. While the names have changed, the procedures are much the same as before but incorporate many of the lessons learned over many years about how best to organise such hearings so that they meet the expectations of those making representations while not adding undue delay to the process. An 'independent' inspector receives or hears representations, which are mostly objections and counterproposals. There is a statutory right to be heard, although this tends to be exercised most by those who are better organised and resourced.

(a)

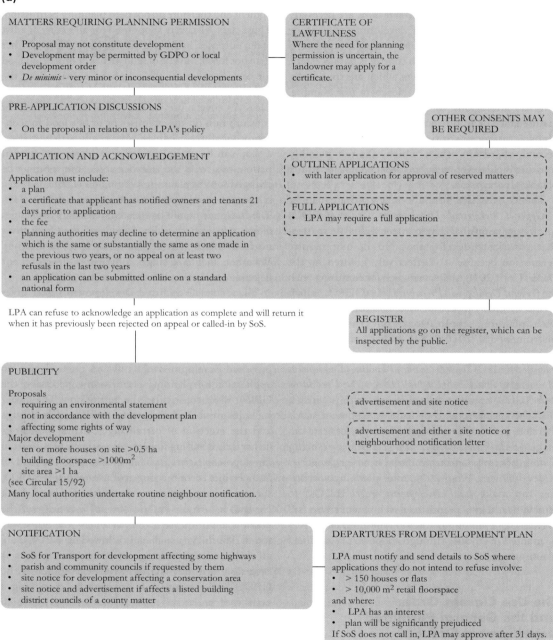

Figure 5.1a and b **The planning application process in England**

(b)

CONSULTATION

In various circumstances consultation is required with:
- neighbouring authorities
- Health and Safety Executive
- Sports Council
- highway authority
- The Coal Authority
- Environment Agency (NRA)
- HMBC (English Heritage)
- The Theatre Trust
- waste regulation authorities
- regional planning bodies
- Secretaries of State for Transport and National Heritage

LPA will have long list of local consultees.

There is formally 21 days to respond.

See Part 3 of GDPO.

PREPARATION OF REPORT

Planning officers will prepare a report on the application, undertake discussions and negotiations with the applicant and other interests, consider consultation returns and policy context, undertake site visits and request further information or changes to the proposal.
Reports are considered by planning committees and sometimes area committees or parish councils.
A majority of decisions are delegated to planning officers.

DECISION

Application is determined in accordance with the development plan, unless material considerations indicate otherwise. Decision should be made within eight weeks for other than major applications.

REFUSED

LPA must give clear and precise reason with reference to development plan policies.

GRANTED

Development to be begun within a specified period of three years from approval of reserved matters unless otherwise specified.

APPEAL

Made to SoS within six months – must include:
- original application
- plans and correspondence
- notices
Determined by Inspectorate by
- written representations
- informal hearing (no cross-examination)
- public local inquiry (inquiry procedure rules apply)

SoS 'recovers' some
- appeals for own decision e.g. if >150 houses or of significant controversy, or for other reasons

DECISION

Inspector makes most decisions but may report to SoS if 'recovered'.

CHALLENGE

Appellant can seek 'statutory review' in the High Court within six weeks on the grounds that:
- decision not within powers of the Act
- procedural requirements not met.
Decision may only be to quash or uphold previous decision.

For a more comprehensive explanation, see Moore and Purdue (2012)
For variations in Scotland, see McAllister *et al.* (2013) and Collar (2010)
For variations in Northern Ireland, see Dowling (1995)
Recent information is available on the recently established planning practice website.

Figure 5.1a and b *(Continued)*

The cynic may be forgiven perhaps for commenting that the freedom given by the UCO and the GPDO has been so hedged by restrictions, and is frequently so difficult to comprehend, that it would be safer to assume that any operation constitutes development and requires planning permission (though it may be noted with relief that painting is not normally subject to control, unless it is 'for purpose of advertisement, announcement or direction'). The legislators have been helpful here. Application can be made to the LPA for a *certificate of lawfulness of proposed use or development* (CLOPUD). This enables a developer to ascertain whether or not planning permission is required. However, a number of changes introduced from 2008 onwards have given much more freedom to developers to extend houses, schools, shops and other business spaces, but some of the changes – particularly those introduced in 2013, discussed below – have provoked much controversy.

Modifications to the Orders are made from time to time – there were eight changes between 2008 and 2013 – either to accommodate changing technology or with the stated intention of 'lifting the burden of regulation'. For example, much of the justification for the changes in 2013 was focused on 'giving a boost to the economy' by easing the process of building extensions to premises. Changes occasioned by developments in technology have included those to deal with the need to accommodate telecommunications equipment for 3G and 4G networks, and to meet the growth in demand for sustainable energy, such as through the installation of micro-generation equipment or recharging points for electric vehicles. Deregulation can be difficult, because changes of use can have dramatic effects on amenity, traffic generation and the quality of places,[5] and as well as new uses having to be accommodated in the system, some fall out of fashion. In England, the UCO refers anachronistically to 'dance halls' but does not adequately recognise the large city pubs and clubs that are associated with 'binge drinking'. Pubs were classified under A2: food and drink, alongside restaurants, and thus large city bars could be developed in any property with the A2 use despite their more extensive and difficult impacts and association with antisocial behaviour. Pubs could be converted into fast food restaurants within the A3 use class, and often were.

In 2001, the Department of Transport, Local Government and the Regions commissioned a review of the UCO (and Part 4 of the GDPO, which deals with temporary uses) (Baker Associates 2001). The research found that the main concerns were with the food and drink class, and especially noise from bars. The main recommendations were to combine A1 (shops) with A2 (financial and professional services) in one 'mixed retail' class. This class was also recommended to include food and drink premises and pubs and bars if they fell below a threshold of 100 square metres. Above that threshold these uses would have their own classes. The government subsequently consulted on a range of options for change in 2002. The outcomes announced in 2005 were new classes A4 (pubs and bars) and A5 (takeaways),[6] but this will include all premises that fall within that category, irrespective of size. The threshold idea was rejected. The changes also put Internet cafes in Class A1 and retail warehouse clubs and nightclubs became *sui generis*.

Concerns about the complexity and difficulty of the interpretation of the GPDO brought it under review by government. Research commissioned from Nathaniel Lichfield in 2003 provided a blow-by-blow account of the operation of the GPDO, with separate sections on the many categories of permitted development. The research did not fully address telecommunications and temporary uses, though they both figure as belonging to that type of development seen as giving most problems (they are being addressed in other ways). The main issues raised by consultees were the inconsistencies and difficulty of interpretation of the GPDO, the adverse impacts which arise from inadequate control, particularly in sensitive areas, and the failure of the system of permitted development overall to contribute to government policy, not least, in achieving more sustainable development.

It would be a matter of debate whether subsequent changes have addressed these criticisms. Whilst changes in 2008 and 2010 lessened the level of control over extensions to domestic and commercial premises, it was the changes introduced by SI 2013/1101 in May 2013 which perhaps most clearly exemplify the nature

Table 5.2 Summary and comparison of the Use Classes Orders

England and Wales (Town and Country Planning (Use Classes) Order 1987 (SI 1987/764), as amended)			Scotland (Town and Country Planning (Use Classes) (Scotland) Order 1997 (SI 1997/3061), as amended)			Northern Ireland (Planning (Use Classes) Order (Northern Ireland) 2004 (SI 2004/458), as amended)		
Class	Use	Development permitted by the GDPO (which may be subject to limitations)	Class	Use	Development permitted by the Permitted Development Order	Class	Use	Development permitted
A1	Shops	To a mixed use as A1 and up to 2 flats. Temporary permitted change (2 years) for up to 150 m² to A2, A3, B1	1	Shops	None	A1	Shops	From a betting office or from food or drink; to a shop and flat
A2	Financial and professional	To Class A1 where there is a display window at ground floor level and to a mixed use for any purpose within Class A2 and up to 2 flats. Temporary permitted change as A1	2	Financial, professional and other services	To Class 1	A2	Financial, professional and other services	From a betting office or from food or drink; to an office and flat
A3	Food and drink	To Class A1 where there is a display window at ground floor level and Class A2. Temporary permitted change as A1	3	Food and drink	To Classes 1 and 2			
A4	Drinking establishments	To A1, A2 and A3 plus similar temporary change as A1						
A5	Hot food takeaways	To A1, A2 and A3 plus temporary change as A1						
B1	Business	To B8 (max 500 m²); to C3 (up to 30/05/16); to state-funded school Temporary (2yr) permitted changed to A1, A2, A3	4	Business	To Class 6 (max 235 m²)	B1	Business	
						B2	Light industrial	To B4
B2	General industrial	To B1 and B8 (max 500 m²)	5	General industrial	To Class 4 or Class 6 (max 235 m²)	B3	General industrial	To B2 or B4 (max 235 m²)

(Continued)

Table 5.2 (Continued)

England and Wales (Town and Country Planning (Use Classes) Order 1987 (SI 1987/764), as amended)			Scotland (Town and Country Planning (Use Classes) (Scotland) Order 1997 (SI 1997/3061), as amended)			Northern Ireland (Planning (Use Classes) Order (Northern Ireland) 2004 (SI 2004/458), as amended)		
Class	Use	Development permitted by the GDPO (which may be subject to limitations)	Class	Use	Development permitted by the Permitted Development Order	Class	Use	Development permitted
B8	Storage or distribution	To B1 (max 500 m²)	6	Storage or distribution	To Class 4	B4	Storage and distribution	To B1 (max 235 m²)
C1	Hotels, boarding houses and guest houses	To state-funded school	7	Hotels and hostels (not including public houses)	None	C1	Dwelling houses	
C2	Residential institutions	To state-funded school	8	Residential institutions	None	C2	Guest houses	To C1
C3	Dwelling houses	To C4	9	Houses	None	C3	Residential institutions	To C1
C4	Houses in multiple occupation	To C3	10	Non-residential institutions	None	D1	Community and cultural uses	
D1	Non-residential institutions	Temporary permitted change to A1, A2, A3, B1 for up to 150 m²	11	Assembly and leisure	None	D2	Assembly and leisure	
D2	Assembly and leisure	To state-funded school; Temporary permitted changes as above	Sui generis	Not classified	From car sales to Class 1; from hot food takeaway to Classes 1 and 2			
Sui generis	Not falling in specified uses above	None except casino to D2						

Notes: 1 The subdivision of residential dwellings into two or more separate dwellings is a change of use
2 A4 and A5 classes were introduced by the Town and Country Planning (Use Classes) (Amendment) (England) Order 2005 (SI 2005/84), subdividing Class A3 established in the 1987 Order

BOX 5.2 SUMMARY OF PERMITTED DEVELOPMENT RIGHTS IN ENGLAND

Permitted development rights are based on the Town and Country planning (General Permitted Development) Order 1995 (SI 1995/418). This was amended by an Order in 2008 affecting domestic development and by one in 2010 affecting business development, but most recently some more wide-ranging changes have been introduced by the Town and Country Planning (General Permitted Development) (Amendment) (England) Order 2013 (SI 2013/1101), which came into force on 30 May that year. The GPDO grants planning permission for certain minor forms of development which are listed. The permissions can be withdrawn by Article 4 directions or conditions attached to planning permissions. The application of the Order can be complex and this is only a brief summary.

- Development within the curtilage of a dwelling house has been limited to 50 per cent of the total area, but, somewhat controversially, this was extended to up to 8 metres from the rear wall in the case of detached dwellings (6 metres in the case of other dwellings) for a period of three years up 30 May 2016
- Change of use of up to 500 square metres within Class B uses
- Change of use from office to residential, subject to certain restrictions
- Subject to certain conditions, change of use of agricultural buildings to a range of uses (A1, A2, A3, B1, B8, C1, D2)
- Change of use from Classes A1 (shops), A2 (financial and professional services), A3 (restaurants and cafes), A4 (drinking establishments), Class A5 (hot food takeaways), B1 (business), D1 (non-residential institutions) and D2 (assembly and leisure) to a 'flexible use' for a period of two years, for areas of up to 150 square metres, subject to various conditions (a measure to assist business)
- For a temporary period up to May 2013, it will be permitted to erect, extend or alter industrial and warehouse premises by 50 per cent or 200 square metres; for offices, shops and some other business premises, the figures are 50 per cent and 100 square metres
- A number of provisions relating to the development of schools and change of use to 'state-funded school', subject to certain conditions
- Minor operations such as painting and erection of walls and fences but not over 2 metres in height
- Temporary buildings and uses in connection with construction, and temporary mineral exploration works
- Caravan sites for seasonal and agricultural work
- Agricultural and forestry buildings and operations (although the local planning authority must be notified in certain circumstances)
- Extension of industrial and warehouse development up to 25 per cent of the cubic content of the original building
- Repairs to private driveways and services provided by statutory undertakers and local authorities (including sewerage, drainage, postboxes), maintenance and improvement works to highways by the highway authority
- Limited development by the local authority such as bus shelters and street furniture
- Certain telecommunications apparatus not exceeding 15 metres in height, and closed-circuit television cameras, subject to limitations
- Restoration of historic buildings and monuments
- Limited demolition works

For full details, see the 'base' Order, the Town and Country Planning (General Permitted Development) Order 1995 (SI 1995/418), and the amending Order, the Town and Country Planning (General Permitted Development) (Amendment) (England) Order 2013 (SI 2013/1101).

of the debate. The essence of these changes is included in the summary in Box 5.2, but the key elements are:

- subject to prior approval, offices may be changed to residential use;
- subject to prior approval, a range of non-domestic premises can be changed to educational use;
- increased limits for home and business extensions;
- increased scope for change of use of agricultural buildings;
- two-year change of use within some classes to help business start ups;
- increased thresholds for business change of use.

There are exemptions for protected areas, such as National Parks, conservation areas and World Heritage Sites.

It was the proposals to increase the limits on the sizes of house extensions which aroused most controversy: according to *Planning* magazine, these changes 'are being implemented in the face of overwhelming opposition' (Garlick 2013: 3). A key justification by government was, 'These measures will bring extra work for local construction companies and small traders' (DCLG 2012a: 2), but the scale of this potential benefit was widely disputed. A concern was the scope for the erection of inappropriate structures, causing adverse environmental impacts and leading to neighbour disputes. This has been addressed by the Statutory Instrument through requiring homeowners to submit details of their proposals to their local council (without payment of a fee), which would then consult neighbours. If objections are raised, the council has to consider whether there would be unacceptable impacts on amenity.

This use of a form of 'prior approval' itself raised some concerns. Prior approval is an approach which had, up to this point, been little used. In such cases, the local authority has to consider whether an application would give rise to problems – of amenity, flood risk, traffic generation. Development cannot start until the applicant is notified that prior approval is not needed, that it has been granted, or until fifty-six days has expired without a decision being made. If the council feels it cannot grant prior approval, it can ask

for a formal planning application. At the time of writing, it is unclear whether this process will be quick and effective, as the government no doubt hopes, whether local authority performance will be variable, or whether the list of impacts to be examined might be too narrow to be compatible with a rounded planning assessment. The fact that the process is a relatively unfamiliar one to many planners may mean that there will need to be an initial period of familiarisation.

These changes mean that the range of developments that require planning permission will be reduced, and there have inevitably been fears expressed that some adverse impacts on amenity and the environment could result. Many of the changes are being introduced for an initial period of three years, and there is a government promise to track their impact, but it remains to be seen how easily this will be achieved when much of the development concerned will not need to be notified to the planning office.

Withdrawal of permitted development rights

The development rights that are permitted by the GPDO can be withdrawn by a Direction made under Article 4 of the Order (and hence it is known as an *Article 4 Direction*). The effect of such a direction is not to prohibit development, but to require that a planning application be made for development proposals in a particular location. The direction can apply either to a particular area (such as a conservation area) or, unusually, to a particular type of development (such as caravan sites) throughout a local authority area. Article 4 Directions should be made only where there is clear justification, where it is 'necessary to protect local amenity or the wellbeing of the area'.[7] Whilst Article 4 Directions are confirmed by local planning authorities, the Secretary of State must be notified, and has wide powers to modify or cancel most Article 4 Directions at any point.

The most common use of an Article 4 Direction is in areas where special protection is considered desirable, as with a dwelling house in a rural area of exceptional beauty, a national park or a conservation area.

Without the direction, an extension of the house would be permitted up to the limits specified in the GPDO. The majority of Article 4 Directions in fact relate to 'householder' rights in conservation areas. They are also used in national parks and other designated areas to control temporary uses of land (such as camping and caravanning) which would otherwise be permitted (Roger Tym and Partners 1995).

In a situation analogous to the issuing of Article 4 Directions, when the intention was announced to introduce new permitted development rights to allow the change of use from B1(a) offices to C3 residential, the Secretary of State confirmed that there would be an opportunity for local authorities to seek an exemption in certain circumstances:

> We recognise that, as with all permitted development rights, there may be unique local circumstances which should be taken into account. The chief planner is today writing to local planning authorities giving them the opportunity to seek a local exemption where this can be justified on economic grounds.
>
> We will only grant an exemption in exceptional circumstances, where local authorities demonstrate clearly that the introduction of these new permitted development rights in a particular local area will lead to (a) the loss of a nationally significant area of economic activity or (b) substantial adverse economic consequences at the local authority level which are not offset by the positive benefits the new rights would bring.
>
> (Hansard, 24 January 2013, Written Statements, c. 17WS)

When the revisions to the GPDO were issued, it included exemptions (SI 1101/ 2013, art. 3(2)) for areas in seventeen local authorities, mostly areas in London and the South East. This represented a small proportion of local authorities which applied for exemption, but DCLG did not give reasons to applicant local authorities or more generally as to why applications were refused.

Since the Article 4 Direction involves taking away a legal right, compensation may be payable, although there are time limits which apply for claiming it. In an amendment to Appendix D of Circular 9/95 *General Development Order Consolidation* in June 2012, the circumstances in which compensation might be payable were specified:

> Local planning authorities may be liable to pay compensation to those whose permitted development rights have been withdrawn if they:
>
> • refuse planning permission for development which would have been permitted development if it were not for an article 4 direction; or
> • grant planning permission subject to more limiting conditions than the GPDO would normally allow, as a result of an article 4 direction being in place.
>
> (para. 6.2)

The Lichfield (2003) research mentioned previously recommended that this right be removed. The report also generally advocated a different approach: removing permitted development rights most often removed by Article 4 Directions and then allowing local authorities to bring in permitted development through local development orders.

Local development orders

Provisions for the introduction of local development orders were made by amendments to the Town and Country Planning Act 1990 introduced in the 2004 Act. The purpose of LDOs is to allow local planning authorities to extend 'national' permitted development rights for all or part of their area for specific developments or general classes of development. It is another idea intended to speed up the planning system and provide more certainty for business, although one that was initially received with considerable scepticism, including opposition from 'the Planning Inspectorate, CPRE, the Civic Trust and the Law Society' (Land Use Consultants and Wilbraham and Co. 2003: 12). The main concerns were the potential fragmentation of the system, with local planning authorities making different requirements and thereby creating confusion and, given that standards will vary among local authorities, undermining confidence in the system. The Home Builders' Federation, among others, supported the idea in principle on the grounds that it right speed up the system by removing many smaller applications and others where there were agreements about their implementation.

The effect of an LDO is to grant planning permission 'in advance' so as to speed up the application process. Initially, the focus was on the implementation of policies that had been adopted in the development plan, but the Planning Act 2008 removed the requirement for LDOs to achieve policies set out in adopted LDDs. Like the simplified planning zone, it borrows from the approach to development regulation in continental countries, where the 'regulation plan' determines the grant of a permit and removes the planning authority's discretion once the order is made. Permission can be granted for a specific form of development on one site, or for any development of a particular type within the authority's area, and it can specifically exclude any type of development or location. No separate hearing is required in the procedure, and the order is adopted by resolution of the local planning authority, although, as always, with provision for government to intervene if needed. Also, as in the early days of *simplified planning zones* (SPZs), LDOs have been suggested as a means to facilitate house building. The links to the government's concerns about the rate of house building (see Chapter 6) are obvious. SPZs were never popular. Developers, investors and planners alike found it a difficult concept in a system built on negotiation and compromise at the time a proposal comes forward.

Whilst the provisions for LDOs detailed in section 61A of the 1990 Act appear to be very flexible, Killian and Pretty, writing four years after their introduction, were 'not aware of any confirmed LDOs' (2008: 37), and PAS in a guidance document on LDOs produced in 2011 describes them as 'a little used tool in the planning toolbox' (p. 3). The Coalition government's enthusiasm for simplifying planning as a means of aiding economic development, together with the emerging 'localism' agenda, led to LDOs acquiring a

new impetus. With support from PAS, a group of local authorities piloted the use of LDOs and over a two to three year period a number have been developed and adopted. Several were associated with recently designated enterprise zones and designed as part of the process to promote economic development – for example, Sheffield, Blythe, Great Yarmouth and Lowestoft – but others have a different focus. One of the first of this recent wave and an outcome of one of the PAS pilots was for Carnon Downs, a village in Cornwall. It extended permitted development rights for minor residential developments, but these had to be implemented in line with policies in a recently developed design guide. In this way it was hoped to achieve greater speed and efficiency but without any loss of quality in development. Another example not related to regeneration can be found in Norwich, where an LDO has been put in place to facilitate the replacement of doors and windows in flats. The following paragraph from the order helps highlight the sorts of benefits which may arise from the use of an LDO:

> The aim of the LDO is to assist landlords, leaseholders and freeholders by speeding up the process, providing certainty of outcome and reducing the cost of replacing windows and doors in flats. The proposal will also result in financial savings to the council as it seeks to implement the window replacement programme across its housing stock. In addition there would be scope for more time and resources to be spent on replacement window applications within conservation areas and on listed buildings so a higher quality design can be achieved within Norwich's historic environments.
> (Norwich City Council 2012: 1)

So LDOs are seen as providing efficiency gains for applicants and LPAs alike, as well as a more general benefit of smoothing and speeding up the development process. By removing a number of minor applications from the LPA's workload, they can save staff time and, as was the case in Norwich, enable staff to be deployed on matters which are seen as having greater importance. However, diminution of control is often accompanied by fears that design and environmental standards will be compromised, whilst linking LDOs to design guidance gives rise to concerns that simplification in the process will be replaced by complexity in the interpretation of detailed guidance.

As part of the Localism Act 2011, planning at the neighbourhood level was given greater prominence. In an important change to the planning system, communities could use neighbourhood planning to permit the development they want to see – in full or in outline – without the need for planning applications. These instruments are called *neighbourhood development orders* (NDOs) but their parallel with LDOs is readily apparent. At the time of writing, there was little experience of the use of NDOs.

Special development orders

While the GPDO is applicable generally, special development orders relate to particular areas or particular types of development. SDOs (like other orders) are subject to parliamentary debate and annulment by resolution of either House. Whilst they were designed to promote and ease the passage of economic development, they have provided an opportunity for testing opinion on controversial proposals, such as the reprocessing of nuclear fuels at Windscale. However, most of the SDOs made in England and Wales were to facilitate the operation of urban development corporations. In these cases, the order granted permission for development that was proposed by the corporations and approved by the Secretary of State. The use of the SDO procedure causes considerable controversy, since it involves a high degree of central involvement in local planning decisions. One very contentious case in 1982 was the granting of permission for over a million square feet of offices and homes at the eastern end of Vauxhall Bridge. At that time the DoE said that 'the purpose of making fuller use of SDOs would not be to make any general relaxation in development control, but to stimulate planned development in acceptable locations, and speed up the planning process' (Thornley 1993: 163). In practice, central government has not

made use of the orders in recent years and has instead opted for other means to shape major decisions. In 1999, SDOs became subject to the provisions for environmental assessment. In 2010, the Conservation of Habitats and Species Regulations (SI 2010/490) limited the power of SDOs likely to have a significant impact on European sites.

Planning application process

All planning authorities provide guides on the planning application process and readers should make reference to them for the finer points. For many minor applications it is a straightforward process, but in some cases it can become very complex and time-consuming. Figure 5.1 gives an overview of the process in England, and it is much the same elsewhere. Many applications will begin with pre-application discussions with the local authority, and the importance of pre-application discussion could be seen as being more important in a regime of 'development management'. The 2004 Act introduces provisions that will allow planning authorities to charge for this service. It is especially important for the local authority to ensure that the application is complete and meets its requirements so that there is minimum delay in processing.

These more routine aspects of the development management process, the planning application form and procedures for its acknowledgement and registration came under close scrutiny in 2003 as part of the government's drive to improve efficiency. In 2004, Arup with Nick Davies Associates reported (at great length) on a wide-ranging study for the ODPM of how local planning authorities deal with the receipt of applications. The outcomes include recommendations for a standard application form and lots of guidance for local authorities intended to provide more certainty for applicants and consistency among authorities. This was largely achieved through the development of the Planning Portal and the introduction of a standard planning application form. This was the national attempt to introduce a single, universal planning application form (but with many variants – there are twenty-five of them in all). Together with the launch of the Planning Portal, developers can submit planning applications electronically using the national form to all local planning authorities, which receive them electronically and then deal with them under their own systems. This was first introduced in April 2008 and rolled out nationally by the government through all the local planning authorities and is now the basis for the almost entire electronic delivery of the development management process.

On receipt of the application and fee, the authority will acknowledge and begin publicity, notification and consultation procedures, all of which will vary depending on the type of application. A review of the arrangements for publicising planning applications in England (Arup 2004) provides a summary of arrangements and practices in planning authorities, together with many recommendations for improvement. This is considered in Chapter 13, with comparisons of practice elsewhere in the UK and alongside other forms of public involvement in planning.

As shown in Figure 5.1, the local planning authority will consult with many bodies, some of which are statutory consultees, which means that they have to be consulted by law. Research on statutory and non-statutory consultation found some confusion among planning authorities about who should be consulted for what purpose. Not all consultees have had the capacity and/or been willing to cooperate effectively in this process, and so from 2005 there is a statutory duty to respond to consultation within twenty-one days. Circular 8/03 also enables a planning authority to forgo consultation with certain statutory consultees if it believes that the development is subject to standing advice issued by the relevant consultees.

Many planning applications will also require other consents from the authority and other agencies, notably building regulations approval. Since the mid-1990s, attention has been given to the idea of creating a 'one-stop shop' approach providing a more user-friendly service for those who will be seeking more complicated consents.[8] In 2004, ODPM reported on a more fundamental approach to the *Unification of Consent Regimes* (Halcrow Group 2004). The review concentrated on the potential unification of planning, listed building and conservation area consents, but in

a mammoth report (about 350 pages) provides a Cook's tour of other regimes, including enforcement, building regulations and, hazardous substances consents. The report identifies many benefits of the existing, largely separate, regimes, but also many potential benefits from unification. In a less compendious review by Adrian Penfold (BIS 2010b), similar concerns were voiced – that complex consent regimes were a brake on development – and Penfold suggested a number of changes to increase certainty, speed up processes, reduce duplication and minimise cost. Rather than opting for a major overhaul unifying regimes, changes to management practices and cultures, along with some simplification, were proposed, with particular attention being drawn to heritage, highways and environment consents. The government response (BIS 2010c) welcomed the recommendations and promised to implement them in a phased approach. In a second report (BIS 2013) on progress with implementation, it was stated that progress had been made on changing management practices in key agencies and that draft legislation had been tabled to simplify environmental and heritage consent regimes.

On the basis of consultation returns, the relevance of national and local policies, previous decisions and a site visit, the planning officer will prepare a report for consideration by the planning committee or under delegated procedures, with a recommendation on the decision to be made.[9] Reports to committee, along with the committee agenda, minutes of previous meetings and consultation returns are public documents. The applicant may be able to make a presentation to committee but this is at the discretion of the authority. Decision notices are sent to the applicant, who can appeal against refusal or conditions imposed. Amendments to the GDPO in 2000 and 2003 (applying to England) introduced requirements for decisions to include an explanation of the reasons for any grant of planning permission and a summary of the policies and proposals that were relevant to the decision.[10]

Most applications in many authorities will be decided by the planning officer under delegated powers, subject to their meeting criteria such as being in accordance with development plan policies and below certain thresholds. In 2012, on average, planning authorities delegated 90 per cent of applications, with some delegating 99 per cent of decisions. However, two authorities delegated only 1 per cent of decisions.[11] When elected members consider applications, they may not always agree with officers, and there are some celebrated cases where members have decided applications against the advice of their officers, such as those of Ceredigion and North Cornwall.[12] This is one category of case that has led to planning authorities' decisions being subject to judicial review.

More complex applications will require negotiations between the applicant (or their agent) and the authority. The officer will be seeking to ensure that the application meets policy and will be working from past experience of committee decisions. The discretionary nature of the British planning system allows for negotiation prior to the final decision. In theory this offers scope to ensure that the final development is closer to meeting the needs of all parties, so long as officers and applicant recognise the benefits of negotiation to achieve better outcomes (Claydon 1998). In practice, in the past it appears that local authority officers have been less prepared to make good use of the opportunity for negotiation than developers (Claydon and Smith 1997), though it should be expected that the gradual diffusion of the 'development management' culture will change this.

In complicated cases it is sometimes convenient for an applicant or the LPA (or both) to deal with an application in outline. *Outline planning permission* gives the applicant permission in principle to carry out development subject to *reserved matters*, which will be decided at a later stage. This device enables a developer to proceed with the preparation of detailed plans, with the security that they will not be opposed in principle. In a few cases there will need to be an environmental impact assessment – the procedures for which are described in Chapter 7.

The development plan in the determination of planning applications

Crucial to the development control process is the concept of *material considerations*. These are exactly

what the term suggests: considerations that are material to the taking of a development control decision. The primary consideration is the development plan.[13] Plans have always been important considerations in development control, but during the 1970s and 1980s many local planning authorities did not have adopted statutory local plans and even if they did, they were not always given the weight they deserved in decision-making by either local or central government. In the case of central government, this was also the product of the then current political imperative of deregulation, reflected in Circular 14/85 *Development and Employment*, which stated, 'The development plan is one, but only one, of the material considerations that must be taken into account in dealing with planning applications.' Thus, the status of the plan became ambiguous.

In 1989, the government began the move to a plan-led system and asked all planning authorities to ensure that they had an adopted up-to-date local plan in place. In 1991, statutory force was lent to the role of the plan in decision-making through the insertion of section 54A into the 1990 Act.[14] This may sound strange to those new to planning, and it may be appropriate to ask whether there can be any other sort of planning system. The implications of section 54A (which is now superseded by section 38(6) of the 2004 Act) have been the subject of much debate. In the light of experience and court rulings, the meaning has been clarified in various revisions of policy guidance for England, Wales and Scotland. The current guidance is given in Box 5.3.

Section 54A certainly had a major impact on the planning system. There is much more emphasis on the preparation of statutory plans to ensure that there is an adequate framework of policy against which to test applications. The 'presumption in favour of development' dating back to the beginnings of planning control (Harrison 1992) was effectively changed to a presumption in favour of the development plan, or more accurately in the words of Malcolm Grant,

it is if anything, a presumption *in favour* of development that accords with the plan; and a presumption *against* development that does not. In each case, the development plan is the starting point, and its provisions prevail until material considerations indicate otherwise.

(Grant 1997: P54A.07, emphasis in original)

The provisions of the NPPF, which introduced a 'presumption in favour of sustainable development' (para. 14) also reiterates that this 'does not change the statutory status of the development plan as the starting point for decision making' (para. 12). But in most cases other material considerations will also play a part in the decision, and this has always been the case. Whether or not other material considerations outweigh the development plan is a matter of judgement for the decision-makers.[15] Even with much more comprehensive plan coverage, many issues raised by planning applications will not be addressed in policy, and there is a limit to the extent to which governments at any level can, or wish to, commit policies to paper. The more this is done, the more inflexible will planning become, the less will it be able to adapt to changing circumstances, the greater will be the likelihood of conflict between policies, and the more confusing the situation will be. The discretionary 'hallmark' of the British development control system mentioned at the start of this chapter is also, in comparison with systems elsewhere, a great advantage. But this only applies so long as there are effective safeguards to ensure that discretion is exercised in the proper way.

Legal niceties aside, how do planning authorities actually decide planning applications? On this central question, findings from research in the 1980s on the role of development plans are probably still most enlightening. Davies *et al.* (1986b) found that many considerations were not covered by plans, and policies were typically expressed in general ways and needed 'translation' into operational terms for each application. Supplementary guidance and other planning documents, including design guides, development briefs, informal local plans and 'policy frameworks', were important. With some caveats, notably the more comprehensive nature of many plans in the 1990s, these findings are probably no less valid today. The same authors found in a study of appeals that 'Inspectors nearly always dismissed appeals, and supported the local authority, on proposals for which there was relevant cover in the development plan.' On the other hand, they 'more often allowed appeals which turned on practical appeal considerations lacking firm local policy coverage, but in which national policies were invoked in favour of the appellant'. The message here is that development control and appeal decisions tend to abide by policy, where it exists. And these findings predate the so called 'plan-led system'. It suggests that greater coverage of statutory plans has been more important than the statutory requirement to make it the starting point for decision-making. The introduction of the NPPF, and its presumption in favour of sustainable development, has been said to have acted as a prompt to LPAs to speed up progress on completion of development plans. This is because, where the development plan is absent, silent or relevant policies are out-of-date, permission should be granted unless any adverse impacts of doing so would significantly and demonstrably outweigh the benefits, when these are assessed against the policies in the Framework taken as a whole, or when specific policies in the Framework indicate development should be restricted.

Other material considerations

Since planning is concerned with the use of land, purely personal considerations are not generally material (though they might become so in a finely balanced case). The courts have held that a very wide range of matters can be material.[16]

The list of possible considerations begins with the siting and appearance of the proposed buildings; the suitability of the site and its accessibility; the relationship to traffic and infrastructure provision; landscaping and the impact on neighbouring land and property. But many other matters may be relevant: environmental impacts; the historical and aesthetic nature of the site; the economic and social benefits of the development; considerations of energy and 'sustainable development'; the impact on small businesses; previous appeal decisions; the loss of an existing use; whether the development is likely to be carried out; and, in a few cases, financial considerations, including the personal

BOX 5.3 THE PLAN-LED SYSTEM

For more than twenty years, the government has been committed to a plan-led system of development management. Currently, the legal basis for this is to be found in section 38(6) of the Planning and Compulsory Purchase Act 2004, which says:

> If regard is to be had to the development plan for the purposes of any determination to be made under the planning Acts the determination must be made in accordance with the plan unless material considerations indicate otherwise.

This was amplified in a supplement to PPS 1, *The Planning System: General Principles*, issued in 2005, which noted:

> If the Development Plan contains material policies or proposals and there are no other material considerations, the application should be determined in accordance with the Development Plan. Where there are other material considerations, the Development Plan should be the starting point, and other material considerations should be taken into account in reaching a decision. One such consideration will be whether the plan policies are relevant and up to date.

The NPPF has maintained this position but with some specific caveats:

> The planning system is plan-led. Planning law requires that applications for planning permission must be determined in accordance with the development plan, unless material considerations indicate otherwise. This Framework is a material consideration in planning decisions. In assessing and determining development proposals, local planning authorities should apply the presumption in favour of sustainable development. Where a planning application conflicts with a neighbourhood plan that has been brought into force, planning permission should not normally be granted.
>
> (paras 196–8)

Noting that new or emerging government policy is a 'material consideration' is not new in itself. Indeed the 2005 document quoted above noted (para. 13) that:

> The Courts have also held that the Government's statements of planning policy are material considerations which must be taken into account, where relevant, in decisions on planning applications. These statements cannot make irrelevant any matter which is a material consideration in a particular case. But where such statements indicate the weight that should be given to relevant considerations, decision-makers must have proper regard to them. If they elect not to follow relevant statements of the Government's planning policy, they must give clear and convincing reasons (*E C Grandsen and Co Ltd v SSE and Gillingham BC* 1985).

circumstances of the occupiers. However, whether or not any of these considerations is material depends on the circumstances of each case. Very few considerations have been held by the courts to be immaterial, but they include the absence of provision for planning gain; and to make lawful something that is unlawful (Moore 2000: 206; see also Thomas 1997).

Planning Policy Guidance and its successor, Planning Policy Statements (and their equivalents), together with circulars have been important material considerations. Although they had no formal statutory force (they were not legally binding), the local planning authority had to have regard to them. Where the local authority did not follow national guidance, it had to give 'clear and convincing reasons'.[17] Changes to national policy that postdate the development plan were particularly important. For example, for out-of-town shopping centres, it was explicitly advised (in PPG 6, para. 1.16) that 'key considerations should be applied', including the likely impact of the development on the vitality and viability of existing town centres, their accessibility by a choice of means of transport, and their 'likely effect on overall travel patterns and car use'. But for many topics guidance could be found to justify alternative positions. And the courts have found that it may be expressed in policy, 'previous decisions, written parliamentary answers, and even after dinner speeches' and conference speeches (Read and Wood 1994: 13), whilst 'ministerial statements' can be seen as an intention to establish a position, through legislation or other methods.

The NPPF, in addition to attempting to condense the volume of guidance by 'replacing over a thousand pages of national policy with around fifty, written simply and clearly', also gave greater force to such guidance where an up-to-date plan had not been adopted. After the publication of the NPPF in March 2012, local authorities were given twelve months to bring their plans into conformity with the new guidance. However, by 1 April 2013, 185 local authorities had yet to adopt an updated local plan. For these local authorities, the 'presumption in favour of sustainable development', which was placed at the heart of the guidance, increases pressure for the granting of permission.

Two considerations warrant further discussion: the design and appearance of development, and amenity.

Good design

Questions of design in planning are much more fully considered in Chapter 9, but in assessing a planning application issues of design have always been important. Much of the built heritage may be worth preserving because it is well designed. As the quotation opening this chapter pointed out, it is therefore of more than contemporary concern that new buildings should be well designed. Nevertheless, the extent to which 'good' design can be fostered by the planning system (or any other system) is problematic, as good design is an elusive quality which cannot easily be defined. However, local authorities have to pass judgement on the design merits of thousands of planning proposals each year, and there is continuous pressure for higher design standards to be achieved.

There is a long and inconclusive history to design control. A 1959 statement by the Ministry of Housing and Local Government (MHLG) stressed that it was impossible to lay down rules to define good design, but policy should avoid stifling initiative or experiment in design, whilst 'shoddy or badly proportioned or out of place designs' should be rejected – with clear reasons being given. Since then, design has continued to be a significant factor in assessing development proposals, for both public and planning authorities. Policy guidance issued by government has continued to counsel the avoidance of prescriptive approaches, but at the same time emphasised the value of sound planning policies and design guidance. The importance of new development reflecting and maintaining 'local distinctiveness' emerged as a theme in PPG 1 *General Policies and Principles* and its successor PPS 1 *Delivering Sustainable Development*, which stressed the importance of design in 'making places better for people', achieved through robust policies' (pp. 14–15).

In 2006, government could be seen as having given greater importance to matters of design by the introduction of a requirement that most planning applications be accompanied by a *design and access statement*. Circular 1/06 *Guidance on Changes to the Development Control System* described them as 'a short report accompanying and supporting a planning application to illustrate the process that has led to the development proposal, and to explain and justify the proposal in a structured way' (p. 11) and noted that, without one, LPAs should not entertain the application. It was hoped that the statements would improve the quality of applications, by requiring applicants to think more carefully about design (and access) aspects. The Coalition, in its quest to simplify planning processes, consulted on simplifying the requirements for design and access statements in early 2013. The Killian Pretty Review of the planning application process in 2008 had recommended a revised and more proportionate approach to design and access statements and this led to revised guidance and regulations introduced in 2010 aimed at streamlining the process. However, the response to the consultation was largely in favour of simplification and consequently, from 25 June 2013, only major applications and listed building consent applications had to be accompanied by design and access statements.

The NPPF continues to stress the importance of good design and in many respects echoes the tone of PPS 1 – it also calls for 'robust and comprehensive policies' (para. 58) – and views good design as a 'key aspect of sustainable development' (para. 56). Whilst the general tone of the NPPF is seen as being favourable towards development, it states, 'Permission should be refused for development of poor design that fails to take the opportunities available for improving the character and quality of an area and the way it functions' (para. 64). Aside from emphasising the importance of design, the NPPF eschews the prescription it advises local authorities to avoid, other than suggesting the use of design codes and the development of appropriate policy frameworks.

Many of the more difficult aesthetic decisions have been made by inspectors. Durrant's (2000) explanation of the reasoning that an inspector makes in cases of dispute over quality of design reveals the very subjective nature of the task: in his case including an example of allowing a twenty-storey 'glass mountain'

adjacent to a grade 1 listed parish church on the south bank of the Thames at Battersea. Durrant argues that the reasoning process has two principal components: the context (both aesthetic and functional) and the scale of buildings, but at the appeal stage the options available to the inspector's decision are really only yes or no.

The design qualities of the most 'significant' developments come under particular scrutiny through the Design Review Committee of the Commission for Architecture and the Built Environment (CABE), and the Design Commission for Wales. CABE, now part of the Design Council, scrutinises about 500 projects a year, 100 of which are discussed in the committee. 'Significant' for CABE means that they are prominent, may affect an important site, or are out of the ordinary. But CABE does not try to replicate the job of the local planning authority in testing designs against national and local policy and design guidance. Rather CABE, in this and other activities, seeks to change the development process overall, so that improvements can be made to the quality of proposals.

Design is an important consideration in planning decisions, and not just for the aesthetic qualities of buildings, but also for social and economic goals. For such reasons, consideration of design continues to be an important aspect of the development management function. The existence of clear design policies, accompanied by constructive pre-application discussions, can do much to speed up the progress of development proposals and achieve better-quality outcomes.

Amenity

'Amenity' is one of the key concepts in British town and country planning, but nowhere in the legislation is it defined. The legislation merely states that 'if it appears to a local planning authority that it is expedient in the interests of amenity', it may take certain action, in relation, for example, to unsightly neglected wasteland or to the preservation of trees. It is also one of the factors that may need to be taken into account in controlling advertisements and in determining whether a discontinuance order should be made. It is a term widely used in planning refusals and appeals. Indeed the phrase 'injurious to the interests of amenity' has become part of the stock-in-trade jargon of the planning world. Rather than legislation, maybe the dictionary should be the source to which we should refer. The Oxford English Dictionary defines amenity as 'the quality of being pleasant or agreeable', so refusing an application on grounds of amenity tends to suggest that going ahead with the development in the form proposed would make things unpleasant and disagreeable. However, amenity is easier to recognise than to define, and it can fall into the category of 'you will know it when you see it'. A planning officer will form a view of such matters through experience but, as with another somewhat nebulous concept, 'quality of life', perceptions can vary: there is considerable scope for disagreement on the degree and importance of amenities: which amenities should be preserved, in what way they should be preserved, and how much expense (public or private) is justified.

Apart from problems of cost, there is the problem of determining how much control the public will accept. Poor architecture, ill-conceived schemes, mock-Tudor frontages may upset the planning officer, but how much regulation of this type of 'amenity injury' will be publicly acceptable? And how far can negative controls succeed in raising public standards? Here emphasis has been laid on design bulletins, design awards and such ventures as those of the Civic Trust, a body whose object was 'to promote beauty and fight ugliness in town, village and countryside'. Nevertheless, planning authorities have power not only to prevent developments which would clash with amenity (for example, the siting of a repair garage in a residential area) but also to reject badly designed developments which are not intrinsically harmful.

Planning conditions and obligations

A local planning authority can grant planning permission subject to conditions, and almost all permissions are accompanied by conditions. This can be a very useful way of permitting development which would otherwise be refused. Many conditions are simple, requiring for example, that the materials to be used are agreed with the local authority before development starts. But there are many more complex permutations. Thus a service garage may be approved in a residential area on condition that the hours of business are limited. Residential development may be permitted on condition that landscape works are carried out in accordance with submitted plans and before the houses are occupied.

The power to impose conditions is a very wide one. Section 70 of the Town and Country Planning Act 1990 allows planning authorities to grant permission subject to 'such conditions as they think fit'. However, this does not mean 'as they please'. The conditions must be appropriate from a planning point of view:

> the planning authority are not at liberty to use their power for an ulterior object, however desirable that object may seem to them to be in the public interest. If they mistake or misuse their powers, however *bona fide*, the court can interfere by declaration and injunction.
>
> (*Pyx Granite Co Ltd v Minister of Housing and Local Government* 1981)

DoE Circular 11/95 *The Use of Conditions in Planning Permissions*,[18] stresses that:

> If used properly, conditions can enhance the quality of development and enable many development proposals to proceed where it would otherwise have been necessary to refuse planning permission. The objectives of planning, however, are best served when that power is exercised in such a way that conditions are clearly seen to be fair, reasonable and practicable.
>
> (para. 2)

As might be expected, there is considerable debate on the meaning of these terms. Circular 11/95 elaborates specifically on the meaning of six tests: conditions should be necessary, relevant to planning, relevant to the development to be permitted, enforceable, precise and reasonable in all other respects.[19] Numerous court judgments provide guidance on how the tests should be applied. To meet the test of being necessary, the local authority should ask whether permission would be refused if the condition were not imposed. Relevance to planning and to the development may be difficult to judge. While planning conditions should not be used where they duplicate other controls, such as those of pollution control, they may be needed if the other method of regulation does not secure planning objectives. At one time, development may have been permitted subject to means of access for people with disabilities being agreed, but this is now covered by other legislation, the Disability Discrimination Act 2005. Conditions should not be imposed on one site to seek to improve conditions on a neighbouring site, for example, where existing car parking is insufficient. But it may be appropriate to impose conditions to address problems elsewhere as a result of the new development, for example, increasing congestion on another part of the site. And it is possible to impose conditions on the use of land not under the control of the applicant. The enforceability test requires that the local planning authority should be able to monitor and detect whether the applicant is complying with it. Enforceability is also closely related to precision in drafting of the condition. Both the authority and the applicant need to be able to understand exactly what is required by a condition.

The reasonableness test requires that the condition is not unduly restrictive. In particular it should not nullify the benefit of the permission. A condition may also be unreasonable if it is not within the powers of the applicant to implement it, for example, where it relates to land in the ownership of a third party. A striking example of a condition which was quite unreasonable was dealt with in the *Newbury* case. There the district council gave permission for the use of two former aircraft hangers for storage, subject to the condition that they be demolished after a period of ten years. The House of Lords held that since there was no connection between the proposed use and the condition, it was ultra vires. In granting permission for development at Aberdeen Airport, the planning authority sought to impose a number of conditions to minimise the impact on the local area. One condition

restricted the direction of take-off and landing of aircraft, but this was found to be both unreasonable and unnecessary, since the Civil Aviation Authority (and not the airport) controls flight paths (McAllister and McMaster 1994: 136–7).

Up to 1968, conditions were also imposed to give a time limit within which development had to take place. The 1968 Act, however, made all planning permissions subject to a condition that development is commenced within five years, which in 2004 was reduced to three years. If the work is not begun within this time limit, the permission lapses, and it need not be renewed if the circumstances have changed. The purpose of this provision is to prevent the accumulation of unused permissions and to discourage the speculative land hoarder. This has sometimes been seen as a constraint on housing supply. Accumulated unused permissions could constitute a difficult problem for some planning authorities: they create uncertainty and could make an authority reluctant to grant further permissions, which might result in, for example, too great a strain on public services. The provision relates, however, only to the beginning of development, and this has in the past been deemed to include digging a trench or putting a peg in the ground.[20]

Conditions can cover a multitude of matters, so comprehensive guidance on their use is difficult, beyond the general principles already set out. However, Circular 11/95 gives examples in two annexes of acceptable and unacceptable conditions, which amplify this guidance. It also usefully gives some advice as to how to deal with conditions in practice. It highlights the fact that pre-application discussion can remove the need for conditions, as the applicant can reformulate a proposal in line with the authority's requirements. On the grounds of efficiency and effectiveness, the use of a set of model conditions on common topics can be compiled, but these need to be used with caution: each condition attached to a planning permission needs to be justified and precise, so the need for a case by case consideration remains. Whilst the NPPF made brief reference to planning conditions (p. 47), it did not withdraw Circular 11/95 and it remained in force. The online planning practice guidance, launched in March 2014, gives guidance on the use of conditions.

In addition to the imposition of conditions, local authorities have increasingly relied on what are variously known as 'planning agreements', 'planning obligations' or 'section 106 agreements' as a means of exercising an additional influence over the nature of planning outcomes. The term 'section 106 agreement' refers to the relevant section of the 1990 Act, whilst, generally, a planning *agreement* is the legal agreement entered into as part of the process and a planning *obligation* is the subject of that agreement.[21] Circular 5/05 *Planning Obligations* (revoked by the NPPF) gives a good explanation of the nature and purpose of this mechanism:

> Planning obligations (or 'S106 agreements') are private agreements negotiated, usually in the context of planning applications, between local planning authorities and persons with an interest in a piece of land (or 'developers'), and intended to make acceptable development which would otherwise be unacceptable in planning terms. Obligations can also be secured through unilateral undertakings by developers. For example, planning obligations might be used to prescribe the nature of a development (e.g. by requiring that a given proportion of housing is affordable); or to secure a contribution from a developer to compensate for loss or damage created by a development (e.g. loss of open space); or to mitigate a development's impact (e.g. through increased public transport provision). The outcome of all three of these uses of planning obligations should be that the proposed development concerned is made to accord with published local, regional or national planning policies.
>
> (p. 9)

Whilst conditions and obligations share the purpose of working to make the unacceptable acceptable, a key difference is that 'the imposition of a condition which satisfies the policy tests of DoE Circular 11/95 is preferable because it enables a developer to appeal to the Secretary of State' (Circular 1/97, para. B20). As a section 106 agreement is a separate legal agreement, a right of appeal does not exist.

Planning authorities have had power to make 'agreements' (with the approval of the Secretary of State) since 1932, but it was not until the property boom of the early 1970s that they became widely used. At that time they were known as S52 agreements, from the relevant clause in the 1971 Act. Under this Act, powers were very loosely described, but Circular 22/83 *Planning Gain* offered some 'tests' of reasonableness for the agreements. Such tests have continued to be offered under successor legislation and guidance, and those set out in the NPPF are not dissimilar to those offered in Circular 22/83. They comprise three tests, all of which need to be met: obligations must be necessary to make the development acceptable in planning terms, directly related to the development, and fairly and reasonably related in scale and kind to the development (p. 47).

Such planning agreements have also attracted the term 'planning gain', which indicates a relationship to the concept of betterment, referred to in the introduction to this chapter and discussed in more detail in Chapter 2. It is this concept of 'gain' which has proved to be a source of contention over the use of planning agreements. The fact that agreements can, as set out in Circular 5/05, require developers to make cash or in kind contributions has led to disquiet. The Law Society Gazette (1988), in discussing the use of S52 agreements, put it in these terms: 'It can hardly have escaped anyone's notice that there has been considerable publicity recently about the activities of certain councils in "selling" planning permission to developers.' In this it was echoing the views of the Committee on Standards in Public Life (1997). This is a view that was contested by local authorities. The evidence from a number of studies was that the majority of agreements were legitimate (Byrne 1989; Grimley J. R. Eve 1992; Rowan-Robinson and Durman 1992). These studies effectively demolished the argument that there was widespread extortion by way of planning gain, though the range of infrastructure and community facilities secured by planning authorities through planning obligations steadily widened. In Scotland, research led to the conclusions that:

> most agreements are useful adjuncts to the development control process; abuse of power does

not present a problem; and for the most part, the benefits secured by agreements have been related to the development proposed: where they have not, the benefits have been of a relatively minor order.
(Rowan-Robinson and Durman 1992: 73)

The range of matters covered by section 106 agreements grew significantly, particularly in the 1990s. Examples are shown in Box 5.4, whilst Box 5.5 illustrates some of the issues arising in the negotiation of planning obligations. DoE Circular 7/91 *Planning and Affordable Housing*, confirming that local authorities could negotiate with developers for the provision of social housing through section 106, represented a major extension of the arena of planning agreements, taking it well beyond the provision of facilities required by

BOX 5.4 EXAMPLES OF FACILITIES SECURED BY PLANNING AUTHORITIES THROUGH PLANNING AGREEMENTS

Residential developments

Direct consequences of development
- Offsite highways
- Parking
- Landscaping
- Open space
- Sports facilities
- Community centres
- Schools
- Health services
- Public transport facilities
- Waste and recycling facilities
- Emergency services
- Childcare facilities
- Affordable housing
- Social rented housing
- Key worker housing
- Sheltered housing

Commercial developments

- Offsite highways
- Parking
- Landscape
- Open space
- Public transport
- Green transport plans

- Housing via mixed use policies

Contributions to community needs

- Construction, training and recruitment initiatives
- Town centre improvement
- Public art
- Countryside managements
- Contributions to cultural plans, theatres, museums, etc.

- Training and recruitment initiatives
- Town centre improvement
- Public art

Source: GVA Grimley *et al.* 2004: 16

the proposed development. In addition, the judgment by the House of Lords in *Tesco Stores v. The Secretary of State for the Environment* (1995) meant that local authorities could demand obligations beyond those implied by the circular. By this time there was some support for planning obligations, since they allowed the development industry to overcome development constraints, so long as the costs were offset by the potential profits to be made. This represented a major change in opinion since the Property Advisory Group (1981) declared the pursuit of planning gain to be unacceptable (RICS 1991; Rowan-Robinson and Durman 1992). For profitable developments such as major retail stores, substantial payments could be made on the promise of future profits, which are safeguarded by the planning system, which would effectively 'protect' the development from further competition. By the 1990s, this fundamental change in the roles of the private and public sectors in land development had become accepted. It has become virtually unanimously accepted that the public sector is financially unable to meet all the associated costs of development, and developers are willing to shoulder them as part of the development value created through planning permission (they may be passed on to landowners or users) – a tacit acceptance of the concept of betterment.

Not surprisingly, the increasing range of matters covered by planning obligations was reflected in their increasing use by planning authorities (Campbell *et al*. 2000) and in the value of obligations agreed. A study published by DCLG (2008b) found that, by 2006, 6.4 per cent of permissions were accompanied by planning agreements with a value of £4 billion. Compared with a study carried out two years earlier, the number of agreements was continuing to rise, especially for major residential permissions. This increase was being driven both by developments in policy – an increasing awareness amongst local authorities of what was possible – and by moves in land and property prices – increasing values giving greater scope to negotiate obligations. Practice in local authorities had developed rapidly: 80 per cent of authorities were found to have policies or supplementary guidance in place covering planning agreements;

standard charges were widely used, particularly for affordable housing, open space and school places; and a majority of authorities had staff dedicated to monitoring obligations, but only a 'small minority' had staff dedicated to negotiating section 106 agreements. The latter point is significant, in that the study also found that the availability of staff expertise was also a factor in driving the increase in use of planning obligations. An updating study analysing trends in 2007–8 (DCLG 2010a) showed a continuing increase in the number and value of obligations and in the number of authorities having a policy framework for obligations; the use of 'tariffs' and standard charges also showed an increase.

However, there had been continuing criticism of the operation of planning agreements. The 2001 Green Paper *Planning: Delivering a Fundamental Change* identified that they could be complex and difficult to agree, with the consequence that the negotiation process added a source of delay to the planning applications system. There remained uncertainty for the local authorities and developers as to what was acceptable; this uncertainty was increased by the differing levels of use of obligations by different authorities in different parts of the country. And whilst guidance since Circular 1/97 *Planning Obligations* had made it clear that agreements could not be a means of 'buying and selling' planning permission, it remained a matter of debate as to how far the increase in the use of section 106 was driven by developments in policy and how far by local authorities' quest for additional funding.

The 2001 Green Paper could be seen as the start of an almost continuous process of discussion about possible changes to the system of planning agreements, in a quest for a system offering greater simplicity, clarity and speed. It is a matter of debate whether any of these objectives have yet been achieved by the changes discussed and introduced. The proposal in the 2001 Green Paper was to replace individual agreements by a tariff payable by developers. The details of the tariff – what would be payable for different types of development – would be set by the LDF and thus known in advance. However, the House of Commons Select Committee, examining the proposals in 2002,

BOX 5.5 PLANNING OBLIGATIONS – VODAPHONE IN NEWBURY

At the time of the application (1998) Vodaphone was an expanding company employing 3,000 people in fifty-eight buildings in Newbury. For business reasons it wished to consolidate and expand its operations on a new site, preferably in Newbury. The site they identified was not allocated on the local plan, but the application was accompanied by a draft section 106 agreement addressing the main impacts of the proposed development.

The process of negotiating the section 106 was conditioned by a number of factors, including the significance of the company for the town – it then accounted for about 7 per cent of all local jobs and after the new development would account for 10 per cent. It therefore had some local support. The lack of any policy framework for planning obligations in the local plan was also significant. This was felt to have limited the measures negotiable to those related to the most immediate impacts of the development. Had there been a policy in place, the company would have factored a greater burden of obligations into its calculations. In the absence of a policy, the process of negotiation had to be, and to be seen to be, open and legitimate. From the point of view of the council, 'This area is occupied by a large number of influential and wealthy individuals. The Council has been taken to judicial review on a number of occasions', whilst from the point of view of the company it had to be clear that they were not in any sense 'buying' planning permission.

The agreement as finally set out at the end of 1999 included a number of elements:

- the preparation, approval and implementation of a green travel plan to increase the use of methods other than the private car for employee travel;
- off-site highway works (cycle routes, improvements to bus services, traffic management measures);
- contributions to repair costs of a grade 1 listed building adjoining the site;
- environmental works;
- contributions towards nursery, school and college costs;
- the cost of studies of secondary effects of the development on drainage, the housing market and traffic;
- measures designed to limit the occupation and use of the building.

The above measures are thought to have cost just over £12 million, with most of the sum being related to traffic-related measures. In total this represented about 10 per cent of the estimated cost of the project.

Source: Adapted from Campbell *et al.* 2001: 21–27

described them as 'sketchy' and identified a number of concerns, including the complexity involved in setting the tariff, uncertainty over how income from the tariff should be used and the element of inequity in that areas with the highest land values would benefit the most.

A consultation took place at the end of 2003 on *A New Approach to Planning Obligations* so as to allow provisions for changes to planning obligations to be inserted into the Planning and Compulsory Purchase Act 2004, and in particular to allow for a 'standard (planning) charge'. In January 2004, the ODPM provided a statement on its response to the consultation returns in *Contributing to Sustainable Communities: a New Approach to Planning Obligations*. In the meantime, the Barker *Review of Housing Supply* was published with a specific recommendation for a *planning gain supplement* (PGS), which was quickly followed by commitment by the

Chancellor – who jointly commissioned Kate Barker's report – in the 2003 Budget Report to consider carefully these proposals. In the meantime, ODPM went ahead with consultations on revisions to Circular 1/97 and published a revision for consultation at the end of 2004, together with proposals for further good-practice guidance. The revised circular – 5/05 – was issued in July 2005. A whole raft of consultations was issued on the proposed planning gain supplement towards the end of 2006, by both DCLG and the Treasury. However, the number of consultations issued may have given some indication of the difficulties, as the Pre-Budget Report for 2006 issued in the same month noted that 'the Government now proposes that a workable and effective PGS would not be introduced earlier than 2009' (p. 69). The Pre-Budget Report for 2007 confirmed that the PGS would not be going ahead:

> Following discussions with key stakeholders, the Government will legislate in the Planning Reform Bill to empower Local Planning Authorities in England to apply new planning charges to new development, alongside negotiated contributions for site-specific matters. Charge income will be used entirely to fund the infrastructure identified through the development plan process.
>
> (HM Treasury 2007b: 103)

That charge is the Community Infrastructure Levy (CIL), for which the government legislated in the Planning Act 2008 (Part 11). The findings of the 2010 study of the use of planning obligations became part of the argument for its introduction.

The CIL became operative on 6 April 2010 and was described as 'a new charge which local authorities will . . . charge on most types of new development in their area. The proceeds of the levy will be spent on local and sub-regional infrastructure to support development of the area' (DCLG 2008c: 2). The CIL has echoes of the proposals for a tariff put forward in 2001 and the justifications offered for its introduction have echoes of the discussions about betterment (DCLG 2010c: para. 8). The CIL was seen as serving the development needs of the wider area, whilst section 106 was seen as still being relevant to dealing with more local impacts. However,

it was also seen as important that the two instruments were complementary and as a result the use of section 106 was seen as needing to be 'scaled back'.

The introduction of the CIL proceeded relatively slowly, though this is partly explained by changing governments and changes to regulations – five sets issued between 2010 and 2013[22] – but also because some of the process involved in setting the charge can be complex. Councils need to specify the projects for which money is needed from the CIL (known as a Regulation 123 list), they need to publish a charging schedule which is subject to an independent examination, and the process depends on having an up-to-date development plan (thought not necessarily an approved core strategy). Some examples of projects funded by CIL are shown in Table 5.3. By mid-2013, only a small number of authorities had gone through all processes and introduced a levy, whilst about fifty were currently consulting on proposals. Introduction of the CIL will bring some changes to the use of section 106. These include putting the Circular 5/05 tests on a statutory basis for developments which are capable of being charged CIL; ensuring the local use of CIL and planning obligations does not overlap; and limiting pooled contributions towards infrastructure which may be funded by CIL. Developer contributions towards affordable housing will continue to be made through section 106 agreements.

A key concern in setting levels of CIL rates and what is asked for through planning obligations is their impact on project viability. These items need to be set at a level which will not choke off development and a judgement has to be made as to what is an appropriate level. This is usually done with the assistance of consultants. For developers, CIL and any planning obligations become an element in development appraisals, along with construction costs, requirements for affordable housing and profit for developer and landowner. However, the final value of a development and, to some extent, development costs can be hard to predict, and profit requirements will vary with risk, so deciding on these matters is far from an exact science. These are not concerns that have arrived with the introduction of CIL. Developers have often argued that section 106 affects profits and viability, but the depressed state of

Table 5.3 Example of content of a community infrastructure plan (London Borough of Redbridge)

Type of facility	Cost £ million
Early education – 1 children's centre	1.0
Primary schools – 4.5 new schools	69.6
Secondary schools – an academy + 2 new secondary schools	96.4
Transport improvements	10.1
Leisure centres – a new centre + pool and courts	15.8
Library modernisation	0.7
Open space improvements to 54 ha.	5.4
Health (NHS/PCT/CCG responsibilities) – 11GPs, 50 hospital beds + specialist care space	21.9
Further education (+1286 places)	6.9
TOTAL	227.8

The above sums are to provide the infrastructure needed for an expected 18,000 new residents over a ten-year period; they are not to remedy any existing deficiencies.

Source: London Borough of Redbridge 2011: 1–2

the housing market following the financial crisis in 2008 gave new prominence to these arguments. Developers argued – with some support from stagnating house building rates – that planning agreements that had been entered into before the crisis no longer reflected current financial circumstances, and the level of contributions demanded were making developments unviable. In August 2012, DCLG consulted on a proposal to renegotiate section 106 agreements from before 6 April 2010 (the date the CIL came into force) 'to make them more reflective of the current market and help unlock stalled development' (DCLG 2012b: para. 4). The response came later in 2013 and resulted in three new sections being inserted into the 1990 Act (through the Growth and Infrastructure Act 2013) which introduced a new application and appeal procedure for the review of planning obligations on planning permissions which relate to the provision of affordable housing.[23]

Fees for planning applications

Fees for planning applications were introduced in 1980. This represented a break with planning traditions, which had held (at least implicitly) that control of development is of general communal benefit and directly analogous to other forms of public control for which no charges are made to individuals, but the Thatcher administration had a very different view. The 1980 Bill provided additionally for fees for appeals, but this was dropped in the face of widespread objections from both sides of the House. The 2004 Act enabled regulations to be made on fees for pre-application discussions and fees for call-in and appeals recovered by the Secretary of State. Previous attempts by authorities to charge fees for pre-application discussions were halted by a decision of the House of Lords, but now many but not all local authorities have a scale of charges for pre-application discussions.

The application fee structure has been subject to regular changes, and providing a detailed schedule here is therefore not appropriate. At the time of writing, the most recent change of fees had been put in place in 2012 and, perhaps not surprisingly, the overall effect had been to increase the cost of making an application: for illustrative purposes, the cost of an application for a dwelling under the 2012 Regulations was £385, compared with £190 six years earlier. The government's philosophy underpinning the charge of application fees was set out in the consultation document preceding the 2012 Regulations, in the following terms:

> It is an established principle that local authorities should pay for activities that are purely or largely for the wider public good. The intention of development management is above all to promote the public good: since managing local development helps to secure the long-term benefits of sustainable, well-designed communities. Yet planning decisions often bring private benefit to the applicant as well; in particular, a property with planning permission may be much more valuable than it would be without. The power granted to authorities to charge planning application fees reflects the possible private benefit implicit in a planning permission. An applicant should expect to pay a fee for an application that could bring a measure of gain. The fee payable reflects the overall cost of handling, administering and deciding the application, including related overheads.
>
> (DCLG 2010c: 8)

For some time it has been government policy that local authorities achieve cost recovery through fees, but progress towards this goal varies between localities and overall it has been estimated that about £9 of every £10 of costs is recovered. The source of this estimate is a study by Ove Arup (2010), which considered rates of cost recovery and recent changes affecting costs, such as alterations to the GPDO. Fee income makes a significant contribution to the cost of the planning service and therefore levels of income impact on the ability to maintain or offer improvements to services. Because local authorities have not traditionally operated as trading entities – partly because of the mix of 'missions' identified in the above quotation – it is difficult to identify the appropriate costs to set alongside application fee income. This was identified as one of the barriers to progress on another objective identified in the consultation paper, the introduction of locally set fees, an approach that is not uncommon in other countries (TCPA 2011b). The economic downturn from 2008 affected the number of applications and

therefore fee income, which was partly reflected in a reduction in staffing numbers. Changes to the GPDO giving greater permitted development rights for householders resulted in a reduction in the number of applications but an increase in the demand for lawful development certificates, which attract a lower fee income. Deciding on the level of application fees, whether nationally or locally, has also to consider the impact of the level of fees on the willingness to develop and submit proposals: setting fees too high may discourage developers, whilst setting fees too low may risk compromising the quality of the service offered.

Planning appeals

An unsuccessful applicant can appeal to the Secretary of State, and a large number in fact do so. Appeals are allowed on the refusal of planning permission, against conditions attached to a permission, where a planning authority has failed to give a decision within the prescribed period, on enforcement notices and other matters as discussed below. Appeals decided during 1998–9 (England and Wales) numbered 12,877, of which about one-third were allowed. By 2012–13 the number had reached 15,479, with the proportion allowed remaining at around one-third. However, numbers of appeals fluctuate over time and Figure 5.2 illustrates trends in the number of appeals. Whilst the number of appeals is related to numbers of applications, itself reflecting the buoyancy of the development market, it is also affected by administrative decisions. For example, the decision in September 2003 to reduce the time limit for making an appeal from six to three months is thought to account for about half of the increase in appeals at that time. The sharp rise in appeal numbers and the resulting backlog in dealing with appeals led to a government U-turn in January 2005, when the time limit was put back to six months. By 2004, the Planning Inspectorate was failing to meet any of its performance targets for appeals and appellants would typically wait one year for their appeal to be dealt with. However, performance by the Inspectorate had, through a range of management changes, reached the position in 2013 where householder appeals were decided in an average of seven weeks, whilst more complex cases were taking around twenty weeks.

Although the appeal is made to the Secretary of State, the vast majority are considered by inspectors 'standing in the Secretary of State's shoes'. The same applies in the other countries of the UK, although there have been some distinctive arrangements in Wales.[24] Until 1969, the ministry responsible for planning dealt with all appeals. In view of increasing delay in reaching decisions and the huge administrative burden, the Planning Act 1968 introduced a system whereby decisions on certain classes of appeal were 'transferred' to professional planning inspectors who had previously only made recommendations to the minister. Over time, the range of decisions transferred to inspectors has been extended such that virtually all are now decided by the Planning Inspectorate. Matters of major importance may be 'recovered' for determination by the Secretary of State. In fact, less than 1 per cent of all appeals are recovered, although it can be argued that the significance is much greater than the figure suggests. Even where decisions are recovered, it is the senior civil servants in the department rather than the minister who make most decisions.[25]

Wide powers are available to the Secretary of State and inspectors. These include the reversal of a local authority's decision or the addition, deletion or modification of conditions. The conditions can be made more onerous or, in an extreme case, the Secretary of State may even go to the extent of refusing planning permission altogether, if it is decided that the local authority should not have granted it with the conditions imposed.

Before reaching any decision, the inspector or Secretary of State needs to consider the evidence and this can be done in three ways: by inquiry, hearing or written representation. Most appeals are considered by written representation, and this proportion has been increasing, from 73 per cent of all planning appeals in England in 1998–9 to 91 per cent in 2012–13. There has been a corresponding reduction in other procedures to 7 per cent for hearings and 2 per cent for inquiries. The procedures are governed by the rules of natural

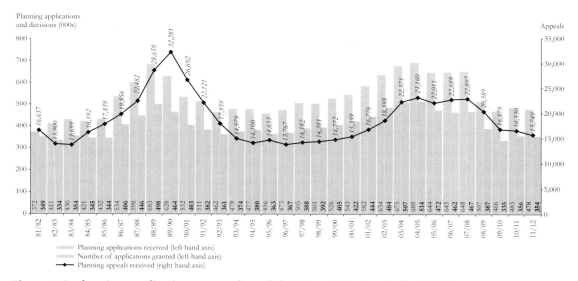

Figure 5.2 Planning applications, appeals and decisions in England 1981–2012

Sources: DETR/ODPM/DCLG Statistical Releases and Planning Inspectorate Annual Statistical Reports

justice and by inquiry procedure rules, which have been updated in England.[26] Although an appellant may select which procedure it considers to be the most appropriate, in the majority of cases the Planning Inspectorate has the power to determine by which procedure the appeal will progress. The efficiency of procedures leading up to and during inquiries has been strongly criticised (Graves *et al.* 1996; O'Neill 1999), but the Inspectorate's annual reports suggest performance targets are largely met.

Inquiries are *adversarial* debates conducted through the presentation and questioning (cross-examination) of evidence. The proceedings are managed by inspectors, but advocates, often barristers, play a dominant role in the proceedings, which leads the proceedings a courtroom atmosphere. Such an approach has benefits in safeguarding the principles of *open, impartial and fair* consideration of the issues. Nevertheless, it is widely acknowledged as unnecessary for certain less complex appeals, especially where one party is not professionally represented. Thus, the hearing procedure has been created; this proceeds in an *inquisitorial* way, with the inspector playing an active role in structuring a round-table discussion and asking questions, but with no formal cross-examination. But the most popular and straightforward procedure is through 'written reps'.

Over the years, the mechanisms for considering appeals have been streamlined. The substantial increase in the number of appeals in the late 1980s led to reviews of the process. The first in 1985 introduced rules to govern the written representation procedure in a similar way to the rules for inquiries, which were also strengthened. Further minor changes were made in 1992 and further substantial revisions in 2000, aimed at speeding up the process, providing statutory rules for the hearings process, and reducing the time allowed for submission of statements. Subsequent changes have continued the quest for speed and greater simplicity, exemplified by the introduction of a Householder Appeal Service in 2009 following consultation in 2007. This service aims to deal with householder appeals – for matters such as extensions and alterations – in a 'simple and proportionate manner . . . so that decisions can be made more quickly' (PINS 2012: 1). Householders

have twelve weeks to submit their appeal and electronic means are preferred, again to ease and speed up processes. There has also been an attempt to encourage negotiation and resubmission of applications. In 2012, further consultation was undertaken on changes to appeal procedures, 'to further support the delivery of a reformed planning system' (DCLG 2012c: 4). Any proposals which may be translated into new guidance will probably focus on continuing to make the system faster and more transparent.

The 2012 consultation document identified the fact that if these objectives are to be realised, then it will not only require action by the Inspectorate. It identifies that all parties will need to be more cooperative and less adversarial, and all must act in a way which avoids wasting time – for example by an early sharing of evidence and a timely provision of relevant material. This continues an approach promoted by the Inspectorate some years earlier, which placed a stronger emphasis on the appellant and local authority agreeing the matters in dispute beforehand and keeping evidence concise.[27] This aspiration is to be encouraged by a strengthening of the processes by which costs are awarded.

The state of the appeals process is critically important for the system as a whole, both in terms of planning policy and how the system should be operated. Although each appeal is considered on its own merits, the cumulative effect is to operationalise policy. It is here that the sometimes vague, sometimes contradictory, messages in government policy must be resolved. The wider effect of appeal decisions may be difficult to assess, but clearly they have a very real influence on other decisions made by planning authorities, and are a route for the imposition of central government policy on local authorities. Inspectors pay particular attention to national policy, which is the determining factor in many appeals (Rydin *et al.* 1990; Wood *et al.* 1998).[28] In addition, discussions on the appeal system can be seen to give a pointer to the government's overall philosophy on decision-making in planning, as explained by Shepley (1999: 403), which recognises the need to make decisions more quickly, more cheaply, and earlier in the development process.

Call-in of planning applications

The power to 'call in' a planning application for decision by the Secretary of State is quite separate from that of determining an appeal against an adverse decision of a local planning authority. The power is not circumscribed: the Secretary of State may call in *any* application. Answers to House of Commons written questions confirm that call-in will be 'used when matters are of national significance'. 'Such cases may include, for example, those which, in the opinion of the Secretary of State may conflict with national policies on important matters, could have significant effects beyond their immediate locality, give rise to substantial regional or national controversy, raise significant architectural or urban design issues or may involve the interests of national security or of foreign governments' (Hansard, HC Deb, 30 April 2012, Vol. 543, c. 1234 and Written Answers, 16 June 1999, c. 138). In the period from 2002 to 2006, 1,326 decisions were taken by the Secretary of State personally, compared with 72,040 by the planning inspectorate. On 114 occasions, the Secretary of State went against the Inspector's recommendations.

Following the closure of Regional Offices, DCLG set up the National Planning Casework Unit (NPCU) to manage planning decisions on behalf of the Secretary of State for Communities and Local Government, including call-ins. An inspector is appointed to carry out an inquiry into a proposal and the Secretary of State has to take the inspector's findings into account when making the decision. The approach by the SoS has to be in line with the Ministerial Code, and this was codified in relation to planning decisions by *Guidance on Planning Propriety Issues* issued by DCLG in 2012, which stated, 'Planning ministers are under a duty to behave fairly ("quasi-judicially") in the decision-making procedure. They should therefore act and be seen to act fairly and even-handedly' (DCLG 2012g: 1).

The 2006 *Barker Review of Land Use Planning Final Report* proposed keeping the Ministerial role to a minimum, and reducing the number of applications considered for calling in. The Labour government accepted Barker's recommendation. On 30 March 2009, the Labour government published a new direction stating the type of application for which the Secretary of State has to be consulted. In summary, the types of application are those for green belt development, development outside town centres, World Heritage Site development, playing field development and flood risk area development. Full details – for example, size limits – are given in Circular 2/09.

In the past, certain types of development have tended to invite central government involvement. In the light of the government's commitments to increasing the delivery of new housing but in a sustainable way, new settlements and other very large housing developments figured prominently; so have applications involving the green belt, large-scale minerals proposals and development affecting buildings of national significance. Mineral workings have often raised problems of more than local importance, and the national need for particular minerals has to be balanced against planning issues. It is argued that such matters cannot be adequately considered by local planning authorities (who will invariably face massive local opposition) and such cases involve technical considerations requiring expert opinion of a character more easily available to central government. A large proportion of applications for permission to work minerals have been called in.

On important questions of design, CABE has, in its terms of reference, the power 'to call the attention of any of our departments of state . . . to any project or development which [it considers] may appear to affect amenities of a national or public character'.[29] Inevitably, the Secretary of State has the job of balancing local concerns with national policies and priorities.

Variations within the UK

Although the basic structures of the systems in the four countries are similar, there are differences in the detail and in how each system works. Changes introduced by the UK Coalition government have seen a greater divergence between the system in England and the other three countries – for example, England is the only country that has provision for Neighbourhood Development Orders, although in 2012 Wales introduced a power for local planning authorities to make Local Development Orders. England, Scotland and Northern Ireland each have their own primary planning legislation, although this will not be fully in force in Northern Ireland until 2015. However, the principles by which development management operates are similar across the whole of the UK, although there is clearly scope for greater differences to emerge as devolution develops.

Currently, the most substantial difference is to be found in Northern Ireland, where development management is the responsibility of the Planning and Local Government Group of the Department of Environment for Northern Ireland. Applications are made to one of the six local area planning offices, but the decision to grant or refuse planning permission is the responsibility of the Department. Decisions are made following statutory consultation with the local district or borough council, and the Department considering the views of elected councillors. As part of local government reform in Northern Ireland, the bulk of planning functions will no longer rest with the Department. Instead these powers will be devolved to the eleven district councils which, through their elected representatives, will be responsible and accountable for most planning decisions. The transfer of most planning functions to new district councils does not just involve the redistribution of functions but includes the creation of a new planning system for Northern Ireland. It will take place when the eleven local authorities formally come into being in April 2015.

Appeals in Northern Ireland are heard by the Planning Appeals Commission, an independent appeals body which operates under the Planning (Northern Ireland) Order 1991 (SI 1991/1220). Decisions are transferred out from the political arena, unlike elsewhere in the UK, where the appeal bodies make decisions in the name of the relevant Ministers. In Northern Ireland, the PAC must reach its decision on the basis of the reports made by the Commissioners. Commissioners are appointed by the First and Deputy First Ministers under Article 110(2) of the 1991 Order. Commissioners are not Civil Servants but are appointed following open public competition. Commissioners' decisions are final but they are open to

challenge by application to the High Court for judicial review.

In Scotland, for the purposes of planning applications, developments are put into one of three categories: local, major or national. These are specified in the Town and Country Planning (Hierarchy of Developments) (Scotland) Regulations 2009 (SI 2009/51) and the different categories require different levels of supporting information – for example, national and major projects require community consultation. Local developments include changes to individual houses and, for example, smaller developments for new housing and retail. Major developments include developments of fifty or more homes, certain waste, water, transport and energy-related developments, and larger retail developments. National developments are mainly large public works (for example, the replacement Forth crossing) and are identified in the National Planning Framework (NPF). Inclusion in the NPF is deemed to have established the need for the development, so objections can only be made on matters of detail.

Enforcement of planning control

If the machinery of planning control is to be effective, some means of enforcement is essential. Under the pre-war system of interim development control, there were no such effective means. A developer could go ahead without applying for planning permission, or could even ignore a refusal of permission. The developer took the risk of being compelled to 'undo' the development (for example, demolish a newly built house) when, and if, the planning scheme was not approved, but this was a risk that was usually worth taking. If the development was inexpensive and lucrative (for example, a petrol station), the risk was virtually no deterrent at all. This flaw in the pre-war system was remedied by the strengthening of enforcement provisions.

These are required not only for the obvious purpose of implementing planning policy, but also to ensure that there is continuing public support for, and confidence in, the planning system. To quote PPG 18 *Enforcing Planning Control* (1991):

the integrity of the development control process depends on the LPA's readiness to take effective enforcement action when it is essential. Public acceptance of the development control process is quickly undermined if unauthorised development, which is unacceptable on planning merits, is allowed to proceed without any apparent attempt by the LPA to intervene before serious harm to amenity results from it.

(para. 4)

Enforcement provisions were radically changed by the Planning and Compensation Act 1991 following a comprehensive review by Robert Carnwath, QC, published in 1989. Provisions of this new enforcement regime were summarised in DoE Circular 10/97 *Enforcing Planning Control*.[30] The 1991 Act provided a range of tools in addition to the long-standing provision for enforcement notices.

Development undertaken without permission is not an offence in itself, but ignoring an *enforcement notice* or *stop notice* is an offence, and there is a maximum fine following conviction of £20,000. (In determining the amount of the fine, the court is required to 'have regard to any financial benefit which has accrued'.[31]) There is a right of appeal against an enforcement notice. An appeal also contains a deemed application for development, for which a fee is payable to the planning authority. Appeals can be made on several grounds, for example that permission ought to be granted, that permission has been granted (e.g. by the GPDO), and that no permission is required. There is also a limited right of appeal on a point of law to the High Court. New procedures for enforcement appeals came into effect from December 2002 and brought them into line with changes made to the planning appeals procedure in 2000, for example, in the use of hearings rather than inquiries, simultaneous submission of evidence and new stricter timetables. Further guidance was issued by PINS in 2011 relating to procedures by which appeals are heard – written representations, hearings or local inquiries – with the Inspectorate ultimately determining which procedure is to be used.

Enforcement can be a lengthy process. For example, South Hams District Council issued an enforcement notice in January 1990 for the removal of a house built without consent. In 1993, the owner was fined £300 for breaching the enforcement notice. In 1995, he was jailed for three months for contempt of a court order requiring demolition. He had demolished only the upper storey and grassed over the lower half.[32]

Where it is uncertain whether planning permission is required, an LPA has power to issue a *planning contravention notice*. This enables it to obtain information about a suspected breach of planning control and to seek the cooperation of the person thought to be in breach. If agreement is not forthcoming (whether or not a contravention notice is served), an enforcement notice may be issued, but only 'if it is expedient' to do so, 'having regard to the provisions of the development plan and to any other material considerations'. In short, the local authority must be satisfied that enforcement is necessary in the interests of good planning.

In view of government commitments to fostering business enterprise, planning authorities were advised in PPG 18 to consider the financial impact on small businesses of conforming with planning requirements. 'Nevertheless, effective action is likely to be the only appropriate remedy if the business activity is causing irreparable harm' (para. 17). Development 'in breach of planning control' (development carried out without planning permission or without compliance with a planning condition) might be undertaken in good faith, or ignorance. In such a case, application can be made for retrospective permission. It is unlikely that a local authority would grant unconditional permission for a development against which it had served a planning contravention notice, but it might be willing to give conditional approval.

The 1991 Act also introduced a *breach of condition notice* as a remedy for contravention of a planning condition. Provided that the planning condition is clearly and precisely worded, this is a simple and effective tool to use. A breach of condition notice cannot be appealed, and a failure to comply constitutes a criminal offence, for which the recipient of the notice can be prosecuted in the magistrates' court, as it is a summary offence.

Where there is an urgent need to stop activities that are being carried on in breach of planning control, an

LPA can serve a *stop notice* or, as introduced by Circular 2/05, a *temporary stop notice*. The stop notice can be served as soon as building works start or unauthorised use begins, and there is no way to delay its effect. This is an attempt to prevent delays in the other enforcement procedures (and advantage being taken of these delays) resulting in the local authority being faced with a *fait accompli*. Development carried out in contravention of a stop notice constitutes an offence and attracts the same penalties as failing to comply with an enforcement notice, as it is issued 'on the back of' an enforcement notice. The temporary stop notice is a 'stand-alone' notice and its effects last for twenty-eight days from the day it is displayed. In 2013, government consulted on changes to the use of temporary stop notices. The changes were introduced by the Town and Country Planning (Temporary Stop Notice) (England) (Revocation) Regulations 2013 (SI 2013/830) and the new provisions remove the restriction on LPAs' ability to serve temporary stop notices on caravans which are used as a main residence, where there is a suspected breach of planning control. This measure, which is concerned with control of unauthorised activity by gypsies and travellers, means that LPAs will determine whether the use of a temporary stop notice in these circumstances is a proportionate and necessary response.

Section 215 of the Town and Country Planning Act 1990 provides a local planning authority with the power, in certain circumstances, to take steps requiring land to be cleaned up when its condition adversely affects the amenity of the area. If it appears that the amenity of part of its area is being adversely affected by the condition of neighbouring land and buildings, it may serve a notice on the owner requiring that the situation be remedied. These notices set out the steps that need to be taken, and the time within which they must be carried out. LPAs also have powers under section 219 to undertake the clean-up works themselves and to recover the costs from the landowner. In a guide to best practice in the use of section 215, published in 2005 by the ODPM, it was shown that they are effective in securing compliance, with few cases of appeal or inaction leading to LPAs using powers under section 219. It also showed that the threat of issue of a notice is by itself effective, with 20 per cent of the notices approved not needing to be served. Section 215 notices are possibly particularly useful in the context of regeneration programmes.

The NPPF replaced PPG 18, but offers little by way of detailed guidance on matters of enforcement. It does assert that enforcement activity should be 'proportionate' and suggests that LPAs should consider developing and publishing a local enforcement plan (an idea which to some extent echoes advice issued by the Planning Officers Society (POS) in 2008), but only devotes one short paragraph to the matter. However, the Localism Act 2011 does make some specific provisions related to enforcement. These are section 123 'Retrospective planning permission', which strengthens the hand of the planning authority and weakens the hand of the person who has undertaken development without planning consent; section 124 'Time limits for enforcing concealed breaches of planning control' (e.g. concealing the development of a house in the countryside); and sections 125–7, dealing with time limits, advertisement control and other technical matters.

The provisions for enforcement can be complex and there are many difficulties in their operation. In the guidance note *Planning Enforcement – An Overview for the LGA* produced in 2010, the POS described enforcement as a 'Cinderella service', partly because it had not attracted adequate attention or resources. As a consequence, 'backlogs of casework have built up in some areas, with unresolved cases running on for years' (p. 1). In addition to problems of inadequate resources, a shortage of people with appropriate knowledge and skills was identified as a problem. The higher profile that had, at times, been given to contraventions of planning control had in fact increased community awareness of planning control as a means of shaping communities. Government had previously recognised some of these issues in 2006, when, in a review of enforcement, it had recommended that LPAs should use some of their planning delivery grant to enhance the level of resource allocated to enforcement and that they should make efforts to establish a career structure in enforcement, contributing to the improvement of the status of the work. However, in 2010,

John Silvester, spokesman for the POS, expressed professionals' continued dissatisfaction when he said, 'Despite many promises Government has not improved the prevailing planning enforcement system that dates from 1991' (p. 4).

Revocation, modification and discontinuance

The powers to control development available to a local authority go further than simply granting or refusing permission. A local planning authority in England may make a change to a planning permission if it is satisfied that that change is not material (Town and Country Planning Act 1990, section 96A). In determining whether a change is material it must have regard to the effect of the change and any previous changes already made under section 96A to the original planning permission.

A planning authority also has power to make orders revoking or modifying planning permissions already granted, but such orders become effective only if confirmed by the Secretary of State, except where they are unopposed. This power is granted under section 97 of the 1990 Act, which states:

> If it appears to the local planning authority that it is expedient to revoke or modify any permission to develop land granted on an application made under this Part, the authority may by order revoke or modify the permission to such extent as they consider expedient.

There is a liability to pay compensation, under section 107 of the Act, in respect of expenditure rendered abortive by the order and for any other loss or damage directly attributable to the revocation or modification. Section 189 of the Planning Act 2008 made changes to the 1990 Act in respect of entitlement to compensation. The cases where planning consent is revoked do not, however, normally cover recent consents. More often, they relate to old consents that have been started but not completed.

The planning authority may, in the interests of proper planning of their area, make an order under section 102 of the Town and Country Planning Act 1990 (Town and Country Planning (Scotland) Act 1997, section 71) requiring the discontinuance of any existing use of land, or the modification of such existing use by the imposition of conditions, or the removal or alteration of buildings or works on the land. Such an order is subject to confirmation by the Secretary of State. Compensation is payable for any reduction in the existing use value of the land and for any disturbance which is the result of such a requirement (see Box 5.6 for an example).

In cases where the planning permission may not have been properly granted in terms of procedure, it may be better financially for the local council to have the planning permission quashed at judicial review, rather than revoke the planning permission and have to pay compensation. The rules about bringing a judicial review are strict, however, and a claim must be made within legal time limits.

British planning legislation does not assume that existing non-conforming uses must disappear if planning policy is to be made effective. This may be an avowed policy, but the Planning Acts explicitly permit the continuance of existing uses.

Purchase and blight notices

A planning refusal does not of itself confer any right to compensation. On the other hand, revocations of planning permission or interference with existing uses do rank for compensation, since they involve a taking away of a legal right. In cases where, as a result of a planning decision, land becomes 'incapable of reasonably beneficial use' the owner can serve a *purchase notice* upon the local authority requiring it to buy the property. In all cases, ministerial confirmation is required. The circumstances in which a purchase notice can be served include:

- refusal or conditional grant of planning permission
- revocation or modification of planning permission
- discontinuance of use.

BOX 5.6 A REVOCATION OF PLANNING CONSENT

In 1993, Alnwick District Council granted planning permission to Northumberland Estates for a supermarket near Alnwick. A protest campaign was launched two years later when it emerged that Safeway had bought the land. Protestors feared that Safeway would close its existing branch in Alnwick and consolidate operations on the new site. The Secretary of State in 1997, John Gummer, revealed that he proposed to revoke the permission. The council challenged the decision in the High Court but lost. Safeway submitted a claim for £4.6 million in compensation for the loss of its planning consent. The council feared that it might be bankrupted. However, in the end it was all settled amicably with the Duke of Northumberland buying back the land that he had sold to Safeway. Safeway agreed to forego £2.6 million of their compensation claim and accepted £2 million, which was paid by the council's insurers, Zurich Municipal.

Source: House of Commons Library Note SN/SC/905, May 2013

In considering whether the land has any *beneficial use*, 'relevant factors are the physical state of the land, its size, shape and surroundings, and the general patterns of land-uses in the area; a use of relatively low value may be regarded as reasonably beneficial if such a use is common for similar land in the vicinity' (DoE Circular 13/83).

A purchase notice is not intended to apply in a case where an owner is simply prevented from realising the full potential value of the land. This would imply the acceptance in principle of paying compensation for virtually all refusals and conditional permissions. It is only if the existing and permitted uses of the land are so seriously affected as to render the land incapable of reasonably beneficial use that the owner can take advantage of the purchase notice procedure.

There are circumstances, other than the threat of public acquisition, in which planning controls so affect the value of the land to the owner that some means of reducing the hardship is clearly desirable. For example, the allocation of land in a development plan for a school or for a road will probably reduce the value of houses on the land or even make them completely unsaleable. In such cases, the affected owner can serve a blight notice on the local authority requiring the purchase of the property at an 'unblighted' price. These provisions are restricted to agricultural or residential owner-occupiers, or business owner-occupiers, if the value for business rates is £34,800 or less (set in 2010 and subject to periodic revisions). You cannot serve a blight notice if you are an investment property owner. You are also usually expected to have made 'reasonable endeavours' to sell your property and to have been unable to do so, except at a price substantially lower than its value under normal market conditions because of certain defined planning actions. These include land designated for compulsory purchase, or allocated or defined by a development plan for any functions of a government department, local authority or statutory undertaker, and land on which the Secretary of State has given written notice of his or her intention to provide a trunk road or a special road (i.e. a motorway).

The subject of planning blight takes us into the much broader area of the law relating to compensation. This is an extremely complex field, and only an indication of three major provisions can be attempted here.

First, there is a statutory right to compensation for a fall in the value of property arising from the use of highways, aerodromes and other public works which have immunity from actions for *nuisance*. The depreciation has to be caused by physical factors such as noise, fumes, dust and vibration, and the compensation is payable by the authority responsible for the works. Second, there is a range of powers under the heading 'mitigation of injurious effect of public works'. Examples include sound insulation; the purchase of owner-occupied property which is severely affected by construction work or by the use of a new or improved highway; the erection of physical barriers (such as walls, screens or mounds of earth) on or alongside roads to reduce the effects of traffic noise on people living nearby; the planting of trees and the grassing of areas; and the development or redevelopment of land for the specific purpose of improving the surroundings of a highway 'in a manner desirable by reason of its construction, improvement, existence or use'. Third, provision is made for *home loss payments* as a mark of recognition of the special hardship created by compulsory dispossession of one's home. Since the payments are for this purpose, they are quite separate from, and are not dependent upon, any right to compensation or the *disturbance payment* which is described below. Logically, they apply to tenants as well as to owner-occupiers, and are given for all displacements whether by compulsory purchase or any action under the Housing Acts. These provisions were slightly extended in the Planning and Compensation Act 1991.

Additionally, there is a general entitlement to a *disturbance payment* for persons who are not entitled to compensation. Local authorities have a duty 'to secure the provision of suitable alternative accommodation where this is not otherwise available on reasonable terms, for any person displaced from residential accommodation' by acquisition, redevelopment, demolition, closing orders and so on.

Development by the Crown, government departments and statutory undertakers

Part 7 of the 2004 Act brought an end to Crown immunity from planning control (the actual change came into effect in 2006). Because the Crown is generally not bound by statute, development by government departments did not require planning permission.

However, since 1950, there have been special arrangements for consultations. Increased public and professional concern about the inadequacy of these led to revised, but still non-statutory, arrangements culminating in DoE Circular 18/84. This said that, before proceeding with development, government departments will consult planning authorities when the proposed development is one for which specific planning permission would, in normal circumstances, be required. In effect, local authorities should treat notification of a development proposal from government departments in the same way as any other application. Where the local authority was against the development, the matter was referred to the Secretary of State.

Circular 2/06 stated that:

> from 7 June 2006, the planning Acts will apply to the Crown, subject to certain exceptions. The Crown has hitherto been immune from the planning system, but successive governments have had a policy that immunities enjoyed by the Crown should be removed where they are not necessary. This will put the position of the Crown onto a statutory basis and ensure full compliance with the Environmental Impact Assessment Directive.
>
> (p. 3)

The 'exceptions' are mainly concerned with national security and defence (where there are limits on information that has to be provided), urgency and enforcement, together with new permitted development rights (to put the Crown on a similar footing to local authorities and statutory undertakers) and use classes (concerned with secure residential institutions).

Development by private persons on 'Crown land' (i.e. land in which there is an interest belonging to Her Majesty or a government department) has required planning permission in the normal way, although there are limitations on the ability of the planning authority to enforce in these cases.

Development undertaken by statutory undertakers is subject to planning control but it is also subject to special planning procedures. Where a development requires the authorisation of a government department (as do developments involving compulsory purchase orders, work requiring loan sanction, and developments on which government grants are paid), the authorisation is usually accompanied by *deemed planning permission*. Much of the regular development of statutory undertakers and local authorities (e.g. road works, laying of underground mains and cables) is *permitted development* under the GPDO. Statutory undertakers wishing to carry out development which is neither permitted development nor authorised by a government department have to apply for planning permission to the local planning authority in the normal way, but special provisions apply to *operational land*. The original justification for this special position of statutory undertakers was that they are under an obligation to provide services to the public and could not, like a private firm in planning difficulties, go elsewhere.

Development by local authorities

Until 1992, planning authorities were also deemed to have permission for any development which they themselves undertook in their area, as long as it accorded with the provisions of the development plan; otherwise they had to advertise their proposals and invite objections. The only requirement was for the local authority to grant itself permission by resolution. These 'self-donated' planning permissions were problematic. Although local authorities are guardians of the local public interest, they can face a conflict of interest in dealing with their own proposals for development. Pragmatic consideration of the merits of a case involving their own role as developers can easily distort a planning judgment. Examples include attempts by authorities to dispose of surplus school playing fields with the benefit of permission for development, and competing applications for superstore development when one of the sites is owned by the authority itself. The local authorities' position was not helped by judgments against them that found many irregularities in the necessary procedures (Moore 2000: 311).

Because of these difficulties, new regulations were issued in 1992 which require planning authorities to make planning applications in the same way as other applicants, and generally follow the same procedures, including publicity and consultation. There must be safeguards to ensure that decisions are not made by members or officers who are involved in the management of the land or property, and the planning permission cannot pass to subsequent land and property owners. Local authorities continue to enjoy permitted development rights, which can be important in certain areas such as schools. Where other interests propose development on local authority-owned land they must apply for permission in the normal way. The new procedures did not go as far as some had hoped and criticism continues, and inevitably so, since the accusation of bias is always possible while local authorities are able to grant themselves planning permission. The Scottish Local Government Ombudsman has complained about 'the ease with which planning authorities breach their own plans, particularly considering the time, effort, and consultation which goes into them'. One solution would be for the Secretary of State to play a role in all applications in which the local authority has an interest (as proposed by the Nolan Committee on Standards of Conduct in Local Government).

Control of advertisements

The need to control advertisements has long been accepted. Indeed, the first Advertisements Regulation Act 1907 preceded by two years the first Planning Act. But, even when amended and extended (in 1925 and 1932), the control was quite inadequate. Not only were the powers permissive, but also they were limited. For instance, under the 1932 Act the right of appeal (on the ground that an advertisement did not injure the amenities of the area) was to the Magistrates' Court – hardly an appropriate body for such a purpose. The 1947 Act set out to remedy the deficiencies. There are, however, particular difficulties in establishing a legal code for the control of advertisements. Advertisements may range in size from a small window notice to a massive hoarding, in the form of a poster, a balloon or even lasers; they vary in purpose from a bus stop sign to a demand to buy a certain make of detergent; they could be situated alongside a cathedral, in a busy shopping street or in a particularly beautiful rural setting; they might be pleasant or obnoxious to look at; they might be temporary or permanent, and so on. The task of devising a code which takes all the relevant factors into account and, at the same time, achieves a balance between the conflicting interests of legitimate advertising and 'amenity' presents real problems. Advertisers themselves frequently complain that decisions in apparently similar cases have not been consistent with each other. The official departmental view has been that no case is exactly like another, and hard and fast rules cannot be applied: each case has to be considered on its individual merits in the light of the tests of amenity and – the other factor to be taken into account – public safety.

Current policy guidance is to be found in the NPPF, where it states (para. 67) that 'Poorly placed advertisements can have a negative impact on the appearance of the built and natural environment' but adds that 'Only those advertisements which will clearly have an appreciable impact on a building or on their surroundings should be subject to the local planning authority's detailed assessment.' More detailed guidance is to be found in Circular 3/07 and the Town and Country Planning (Control of Advertisements) (England) Regulations 2007 (SI 783/2007), which established three categories of advertisement:

- Advertisements which the rules exclude from the planning authority's direct control and which can be displayed 'as of right'. These advertisements are referred to under Schedule 1 of the Regulations and involve nine classes which include traffic signs, national flags, advertisements within buildings and parliamentary and local government election signs. Detailed descriptions of each class, its conditions, limitations and interpretation are given in the Regulations. Subject to the conditions being met those advertisements are excluded from the LPA's control.
- Advertisements for which the rules give a 'deemed consent' so that the planning authority's consent is not needed, provided the advertisement is within the rules. There are sixteen separate classes

identified, including such items as signs on businesses, functional signs for local authorities and transport undertakings, signs on balloons and flags, and some temporary signs. These are detailed in the Regulations and are in most cases subject to size limitations and standard conditions.

- Advertisements for which the planning authority's 'express consent' is always needed. As a general guide the following types of sign would require express consent (an indicative, not a comprehensive list): poster hoardings, externally illuminated signs, advance and directional signs, advertisements above 4.6 metres from the ground or above the bottom of any first floor window. Applications for permission should be made to the LPA in the normal manner, accompanied by drawings indicating the location, size, position, colour and materials of any sign, together with details of the size and type of lettering and means of illumination, if any. The LPA has three courses of action:

(a) grant consent thereby requiring compliance with the standard conditions laid down by the Regulations and any other optional conditions which may be appropriate;

(b) issue a split decision approving certain signs, if more than one sign has been applied for, and refusing any unsatisfactory sign or signs giving reason(s); and

(c) refuse consent giving reasons.

The five standard conditions incorporated in the Regulations are concerned with safety, maintaining the advertisement in good condition and ensuring that the consent of the owner of the site where the advertisement is displayed.

Some minor amendments were made to the Regulations by the Town and Country Planning (Control of Advertisements) (England) (Amendment) Regulations 2011 (SI 2011/2057), affecting advertisements in telephone boxes and inserting a new class of deemed consent for advertisements on electric vehicle charging points.

There are some areas – such as conservation areas, National Parks or Areas of Outstanding Natural Beauty – which can be regarded as especially vulnerable to the visual effects of outdoor advertisements. All planning authorities have three additional powers which enable them to achieve a stricter control over advertisements. These powers are:

(a) to define an Area of Special Control of Advertisements;
(b) to remove from a particular site or a defined area the benefit of the deemed consent normally provided by the rules; and
(c) to require a particular advertisement, or the use of a site for displaying advertisements, to be discontinued.

The way in which the planning authority proposes to use the first and second of these powers must be formally approved by the Secretary of State before it is effective. There is the normal right of appeal to the Secretary of State against the planning authority's use of a discontinuance notice.

The NPPF gives some further guidance on the use of these powers:

> Before formally proposing an Area of Special Control, the local planning authority is expected to consult local trade and amenity organisations about the proposal. Before a direction to remove deemed planning consent is made for specific advertisements, local planning authorities will be expected to demonstrate that the direction would improve visual amenity and there is no other way of effectively controlling the display of that particular class of advertisement.
>
> (para. 68)

Where an order establishing an Area of Special Control is in force, the LPA has a duty, under regulation 20(4), to consider at least once in every five years whether it should be revoked or modified.

Control of mineral working

The NPPF was trumpeted as reducing planning guidance dramatically, but in the case of minerals planning, there seems to be quite a lot of guidance which escaped the axe. There are four pages of minerals guidance in the NPPF itself, plus another ten or so pages in a separate 'technical guidance document'; also, a number of Minerals Planning Guidance notes (MPGs) are still in force.[33] Part of the explanation for this may be found in an evaluation of minerals policy guidance published in 2011 but which took into account the nature of the emerging NPPF. It stated:

> An adequate, steady and secure supply of minerals is important to sustaining industries that depend upon them as essential raw materials. However, in a densely populated country like England, minerals extraction is almost always controversial and opposed by host communities. Managing the nation's finite mineral resources through a strategic, plan-led approach is, therefore, crucial. A clear and unambiguous national minerals policy, allied to other national policy objectives, is essential and the key to maintaining sustainable, long-term continuity of minerals supply.
>
> (Bloodworth 2011: vii)

Paragraphs 142–9 of the NPPF echo these sentiments and emphasise that LPAs need to develop a suitable policy framework for mineral safeguarding and extraction, including suitable environmental safeguards for mineral workings and post-extraction remediation. Planning for aggregates (stone) should be based on a *local aggregate assessment* which seeks to balance demand and supply, undertaken with the assistance of an *aggregate working party*: these are cross-sector working groups which have been in existence for some time in some form in stone extraction areas. So, as in other policy areas, management of minerals developments is also plan-led, and applications for extraction are determined in line with plan policies. These policies will reflect the national framework noted above. There are also a number of specific matters, such as the necessity to consider how to meet the need for stone extraction for restoration of historic buildings and a prohibition on peat extraction from new or extended sites (for reasons of environmental protection).

Powers to control mineral workings stem from the definition of development, which includes 'the carrying out of . . . mining . . . operations in, on, over or under land' (Town and Country Planning Act 1990, section 55(1)). However, a special form of control is necessary to deal with the unique nature of mineral operations. Unlike other types of development, mining operations are not the means by which a new use comes into being; they are a continuing end in themselves, often for a very long time. They do not adapt land for a desired end use. On the contrary, they are essentially harmful and may make land unfit for any later use. They also have unusual location characteristics: they have to be mined where they exist, which can be in areas of significant environmental quality. For these reasons, the normal planning controls need to be supplemented.

Two major features of the minerals control system are that it takes into account the fact that mineral operations can continue for a long period of time, and that measures are needed to restore that land when operations cease. It is, therefore, necessary for MPAs to have the power to review and modify permissions and to require restoration. Under current legislation, MPAs have a duty to review all mineral sites in their areas. This includes those which were 'grandfathered' in by the 1947 Act. These old sites, of which there may be around a thousand in England and Wales, often lack adequate records. They present the particular problem that they can include large unworked extensions which are covered by the permission; if worked, these could have serious adverse effects on the environment. The provisions relating to these sites are even more complicated than those pertaining to the generality of mineral operations, and they have been significantly altered by the Environment Act 1995. Details are set out in MPG 14 *Environment Act 1995: Review of Mineral Planning Permissions*, one of the MPGs not deleted by the NPPF.

Policies for restoration (and what the Act quaintly calls 'aftercare') have become progressively more stringent, mainly in response to what the Stevens Report (1976) referred to as a great change in standards and attitudes to mineral exploitation. The lengthy – 75 pages – guidance note, MPG 7 *Reclamation of Mineral Workings*, set out policy for this activity. More recently, the *Technical Guidance to the National Planning Policy Framework*, published in 2012, noted that 'planning authorities should provide for restoration and aftercare at the earliest opportunity to be carried out to high environmental standards. This should include through provision of a landscape strategy, restoration conditions and aftercare schemes as appropriate' (p. 19). It perhaps emphasises how important restoration has become that the note pays attention to the financing of such works (by the operator or the landowner) by requiring applicants to 'demonstrate with their applications what the likely financial and material budgets for restoration, aftercare and after-use will be, and how they propose to make provision for such work during the operational life of the site' (p. 22).

The extraction of minerals is one of the most obvious examples of a 'locally unwanted land use' (LULU) and one that has a disproportionate effect on particular locations (Blowers and Leroy 1994). But minerals extraction may also bring economic benefits especially in more remote rural locations. Management of minerals development seeks to reconcile these conflicting interests, and reviews of minerals planning guidance have progressively taken more account of the need for sustainable development. Nevertheless, a major limitation of the control of minerals exploitation is the emphasis on finding suitable locations, albeit in the interests of mitigating environmental impacts.[34] However, much less attention is given to managing the demand for these resources.

Major infrastructure projects

Decision-making on major infrastructure projects has proved to be very difficult and the process has tended to be long-winded. This has been exacerbated in the UK by limited national policies or strategies concerning investment in roads, bridges, airports and the like. From time to time, this problem reaches public attention, or rather the inquiry part of the process does. The inquiry into Terminal 5 at Heathrow Airport is perhaps the 'classic' case, where the inquiry sat for 525 days, heard 700 witnesses and received

6,000 documents. The terminal finally opened in 2008, thirteen years after the inquiry began and twenty-six years after the scheme was first mooted. It is understandable, therefore, that the government should seek to improve on performance in dealing with major projects.

Approval for major infrastructure projects such as railways, light rail systems and bridges has been given in different ways. Before 1992, most projects were approved through private members bills or the hybrid bill procedure and Act of Parliament.[35] The Channel Tunnel Rail Link was approved in this way in two years. The procedure involved select committees in each House hearing petitioners' requests for amendments to the scheme. Following the Act, planning permission is still required from local planning authorities for detailed 'reserved matters'.

The ODPM consulted on new parliamentary procedures for processing major infrastructure projects in 2001. The proposals included more up-to-date statements of government policy on infrastructure, an improved regional policy framework (coming forward in regional spatial strategies), a procedure to allow Parliament to give approval to the project in principle, improved inquiry procedures, and changes to compulsory purchase and compensation provisions. The Planning Act 2008 introduced a new system for approving major infrastructure projects of national importance. These included railways, large wind farms, power stations, reservoirs, harbours, airports and sewage treatment works. Decisions on these were to be taken by a new independent *Infrastructure Planning Commission* (IPC), based on new *national policy statements* (NPS). Commissioners, who were to be independent of government and all other interests, would examine the evidence for and against each project. They would act in accordance with government policy as set out in national policy statements, which were to consider national priorities and explain the case for investment in energy, transport, water and waste infrastructure. An important element of the process was to be pre-application consultation by the promoters of projects, with the IPC judging whether such consultation had met required standards. The hearing and decision-making process by the

Commission were to be timetabled – for example six months to carry out an examination, three months to make a decision. The effect would be that the SoS would no longer have the final say on major infrastructure decisions: instead it would be in the hands of an independent and 'objective' Commission.

However, the new Commission had hardly got to work when, following the election in 2010, the new government announced its intention to abolish the IPC:

> It will be replaced with a new rapid and accountable system where Ministers, not unelected commissioners, will take the decisions on new infrastructure projects critical to the country's future economic growth.
>
> (Statement by Greg Clark, 29 June 2010)

This intention was made good by the Localism Act 2011, which recognised that some planning decisions are so important to the overall economy and society that they can only be taken at a national level. The Act also provided for national policy statements, which will be used to guide decisions by ministers and can be voted on by Parliament. It was asserted that 'Ministers intend to make sure that major planning decisions are made under the new arrangements at least as quickly as under the previous system' (DCLG 2011b: 14).

The mechanics of the current system are little changed from those proposed in the 2008 Act, apart from where the final decision is taken: these are summarised in Figure 5.3. Any developer wishing to construct a nationally significant infrastructure project (NSIP) must first apply for development consent. For such projects, the relevant Secretary of State appoints an 'examining authority' to examine the application. The examining authority will be from the Planning Inspectorate, and will be either a single Inspector or a panel of three or more Inspectors. Ironically, the IPC was, before its abolition, based in the same building as the Inspectorate in Bristol. Once the examination has been concluded, the examining authority will make a recommendation to the Secretary of State, who will make the decision on whether to grant or refuse consent. In mid-2013, about three-quarters of the

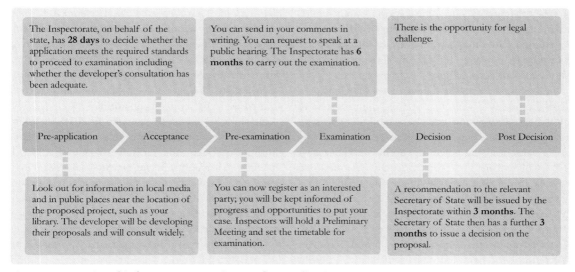

Figure 5.3 National infrastructure projects – the application process
Source: PINS Advice Note 8.1, April 2012, p. 3; guidance for the public

projects in process or where decisions had been reached were concerned with energy generation. An example would be Hinkley Point C nuclear power station in Somerset, where the application was made on 31 October 2011 and the decision letter was issued on 19 March 2013. The process for infrastructure planning is discussed further in Chapter 12.

Caravans

During the 1950s, the housing shortage led to a boom in unauthorised caravan sites. The controversy and litigation that this prompted led to the introduction of special controls over caravan sites (by Part I of the Caravan Sites and Control of Development Act 1960).[36] The Act gave local authorities new powers to control caravan sites, including a requirement that all caravan sites had to be licensed, with licences dealing with matters such as standards for sites, access, services and safety. These controls over caravan sites operate in addition to the normal planning system. Planning permission has to be obtained for caravan (and camping) sites which are occupied for more than twenty-eight days in

a year, in addition to the appropriate licence. Most of the Act dealt with control, but local authorities were given wide powers to provide caravan sites.

Holiday caravans are subject to the same planning and licensing controls as residential caravans. To ensure that a site is used only for holidays (and not for 'residential purposes'), planning permission can include a condition limiting the use of a site to the holiday season. Conditions may also be imposed to require the caravans to be removed at the end of each season or to require a number of pitches on a site to be reserved for touring caravans.

Meeting the needs of one group of caravanners has proved particularly difficult: gypsies and travellers. This 'difficulty' has been summed up by Home (2002: 345) as stemming from a number of factors:

> The settled community's general antipathy towards them, the perceived inconsistency of the nomad seeking a settled base, the difficulties of incorporating caravans (essentially moveable property) under land use regulation, and the reluctance of the planning system to accord a special exemption from countryside protection policy to a minority group.

The fact that 'gypsy' is in some ways a cultural concept makes exact definition (and counting[37]) difficult, but legislation tends to use terms such as 'nomadic' and 'travelling showmen', but the amount of time spent travelling can be variable.

Following on from the Caravan Sites and Control of Development Act 1960, when the commons were closed to gypsies and travellers and there was an increasing loss of traditional stopping places, few local authorities used the power to supply sites. Between 1970 – when the duty on local authorities contained in the Caravan Sites Act 1968 to provide adequate sites for gypsies 'residing in or resorting to' their areas was brought into force – and 1994 when that duty was repealed, some 350 sites were built in England. Consultation on the Reform of the Caravan Sites Act 1968 in 1992 heralded a marked shift in policy, detailed in Circular 1/94. In addition to repealing the obligations imposed on local authorities by the Caravan Sites Act, it provided stronger powers to remove 'unauthorised persons', though the DoE circular espouses a policy of tolerance towards gypsies on unauthorised sites. Between 1994 and 2006, the government put emphasis on the private provision of sites by gypsies and travellers themselves, but it proved very difficult to obtain planning permission. New policy advice was issued in February 2006, in Circular 1/06. According to the circular:

> A new Circular is necessary because evidence shows that the advice set out in Circular 1/94 has failed to deliver adequate sites for gypsies and travellers in many areas of England over the last 10 years. Since the issue of Circular 1/94, and the repeal of local authorities' duty to provide gypsy and traveller sites there have been more applications for private gypsy and traveller sites, but this has not resulted in the necessary increase in provision.
>
> (p. 4)

The circular sets the provision for gypsies and travellers in the wider context of the objective of everyone having access to a decent home, and the general approach to having a plan-led system. The role for planning was both to assess levels of need – a duty also placed on local authorities by the Housing Act 2004 – and to identify suitable sites to meet this need. This assessment of need also had a regional perspective, within the RSS. As a result, a number of LPAs developed *development plan documents* (DPDs) for gypsies and travellers. The effect of the change was to secure an increase in the numbers of sites.

However, following the election in 2010, the view of the new SoS Eric Pickles was that 'the current planning policy for traveller sites does not work' (DCLG 2011c: 3), although in terms of provision of sites evidence suggested otherwise.[38] Perhaps the real reason for his dissatisfaction is to be found in another statement, that 'there is a widespread perception that the system is unfair and it is easier for one group of people to gain planning permission, particularly on sensitive Green Belt land' (ibid.). Nevertheless, new legislation was introduced in March 2012, in the form of 'Planning policy for gypsy sites' (DCLG 2012f). This is to be incorporated into the NPPF when it has been 'tested'. The guidance maintains the plan-led approach to policy and continues the emphasis on an evidence-based form of policy, but seeks to 'promote more private traveller site provision while recognising that there will always be those travellers who cannot provide their own sites' (p. 1). A concern has been expressed about the impact of the new guidance in the context of localism and the abolition of RSS, with research suggesting that 'Councils' targets for additional residential pitches fell by 52 per cent from the 2,919 in the three published and emerging RSSs to the 1,395 recognised by the authorities themselves' (Hargreaves and Brindley 2011: i). In 2013 there was increased ministerial intervention in the issue of traveller sites when it was announced that 'The Secretary of State wishes to give particular scrutiny to traveller site appeals in the Green Belt' and will consider 'recovering' appeals for his own decision rather than leaving them in the hands of the Inspectorate (DCLG 2011c: 3).

As with many areas of planning policy, it is issues of implementation that are of particular sensitivity, so debate focuses not only on the question of whether local plans are identifying sufficient sites (a source of much debate in itself) but whether these identified sites eventually appear as available pitches for travellers. To assist here, there have been at various times government funds available to aid with the development of sites. Currently, the Gypsy and Traveller Sites Grant scheme, operated by the Homes and Communities Agency (HCA), provides funding for local authorities and registered social landlords to create new sites and refurbish existing sites. From taking over the scheme from DCLG in 2009, HCA invested £16.3 million in twenty-six schemes, providing eighty-eight new or additional pitches and making improvements to existing pitches, but in the view of many gypsy groups this level of provision fails to keep up with needs.

Dealing with sites and enforcement processes takes LPAs into difficult legal territory, not least because how gypsy and traveller issues are handled frequently moves beyond issues of land use into the areas of equality and human rights. Whether this situation will be eased by the cancellation in 2013 of the 2005 guidance *Diversity and Equality in Planning* remains to be seen. Some of these issues are illustrated in Box 5.7, which outlines the proceedings at the Dale Farm site in Billericay in Essex. Dealing with this aspect of planning has been and seems likely to remain one of the more contentious areas of planning and development management.

BOX 5.7 DALE FARM: THE MOST CONTENTIOUS TRAVELLER SITE?

A 2007 report for DCLG noted that 'No report on site provision and enforcement could be complete without reference to what is probably the most infamous unauthorised site in the country – at Dale Farm in Basildon, Essex' (p. 46). Not only was the process of ending the unauthorised use one which was long and protracted, but it illustrated some of the issues which arise in dealing with these matters. However, it should be noted that Dale Farm is not typical in either its size – possibly the largest site in the country – or the scale of international attention it attracted, but it does serve to cast light on the almost febrile nature of the issues.

The site has a long and contentious planning history. It is a six-acre site in the green belt and had previously been used, without planning permission, as a scrapyard. Somewhat confusingly, it is located next to an authorised site which provides thirty-four pitches. Although there was some evidence of traveller use in the 1980s, development started in earnest in 2001, which was when Basildon Council first served enforcement notices. Appeals against the notices were dismissed by the SoS but the time frame for compliance was extended to two years, to allow time for residents to find suitable alternative accommodation. Between the end of this period in June 2005 and June 2009 there were a number of court actions, ending at the High Court, where the planning enforcement action was ruled to be legal. However, it took a further two years for the enforcement action to be put into effect, in October 2011.

However, the planning issue of protecting the Green Belt was not the only sensitive element. At one point the representative of the UN High Commissioner for Human Rights offered to mediate. Amnesty International noted on a web page dedicated to Dale Farm that it was 'outraged at Basildon Council's decision to ignore the advice of Amnesty, a wide range of UN and Council of Europe bodies and experts, and other UK-based civil society organisations about the human rights impact of these evictions'. The DCLG 2007 report noted that:

> opposition amongst parts of the settled community towards site residents has become ever fiercer, with parents from the settled community withdrawing their children from the school attended by children from Dale Farm, and the view regularly expressed in letters to the local press that Gypsies and Travellers living on the site are somehow 'above the law'.

(p. 46)

Essex Police also opened a website covering the process, where it was noted that twenty-five people were arrested during the eviction proceedings, all of whom came from elsewhere to 'support' the site residents.

Basildon Council's minutes of a key meeting in May 2011 noted that 'The main issue was proportionality, weighing the interest of the community at large in upholding planning law and policy against the hardship this would cause to the occupiers of the site.' The minutes went on to note that a key piece of legislation was the Equality Act 2010, which consolidated existing anti-discrimination legislation on race, disability and sex.

Amidst a storm of media coverage, the residents were evicted from the site, but many had no alternative accommodation to go to. According to the *Guardian* of 28 September 2011, this was partly explained by the fact that funds allocated to the Homes and Communities Agency to provide traveller sites had been diverted to other uses, a view endorsed by the Equality and Human Rights Commission and Lord Avebury, Chair of the all-party parliamentary group on Gypsies, Roma and Travellers. It was reported that many of the former residents were parked illegally on the lane leading to the former site. However, in 2013, Basildon Council gave permission for a new site half a mile from the previous site to be built using the HCA grant mechanism.

The whole proceedings proved to be costly, with estimates of monetary cost varying between £5 million and £18 million. No doubt it was also costly in terms of time for the council and the site residents. It also gave bad publicity to travellers, many of whom lived almost anonymously in settled communities on a long-term basis, and to the whole process of site provision for travellers, which has been contentious and effectively unresolved for fifty years.

Sources: DCLG 2007; Equality and Human Rights Commission 2009; Amnesty International, BBC and Essex Police websites; Basildon Council Minutes (13 December 2007; 14 March 2011; 17 May 2011)

Telecommunications

One area of development management work that expanded rapidly and with some controversy has been telecommunications. The expansion of masts to service mobile phone networks was a particular concern, at first because of their visual impact, but also because of potential health effects. In England, PPG 8 *Telecommunications*, first issued in 1988 and revised in 1992 and 2001, was the main source for policy on telecommunications in England. Similar guidance applies in the other parts of the UK. As with other planning guidance and policy, this advice was replaced by the NPPF. However, this largely summarises the earlier guidance, retaining the positive approach towards telecommunications and restricting the grounds on which planning authorities can reject applications.

Government policy has been strongly behind the expansion of telecommunications, and this has presented many challenges for the planning system,

not least in its effect on spatial development patterns (see, for example, Graham and Marvin 1999). The erection of masts and related equipment has been the main talking point, but government policy is clearly influenced by the economic argument for efficient communications networks and fostering competition among rival networks. There is a tension between the central government's policy (then led by the DTI) for expanding mobile telecommunications, which has brought in consider-able revenue, and its local implementation in the face of concerted opposition in some places from the public. It should also be said that 'the public' faces two ways on this, with people wanting the convenience of mobile communications while not always accepting the infrastructure that provides it.

In England, masts under 15 metres in height[39] are permitted development (with some exceptions),[40] but there is a *prior approval procedure* which gives the planning authority fifty-six days to say whether it wishes to approve details of the siting and appearance of the development. The local authority consults in the same way as for planning permission, and it may refuse or add conditions to the approval. If the authority fails to notify the applicant within the fifty-six day period, the application is deemed granted.

Local development frameworks should include general policies for telecommunications-related development and may allocate sites for large masts. The planning authority should also encourage different operators to share facilities, though competition between the networks limits their willingness to cooperate. There is also an obligation on the developer to site the mast so that it has least effect on the external appearance of buildings. Where this is not followed, the planning authority may serve a breach of condition notice on the basis that a condition of the permitted development right has not been complied with. Masts over 15 metres in height require planning consent.

As noted above, health has been a major concern in the siting of mobile phone masts. This was particularly prominent in the 1990s but health considerations are barely mentioned in the 1992 version of PPG 8. The Independent Expert Group on Mobile Phones ('the Stewart Report') conducted an assessment of the health effects and concluded in its report in 2000 that 'there is no general risk'. Nevertheless, the Group recommended a precautionary approach and the removal of permitted development rights. This was not accepted by government in England, although the period for prior approval was extended to fifty-six days from forty-two. While health and related concerns, such as the perception of risk, can be material considerations as a matter of law, the government's view is that the planning regime is not the appropriate place for determining health safeguards. It is seen as the responsibility of central government to decide what measures are necessary to protect public health. Hence, as a matter of policy, if a proposed base station meets the recognised guidelines for public exposure to non-ionising radiation, it should not be necessary for a planning authority, in processing an application, to consider further the health aspects and concerns about them. All radio base stations in the UK are built to comply with the International Commission on Non-Ionising Radiation Protection (ICNIRP) guidelines for exposure to radio waves. The mobile operators have committed to present an ICNIRP certificate with each planning application.

Permitted development rights were removed in Scotland and the impact was evaluated in 2004 (Lloyd and Peel). Amongst the planning authorities there was a consensus that the enhanced level of control had resulted in better siting and design, whilst there was no evidence that it had slowed up telecommunications infrastructure roll-out. However, the report revealed continuing concerns about health matters:

> There is frustration that the planning system is confined to matters of siting and design. The findings suggest that the public remains unconvinced in general terms as to the protection afforded communities by the planning system with respect to telecommunications developments. More needs to be done to convince the public of the openness and transparency of the planning decision-making process.
>
> (p. 3)

So, whilst familiarity and the passage of time may have meant that the issue of telecommunications development now has a somewhat lower profile, concerns have not completely disappeared. However, government policy has, in general terms, been quite consistent, and current policy is well summed up in paragraph 46 of the NPPF:

> Local planning authorities must determine applications on planning grounds. They should not seek to prevent competition between different operators, question the need for the telecommunications system, or determine health safeguards if the proposal meets International Commission guidelines for public exposure.

Efficiency and resourcing of development management

There has been a succession of attempts on the part of central government to 'streamline the planning process' and to make it more 'efficient', though the reasons have varied (further explanation is given in Chapter 3). In the early 1970s, the concern was with the enormous increase in planning applications and planning appeals which, of course, stemmed from the property boom of the period. By 1981, government concern was with the economic costs of control, with cutting public expenditure and with 'freeing' private initiative from unnecessary bureaucratic controls. During the 1990s, the emphasis was on speeding up and raising standards of the 'planning service' so as to achieve better efficiency and value for money. In 2001, the Green Paper *Planning: Delivering a Fundamental Change* set four objectives for a reformed system of development control: to be responsive to the needs of all its customers and offer a new culture of customer service; to deliver decisions quickly in a predictable and transparent way; to produce quality development; and to genuinely involve the community. Development control was subject to review following the Green Paper and the 2004 Act but the Killian Pretty Review *Planning Applications: A Faster and More Responsive System* (2008) produced seventeen recommendations which were aimed at making the applications system 'customer-focused, fair, proportionate and transparent' (p. 5). The Conservative discussion paper *Open Source Planning* stated that it would 'accept the recommendations of the Killian Pretty Review where they are applicable to our new Open Source planning system' (p. 21). This summary of policy initiatives highlights an ongoing concern with 'efficiency' in dealing with planning applications, and the nature of these concerns and proposed solutions are set out in the brief history which follows below.

A good starting point is the analysis of an inquiry chaired by George Dobry, QC in 1975. In the 1970s there was lengthening delay in the processing of planning applications during a property boom. Dobry was quick to point out that 'not all delay is unacceptable: it is the price we must pay for the democratic planning of the environment'. He also took account of increasing pressure for public consultation and participation in the planning process, and the 'dissatisfaction on the part of applicants because they often do not understand why particular decisions have been made' and general concerns that the system was not doing enough to protect a good environment or promote high-quality development.

Dobry, like his successors, had the difficult task of reconciling apparently irreconcilable objectives: to expedite planning procedures while at the same time facilitating greater public participation and devising a system which would produce better environmental results. His solutions attempted to provide more speed for developers, more participation for the public *and* better-quality development and conservation. His solution was to divide applications into minor and major, such that minor applications could be dealt with more expeditiously through a simpler process, though with the opportunity for some participation and with a safety channel to allow them to be transferred to the major category if this should prove appropriate.

Dobry's scheme was a heroic attempt to improve the planning control system to everyone's satisfaction (Jowell 1975). Inevitably, therefore, it disappointed everyone, not least because his overriding concern for expediting procedures forced him to compress 'simple' applications into an impracticable timescale. In the meantime, the boom had collapsed and a new Labour government had other concerns. The government rejected all Dobry's major recommendations for changes in the system, though it was stressed that their objectives could typically be achieved if local authorities adopted efficient working methods. Dobry's view that 'it is not so much the system which is wrong but the way in which it is used' was endorsed, and his *Final Report* was commended 'to students of our planning system as an invaluable compendium of information about the working of the existing development control process, and to local authorities and developers as a source of advice on the best way to operate within it'.

Following the next change of power in 1979, the incoming Conservative government quickly picked up the theme of planning delay and lost no time in preparing a revised development control policy. A draft circular created alarm in the planning profession, partly because of its substantive proposals, but also because of its abrasive style. 'The Most Savage Attack Yet', expostulated *Municipal Engineering*, while *Planner*

News remonstrated that the results of the circular 'could be disastrous'. The revised circular, as published (22/80), was written with a gentler touch, but much of the message was very similar. The emphasis was on securing a 'speeding up of the system' and ensuring that 'development is only prevented or restricted when this serves a clear planning purpose and the economic effects have been taken into account'. This might seem a little superfluous, as the Dobry Report had pointed out (para 1.32) that a role of planning is 'positive encouragement, and help for, development' (more recently in 2011, he noted in a letter to *The Times* that, in English planning law since 1946, 'the presumption is yes'). It was at this point that the infamous target eight-week period for deciding on planning applications was instigated, with regular publication of comparative performance figures. Quarterly figures have been published since 1979, and are used by both the government and the development industry to bolster criticisms of the system.

The policy 'to simplify the system and improve its efficiency' (to use the words of the 1985 White Paper *Lifting the Burden*) continued, with revised circulars, new White Papers, and the introduction of planning mechanisms which reduced or bypassed local government control, such as simplified planning zones and urban development corporations. However, towards the end of the 1980s, a greater emphasis on 'quality' emerged, as environmental awareness and concern increased. A change in direction was signalled by the 1992 Audit Commission report on development control, significantly entitled *Building in Quality*. Though the major emphasis was still on the process of planning control rather than its outcome, there was a very clear recognition of the importance of the latter. The report noted that there had been a preoccupation with the speed of processing planning applications, 'ignoring the mix of applications, the variety of development control functions, and the quality of outcomes' (para. 35). But there had been no 'shared and explicit' concept of quality and added value. What that 'added value' may be is dependent upon the authority's overall objectives: 'in areas under heavy development pressure or in rural areas, environmental, traffic, or ecological considerations may be paramount' (para. 75); in Wales,

'the impact of the development on the Welsh language can be a consideration' (para. 75).

The effect of *Building in Quality* was to redress the balance somewhat from the emphasis on lifting the burden of regulation, but the importance of meeting the eight-week target (and for major applications, the thirteen-week target) remained. There was a considerable overall improvement in performance in the first part of the 1990s: from 46 per cent of applications decided within eight weeks in 1989–90 to 65 per cent in 1993–4. Performance figures – by type of authority, by type of application – are regularly available on the government website, but performance has tended to fluctuate: 2013 rates of 5 per cent of major applications dealt with in thirteen weeks and 6 per cent of minor applications dealt with in eight weeks both represent falls compared with peaks in the mid-2000s.

A review of progress on *Building in Quality* in 1999 pointed to the value of increasing delegation of decision-making to officers for those authorities whose performance has improved. The reduced number of applications from the peak in 1988–9 (illustrated in Figure 5.2) and the increasing coverage of local development plan policy were also significant factors. However, at some points, improvements have been made in the face of a sharp increase in the number of applications, and there is still great variation in performance, all of which makes for difficulties in producing a generalised diagnostic. The varying conclusions of two reports published in 2002 gave a very good impression of the complexity of the problem. The 2002 Audit Commission Report on *Development Control and Planning* identified 'intractable barriers' to improvement:

These include resource limitations, competing priorities within local government and the inherited complexities of the planning system. But the slow pace of change is also symptomatic of a wider malaise: there has been a reluctance to accept the need for improvement in many cases. In the area of customer service, for example, there is evidence that planning has failed to keep pace with improvements in other council services.

(p. 7)

However, the second report paints a quite different picture of authorities struggling to meet rapidly increasing demands with declining resources. *Resourcing of Planning Authorities*, a report by Arup with the Bailey consultancy for the ODPM, notes 'the overwhelming finding is that resources have declined significantly over the past five years and performance has generally worsened, albeit in different functions in different authorities' (2002: 13).[41] In the light of their findings, and taking a 'highly conservative assumption', the researchers concluded that:

> to achieve the equivalent of 1996/97 levels of gross expenditure would therefore require increases on the 2001/02 levels of gross expenditure of 37 per cent for unitary authorities and district authorities and 23 per cent for county authorities . . . The increase would equate to between four and five additional staff, on average, per authority.
>
> (p. 16)

The report was prepared in time to feed into the Green Paper modernising planning process and was important in bringing forward the Planning Delivery Grant. The grant aimed to improve performance and resources for local planning authorities (and latterly other planning bodies). Perhaps counter-intuitively, the grant was targeted at authorities that performed well. In the first year of operation (2002–3) planning authorities received between £75,000 and £475,000, with a mean of £129,000, with the nine authorities that had always met government targets getting the most money and 152 authorities receiving the 'basic' £75,000. An evaluation of the Planning Delivery Grant (PDG) by Addison and Associates published by the ODPM in 2004 concluded that the PDG had provided a demonstrable incentive, considerably raising the profile of the planning function within local authorities and focusing attention on the effectiveness of the planning service. Money was spent on various items, with 46 per cent going on the retention and recruitment of staff, 22 per cent on ICT, whilst training was important, in particular to bring non-planning graduates up to speed in the system. However, in a policy paper the Conservative Party signalled that they would end the PDG if

returned to government and on 10 June 2010, the SoS announced that no further payments of grant would be made, as part of the programme of reduction of public expenditure and because 'we believe that the housing and planning delivery grant has proved to be an ineffective and excessively complex incentive' (Hansard, Written Statements, 10 June 2010, c. 16WS).

Throughout the 2000s, reviews of the applications process continued to be produced. For example *Key Lessons for Development Control* (DCLG 2006b: 14) noted that there was a need for 'improved processes and procedures, increased delegation, formulation of improvement strategies, investment in enhanced ICT and pro-active management of cases, especially major applications, both pre-application and during the application process'. The more extensive Killian Pretty Review of 2008 took a different approach to achieving similar objectives. Its seventeen recommendations can be grouped into five areas:

1. The application process is made more proportionate with more permitted development and streamlined processes for small-scale development and streamlined information requirements.
2. The process is improved particularly in relation to pre-application and post-decision stages.
3. Engagement is made more effective by improvements in the way stakeholders are involved in the process.
4. Changes in culture are encouraged by replacing time-based performance targets with a measure of customer satisfaction and by seeking ways to reward better quality applications.
5. Unnecessary complexity is removed by making the national policy and legislative framework clearer, simpler and more proportionate.

Clearly, some of the Coalition government's changes, such as the reduction of information requirements and changes to permitted development, echo these recommendations.

In parallel with these attempts to improve the performance of the applications process, local government was, for a time, faced with more general exhortations and requirements to improve its 'customer focus' across service areas, including planning. Two initiatives merit mention. The first was a component of the Citizen's Charter programme launched by John Major's government. One of the most important features of the Citizen's Charter was its stated aim of setting out people's entitlements to public services with charters for individual public services intended to outline the standards of service that people could expect to receive. An example is shown in Box 5.8. However, in the view of the House of Commons Public Administration Select Committee (2008: 2) 'the Citizen's Charter programme was rather confused – promises contained in the charters were often vague and aspirational, confounding the aim of defining a tangible set of entitlements to public services that people could readily understand and use'. Typical 'Charters' for planning would include standards for acknowledgement of receipt of applications and named officers to deal with matters. The second

programme was *Best Value*. Again this applied across local government and aimed to introduce a process of continuous improvement to services. For planning there were a number of targets to be met, some for cost and efficiency, some for outcomes (including time-based targets for dealing with applications) and some for 'quality' based on comparing performance with a checklist of good practice. The last set of Best Value Performance Indicators was issued in 2008, when, along with various other measures used by central government to manage performance of local government, they were replaced by a single set of indicators, the *National Indicator Set*. This itself was replaced by the incoming Coalition government in 2010 with the *Single Data List*, a reduced list of requirements. However, performance data for the processing of applications continues to be collected quarterly.

BOX 5.8 SERVICE CHARTER FOR DEVELOPMENT CONTROL, MID SUSSEX DISTRICT COUNCIL

The Development Control Department of Mid Sussex District Council published details of its Service Charter which sets out the standard of service delivery they aim to provide for their planning services, including the processing of planning obligations. Some of these commitments are to:

- reply to letters within 10 days or acknowledge them within 3 days explaining the reason for delay;
- respond to requests for pre-application advice by letter, on-site meeting, off-site meeting or telephone within 10 days;
- make available copies of planning applications for inspection in the office within 2 days of registration;
- have regard to all representations received on planning applications within the time periods available;
- notify interested parties of the planning decision if their representation is accompanied by a stamped addressed envelope;
- monitor a 10% sample of implemented permissions;
- return all telephone calls by the end of the following day; and
- make any further necessary responses to telephone calls within 10 days.

The council provide a summary leaflet in hard copy and via their website setting out these standards as well as making the full report available from the Environment Directorate on request. The leaflet also provides a clear demarcation map of the three district planning teams, setting out the wards that are included within them, and contact phone numbers for each team.

Source: DCLG 2006f: 28

The use of quantitative performance measures alone attracted criticism, with a frequent feeling being that it tended to 'crowd out' important qualitative aspects of performance. In some respects, this approach is more in tune with 'development control' than 'development management', but the final element to note is one which can be seen as being linked to the emergence of a development management approach. *Planning performance agreements* (PPAs) were introduced in 2001 in response to the challenges of delivering and achieving high-quality outcomes, for large, complex development proposals through the planning system. PPAs are a project management tool. There are two main reasons for using them – to improve the quality of development proposals and to improve the decision-making process through collaboration. PPAs bring together the LPA, the applicant and other stakeholders to work together in partnership through the different stages of the planning application process. They differ from other pre-application-focused approaches in that they are collaborative, not a unilateral statement of what the LPA will offer. An evaluation of PPAs by Tribal (2010) found that one of their key benefits is the removal of the application from the statutory determination deadlines (data is collected separately on PPAs), which eliminates excessive haste in handing what are often complex cases. However, many cases were in fact determined more quickly than would have been the case without a PPA being in place. The study concluded that, overall, 'PPAs are becoming an increasingly important part of good development management practice' (p. 11).

Performance of LPAs in dealing with planning applications is unlikely to slip out of the headlines, and 'bad news'/poor performance is likely to dominate coverage. So, for example, in mid-2013, we could read a headline 'Six councils face "special measures"', referring to councils whose performance in dealing with planning applications fell below a threshold whereby developers could submit major applications directly to the Planning Inspectorate.[42] However, if LPAs are given adequate resources, experience in the last twenty years suggests that there are steps that LPAs can take to improve performance, quantitatively and qualitatively. Maybe the collaborative approach

exemplified by PPAs and inherent in the development management approach can be a force for aiding improvement in coming years.

Further reading

Legal texts

The law and procedure of development control is explained fully in several textbooks. Valuable works include Moore and Purdue (2012) *A Practical Approach to Planning Law*, Duxbury (2012) *Telling & Duxbury's Planning Law and Procedure,* plus of course the regularly updated *Encyclopaedia of Planning Law and Practice*; for Northern Ireland: Dowling (1995) *Northern Ireland Planning Law*; for Scotland: Collar (2010) *Planning Law* and McAllister *et al.* (2013) *Scottish Planning Law*. Much material – both texts of legislation and some guidance – is available on government websites, and the Planning Portal (www.planningportal.gov.uk) describes itself, with some justification, as 'the first port of call for anyone wanting to find out about the planning system in England and Wales.' Also on this site is the developing *Planning Practice Guidance*, which has importance, as it can be a material consideration. It covers a range of topics – alphabetically from advertisements to water supply – and can be a useful starting point on a number of the topics in this chapter and listed below, but it has attracted mixed views – see Carmona (2013a) for example.

Use Classes and Development Orders

For an interesting introduction to the evolution of the Use Classes Order, see Home (1992). The Lichfield (2003) Study for the ODPM considers each part of the GPDO in detail. See also Edinburgh College of Art *et al.* (1997) *Research on the General Permitted Development Order and Related Mechanisms* and also BDP Planning and Leighton Berwin (1998) *The Use of Permitted Development Rights by Statutory Undertakers*. On Article 4 Directions, see Roger Tym and Partners (1995) *The Use of Article 4 Directions by Local Planning Authorities* and Larkham and Chapman (1996) 'Article 4 Directions and development control'. The standard legal text is Grant

(1996) *Permitted Development*. Halcrow Group's (2004) report, *Unification of Consent Regimes*, is very informative in placing the planning and listed building regimes alongside others.

The development plan as a consideration

The role of plans in appeal decisions is considered in Bingham (2001) 'Policy utilisation in planning control'. The impact of the introduction of section 54A (now section 38(6) of the 2004 Act and section 25 of the Scottish Act) is reviewed by Gatenby and Williams (1996) 'Interpreting planning law', and this early assessment is still very relevant. See also their earlier article (1992) 'Section 54A: the legal and practical implications'. Other sources include MacGregor and Ross (1995) 'Master or servant?', Purdue (1991) 'Green belts and the presumption in favour of development', Harrison, M. (1992) 'A presumption in favour of planning permission?' and Herbert-Young (1995) 'Reflections on section 54A and plan-led decision-making'.

Other material considerations

The ODPM statement *The Planning System: General Principles*, published alongside PPS 1, presented a succinct statement on material considerations and this has not been overwritten by the NPPF. See *Scottish Planning Policy* and PAN 40 for Scotland, and PPS 1 for Northern Ireland. A categorisation of considerations, drawing on work by Lyn Davies, is summarised in Thomas (1997) *Development Control: Principles and Practice*.

Design

There are two very good starting points for considering the role of design as a factor in planning: the DETR and CABE (2000) *By Design: Urban Design in the Planning System: Towards Better Practice* includes checklists of design considerations and a list of other references, and Carmona's two-part article in *Planning Practice and Research* (1998, 1999) 'Residential design policy and guidance: prevalence, hierarchy and currency'. On design and access statements, see the guidance produced by CABE and

Punter (2010) 'Planning and good design: indivisible or invisible? A century of design regulation in English town and country planning'. Further suggestions are given at the end of Chapter 9.

Amenity

Despite its significance there have been few studies of amenity in the planning process. For a discussion of statutory provisions, see Sheail (1992) 'The amenity clause'. For an unusual historical study of the development of the notion of amenity, see Millichap (1995a) 'Law, myth and community'. Amenity is an important element to residents – for a reflection on how this importance might be measured, see Gibbons *et al.* (2011).

Appeals

The Planning Inspectorate's annual statistical reports give a wealth of detail on numbers of appeals and outcomes. Circular 5/00 *Planning Appeals Procedures* provides details of the various elements of process. In Scotland, the latest position is given in the Town and Country Planning (Appeals) (Scotland) Regulations 2013 *(SI 2013/156)*. The former Chief Planning Inspector gave an account of 'Decision-making and the role of the Inspectorate' (Shepley 1999). An example of the analysis of appeals data is given by Wood *et al.* (1998) 'The character of countryside recreation and leisure appeals'.

Enforcement

Two standard legal texts are Millichap (1995b) *The Effective Enforcement of Planning Controls* and Bourne (1992) *Enforcement of Planning Control*, which are still relevant although they are superseded by recent changes. The DETR published Circular 10/97 *Enforcing Planning Control: A Good Practice Guide*. A useful complement is the publication by the Planning Officers Society *Practice Guidance Note 5: Towards Proactive Enforcement*; the POS have also published a short and useful overview for the Local Government Association (available via the POS website www.planningofficers.org.uk). The operation of the procedures in Scotland was investigated by Edinburgh College of Art *et al.* (1997), whilst the Scottish government publication Circular 10/09 *Planning Enforcement* provides a comprehensive review of procedures.

Advertisements

The regulations are detailed and explained in Circular 3/07, which is complemented by *Outdoor Advertisements and Signs: A Guide for Advertisers,* also published in 2007 by DETR. The Planning Inspectorate's Good Practice Advice Note 17 *Advertisement, appeals and related issues – England* is also a useful source of information. The position in Scotland is defined in the Town and Country Planning (Control of Advertisements) (Scotland) Amendment Regulations 2013 (SI 2013/154). The fullest exposition of the law of advertisement control is given in Mynors (1992) *Planning Control and the Display of Advertisements.*

Minerals

In England policy is set out in the NPPF and an accompanying supplementary note *Technical Guidance to NPPF* (which also deals with flood prevention). A detailed (two volume) evaluation of minerals policy and guidance was published by DETR in 2011. In Wales, policy is set out in *Minerals Planning Policy Wales* and in supporting advice notes (MTANs); in Scotland it is set out in the national planning framework and supported by a number of PANs. An excellent review of mineral resource planning and sustainability is given by Owens and Cowell (1996) *Rocks and Hard Places.*

Caravans and gypsies

The central text is the research by Niner (2002) at the Centre for Urban and Regional Studies, Birmingham University, *The Provision and Condition of Local Authority Gypsy and Traveller Sites in the English Countryside.* Another account is given by Morris (1998) 'Gypsies and the planning system'. See also Gentleman (1993) *Counting Travellers in Scotland.* Official policy is given in *Planning policy for traveller sites,* published in 2012. A valuable source of background material is available in *The Road Ahead: Final Report of the Independent Task Group on Site Provision and Enforcement for Gypsies and Travellers,*

published by DCLG in 2007. A good practice note produced by RTPI in 2007 on planning for gypsies and travellers is also useful.

Efficiency in planning application systems

The Audit Commission (1992) report, *Building in Quality: A Study of Development Control*, and the subsequent (1998) *Building in Quality: A Review of Progress on Development Control* and (2002) *Development Control and Planning* provide valuable information on the genesis of the current approach to managing and measuring performance. Tables of application statistics are available on the gov.uk website. The Killian Pretty Review (2008) *Planning Applications: A Faster and More Responsive System* provided conclusions supported by both the commissioning government and its successor. A review of Planning Performance Agreements, introduced in 2001, was commissioned from Tribal and published in 2010.

Notes

1   The reviews by Kate Barker are possibly the most prominent, with a significant feature being that what would have traditionally been seen as planning matters were the subject of reports commissioned jointly by the Treasury.

2   The provision is intended to enable controls over large increases in floorspace by using basements or building mezzanine floors in existing large retail stores. The Secretary of State issued a consultation document in March 2005 suggesting that the threshold should be 200 square metres.

3   Changes of use in sites of special scientific interest may also require approval from English Nature, although they are allowed by the UCO.

4   Until 1995, the General Development Order contained both permitted development rights and procedural matters (relating to planning applications). In 1995 these were separated (following the Scottish model introduced in 1992). There is therefore now a General Permitted Development Order and a General

Development Procedure Order. See Circular 9/95 *General Development Order Consolidation*. Though these new orders are predominantly consolidations, they contain a number of changes.

5 See, for example, Bell (1993) on problems arising from changes of use.

6 The Town and Country Planning (Use Classes) (Amendment) (England) Order 2005 (SI 2005/84).

7 Circular 9/95, para. 1. The Secretary of State approves Article 4 Directions except those that are specifically related to dwelling houses in conservation areas. PPG 15 *Historic Environment* explains the application of Article 4 Directions in conservation areas in England. The circular was amended in 2012: see *Replacement Appendix D to Department of the Environment Circular 9/95: General Development Consolidation Order 1995*, published by DCLG.

8 In 1998, the DETR published a good-practice guide, *The One-Stop Shop Approach to Planning Consents*. There is also increasing interest in comparisons of practice in the UK with other countries; on development control, see GMA Planning *et al.* (1993).

9 The changes in local authority management structures mean that many decisions are now made by a cabinet rather than committees, but in the case of planning and other regulatory activities local authorities must retain the committee decision-making procedure (although some decisions will be delegated). Some large authorities will divide up the committee into smaller local area committees.

10 The Planning Officers' Society has published a Practice Note on Reasons for the Grant of Planning Permission (see www.planningofficers.org.uk).

11 These and other development control figures are from the quarterly returns on planning application statistics and available on the www.gov.uk website.

12 Essex (1996) reviews these two cases and the general issue of relationships between officers and members in decision-making.

13 During the transition phase following the 2004 Act, the development plan in England meant the unitary development plan or the structure and local plan, including minerals and waste plans (and even old subject plans), depending on the area in question. As the new plan system was put into place, it comprised the regional spatial strategy and the development plan documents in the local development framework. However, RSS was abolished by the Coalition government in 2011 in the Localism Act 2011 (see Chapter 4 for the definition of development plans elsewhere in the UK).

14 Exactly the same provision was made in the Scottish legislation as section 25 of the 1997 Act (formerly section 18A of the 1972 Act). No such provision was made for Northern Ireland.

15 In his judgment in the case of *The City of Edinburgh v. the Secretary of State for Scotland*, Lord Hope said: 'it requires to be emphasised however, that the matter is nevertheless still one of judgment, and that this judgment is to be exercised by the decision taker. The development plan does not, even with the benefit of section 18A, have absolute authority. The planning authority . . . is at liberty to depart from the development plan if material considerations indicate otherwise' (Grant 1997: P54A.05/2).

16 In the oft-quoted *Stringer* case, it was stated that 'any consideration which relates to the use and development of land is capable of being a planning consideration' (*Stringer v. Minister of Housing and Local Government* 1971). Whether a particular consideration falling within that broad class in any given case is material will depend on the circumstances. In another important case (*Newbury*), the House of Lords formulated a threefold 'planning test': to be valid, a planning decision had to (i) have a planning purpose; (ii) relate to the permitted development; and (iii) be reasonable (*Newbury District Council v Secretary of State for the Environment* 1981).

17 The government's statement *The Planning System: General Principles* (2005), like its predecessors, notes that the government's statements of planning policy 'cannot make irrelevant any matter which is a material consideration in a particular case. But, where such statements indicate the weight that should be given to relevant considerations, decision-makers must have proper regard to them. If they elect not to follow relevant statements of the Government's planning policy they must give clear and convincing reasons (*EC Grandsen and Co Ltd v. SSE and Gillingham BC* 1985). Emerging planning policies, in the form

of draft departmental circulars and policy guidance, can be regarded as material considerations, depending on the context. Their very existence may indicate that a relevant policy is under review and the circumstances which have led to that review may need to be taken into account' (paras 13 and 14).

18 Circular 11/95 did not deal with conditions in respect of minerals or waste, which were dealt with in the minerals planning guidance notes and PPG 23 *Planning and Pollution Control*. However, this was replaced by the NPPF, which simply states (para. 110): 'In preparing plans to meet development needs, the aim should be to minimise pollution and other adverse effects on the local and natural environment.'

19 In Scotland, guidance is given in Circular 4/98, and in Wales, it is Welsh Office Circular 35/95, both entitled *The Use Conditions in Planning Permissions*.

20 But the trench digger may be brought up against a further provision: the serving of a *completion notice*. Such a notice states that the planning permission lapses after the expiry of a specified period (of not less than one year). Any work carried out after then becomes liable to enforcement procedures.

21 The statutory provisions relating to agreements were amended by the Planning and Compensation Act 1991. Agreements were replaced by 'obligations' and can now be unilateral – not involving any 'agreement' between a local authority and a developer at all. This provision allows a developer to make an agreement to provide the necessary off-site works even if the local authority is not prepared to be a party to the agreement (a *unilateral undertaking*).

22 These are: the Community Infrastructure Levy Regulations 2010 (SI 2010/948); the Community Infrastructure Levy (Amendment) Regulations 2011 (SI 2011/987); the Local Authorities (Contracting Out of Community Infrastructure Levy Functions) Order 2011 (SI 2011/2918); the Community Infrastructure Levy (Amendment) Regulations 2012 (SI 2012/2975); and the Community Infrastructure Levy (Amendment) Regulations 2013 (SI 2013/982).

23 Affordable housing obligations on sites granted permission in accordance with a Rural Exceptions Site policy are exempt from the review procedure introduced by the Growth and Infrastructure Act 2013.

24 In England appeals are made to the Deputy Prime Minister and in Wales to the Welsh Assembly. In both cases the Planning Inspectorate Executive Agency considers and makes decisions on most (the nature of the Agency is explained in Chapter 3). In Scotland, the Inquiry Reporters Unit considers appeals representing the Scottish Minister for Planning. In Northern Ireland, the Planning Appeals Commission has the same role as the Planning Inspectorate.

25 Cases may be recovered by the Secretary of State where they involve substantial development (over 150 houses or retail development over 100,000 square feet), significant proposed development in the green belt, major mineral planning appeals, where other government departments have an interest or where there is major controversy over the development.

26 Circular 5/00 explains the procedures and gives references to the inquiry, hearing and written representation rules. Procedures in Scotland are governed by *the Town and Country Planning (Appeals) (Scotland) Regulations 2013 (SI 2013/156)*. Procedures in Wales bear significant similarities to those in England but are governed by a number of regulations issued in 2003, with some updating after consultation in 2011, such as the introduction of the householder service.

27 The Inspectorate agreed and published *Better Presentation of Evidence in Chief* with the Local Government Planning and Environment Bar Association (2000).

28 The difficulties of the interpretation of aggregate appeals data (since each decision is made on its merits) have been a subject of continuing debate: for example, see Brotherton (1993). In a different context, the coverage of the appeal in the Court of Appeal by Cala Homes against SoS Pickles' revocation of Regional Spatial Strategies in 2010 left a period of uncertainty over the materiality of RSS for a number of months (see the government statement at www.gov.uk/government/news/government-wins-second-legal-challenge-on-regional-strategies (accessed 30 January 2014)).

29 Until September 1999, this role was undertaken by the Royal Fine Arts Commission, which requested intervention by the Secretary of State on numerous occasions, but not always successfully.

30 The circular is best read in conjunction with PPG 18 and the DETR *Enforcing Planning Control: Good Practice Guide for Local Planning Authorities* (1992). In Scotland, Circular 10/09 *Planning Enforcement* brings together current guidance.

31 *Planning* on 25 April 2014 reported on a number of cases where planning authorities had made use of the Proceeds of Crime Act 2002 to secure larger penalties, with the largest fine being £1.4 million imposed by the court on a landlord who converted a house to twelve flats without planning permission. Such fines represent the increase in value gained by the illegal action.

32 *Independent*, 9 December 1995, p. 9.

33 A number of MPGs were left in place after the general 'cull' of detailed guidance by the NPPF. These are: MPG4 *Revocation, Modification, Discontinuance, Prohibition and Suspension Orders*, which relates to the review and modification of minerals permissions; MPG8 *Planning and Compensation Act 1991 – Interim Development Order Permissions (IDOS): Statutory Provisions and Procedures*, concerned with mineral permissions granted under IDOs on or after 22 July 1943 in respect of development which had not been carried out before 1 July 1948; MPG9 *Planning and Compensation Act 1991 – Interim Development Order Permissions (IDOS): Conditions*, giving advice on the considerations to be taken into account by applicants and minerals planning authorities in preparing and determining the conditions to which registered IDO permissions should be subject; and MPG14 *Environment Act 1995: Review of Mineral Planning Permissions*, giving advice to mineral planning authorities and the minerals industry on the statutory procedures to be followed and the approach to be adopted in the preparation and consideration of updated planning conditions in the review process.

34 The DETR funded a series of research projects on the environmental impacts of minerals exploitation that inform national policy on development control. The most recent reports are Arup Environmental and Ove Arup and Partners (1995) *The Environmental Effects of Dust from Surface Mineral Workings*, Vibrock Ltd (1998) *The Environmental Effects of Production Blasting from Surface Mineral Workings*, ENTEC UK Ltd (1998) *The Environmental Effects of Traffic Associated with Mineral Workings*, and University of Newcastle upon Tyne (1999) *Do Particulates from Opencast Coal Mining Impair Children's Respiratory Health?*

35 An Annex to the ODPM's consultation paper on *New Parliamentary Procedures for Processing Major Infrastructure Projects* gave a useful summary of the ways in which major infrastructure projects may be approved. The examples here are drawn from the paper.

36 This legislation has remained as a separate code and is not consolidated in the Town and Country Planning Act 1990. The Caravan Sites Act 1968, which required local authorities to provide caravan sites for travellers, if there was a demonstrated need, and dealt with the protection from eviction of caravan dwellers and gypsies, was similarly separate. The Criminal Justice and Public Order Act 1994 removed the duty of local councils to provide authorised pitches and gave the council and police powers to move travellers on, subject to certain welfare issues.

37 Niner (2004) has given an upper estimate of 250,000.

38 Research by Dr Jo Richardson and Ros Lishman of De Montfort University for Lord Avebury (*Impact of Circular 01/2006: Supply of New Gypsy/Traveller Sites*, 29 March 2007) reviewed a total of 129 appeal decisions, seventy-five being before 1 February 2006 (the implementation date of Circular 1/06) and fifty-four being after that date. Between the two periods, the number of allowed appeals increased by 20 per cent and the number of dismissed appeals decreased by 20 per cent. CLG's own evidence indicates that in the year ending December 2009, local authorities determined 217 applications for gypsy and traveller pitches, 50 per cent of which were granted. This is a figure that is unprecedented in terms of the period prior to the introduction of Circular 1/06.

39 Under the General Permitted Development Order 1995 (SI 1995/418), Schedule 2, Part 17, telecommunications operators can erect masts up to 15 metres in height under permitted development rights. Railway companies also have permitted

development rights under Schedule 2, Part 11 of the 1995 Order. These rights, unlike those for telecommunications companies, have no height limit. Part 11 permitted development rights can also be used by Network Rail, because railways were originally constructed under Local or Private Acts of Parliament. *Open Source Planning* stated that: 'We believe that all types of mobile phone masts in England (including Network Rail, TETRA and small/pico masts) must be *subject to the same, full planning process* as other forms of development, so giving local communities a greater say on where they are located' (p. 18) but legislation has yet to emerge to put this into effect.

40 Permitted development applies to ground-based masts and those installed on buildings or other structures, and a public call box. Some masts or antennae may be so small that they do not constitute development – for example television aerials have been treated as outside the definition of development (despite their sometimes significant impact on the external appearance of buildings). The exceptions from permitted development for masts under 15 metres include proposed masts on listed buildings, scheduled ancient monuments and where the planning authority has made an Article 4 Direction withdrawing permitted development rights.

41 The study found that the typical cost of dealing with a householder application is about £200, whereas the fee at that time was £95 (so even if the system operates efficiently, it does so making a loss) and that the ideal ratio is about one member of staff 'for every 150 to 200 applications, plus support services'.

42 *Planning*, 12 July 2013. The article referred to six planning authorities which had determined less than 30 per cent of major applications within thirteen weeks over a two year period.

6 Developing planning policies

Planning is about getting the right stuff in the right place at the right time.
(Steve Quartermain, DCLG Chief Planner, at RTPI Conference, 27 February 2014)

It is unrealistic and unreasonable to expect even more homes to be built if it means concreting over large parts of our countryside or failing to put the infrastructure in place to support new developments.
(Andrew Povey, Leader, Surrey County Council, quoted in *Planning*, 28 August 2009)

I am struck by how little people appreciate the severity of the housing supply problem. People do not act in spite of the evidence. Is it ignorance or denial? It is a big problem that is getting worse.
(Neil McDonald, Chief Executive, National Housing and Planning Advice Unit, quoted in *Planning*, 28 August 2009)

Introduction

The quote from Steve Quartermain encapsulates some key challenges for planning policymakers. If we are to have successful communities, decisions need to be made about what is 'the right stuff' – how many houses, how much shopping, new sources of employment, key services – where it should go, and how can it be put in place when it is needed. When we have new residential development, will the jobs and services be there to meet residents' needs without the need to travel significant distances? This chapter looks at some of the factors involved in making decisions about policy to influence such matters, and at the policy guidance that has helped to shape such decisions. It does this by selecting policy areas which are the major users of land identified in plans to accommodate new development.

For some time, the purpose of planning has been seen as promoting sustainable development and the most recent government policy guidance – the National Planning Policy Framework (NPPF) – has underlined the importance of planning in helping to achieve this objective. In doing this, it identifies and describes three roles for planning: an economic role, which involves ensuring that sufficient land of the right type is available in the right places and at the right time to support growth and innovation; a social role, which focuses on facilitating the provision of housing required to meet the needs of present and future generations, along with the necessary supporting services; and an environmental role, contributing to protecting and enhancing our natural, built and historic environment. The third area of policy is covered in a number of other chapters of this book, but in this chapter there is an emphasis on social and economic matters. Chapter 4 has reviewed the *processes* involved in preparing a development plan and developing planning policies, but this chapter looks at some substantive issues in determining policy on housing, employment and other related policy areas, which provide the main sources of demand for the identification of new sites, as well as reviewing the evolution of policy in these areas. Through a review of some of the mechanics of policy development, of the changing emphasis of policy – often shaped at the interface between politics and planning – and the areas of uncertainty about policy development, it is hoped to give a better understanding of the *task* of making policy within the context provided by the nature of planning in the UK, where how the task is approached is itself the subject of government guidance of varying levels of detail.

Observation of, or participating in, the process of preparing a development plan will quickly reveal that questions of how much development – particularly housing – and where it should go are amongst the most contentious issues: this is illustrated by the second and third quotations at the beginning of this chapter. Generally, policies for *where* development should go have emphasised concentration in some form. At one time, the idea of focusing new development in larger, established settlements was favoured on the basis that it not only helped in resisting sprawl and protecting the countryside but that it was also a more efficient foundation for the delivery of services and therefore in the use of public funds (Cloke 1979; Gilder 1984; Moseley 2003). More recently, concentrating new development in larger communities has been seen as a way of helping promote sustainable development, as it is conceived as facilitating the co-location of activities and therefore reducing the need to travel.[1] However, such a policy approach has not always been popular. Those communities which have been earmarked for new development could see it as a threat to the character of the settlement, as adding to the problems of already overburdened services or as delivering some other unacceptable impact. On the other hand, those communities which have been identified as areas of restraint or where no new development should take place can feel that this renders them unable to adapt to change, and the policy is, in effect, a recipe for long-term decline. This is perhaps most dramatically exemplified by Durham's 'D' villages – D for destruction – policy of the 1950s and 1960s.[2]

Debates over *how much* development can perhaps be seen as more technical in nature, with disputes over both the methods used to construct forecasts of future need, the meaning and reliability of base data, and the ability to accommodate satisfactorily the levels of development foreseen often occupying much time at inquiries into development plans. Whilst much of the debate can be focused on resisting additional development, there are always voices in favour. At the local level, these might be focused on the need for more affordable housing, more employment opportunities or a greater choice of shops, whilst at the national level economic arguments about a need for higher rates of development can be more prominent. The latter type of argument was given greater profile by the work of Kate Barker, who in her 2004 *Review of Housing Supply* argued that increased land allocations for housing were an important means of restraining housing price growth, and these types of arguments continue to feature prominently in government exhortations for increased levels of house building. However, planning is very much concerned with the future, so it is inevitable that the somewhat specialised activity of forecasting is part of the process of developing planning policy and therefore a focus for argument. Forecasts are part of what has always been a key component of the plan-making process but have recently possibly achieved a higher profile, given the importance attached to 'evidence' as part of the test of soundness of planning strategies, as discussed in Chapter 4. But planning is a normative as well as a positivist activity: it is concerned with how we might want things to be as well as needing an understanding of where trends might be leading us, and where and how we strike a balance between the two can be at the heart of many debates about how much development we should be planning for.

Whilst this book is not the place to discuss the technical intricacies of forecasting and the reliability of various data sources, it is nonetheless important to gain some insight into the range of factors which enter into consideration when developing policy, as a basis for understanding the nature of those important questions of how much development and where it should go, and the processes through which planning seeks to address these questions. As an aid to achieving this objective, this chapter considers how planning has sought to decide how much housing is needed, what provision should be made for employment activities and what provision should be made for new retail development, along with tracking a number of key changes in the emphasis of policy in these areas. First, it briefly considers the question of the importance of evidence in planning before going on to reviewing how evidence is used, alongside other considerations, in developing policy for these three themes.

Evidence in planning

Faludi and Waterhout (2006a) have suggested that the Domesday Book could be seen as perhaps the first attempt in Britain to collect evidence systematically as a basis for developing policy, but in the planning context Geddes's 'survey before plan' paradigm might seem more relevant. Whilst Geddes's focus was on physical aspects of development, rather than the wider social, economic and environmental palette which we are concerned with today, reference to it does emphasise that developing planning policy has for a long time involved the use of evidence. Not surprisingly, for a considerable period disputes over the meaning of evidence and the validity of the policy conclusions drawn from it have been a significant component of debates over new planning policies, at planning inquiries and in other less formal arenas.

There is some agreement (Faludi and Waterhout 2006a; Davoudi 2006a) that an interest in evidence-based policies and planning developed in the UK and was propelled forward by New Labour's 'emphasis on a pragmatic rather than ideological stance' (Davoudi 2006a: 14) on policy development and by its general 'modernisation' agenda for government (Lord and Hincks 2010), perhaps summed up by a statement from the Cabinet Office that 'good quality policy making depends on high quality information, derived from a variety of sources' (1999: 33).[3] So it should not have been unexpected that changes to planning following the 2001 Green Paper resulted in an enhanced emphasis on the role of evidence in developing policies in the newly introduced local development

frameworks. This can be seen reflected in numerous items of guidance from DCLG and the Inspectorate. For example, PPS 12 stated:

> To be 'sound' a DPD should be JUSTIFIED, EFFECTIVE and consistent with NATIONAL POLICY.
> 'Justified' means that the document must be:
>
> * founded on a robust and credible evidence base
>   . . .
>
> 'Effective' means that the document must be:
>
> * deliverable
> * flexible
> * able to be monitored
>   (p. 24, emphasis in original)

The latter point emphasises that collection of evidence is not a one-off activity, but one which sets the parameters by which the plan will be monitored, and therefore evidence collection and interpretation are more akin to a continuous activity than an occasional burst of effort. This perhaps aligns approaches in planning with the 'new public sector management' which is associated with the New Labour 'modernising' agenda for public services. The NPPF makes many mentions of 'evidence' in its fifty or so pages: indeed, it offers a section of twenty paragraphs on the use of a 'proportionate' evidence base. However, neither the NPPF nor the practice guidance attempts to pin down the meaning of 'proportionate', although, as will be seen later in this chapter, it does offer guidance on what local planning authorities (LPAs) might do to collect evidence for various policy areas. But, the nature of the guidance offered by the NPPF is not wildly different from that in PPS 12 quoted above: a sound plan will still be justified, effective and consistent with national policy, but it also now needs to be 'positively prepared'.

However, as has been noted earlier, there is more to policy than evidence, so what part evidence plays in policy development and how it is used are key questions. A 'rational' approach, where evidence directly drives policy, is generally felt to be rare (Weiss 1977), though evidence might well be used more often instrumentally as ammunition to support a position. In a planning context, evidence, in the sense of statistical data, is likely to be used interactively with other inputs – the Cabinet Office (1999) has identified numerous sources such as consultation inputs, evaluation evidence on previous approaches and 'experience' of both professionals and communities affected by policy. Evidence might be seen as being used indirectly in what is sometimes termed the 'enlightenment' model, where it provides contextual knowledge and influences conceptual thinking. Davoudi (2006a: 18) terms this 'evidence-informed' policymaking, while Davies et al. (2000: 11) use the terms 'evidence-influenced' and 'evidence-aware', regarding them as 'the best that can be hoped for'. So, evidence-based planning and policy development is an activity which generally – or perhaps universally – is not one where evidence collection and analysis will deliver a policy position, but one where evidence has to be interpreted and balanced with other inputs, such as community wishes, political imperatives and professional judgements. For planning, there are clearly risks in giving too much prominence to evidence in policy processes, with a major one being that it will induce planning to cross the line from being a technical to becoming a technocratic activity. As Davoudi (2006a: 15) puts it, the expert helping with evidence moves from being 'on tap' to 'on top', which is contrary to the collaborative and inclusive mission that planning espouses.

So, whilst evidence has assumed a more prominent role in planning, this has not eased the process of policymaking. Aside from the almost philosophical questions alluded to above, there are significant practical difficulties. Research by Lord and Hincks into the use of evidence in plan making following the 2004 Act (2010: 486) found that 'the use of evidence has been a feature of the reformed planning system to which planning authorities had found it difficult to adapt'. These difficulties ranged from barriers in understanding the 'new model' of planning to a lack of skills and the capacity to carry out the tasks necessary. Beyond such practical difficulties, questions of how evidence is interpreted can be a challenge, as

will be illustrated in some of the policy-specific material that follows and as described somewhat colourfully by Faludi and Waterhout (2006b: 71):

> The mud pools of information that are usually available allow one to adduce evidence to support various, and maybe even opposite, policy directions. Policy-makers can pick and choose, trying to prove their point.

Clearly this question of multiple interpretations is an issue for those developing plans and policies but also for an interested public. Such multiple interpretations are often a contributory factor in fuelling debate over changes to policy at planning inquiries.

PLANNING FOR HOUSING

In the wake of 1995 government forecasts that there would need to be planned provision to accommodate an extra 4.4 million households between 1991 and 2016, the Town and Country Planning Association produced a report in 1996 entitled *The People: Where Will They Go?* In the introductory chapter, the editors, Breheny and Hall stated:

> There can be no doubt about the major planning issue for 1996 and beyond: it is how much housing we are going to need in the UK over the next twenty years or so, and where are we going to put it. These are amongst the oldest and most basic questions for planners. Ebenezer Howard posed them in 1898, at the centre of his Three Magnets diagram: *The People – Where Will They Go?*
>
> (p. 4)

When the first edition of this book was published the 1961 census results had just become available and the population of Great Britain was shown to be just over 51 million; the results of the 2011 Census showed the population to be over 61 million. However, this increase of about 10 million in population was almost matched by the increase in the number of households, from 16.3 million to 25.5 million. Part of the reduction in average household size (from 3.1 to 2.4 persons) can be attributed to a reduction in the proportion of families with children and to fewer children in those families with children, but another significant factor was the rise numbers of single-person households, up from 12 per cent in 1961 to 29 per cent in 2011 (Macrory 2012; *Social Trends 41*). The forecast growth of 4.4 million households between 1991 and 2016 was largely predicated on a continuation in the rise in the numbers of single-person households.[4] The scale and nature of these trends give an idea of the sorts of challenges faced by planning in meeting housing needs – not only to find sites for growing numbers of houses needed but also to try to match the nature of the stock to the changing needs of the population.

Not surprisingly, an objective of successive governments has been to boost housing supply. At the time of the election of the Coalition government, the National Housing and Planning Advice Unit suggested that over 290,000 houses per year needed to be built for the following thirty years to meet housing needs. This is a little below the total that was achieved in the period of post-war reconstruction but around twice the number achieved recently. PPS 3 *Housing*, the fourth edition of which was published in June 2011, a year after the election of the Coalition government, was developed in response to the recommendations of the *Barker Review of Housing Supply* and included amongst its objectives the achievement of 'a wide choice of high quality homes, both affordable and market housing, to address the requirements of the community' and improving 'affordability across the housing market, including by increasing the supply of housing' (p. 6). The NPPF seeks to deliver 'a wide choice of high quality homes' and to 'boost significantly the supply of housing' (p. 12). The mechanisms through which this might be achieved include having a full understanding of housing need and demand, and of housing land supply. Housing need can be thought of as households which cannot meet their needs in the housing market without support, and it underpins the development of policy on affordable housing, whilst housing demand is the product of the interaction of a range of economic and demographic factors. Whilst the provision for new housing in a development plan will mainly be through the identification of suitable sites, there is also a role for the improvement of previously unfit housing and for the conversion of otherwise redundant industrial or commercial property in providing for housing demand. Recent changes to the Use Classes Order have made the conversion of offices to housing somewhat easier.[5]

Under the previous Labour administration, targets for the scale and location of new housing development were set at regional level, whilst details of site selection were dealt with at the local authority level. The Coalition government abolished regional housing targets, but local planning authorities still have to ensure that their local plans identify sufficient land to satisfy housing demand. The NPPF requires that assessment of future housing requirements in local plans should have regard to current and future demographic trends and profiles, and take into account evidence, including the government's latest published household projections. The household projections therefore provide an important part of the evidence base for the assessment of future requirements for housing. The NPPF does not state that local planning authorities must use these government projections as the basis for their policy, but the experience of examination of core strategies by inspectors suggests that they are a good starting point (PAS 2013b).

These population projections make clear at several points that they represent an extrapolation of a set of observable trends and that the value of these projections in developing plans and policies for the future is limited by the extent that assumptions made about past relationships continue to be applicable in the future. As has been stated earlier, planning always contains a substantial normative element and has never been solely about constructing policies to fulfil projections. However, projections of past trends do represent a test of sorts for normative policy prescriptions, in that they require planners to answer questions about how and why the future may differ from the circumstances in the past which have produced the trends on which projections are based, and which they are seeking to amend with their policies.

Demographic forecasts are, in concept, relatively simple. Starting with figures for the resident population, birth and death rates – ratios which have been relatively stable and predictable over time – are applied, and then assumptions about the impact of migration into and out of the area are added. These latter items are both more difficult to predict and can raise some controversial issues. In many regions of the UK, there are both large inflows and outflows of population from other regions of the country, making the prediction of the net figure in which the planning authority would be interested that much more problematic, whilst more recently in particular parts of the country there has been a significant level of net in-migration from other countries. There is undoubtedly a relationship between the economic prospects of a locality and levels of migration, but this is difficult to quantify; in the case of international migration there are also unpredictable political variables. Locally, controversy may arise over the relationship between in-migration and house prices and environmental factors. So, in an attractive rural area, there may be significant numbers of people wishing to move in for retirement or commuting purposes, but catering for such a demand may be seen as posing unacceptable environmental threats, whilst restricting supply would be likely to result in excess demand and upward pressure on house prices, which would create an affordability problem for 'locals'. In the case of migration, the absence of regional forecasts gives relevance to the 'duty to cooperate', as the rate of provision for migration in one authority's forecasts can have an impact on other sending or receiving authorities' need for housing (McDonald 2013).

But, the local planning authority will be more interested in changes in numbers of households than changes in numbers of people. The significance of this distinction, as has been noted earlier, is demonstrated by the fact that the need to accommodate an additional 4.4 million households identified in the 1995 forecasts was driven more by changes in household characteristics than by simple population numbers. Figures for changes in numbers of households are shown in Table 6.1. Moving from demographic to household forecasts introduces a number of other assumptions, and there is also a need to consider other sorts of information related to housing stock and usage. As PAS (2013b: 4) has pointed out, 'It is important to understand the sensitivity of assumptions to minor adjustments. This is particularly the case in relation to household size assumptions where small adjustments can have big consequences for the number of houses to be provided.'

Table 6.1 Illustrative figures for changes in numbers of households

Household estimates and projections for England in 1991, 2001 and 2011 (thousands)

	1991 (Census)	2001 (Census)	2011 (Projected)	2011 (Census)
Couples, no other adults	8,853	9,151	9,597	9,465
Couples, one or more other adults	2,779	2,290	1,925	2,508
(All couples)	(11,631)	(11,441)	(11,504)	(11,973)
Lone-parent households	982	1,438	1,811	1,712
Other multi-person households	1,499	1,341	1,301	1,632
One-person households	5,052	6,304	7,773	6,785
All households	19,164	20,523	22,389	22,102

Source: Holmans 2013

Minor apparent discrepancies in totals are due to independent rounding.

This focus on households – and therefore houses and land take – serves to highlight a range of factors that have become increasingly important in planning for housing in recent years. The first could be summarised as a growing emphasis on the links between economic trends, housing demand and the housing market. Under New Labour, this eventually took the form of seeking a closer relationship between regional economic and planning strategies (RES and RPG/RSS), but prior to this there had been an interest in how, for example, levels of migration or household formation might be linked to local economic prospects. More recently, this can be seen to be reflected in the development of *strategic housing market assessments* (SHMAs), which remain a key component of the process of developing core strategies; it has also perhaps help embed the issue of housing affordability into the planning agenda. The work of Kate Barker, in her studies of planning for housing and planning more generally, placed a new emphasis on the economic impact of planning through, amongst other matters, its impact on the housing land market. This in turn has re-emphasised the importance of plans being based on a clear understanding of land availability through a *strategic housing land availability assessment* (SHLAA). The second group of factors, whilst not divorced from economic considerations, is perhaps more in the realm of the physical and includes matters such as vacant houses – what can be done to bring unused houses back into use – demolitions and clearances of unfit houses – this will include consideration of the impact of policies to improve housing conditions locally – and what is sometimes termed 'sharing and concealed households' or what might have been captured in the past by the term 'overcrowding'. In summary, success in bringing vacant dwellings back into use will reduce the need to provide new houses, whilst initiatives at clearing unfit houses or to tackle overcrowding will increase the need for new houses. In an effort to map some of the significant factors in making provision for housing in development plans, these factors will be reviewed below, along with other factors important in making a choice of sites, such as the emphasis on developing 'brownfield' sites first, and the quantity of land needed, such as the sorts of densities at which housing should be developed.

Links between economic trends and household formation

It seems logical to assume that there is some form of relationship between population and household growth (decline) in a region and its economic fortunes. If a region is developing, it would seem likely to attract in-migrants, and the prosperity associated with growth may encourage local individuals or couples to set up home or move up the property ladder. Conversely, a decline in economic fortunes seems likely to encourage some people to look to other regions for employment – and consequently a home – and to discourage others remaining locally from setting up home or moving home. However, whilst empirical research has identified relationships between an economic indicator like unemployment and, say, migration flows (Rees *et al.* 1996), attempts to move from correlations to explanatory models have proved difficult, and no explanatory/predictive model of migration and economically driven changes in household formation has yet been constructed which has wide and uncritical acceptance. Consequently, whilst logic dictates that planning cannot ignore economic factors, deciding exactly how it takes them into account is problematic.

An illustration of the sort of questions that might arise can be provided by the Yorkshire and Humber Regional Planning Guidance (RPG). Like most regions, it is a region of contrasting economic fortunes: more rural North Yorkshire is generally seen as prosperous, whilst the former industrial areas are less so. North Yorkshire had long attracted in-migrants, but South Yorkshire had suffered out-migration and out-commuting for work. In recognition of its adverse economic circumstances, South Yorkshire had a number of programmes in place to improve and develop its economy. The RPG sought to reduce in-migration into North Yorkshire and reduce out-migration from South Yorkshire. This would involve providing fewer housing allocations than trends would suggest in North Yorkshire, but a figure above trend in South Yorkshire. However, this raises the question of how this will be achieved, given that there are no means to actually *control* people's decisions about where to live. If trends continued, the level of housing provided in North Yorkshire would be below market demand and prices would rise, which would pose a local problem of affordability. In South Yorkshire, if the economic regeneration programmes were not successful in providing levels of local employment to discourage out-migration and out-commuting, then there would be an oversupply of houses, which would result in prices falling or, more likely, building levels not reaching that being provided for. Clearly, in both these examples, over or under provision of housing land would result in outcomes that planning policy would seek to avoid, but raising the questions serves to emphasise that planning has to pay some regard to market factors when developing policy, even if it is not easy to do so on a basis which offers the prospect of a 'right' answer.

The above discussion has been framed by considerations of migration, but it is also important to be aware of how economic factors might influence the propensity of people already living in an area to set up home (or move house) and therefore create a demand for housing. While the house purchase market cannot be taken as a direct indicator of rates of household formation, constraints on the supply of social housing, among other factors, means that there is some relationship between the two variables. In the past, about four out of ten newly formed households have been owner-occupiers; more recent figures see about seven out of ten additional homes as being in the market sector (comprising owner-occupation or private renting without benefits) (Holmans 2013). The scale of the expected future need is illustrated in Table 6.2.

There are well-established links between economic variables and the house purchase market, particularly money market (interest rates and mortgage availability) and labour market (unemployment and job availability) factors. Research for the Council of Mortgage Lenders (Ford and Seavers 2000) emphasised the importance of consumer confidence in influencing a decision to purchase, with negative influences coming from personal household financial factors. Contributory factors to lack of financial confidence include perceived job insecurity and lower levels of income. So, in estimating how many new households will need to be

Table 6.2 Illustration of market/social housing split

Estimated division between market and social sector households in 2011, 2021 and 2031

		Couple households	Lone parent households	Other multi-person households	One person households	Total	Percentage of total
		thousands					%
2011	Market sector	10,102	605	1,158	4,576	16,443	74.4
	Social sector	1,870	1,107	474	2,208	6,659	25.6
	Total	11,973	1,712	1,632	6,785	22,102	100.0
2021	Market sector	10,887	762	1,399	5,009	18,057	74.2
	Social sector	1,972	1,354	599	2,391	6,276	25.8
	Total	12,859	2,116	1,958	7,400	24,332	100.0
2031	Market sector	11,638	8,125	1,526	5,754	19,742	74.2
	Social sector	2,036	1,426	625	2,764	6,851	25.8
	Total	13,674	2,251	2,150	8,518	26,593	100.0

Source: Holmans 2013

Minor apparent discrepancies in totals are due to independent rounding.

provided for by a local plan, a local planning authority will also in some way need to take account of these 'confidence' factors, by paying attention to local and wider considerations of job prospects and economic prosperity. The importance of such factors could be seen as part of the imperative to develop strategic housing market assessments, which are discussed later.

The Barker Reviews

Whilst planning may, for some time, have had an interest in the relationship between economic factors and local housing provision at the local level, at the national level there was also a concern about the interaction between planning, housing supply and national competitiveness. The *Barker Review of Housing Supply* (2004) was a wide-ranging study commissioned by the government because of concern about the long-term upward trend in real house prices in the UK and its effect on the wider economy. Two points should be made at the outset: first, the review was commissioned by the Treasury in association with the ODPM, and not solely by the department responsible for planning and housing; and second, the report had a substantial impact on policy and action, promoting fresh thinking about the relationship between planning and the economic health of the country. Along with Kate Barker's subsequent report, *Barker Review of Land Use Planning* (2006), it has been seen by some as pushing planning towards being driven by the market, which may not be a surprise given that its principal sponsor was HM Treasury. However, for others it concentrates on the economics rather than the more difficult politics of housing supply. Nevertheless, it remains something of a landmark in the development of approaches to addressing the question of planning and housing supply.

The argument presented by the Housing Supply Report was that high house price inflation (2.4 per cent in the UK compared with a European average of 1.1 per cent) was creating problems of affordability and, because of the volatility of the housing market, had exacerbated problems of macroeconomic instability

and consequently had had an adverse effect on economic growth. This was not seen as a current 'crisis' but a long-standing trend which explained why consumers see housing as an investment and hold expectations that price inflation will continue. New house building was significantly lower than the rate of new household formation – by about 30 per cent below what was required at the time of the report – and below targets set out in planning policy (RPG/RSS). High demand (in the housing boom years) did not seem to affect the rate of completions; and there was an increasing rate of refusal of large housing applications. To put it simply, planning was not providing for levels of housing need, an argument which continues to have significant resonance ten years after the report was published. On the economy, the report argued that higher rates of house building would:

* help to reduce volatility in house prices, thereby improving macroeconomic stability and supporting growth;
* improve flexibility and performance of the UK economy via greater labour mobility;
* bring greater access to housing for many households, avoiding unwelcome distributional effect, and the ill-effects of poor housing.

The government recognised that doing nothing was not an option. Barker's views of what could be done ranged widely over the role of the Housing Corporation, the failure of the development industry to innovate, and the impacts of taxation. Deficiencies in the administrative machinery for assessing housing needs and influencing the market were noted, particularly at the regional level, where three bodies dealt with aspects of housing more or less independently: *regional bodies* (assemblies) and *regional spatial strategies* made broad allocations of housing land, *regional housing boards* advised on the demand for, and funding of, social and other 'sub-market' (non-market) housing, through the *regional housing strategy*, and the *regional development agencies' economic strategies* supported regeneration, which is closely linked to housing demand and supply. It hardly needed to be said that 'they often use a different evidence base and operate over different timescales'. However, the Review found that the single most important barrier to the delivery of housing (and thus part remedy to house price inflation) is availability of land through the planning system.

In order to reduce house price inflation to levels experienced in other European countries, Barker recommended that the delivery of new housing would need to almost double, from 150,000 to 295,000 homes per year, though there were numerous cautionary notes about the uncertainty of the calculations, endorsed by other commentators.[6] For this to happen, a more 'effective' planning system was needed, with the implication being that all institutional improvements were needed in planning rather than in the construction or housing finance sectors. This would be:

- a system that responds to market signals;
- decision making procedures that take full account of the wider cost and benefits of housing development, including environmental and amenity costs;
- appropriate incentives for development at the local level;
- clear and timely mechanisms to provide the necessary infrastructure and services to support development and deliver sustainable communities; and
- sufficient resources to enable effective decision-making.

(Barker 2004: 32)

It is interesting to note that all of the recommendations had been voiced in various forms before, but the difference here (as noted above) is that the recommendations were made to HM Treasury as well as the ODPM. The Chancellor responded to this report in the 2004 Budget Statement.

The main themes of the Review were to ensure that more land is allocated for housing and associated development, to give more weight to the economic considerations, costs and benefits of decisions on housing land and to seek ways of distancing the decisions on housing land from the political process. These did not go unnoticed by the main protagonists and

alternative arguments were advanced. They highlighted that the decline in house building since the late 1970s was explained by the drastic reduction in building of social housing and they supported a substantial boost in spending on affordable housing. Of equal concern to opponents was the potential 'relaxation' of planning controls over house building, and democratic control over planning. The suggestion that housing market indicators should trigger the automatic release of housing land was felt to undermine the role of planning in taking account of the full range of considerations, including the environmental interest. Certainly, Barker gave insufficient attention to the question of how the housing crisis could be tackled while also contributing to sustainable development (mentioned in the terms of reference) and particularly within environmental capacities.

Notwithstanding the perceived weaknesses of the Review, it is difficult for planning to sidestep the criticisms both expressed and implied in it. Planning had failed to deliver even its own estimates of land needed for housing. A pointed 'case study' of York and Harrogate is used to illustrate the costs and benefits of housing restraint (p. 38 of the Analysis Report). Both are attractive areas that suffer from a lack of development sites and both are constrained by green belt designation. House price growth in York had exceeded 12 per cent, as 'many of the new dwellings in these areas are bought by newcomers, some no doubt moving from London or the south east, having cashed in their capital gains'. The result was that public-sector workers and others in the tourism business were priced out of the market, which created labour shortages, longer commuting distances, affordability problems and forced developers into targeting other land and property (such as pubs, shops and businesses) for lucrative housing development. The illustration in the report is too kind, however, and does not give the full story. It failed to mention that the local authority had yet to adopt a statutory local plan some thirty-six years after the system was created and fifteen years since the government called on all authorities to adopt local plans with haste (and, at the time of writing, it has yet to adopt an up-to-date local plan, although it has to be acknowledged that factors such as changes to the planning system and local government reorganisation have contributed to the delays). Procrastination over the plan, of course, happens because of resistance to new development, and the unwillingness and inability of the system to make difficult decisions, and that included – at the time – the failure of the Regional Office and the ODPM to sort things out with the local authority. With examples like this it is fair to suggest that planning got off lightly in the criticism. York is only one of many authorities that have failed to deliver housing land. It can be argued forcefully that this is the democratic process in operation, and the abysmal quality of much previous (sub)urban development is a good deterrent to any support for further growth. Local 'NIMBYs' (not in my back yard) are acting 'rationally' in opposing new growth. Weak support from government in providing physical and social infrastructure should also be noted. Unfortunately the costs of not acting are most heavily felt by those who are most in need of housing and jobs. In line with arguments rehearsed earlier, in the case of York, the potential of one of a small number of locations in the north of England that might succeed in the knowledge economy may not be realised.

As underlined by the case study, problems of housing supply are not just associated with one period of time: they are more persistent, and many of the points raised by Barker remain high on the planning and housing agendas, as the problem of house price inflation and shortages of housing are still with us. Whilst, as has been stated, planning authorities had been for some time considering issues such as the links between economic and social aspects and housing land availability, Barker gave something of an additional impetus to these efforts, although it might be argued that her emphasis on market factors had echoes of the permissive planning regime of the 1980s. Mechanisms were introduced both to forge a closer link between planning policies for housing and market factors, and to ensure that land availability constraints on housing development were eliminated. Both of these mechanisms – strategic housing market assessments and strategic housing land availability assessments – now play a part in decisions on housing policy.

Strategic housing market assessments and strategic housing land availability assessments

SHMAs are essentially a means of planning for, and trying to manage, markets. By surveying and setting out how much, and what kind of, new housing development is needed, they try to ensure that markets behave more equitably than in a more laissez-faire system. However, as has been noted, the management of development has not been inherently about being more equitable:[7] it can also be used to protect house prices and prevent development that might otherwise end up as burdensome and unsightly.

From the perspective of free-market economics there is no need to tell developers what type of housing they should build (i.e. terraced, semi-detached or detached) or how many bedrooms those units should contain. Producers of housing are expected to respond to the demands of the consumer and so, if the market is working according to proper economic principles, then it is in the interests of the developer to respond as closely as possible to demand, albeit within the constraints of their surrounding costs. A situation can therefore be imagined, perhaps most closely exemplified by some suburban parts of the United States, in which housing pops up in various different locations according to the propensity of consumers to pay and also according to expectations of what consumers want. Perhaps as a result of a crisis in the price of oil or a structural economic shift which leads to serious levels of job losses in a city, some of this development will become inaccessible or undesirable and may be abandoned. If economic fortunes change, these underused and abandoned areas may (or may not) become redeveloped or converted to new uses in accordance with those processes of creative destruction which are inherent to the pursuit of competitiveness through market mechanisms.

The tradition of planning in the UK still lies some distance from this free-market 'ideal' and to a large extent this has to do with a different cultural attitude towards the countryside (Matless 1998). In contrast with frontier attitudes in the United States, which saw the countryside as something to be parcelled up, sold off, worked and developed, the countryside in the United Kingdom has been, and largely remains, the property of aristocrats and/or commercial elites.[8] Tied up with our legacy of landed gentry and countryside estates is a highly paternal attitude towards parts of the countryside, and historically the English countryside has been held in high regard by commentators from across the political spectrum. For many, the English dream is not so much of a suburban holding as a country cottage in a village with a church and a duck pond. But the preservation of the countryside aesthetic is by its very nature an issue of place making, or place management at least. In this sense it cannot be achieved by the free-market approach outlined above, because that approach leads to places which are made up of a jumble of little territories governed by consumers, rather than to consideration of some of the more 'social' issues arising from the interrelationships between territories which form place.

'Town and country planning' is a notion which derives from this legacy and it necessitates an attempt to manage markets as a means of achieving place goals, which might now be justified either in terms of countryside preservation or by other arguments, such as supporting public transport. The free-market approach to development provides enormous power to consumers to exercise control – or at least those with the means to wield such power – over their private property but, ironically, means that the way places look is often given over to market forces. Thus, it often gives consumers a 'choice' of which (often relatively bland and homogeneous) environment they would like to live in. In relation to this, the UK approach of incorporating a collective concern for place (albeit sometimes minimally restricted to countryside preservation) is seen by some as evidence of a socialist legacy within the planning system (Cheshire *et al.* 2012). But this is a misreading of the cultural forces which have been shown by history (Murdoch and Abram 2002) to have exerted most influence over policy on housing and the countryside. The planning system, as it has worked out in practice over the last thirty years and more, has tended both to encompass and to oscillate between the conflicting objectives of countryside preservation and encouraging a permissive approach to private development.

Set within this context, SHMAs are just the latest of a series of tools designed to provide policymakers with the evidence they require in order to judge how to balance the 'need' for new housing land allocations with the preservation of the countryside. Their origin lies in the decline of council house building rates in the 1980s[9] and in attempts by the then government to promote a more permissive approach to planning for private house building in the hope of taking up the slack. Over time, the generation and passing down of housing need projections evolved as a way of imposing this more permissive approach on local authorities. However, by the year 2000, New Labour's revision of PPG 3 *Housing* was directing local authorities to carry out assessments of housing need 'to assess the range of need for different types and sizes of housing across all tenures in their area' (para. 13). In contrast to a free-market perspective, which would expect this need to be provided through the consumer/producer relationship and therefore driven by those with the greatest ability and willingness to spend, these assessments of need aimed to ensure that the needs of those on lower incomes would also be represented. Thus, the elderly, disabled, students and the homeless should all be considered. PPG 3 promoted mixed communities – a term which is also used by the NPPF (para. 50) – and directed local authorities to identify sites where affordable housing would be expected. In this sense it was expected that social housing providers would collaborate with developers to ensure a mix of housing, without which planning permission would not be granted. However, where public-sector funding for affordable housing was insufficient to meet all the need identified by housing needs assessments, PPG 3 also laid the ground for planners to require developer contributions to address the shortfall.

Housing needs assessments thus became important as the evidence base on which planning could not only build local planning policy but also justify requiring financial contributions from developers towards the provision of sub-market housing. Their extension to cover housing in the private sector also allowed planners to consider whether certain types and tenures of housing were over- or under-provided for by the market in their area and to adopt measures to steer provision towards meeting need more effectively. In effect, then, this was a recognition by government that, in practice, markets did not work in accordance with the textbook free-market approach. It recognised that the historical legacies of housing provision varied by place and that need was affected by complex processes of economic change. Unlike the post-war approach to these issues, however, the solution was not seen as lying in a two-tier, public and private system of housing provision but in an attempt to regulate the market and steer it towards meeting wider social aims. As part of this, Labour's traditional approach towards land value taxation was revived, albeit to a more limited degree than in the past, as once again the increase in land value produced by the granting of planning permissions was seen as offering an opportunity to help meet social aims, which were now defined through assessments of need. As the approach emerged throughout the 2000s, tools that were formerly called housing needs assessments came instead to be termed SHMAs, which reflected the increasingly strategic role they came to play in guiding the structure of mixed-tenure markets.

The processes described above established the principle of market intervention as a means of addressing social need, conceptualised in the first instance as increasing the overall stock of housing for different special-needs groups. However, it was only a short step from thinking about needs thematically in this way to considering them spatially. Thus, in parts of the North and West Midlands of England, housing market renewal emerged as an attempt to manage the provision of different types and tenures of housing across areas with high and low pressure for new development. Through undertaking housing needs assessments, and later SHMAs, local authorities came to realise that the nature of the housing stock differed across their areas. It seemed that the management of new housing development might not just be useful as a means of meeting need but also as a way of directing that need towards particular locations deemed most suitable for accommodating it. Thus, areas were identified where pressure for new housing development was very low. In these housing market renewal areas, public subsidy would be used to promote more 'aspirational', private-sector

development. This approach was effectively the corollary of that taken in more prosperous development areas, where affordable housing would be introduced to create new, mixed-tenure developments.

One of the key characteristics of this managerialist attempt to achieve social aims via market means was the different level of intervention faced by poorer and more affluent areas. Where affluent areas would simply be complemented by new, mixed-tenure development, many of the housing market renewal areas would see significant restructuring, including demolition and redevelopment. The role of SHMAs in this context thus grew beyond a rational process of managing new private-sector housing development to become a rational model for restructuring the housing stock of certain neighbourhoods. It was this requirement to realign housing provision with need as identified in the SHMA, and in many cases to do this through the use of demolition, that prompted residents in some areas to complain that the initiative was not sensitive enough to other ways of achieving more community-led forms of regeneration.

In order for SHMAs to work, they needed to be set within a planning system that prevented development in some, usually rural and more peripheral, locations and concentrated it elsewhere. It is only by creating this extra scarcity in the supply of land that betterment value would arise and that the potential to secure that value as a means of addressing community need would be created.[10] SHLAAs emerged as the mechanism through which this extra scarcity was to be brought about. Like the SHMA, the SHLAA is the latest of a series of terms and tools which aimed to help manage how much scarcity should be imposed on the housing land supply. During the late 1990s, amidst significant increases in the forecast amount of housing that would be needed in the UK, there was a move first by the Conservatives under John Major and then by New Labour through its Urban Renaissance initiative to concentrate an increasing percentage of development on brownfield land. Studies of brownfield land availability, linked at the time to the National Land Use Database, evolved to become what is now known as the SHLAA, which serves as a local authority-wide survey of sites with the potential for new housing.

- identifying areas where, through land assembly, area-wide redevelopment can be promoted.

This emphasis was continued in PPS 3, where it was noted that 'The priority for development should be previously developed land, in particular vacant and derelict sites and buildings' (p. 12). The national target was identified as 60 per cent of new housing to be provided on previously developed land or through conversions, whilst LPAs were advised to include a 'local previously developed land target and trajectory' in their local plan documents. The Urban Task Force promoted the 60 per cent target and devised its own estimates for the various types of recyclable land and also of the number of dwellings that are likely to be accommodated on this land under current policies, but with its overriding aim being that reflected in its report title, *Towards an Urban Renaissance* (1999). Whilst the NPPF maintains an emphasis on the use of previously developed land, *The Plan for Growth* announced that the target for 60 per cent of housing to be on brownfield land was to be removed. This was reflected in the impact statement for the NPPF, where the reference to 'removing the target and the priority for brownfield development' on page 55 led some to fear that the commitment had been watered down. The Central and Local Government Committee on 11 December 2011 echoed this feeling in its comment on the draft NPPF:

> There is a danger, nevertheless, that the removal of the brownfield target and the 'brownfield first' policy – in conjunction with the introduction of the presumption in favour of sustainable development and changes to requirements for allocating land for housing – will result over time in less importance being attached to the use of previously-developed land first where possible. This principle should be strongly stated in the NPPF, and reiterated by requiring local authorities to set their own targets for the use of brownfield land. This would allow for adaptation to particular circumstances and would in addition be a useful mechanism for local accountability.

Nonetheless, LPAs continue to prioritise the use of previously developed land in policy, even if they do not now specify a target level of use. It is too early to say what effect these changes might have, but as Table 6.3 shows, there has been a fairly consistent increase in the proportion of housing built on previously developed land since targets were first mooted in 1998. The 2014 planning practice guidance seeks 'to promote the viability of brownfield sites' by suggesting that local plan policies give particular consideration to matters such as planning obligations and Community Infrastructure Levy charges and the additional costs they can impose on the development of such sites, which are often more expensive to develop.

Since PPG 3, guidance has contained some form of definition of what constitutes 'previously developed land'. While there has been a fair amount of consistency, there have been some changes over time. Box 6.3 shows the definition included in the glossary accompanying the NPPF. It is perhaps worth noting the reference to 'private residential gardens', as this had been an area of some concern in earlier guidance, where, as an unintended consequence, gardens had been in some cases seen as brownfield land. However, this was remedied by a revision to the definition in June 2010.

There is no clear line between vacant, derelict and contaminated land (or neglected, underused, waste and despoiled land). It is all previously developed land (see Box 6.3). There are, however, significant differences between the development potential of the different types of land. In a submission to the Urban Task Force, the House Builders Federation (HBF) noted 'developing these sites is seldom straightforward. House builders frequently encounter obstacles inherent in the planning process as well as others including issues of contaminated land' (HBF 1998, preface). In a subsequent document reporting research on PPG 3 (HBF 2000: 8), the HBF identified potential consumer resistance to dwellings on certain types of brownfield locations, with contaminated land attracting high levels of concern. However, there was also some concern about the location of much brownfield land, which was felt to be in many cases in congested built-up areas.

A further factor to consider, which became associated with focusing more development in 'the compact city', is that of housing density. The second HBF report referred to above identified a number of negative public

Table 6.3   Proportion of new dwellings on previously developed land, England 1989–2011 (%)

Year	New dwellings on previously developed land		Land area changing to residential use that was previously developed[3]
	Including all conversions[1]	Excluding all conversions[2]	
1989	55	52	44
1990	54	51	45
1991	53	50	45
1992	56	53	47
1993	56	53	48
1994	54	51	46
1995	57	54	48
1996	57	54	48
1997	56	53	47
1998	58	55	48
1999	59	56	50
2000	62	59	52
2001	64	61	55
2002	67	64	57
2003	70	67	58
2004	75	72	62
2005	77	74	63
2006	76	73	65
2007	77	74	68
2008	81	78	70
2009	80	77	69
2010	71	67	53
2011	68	64	53

Source: DCLG Table P211, December 2013
Notes: 1 Conversion of existing buildings estimated to add three percentage points up to 2002.
2 As reported by Ordnance Survey, mainly excluding conversions and excluding all conversions from 2003.
3 Excludes land changing to residential use but with no dwellings built.

attitudes connected with increasing density: it was 'built for students and those on benefits and was associated with undesirable parts of town' (2000: 9). The compact city idea was promoted because it was seen as providing opportunities for more effective public transport and increased cycling and walking, sharing of resources, including local energy production; it may reduce development of greenfield sites, and it may provide more social interaction through local provision of services. However, this theory may be difficult to realise in practice. In line with the consumer views, urban life is often characterised by traffic congestion, poor environmental quality and 'town cramming' (Williams 1999: 169). Perhaps partly because of their

BOX 6.3 LAND STATUS DEFINITIONS

Brownfield land or previously developed land (PDL)

Land which is or was occupied by a permanent structure, including the curtilage of the developed land (although it should not be assumed that the whole of the curtilage should be developed) and any associated fixed surface infrastructure. This excludes: land that is or has been occupied by agricultural or forestry buildings; land that has been developed for minerals extraction or waste disposal by landfill purposes where provision for restoration has been made through development control procedures; land in built-up areas such as private residential gardens, parks, recreation grounds and allotments; and land that was previously developed but where the remains of the permanent structure or fixed surface structure have blended into the landscape in the process of time.

Source: NPPF 2012, Annex 2: Glossary, p. 55

Greenfield land

Any land outside the above definition.

Vacant land

Land that was previously developed and is now vacant which could be developed without treatment (see below for definition of treatment). Land previously used for mineral extraction or waste disposal which has been or is being restored for agriculture, forestry, woodland or other open countryside use is excluded.

Vacant buildings

Unoccupied for one year or more, that are structurally sound and in a reasonable state of repair (i.e. capable of being occupied in their present state). Includes buildings that have been declared redundant or where re-letting for their former use is not expected. Includes single residential dwellings where they could reasonably be developed or converted into ten or more dwellings.

Derelict land

Land so damaged by previous industrial or other development that it is incapable of beneficial use without treatment, [which may include] demolition, clearing of fixed structures or foundations and levelling. Includes abandoned and unoccupied buildings . . . in an advanced state of disrepair. . . Excludes land . . . which has been or is being restored for agriculture, forestry, woodland or other open countryside use [and] land damaged by a previous development where the remains of any structure or activity have blended into the landscape in the process of time.

Source: NLUD Data Specification (www.nlud.org.uk)

Contaminated land

Statutory definition

'Contaminated land' is any land which appears to the local authority in whose area it is situated to be in such a condition, by reason of substances in, on or under the land that –

(a) significant harm is being caused or there is a significant possibility of such harm being caused; or

(b) significant pollution of controlled waters is being caused, or there is a significant possibility of such pollution being caused.

Source: Environmental Protection Act 1990, section 78A(2)

Definition for planning policy

Whilst NPPF does not contain a definition of contaminated land, it does recognise that the planning system should assist in its remediation. PPS 23 *Planning and Pollution Control*, which was replaced by NPPF, included the following:

> Where the actual or suspected presence of substances in, on or under the land may cause risks to people, property, human activities or the environment, regardless of whether or not the land meets the statutory definition.

Source: PPS 23 *Planning and Pollution Control*, Annex 2, paras 2.5 and 2.13

location, there is an association between development on sites consisting of previously used land and density of development. Land use change statistics for 2011 noted that overall average density of development was forty-three dwellings per hectare, whilst development on previously developed sites averaged fifty-three dwellings per hectare. So, whilst policy for the reuse of land and minimising land take by increasing density can clearly contribute to some aspects of sustainable development, there are some perceived risks, in the public's eyes at least, to the quality of development that is realised. Planning policy will need to endeavor to strike a balance between these factors.

As an alternative to peripheral or internal expansion of a range of existing settlements, developing completely new settlements is a possibility and one which has been encouraged in national policy at various times. Whilst the New Towns programme was one such national policy initiative, there have been more recent examples. In 2003, the *Sustainable Communities Plan* designated four 'growth areas' as a contribution to tackling housing pressures (at the same time as it was designating housing market renewal pathfinders); this process was subsequently extended by inviting more authorities to apply for 'growth point' status. In 2007, a commitment to building a number of 'eco towns' was announced. According to the *Eco Towns Prospectus* (DCLG 2007a: 4), 'Eco-towns will be small new towns of at least 5-20,000 homes. They are intended to exploit the potential to create a complete new settlement to achieve zero carbon development' and 'places with a separate and distinct identity'. The idea of new settlements is included in PPS 3 as a means of meeting housing need, whilst a supplement to PPS 1 *Delivering Sustainable Development* on eco towns was issued in 2009. The Coalition government has reduced its commitment to eco towns and only one – North West Bicester – of the ten planned under Labour will go ahead meeting the full standards set for eco towns. The

NPPF does continue some lukewarm commitment to new settlements and, somewhat surprisingly, the Chancellor, George Osborne, announced government 'support for a new garden City at Ebbsfleet' in the Budget of 2014 (Budget Report 2014, para. 1.145). However, it should not be forgotten that, in a different era, new settlements contributed significantly to housing Britain's population: as Hall and Ward (1998) noted, in the half century after 1946 Britain built twenty-eight new towns with a combined population of 2,254,300 people at the 1991 census.

There are a number of examples of new settlements which have been promoted by both private and public initiatives. Perhaps the best known is Poundbury in Dorset, a town for around 6,000 and built as Prince Charles's contribution to the development of 'new urbanism'. A much-contested proposal that has yet to go ahead is for what has been called (by the developers) Micheldever Station Market Town, a development of a new community originally for 5,000 people in Hampshire first put forward around 1990 by the then owners, Eagle Star. It has consistently been rejected and the rejection was the subject of a judicial review in 2013. Northstowe, to the north-west of Cambridge, can be seen as a product of prolonged reflection by local authorities on the idea of addressing housing and expansion pressures in Cambridge through the building of a new settlement. It is planned to house around 10,000 people and describes itself as aspiring to be 'an exemplar and vibrant 21st century town enabling more sustainable lifestyle choices and patterns of living', reflecting the fact that it received support from the eco towns programme (see its website at www.northstowe.uk.com). It is being jointly promoted by Gallagher Estates and the Homes and Communities Agency, and in 2014 Gallagher signed a section 106 agreement to contribute £30 million towards the cost of community facilities[18] to allow construction work to start.

How much housing and where?

The task of estimating how much housing needs to be provided for through planning policy has moved to – some might say – a more refined level in recent years. What might have been a straightforward calculation of population and housing numbers has taken on more complexity through a number of social changes affecting the relationship between population and numbers of households and by trends in inter-regional and international migration. A social dimension has been given greater prominence by the growing importance of affordability in housing calculations. Bramley *et al.* (2010: 25) distinguish between a range of relevant concepts which have to be taken into account:

> discussions also generally distinguish 'need' – shortfalls from certain normative standards of adequate accommodation – from 'demand' – the quantity and quality of housing which households will choose to occupy given their preferences and ability to pay (at given prices). The term 'housing requirements' is sometimes used in this context, to refer to the combination of need and demand, particularly where market as well as affordable housing provision is being considered.

Such considerations have been absorbed into the process of planning for housing to a certain extent through the SHMA, but the outputs are no less controversial for all the apparent added sophistication that is being brought to bear. Whilst part of the controversial nature of housing numbers is down to factors such as NIMBYism, debate is fuelled by the range of assumptions that underpin all the calculations.

The policy objectives which guide decisions on the location of new housing provision are similarly controversial and open to challenge. The concept of sustainability in itself tends not to be widely challenged, but the fact that it is nearer to a political than a practical concept means that decisions on which are the best locations for new housing development based on the quest for sustainable solutions are far from universally accepted. Nonetheless, a balance has to be struck between more concentrated and more dispersed patterns of development – including the possible designation of new settlements – in framing planning policies. Such policies also have a social as well as an environmental aspect, in that choices often have to contribute to tackling disadvantage and assisting regeneration.

Additionally, the focus on translating from policy to implementation which emerged when the planning authorities were first obliged to consider matters of land supply has been given greater prominence of late. The SHLAA process outlined above cements the importance of an assessment of availability and developability into the policy development process, adding an extra element into the decision process on the 'how much' and 'where' of planning for housing.

In some senses, the NPPF has continued past approaches to planning for housing, albeit with an additional market 'spin' by its focus on 'delivering a wide choice of high quality homes'. It has made two changes that can be seen as adding both realism and complexity to the task. First, to aid in ensuring deliverability, LPAs have to:

> identify and update annually a supply of specific deliverable sites sufficient to provide five years worth of housing against their housing requirements with an additional buffer of 5% (moved forward from later in the plan period) to ensure choice and competition in the market for land.
>
> (para. 47)

Second, an allowance can be made for windfall sites – that is, sites that had not been included in a local plan but which unexpectedly become available – in the five-year supply of housing land if there is compelling evidence that such sites have consistently been available in the local area and will continue to provide a reliable source of supply in the future. However, such changes do not fundamentally alter the nature of the task of planning for housing, which remains both complex and controversial.

PLANNING FOR ECONOMIC DEVELOPMENT

Planning for economic development can be seen as pre-dating the Planning Act 1947, with the response to

the Depression and the unemployment of the 1930s seeing the launch of regional planning. This, in Peter Hall's words, represents 'a different kind of problem, demanding a different expertise' from regional land use or spatial planning, but there are 'clear interrelationships' between the two types of planning (1992: 63). However, the work of the Barlow Commission can be seen as drawing a link between this regional economic planning and approaches to managing physical growth at the local level. Problems of uneven growth and economic restructuring continue to be a key concern of planning, but more recently the relationship between planning and economic growth at a national level has suffused the planning agenda and acted to shape just how government views the role of planning. Generally, there has been a tendency, from the Thatcher era particularly, to see planning as a problem and impediment to economic development, perhaps notably starting with the view expressed to the Planning Summer School of 1979 by Michael Heseltine that 'thousands of jobs were locked away in the filing trays of planning departments' (Heseltine 1979: 27) but continued by governments since – the work of Kate Barker mentioned earlier illustrates some of the thinking of the Labour government, whilst David Cameron's speech to the 2011 Conservative Party conference, in which he described planners as 'enemies of enterprise', is a more recent example.[19] The result has been that planning has been exhorted to facilitate economic growth through, as we have already seen, ensuring an adequate housing supply and other necessary supporting infrastructure, as well as making adequate provision for economic development and new employment. In some cases, the way to overcome the barriers that planning was perceived to present was to bypass planning completely through initiatives such as urban development corporations and enterprise zones, but the role of local plans in providing for economic development has always been, and remains, significant.

The problems which were present at the launch of regional economic planning in the 1930s have not disappeared – the areas of high unemployment then for the most part remain disadvantaged – but, if anything, the issues posed by the changing nature of work have become of wider significance. In 1999,

TCPA produced something of a companion study to *The People: Where Will They Go?*, referred to above. This study, with the title *The People: Where Will They Work?*, was concerned with the 'changing geography of jobs' (Breheny 1999: 1), with something of the importance of this question being founded on the spatial relationship between the location of work and the location of housing, whilst contemporary changes – the loss of jobs in cities, the continuing north-south divide, an urban-rural shift of employment – were also a major focus. Other issues addressed by the study, such as sectoral changes in employment from manufacturing to services and the accompanying mix of skills required, had identifiable spatial implications for questions such as where growth would take place in future, whether the work-home link would become increasingly decoupled, and to what extent the new geography of jobs is susceptible to public intervention. Clearly, not least because of the link between employment change and future levels of local prosperity, local authorities are centrally interested in this last question and seek to make a suitable provision for economic development and employment through their local economic development activity as well as through their local plan work.

The current framework within which LPAs produce their local plans is set by the NPPF and this continues the rhetoric established in the Thatcher era, stating in paragraph 19 that 'Planning should operate to encourage and not act as an impediment to sustainable growth. Therefore significant weight should be placed on the need to support economic growth through the planning system.' So, the need to promote employment growth should suffuse the plan – it should be part of the vision and the strategy, and this strategy should be matched by a provision that meets needs over the plan period. Specifically, policies should 'support existing business sectors . . . and identify and plan for new or emerging sectors'; with an eye to the future, policies should 'plan positively for the location, promotion and expansion of clusters or networks of knowledge driven, creative or high technology industries'; 'identify priority areas for economic regeneration'; and also 'facilitate flexible working practices such as the integration of residential and commercial uses within the same unit' (para. 21).

Many local authorities would deny that they are anything other than enthusiastic about promoting employment. Whilst some of the concerns, such as poverty or low incomes which are caused by unemployment or by too many low paying or part-time jobs, are not directly addressable by planning policies alone, LPAs would see their planning policies being framed with such issues very much to the fore and be very positive in their stance towards new employment development. Indeed, the RTPI has called the idea that planning is a drag on economic growth 'a myth'.[20] However, whether it is born from local enthusiasm or driven by government exhortation, how an intention to promote or facilitate economic development is translated into planning policy is a key question. Just how can a positive strategy be expressed as an identified range and quantity of sites that meet the future needs of present and future employers?

Developing planning policy for employment and economic development

Whilst some aspects of the practice of developing policy may have changed – for example, in the 1970s there was little explicit link between forecasts of jobs and population, whereas now the links are much more apparent – many of the essential elements have remained. Practice guidance from central government perhaps reached a peak of detail in 2004 with the publication of *Employment Land Reviews – Guidance Note*, which stretched to 112 pages. Previously, guidance had tended to focus on appropriate methods of forecasting employment, with more limited attention paid to matters such as factors affecting land availability. The updated practice guidance issued in 2014 followed the simplification mantra of the NPPF – it amounts to a little over four pages. However, many of the essential components are echoes of the 2004 advice. The 2004 advice note identifies three stages to the planning process: taking stock of the existing situation; creating a picture of future requirements; and identifying a 'new' portfolio of sites. These three elements can encapsulate the steps that most policy development processes will follow.

Assessment of the current situation in both the 2004 and the practice guidance refers to the current market for employment land. In assessing this aspect, inputs from the business community are considered important, but this will need to be supplemented by information on trends in take-up of sites and in land and rental values to help build an understanding of local property markets. Also important will be an assessment of the current supply of employment sites, to identify constraints on availability – such as important infrastructure or ownership factors, contributing to the currently much-loved concept of 'market failure' as a justification for any form of public-sector action – and its suitability for current and future needs. Analyses should have qualitative as well as a quantitative dimension, to try to ensure that the needs of particular types of businesses are able to be met. However, in addition to assessing the situation with regard to land supply, it is likely that LPAs will also be interested in the current situation with regard to employment and unemployment as a backdrop to these considerations – are there local economic issues which could inform policies and priorities?

In attempting to think about future requirements, there has been a consistent stress on the need to consider a range of different property markets. The 2004 guidance usefully identifies the range of potential different market segments that may need to be catered for, and this is reproduced in Box 6.4. However, this omits some types of employment sites, such as those catering for leisure and tourism activities, which are considered briefly later. Whether each local area will need or be able to cater for all these segments is a matter for local determination, but the variety of different types of market segments gives some hint as to the complexity of the task. Both the 2004 and the 2014 guidance identify the same set of factors that need to be considered in developing a forecast. These are labour demand (the sorts of jobs likely to be created in the next fifteen years); labour supply (how many people in the local area are likely to need jobs in the next fifteen years); the past take-up of employment land; and local and contextual information on business trends.

The potential complexity of forecasting labour demand over a fifteen-year period can be illustrated by the fact that the national forecast for the economy requires about twelve pages of A4 to detail its various components (OBR 2011), but it does not look forward for such a period of time, nor does it have to deal with the additional uncertainty posed by looking at smaller geographic areas. Labour supply forecasts are demographically derived – that is, they are a product of the population forecasts discussed earlier – but with additional assumptions about the proportion of people who will be seeking work. With both demand and supply forecasts, spatial planning will require some sort of assumption about patterns of commuting, something which has exhibited significant change over the years. In comparison, data on past rates of the take-up of employment land is easy to collect, but difficulties may be encountered in interpreting what it means, as over a small area trends can be 'lumpy' and shaped by one-off events such as particular businesses experiencing periods of success which may not be replicated in future, or by industry-wide trends. Guidance underlines the importance of the local perspective. Whilst changes in the national economy clearly affect patterns of development, 'national economic trends may not automatically translate to particular areas with a distinct employment base' (Practice Guidance 2014). However, such complexity has to be taken as an additional exhortation to monitor policies regularly and to make provision that is flexible.

The third stage of the process involves deciding on the range and number of sites needed in a plan. This will clearly be determined by a range of local factors and priorities. Employment structure will shape the types of sites required. To take an extreme example, if most new employment is expected in sectors such as tourism and leisure, then that would produce different demands from heavy industry or warehousing, but the mix between less radically different uses will be present in all localities – offices or manufacturing, science parks or more general business parks, all would require different types of space standards and settings. Other trends also need to be taken into account. For example, recent years have seen a rapid increase in the numbers of people working from or at home, which

BOX 6.4 A POTENTIAL CLASSIFICATION OF PROPERTY MARKET SEGMENTS AND TYPES OF SITES

Established or potential office locations: Sites and premises, predominately in or on the edge of town and city centres, already recognised by the market as being capable of supporting pure office (or high technology R&D/business uses).

High-quality business parks: These are likely to be sites, no less than 5 hectares but more often 20 hectares or more, already occupied by national or multinational firms or likely to attract those occupiers. Key characteristics are quality of buildings and public realm and access to main transport networks. Likely to have significant pure office, high office content manufacturing and research and development facilities. Includes 'strategic' inward investment sites.

Research and technology/science parks: Usually office-based developments which are strongly branded and managed in association with academic and research institutions. They range from incubator units with well-developed collective services, usually in highly urban locations with good public transport access to more extensive edge/out-of-town locations.

Warehouse/distribution parks: Large, often edge/out-of-town serviced sites located at key transport interchanges.

General industrial/business areas: Coherent areas of land which are, in terms of environment, road access, location, parking and operating conditions, well suited for retention in industrial use. Often older, more established areas of land and buildings. A mix of ages, qualities and site/building size.

Heavy/specialist industrial sites: Generally large, poor-quality sites already occupied by or close to manufacturing and processing industries. Often concentrated around historic hubs such as ports, riverside and docks.

Incubator/SME cluster sites: Generally modern purpose-built, serviced units.

Specialised freight terminals, e.g. aggregates, road, rail, wharves, air: These will be sites specifically identified for either distribution or, in the case of airports, support services. Will include single-use terminals, e.g. aggregates.

Sites for specific occupiers: Generally sites adjoining existing established employers and identified by them or the planning authority as principally or entirely intended for their use.

Recycling/environmental industries sites: Certain users require significant external storage. Many of these uses, e.g. waste recycling plants, can, if in modern premises and plant, occupy sites which are otherwise suitable for modern light industry and offices. There are issues of market and resident perceptions of these users. Some sites, because of their environment (e.g. proximity to heavy industry, sewage treatment works, etc.), may not be marketable for high quality employment uses.

has to a large extent been made possible by the development and diffusion of communication technologies (Ruiz and Walling 2005). Between 2001 and 2011, the number of homeworkers rose by 24 per cent to around 3.8 million, or about 13 per cent of the workforce. This rise was in large part driven by increasing numbers of self-employed (Live/Work Network 2012). Such a trend would have a significant effect on the amount of (particularly) office space that needed to be catered for, as well as possibly requiring other accompanying policy measures. The approach advised by guidance stresses the need to identify the different sectors where new employment is expected to arise and to associate this with land requirements via assumptions about employment densities – how much space is typically required for a worker by different types of employment – but evidence in the 2004 guidance suggests that the difference between space requirements are between four to six times for different uses, with the extremes being represented by office-based call centres and large-scale warehouses.

Tourism represents a particular type of employment and one which poses some particular issues for planning. Its importance has increased greatly over the years to the point where it contributes around 6 per cent of the GVA to the economy and accounts for around 9 per cent of jobs (ONS 2012a, 2013a). The growth in cultural and heritage tourism – in part as a result of regeneration initiatives, of which the 2012 Olympics is a special example – has meant that the impacts of tourism have become spatially more diffused. However, in encouraging and catering for the development of tourism, planning has to seek to balance the value of the (mainly) economic benefits it brings against the sometimes adverse impacts which follow from increasing visitor numbers. Whilst there may be some uncertainty about placing too great a reliance on tourism as a source of employment – many jobs are part-time and/or temporary and are often relatively low paid – the major area of concern has tended to be potential adverse environmental impacts. The attempts to assuage such concerns have often focused on managing visitor flows, in which planning has a part to play, or in the development of 'sustainable

tourism', although such a concept is regarded by some as something of a chimera (Minhinnick 1993) and by others as partly dependent on visitor behaviour (Krippendorf 1987). Policy guidance on how to plan for tourism is very slight in the NPPF, where the word 'tourism' only appears three times as an adjunct to policy guidance on town centre and rural development. The last planning guidance dedicated to tourism was issued in 1992, although some 'good practice guidance' was issued in 2006. The latter perhaps gave a hint as to why guidance is thin on the ground when it states that those preparing local plans 'should consider whether any policies for tourism are needed beyond what is set out in the core strategy' (DCLG 2006d: 17). However, it goes on to give an example of where specific approaches have been adopted and this is reproduced in Box 6.5, although the location for the example is one where tourism has long been an important source of employment and income. Nonetheless, many areas will give priority to tourism development as part of their planning and economic strategies. For example, Liverpool's Draft Core Strategy (2012) notes that 'A key draw for tourism is the City Centre's unique heritage and waterfront setting, which includes the Royal Liver, Cunard and Port of Liverpool Building, together with numerous other historic buildings' and this is reflected in the weight given to tourism in the plan's strategic objectives.

Given the number of areas of difficulty and the many assumptions that have to be made, it is perhaps not surprising that the conclusion of the 2004 Practice Guidance was that 'quantitative assessment of employment land requirements are not reliable over the time horizons' of development plans (p. 51). However, in spite of the difficulties, estimates and policies have to be constructed. Developing economic forecasts was previously a role undertaken by RDAs. The complexity of this activity and of other matters such as making assessments of employment land markets mean that in many cases outside expertise has become important to LPAs in establishing the foundations for policy. Typically, because of the wide variety of assumptions that are possible, assessments produce a wide range of quantitative results,[21] and for this reason the practice guidance (2004: 60) concludes

BOX 6.5 EXAMPLE OF A LOCAL PLAN TOURISM POLICY FOR BOURNEMOUTH

Bournemouth Local Plan, adopted 2002 and Supplementary Planning Guidance (SPG) on Tourism, adopted in 2004 both emphasise the need to diversify tourism facilities. This need manifested itself in particular through the continued loss of hotels despite planning policies aimed at retention. In its analysis, the Plan has noted that Bournemouth's popularity as a prime holiday destination has been offset by other markets opening up, such as short breaks, activity holidays and business and conference tourism.

The Local Plan has recognised the need to respond to this and for the seaside resort to serve a variety of functions and widen its economic base as a shopping and commercial centre. In response to this, the plan contains policies for the defined Town Centre Tourism Area and Tourism Core Areas as the hub of tourism facilities containing major tourism related facilities such as theatres, cinemas, night clubs, shops and restaurants, the Gardens, Pier and beach; Bournemouth International Centre; and hotels, guest houses and blocks of self catering accommodation.

Source: DCLG 2006d

by saying that assessments 'will need to be updated regularly, at no more than five yearly intervals, as part of the "plan, monitor and manage approach"', a sentiment echoed in the 2014 Practice Guidance. Policies and the assessments on which they are based also need to take account of changes in national and local economic fortunes, and these are an additional reason for keeping assessments, and possibly policies, under review.

However, a focus on quantitative estimates should not obscure either the qualitative considerations that underpin policy or the range of economic and social objectives that guide and inform economic development. The range of types of sites set out in Box 6.4 hints at the many different elements of the employment property market that policy will need to consider and possibly cater for through planning policies. Efforts to foster economic growth through appropriate planning policies will span a whole range of activities – manufacturing, distribution, retail, tourism, sport and leisure – and the extract in Box 6.6 from Liverpool's Draft Core Strategy helps illustrate this. However, another important element of the Liverpool Core Strategy reminds us of the origins of this strand of planning policy. The final section of the document

deals with the strategic objective of 'maximising social inclusion and equal opportunities', echoing the concern in the 1930s, albeit in different language, to develop policies and interventions to help the concentrations of unemployed produced by the Depression. However, this also highlights some of the arguments about the impact of planning in relation to the economy. Such policy approaches, along with other economic, social and environmental objectives of planning, are seen by some as an additional cost that planning imposes on business, to the detriment of the economy as a whole; others, however, would see these as fundamental roles and purposes of planning (Nathan and Overman 2011; Haughton *et al*. 2014).

PLANNING FOR RETAIL DEVELOPMENT

The previous edition of this book described debates in retail planning as 'the very stuff of planning arguments'. This is because the debates encapsulate a couple of the key challenges of how we plan. First, retail trading has been subject to many changes – the emergence of

BOX 6.6 ILLUSTRATION OF THE RANGE OF ELEMENTS INCLUDED IN ECONOMIC POLICIES

Strategic Policy 3: Delivering Economic Growth

1. Development of business sectors with strong growth potential in Liverpool and the City Region will be supported. These include:

    a. Knowledge-based industries
        i. health and life sciences
        ii. advanced science, manufacturing and engineering
        iii. creative, cultural and media industries
        iv. ICT and digital technology
    b. Financial, professional and business services
    c. Port and maritime industry
    d. Airport and aviation-related activity
    e. Tourism/visitor economy, and
    f. Low carbon economy businesses

2. The football clubs of Everton and Liverpool contribute significantly to the City's economy, and proposals for the sustainable development or redevelopment of these clubs will be supported where they are of an appropriate scale, and subject to other relevant planning policies.

Note that this document was not submitted but will go on to form the framework for the Local Plan for Liverpool.

Source: An extract from *Submission Draft: Liverpool Core Strategy 2012*, p. 41

out-of-town trading; the development of Sunday trading and the '24 hour city'; changes affecting the character of city centres such as the developments in financial services and other 'new' uses in centres (mobile phone shops, bookmakers, more coffee shops and restaurants); and the growing importance of the Internet – and all these hard-to-predict trends have to become part of the development of planning policy for retailing. Second, retail perhaps represents an area where the sometimes difficult interface between planning and the market is exposed. Considerations such as competition play a part in developing national policy on shopping and in periods of more laissez-faire approaches to economic policy, such as parts of the Thatcher years, traditional planning objectives became diminished as

the weight given to the views of developers increased. Visible adverse impacts of such approaches contributed to the re-emergence of planning and to the general shape of current policy. In the NPPF (pp. 7–8), the relevant section has the title 'Ensuring the vitality of town centres' and local planning authorities are urged to 'recognise town centres as the heart of their communities and pursue policies to support their viability and vitality' (see Box 6.7 for a consideration of these concepts). One of the means of doing this is to apply a 'sequential test' to proposals to ensure that, as far as possible, what might be considered as town centre uses finish up in the town centre and are only developed in out-of-centre locations if no suitable sites are available in or near the centre. The NPPF also requires LPAs to

BOX 6.7 VITALITY AND VIABILITY OF TOWN CENTRES

'Vitality' and 'viability' have been two key concepts used in planning policy for a number of years. Vitality has been linked by Ravenscroft (2000: 2534) to Jane Jacobs' ideas about what makes a successful place and is reflected in how busy a centre is at different times and in different parts. Viability is related to the capacity of a centre to attract investment to maintain and improve the fabric and to adapt to changing needs. PPG 4 *Industrial, Commercial Development and Small Firms* and PPS 6 *Planning for Town Centres* both gave lists of the sorts of factors which might be monitored through what has become known as a 'health check' to detect the first signs of emerging problems in a centre. However, further tests may be required for a more rigorous diagnosis and prescription. The key elements in a health check are seen as being:

- diversity of uses, by number, type and amount of floorspace
- retailer representation and their intentions to change – expand, contract, leave altogether
- proportion of vacant property
- pedestrian flows (footfall)
- customer and resident views, gained through surveys
- environmental quality of the centre
- perception of safety and occurrence of crime
- ease of access by a choice of means of travel
- potential capacity for growth – opportunities for expansion
- shop rents and patterns of change
- capital value of property in relation to expected rents
- amount of retail and other floorspace in edge-of-centre and out-of-centre locations.

Monitoring such factors can be seen as an input to policy development as well as a measure of plan and policy effectiveness.

Note that the Planning Practice Guidance 2014 gives a slightly edited version of this menu.

Source: PPS 6 *Planning for Town Centres*

'allocate a range of suitable sites to meet the scale and type of . . . development needed in town centres'. How they might go about deciding what is needed and the form of policy guidance best suited to ensuring the vitality of retail centres is considered below, along with a consideration of the changing nature of retail and the changing nature of the planning response.

Trends in retailing

To be successful, retail planning needs both to understand and to some extent to anticipate changes in the sector, so that suitable patterns of land use and land availability can be established. The brief review below highlights some of the sorts of factors that planning has needed to try to anticipate and to respond to, and in doing so it emphasises the degree of difficulty planning has faced, and no doubt will continue to face.

At the time of the Planning Act 1947 there was a clear and long-established pattern of retailing, with a defined hierarchy of uses echoing Christaller's well-known geographic model – main stores and most comparison shopping (items such as household goods and clothing) were to be found in larger town and city centres whilst much of the need for convenience shopping (daily needs, such as food) was met in smaller or local centres. Such a pattern underpinned and was reinforced by planning policy, and reflected a number of established patterns of living and trading. However, a range of changes in society gradually challenged this established pattern. Increased personal mobility meant that people were no longer so strongly tied to this pattern of shopping. Changes in the social economy, particularly the increasing numbers of women going out to work meant that the regular daily shop for convenience items in the local centre was no longer possible. Increasing prosperity meant that levels of spending – and therefore the need for shopping floorspace – increased, but this increase was unevenly distributed between convenience and comparison goods: whilst spending on food has increased by about 50 per cent, spending on items like clothes and furnishings has practically trebled (ONS 2008). These trends were reflected – some might say partly led – by changes in styles of retailing, such as the development of more self-service approaches, the increasing scale of some shops, and the increasing dominance of the High Street by a small number of multiple retailers.

These changes were complemented by a number of changes in the physical nature of shopping provision. The first change, partly induced by post-war reconstruction, could be seen as the modernisation of the central areas of towns, including the introduction of pedestrianised areas. Early examples could be found in Coventry and in some of the new towns, such as Stevenage. Perhaps a more significant trend, in a planning context, was the emergence of out-of-town developments. Early examples were hypermarkets and large-scale food stores, but these were followed by retail warehouses selling 'bulky goods' – furniture, white goods, DIY items – with the 'bulk' being part of the argument for the move out of the centre. The movement outwards from town centres continued with the development of regional shopping centres, with some of the best-known ones being the Metro Centre in the North East, Merry Hill in the West Midlands (see Box 6.8) and, most recently, Bluewater east of London. Smaller-scale variants on these developments – retail parks, factory outlets – have also been seen, each challenging the old established

hierarchical retail model. More recently, retailers have adopted a more subtle approach to retail positioning (Davies and Brooks 1989), seeking to better match their distribution of outlets to local customer profiles and lifestyles. The re-emergence of small local food shops in town and city centres, with Marks and Spencer's 'Simply Food' shops in the vanguard, is an example of this approach. How the space in a town centre is used has also been influenced by changes in Sunday trading laws and moves towards longer opening hours for shops and other facilities, sometimes known as a move to a '24-hour city'.

More recently, a key factor has been the increasing importance of information technology in shaping retail developments. For the retailer, this has enabled them to have a better understanding of customer preferences (through the use of loyalty cards, for example) but also to manage supply chains better – through approaches such as 'just in time' delivery – and consequently make a more effective use of their retail space, because it is less necessary to carry high levels of stock. The emergence of online shopping has been seen to threaten both the future of established retailers and established retail trading patterns. However, many retailers in both the convenience and the comparison sectors have developed what is sometimes known as a 'bricks and clicks' strategy, allowing customers to browse their range of goods and place an order online, to be collected at a time that suits them at a local store. Nonetheless, Internet-based sales had reached more than 10 per cent of total sales by 2012, double the level of 2002 (GVA 2012), which raises the question of whether town centres offer too much shopping floorspace for emerging conditions.

At the time of their emergence, all of these trends have been difficult to anticipate, which makes the task of planning for retail development particularly challenging. Information technology and online retailing are but the most recent of these, with it being difficult to anticipate both the response of customers and of established retailers to the opportunities and the challenges posed.

Perhaps a final development to note is the gradual move of shopping from a mainly functional to a partly leisure activity (Morton and Dericks 2013), itself to some extent a reflection of the increasing prosperity noted above. Whilst the choice of retailers remains an important feature, competition between places takes on a new dimension when a visit to a shopping centre might be seen as both a functional activity and a leisure experience. Indeed some see a transition to increased community/leisure uses as the way forward for town centres (Grimsey 2013). Planning will be concerned for the success of its shopping centres but the task of trying to ensure success takes on new dimensions when it moves beyond control of location and floorspace to areas which are at the limits of planning control and influence, such as a the style of a centre and other factors which help shape its appeal. A desire to differentiate and shape the nature of the 'experience' has emerged, with concerns over the mix of activities in a town centre in various forms becoming prominent, together with a desire to make the visit as pleasant and convenient as possible. The latter can include such practical matters as parking arrangements and charges, and crime prevention strategies, and these are often bound up with an approach known as town centre management. Many of the twenty-eight recommendations of the Portas Review[22] (Portas 2011) on the future of town centres fall into a similar category. The former reflects developments which are felt to adversely affect the atmosphere and liveliness of a shopping area – the proliferation of financial services outlets in the 1980s, increasing numbers of charity shops and, more recently, the increase in the number of branches of bookmakers – as well as a development which is seen to limit the ability of a place to differentiate itself from the competition because they are all dominated by a decreasing number of multiple retailers – the so-called 'clone town' development (NEF 2005). The recent success of small towns and rural areas noted by Guy (2007) can be partly traced to their success in maintaining a representation of independent retailers, thereby differentiating themselves from other competing centres (Powe et al. 2007) and avoiding the 'clone town' image. It could also be argued that increased attention to the future of small towns through policy interventions such as the Market Towns Initiative[23] – very much in tune with the 'town centres

first' policy – also contributed to their relative success in the past, whilst some of the initiatives it spawned continue to support such towns.

Developments in retail planning policies

Guy (2007) identifies three phases in the development of retail planning, with the impetus for the changes between phases being as much political as a response to secular changes. The 1960s and 70s he identifies as an active phase of policy, when local government was leading an approach which supported a clear and established hierarchy of centres through explicit policies. In the 1980s, policy entered a reactive phase, with leadership ceded to retailers and developers, in line with the general neoliberal stance of a government prioritising the views and wishes of the private sector and giving greater weight to market signals. Partly in response to the adverse impacts which were seen as being associated with this more laissez-faire approach, the 1990s saw a return to more active policy and a tightening up of planning policy. This era saw the emergence of a pro-town centre policy as well as the development of an interest in a number of areas related to retail policy, such as the encouragement of competition, the pursuit of sustainable development, and an interest in the impact of retail policy on social inclusion. Whilst current policy guidance espouses a more market-focused language, it continues a priority for town centres and could not be considered a regression to laissez-faire policy.

However, at all phases of policy it is possible to assert that planning was not particularly good at anticipating trends in retailing and was perhaps more following than leading changing development patterns. This perhaps reflects the limited evidence available on which to base or from which to evaluate policy (Delafons 1995; Findlay and Sparks 2005). In spite of this, what has been consistent through the changing phases of policy has been the importance of attempting to assess the trading impact of new development on existing centres, and some of the approaches to making this assessment will be reviewed later. Guy also asserts that there has been an increasing tendency for retail policy to become more centralised, citing a steady increase in ministerial call-ins of retail applications over time, with an outcome of more decisions made centrally being an increase in consistency. The lower level of economic activity taking place since Guy came to this conclusion makes it difficult to judge whether this is becoming an established trend.

Current policy shows a consistent trend from a position emerging in the 1990s. The development of out-of-town shopping centres was identified as a reason for the weakening or even the killing off of traditional town and district centres, with the Merry Hill development in the West Midlands being seen as a seminal example (see Box 6.8). They were also seen as contributing to increasing car travel (and its accompanying pollution)

BOX 6.8 THE MERRY HILL SHOPPING CENTRE

Merry Hill Shopping Centre is the largest retail centre in the Black Country, and comprises 1.5 million square feet of retail floorspace occupied by 200 retail units; it attracts 21 million visits each year. It opened for business and developed between 1985 and 1989 in an enterprise zone on the site of a former steel works adjacent to Dudley and it is now the second most important retail centre in the West Midlands (thirty-fifth in the UK).

A quantitative assessment *The Merry Hill Impact Study* was carried out in 1992–3. It aimed to 'examine the impact of Merry Hill upon the vitality and viability of established shopping centres within the West Midlands' so as to inform future planning policy.

The study concluded that the major centres in the wider catchment area had lost market share between 1989 and 1993, while that of Merry Hill had risen considerably. In particular, Dudley's market share in 1993 was one third of its 1989 level, which is equivalent to an impact of 70 per cent on the pure comparison offer. Significant declines in market share also occurred in Stourbridge and West Bromwich, whilst Birmingham, Wolverhampton and Walsall were within 15 per cent of their 1989 market shares.

The consultants examined qualitative and quantitative indicators of impact, including changes in retail composition and retail floorspace, market perceptions and the views of retailers. They concluded that the effect of Merry Hill on Dudley went deeper than a reduced market share, as the town centre had experienced the loss of vital town centre anchor stores and major multiple retailers. In particular, the increase in vacancy levels combined with the decline in retailing had undermined the vitality and viability of Dudley. In 2011, the Wolverhampton *Express & Star* noted that Dudley had 'more shops standing empty than any other centre of similar size in the country' with almost 30 per cent of shops empty. In the Black Country Core Strategy adopted by the four adjoining local authorities in 2011, Merry Hill and associated developments in Brierley Hill were designated as the top-level 'Strategic Centre' for the area, whilst Dudley town centre was redesignated as a second-level centre 'focusing on its leisure, heritage and tourism role'

The Merry Hill study highlights one of the best-documented and most extreme cases of impact arising as a consequence of the unplanned growth of a major out-of-town shopping centre when Dudley was already a vulnerable centre. In many other cases, the impact of new developments has been less significant, and often less immediate.

Sources: PPS 4 Practice Guidance, p. 49, supplemented by Black Country Core Strategy, Brierley Hill Area Action Plan Baseline Options Report and Wolverhampton *Express & Star*, 8 September 2011

BOX 6.9 AN EXAMPLE OF A RETAIL STUDY FOR RYEDALE DISTRICT

Ryedale District Council commissioned a study to 'provide a robust evidence base on the capacity for additional retail development . . . which can be used to inform the preparation of its Core Strategy'. Ryedale is a largely rural area with the market town of Malton as its main retail centre, but with other centres at Pickering, Helmsley, Kirkbymoorside and Norton. The main findings of the study, based on an analysis of available data and household surveys, included:

Current shopping patterns

The District suffers a high level of leakage of comparison goods expenditure to nearby centres such as York and Scarborough – 70 per cent of total comparison goods spending and 83 per cent of spending on clothes and shoes. It also achieves a low level of retention of convenience goods spending at 66 per cent. Further investigation of qualitative aspects led the consultants to conclude that there was a need for modern retail units to meet the requirements of High Street comparison traders; a new large-format foodstore; and some retail warehouses to accommodate DIY and bulky goods needs unmet in the District.

Qualitative needs

Consideration of strategies to increase floorspace to maintain current levels of trade retention or to increase floorspace at a higher rate to increase retention and address overtrading in some convenience stores is used to calculate future floorspace needs for comparison and convenience goods. These figures are described as 'indicative guidelines' for policy decisions, and they should be used alongside other qualitative factors in reaching decisions.

Opportunities for meeting needs

The report also makes an assessment of the potential of six sites for meeting development needs for new retail space.

Impact assessment

An assessment is made of the impact on the town centres of Malton and Pickering of the development of new edge-of-centre convenience floorspace. Whilst there might be some trade diversion from existing convenience stores, this could be offset by positive benefits such as 'widening consumer choice; generating employment opportunities; addressing identified qualitative needs; attracting new investment to the town centre; and clawing back expenditure leakage'. Overall, the vitality and viability of Malton town centre as a whole would be enhanced whilst the development at Pickering would help minimise any trade diversion to the new development in Malton.

These recommendations are reflected in the retail hierarchy and other development policies included in the Local Plan Strategy adopted in 2013

Sources: Roger Tym and Partners (2011) *Ryedale Retail Capacity and Impact Assessment Update: Volume 1*; Ryedale District Council (2013) *Ryedale Plan – Local Plan Strategy*

require the acquisition and interpretation of a significant amount of evidence, some of which may well involve primary research – for example, to gain information on consumer habits and preferences. This material will need to be used alongside what may be seen as more 'political' inputs. At the local level, there will be a range of priorities including the hierarchy of settlements – which will help define what goes where – and other key strategic priorities – the level of importance given to enhancing sustainability or social inclusion may influence the distribution and nature of new development prioritised in policy. At the national level, changes in the emphasis of economic policy or increasing stress on the importance of consumer choice or enhancing competitiveness can affect the degree of looseness or tightness of policy. The laissez-faire policy stance of the 1980s provided a graphic example of the problems that such a change may generate, but recent debates about the negative impacts on productivity of the 'town centre first' policy (see e.g. Cheshire et al. 2011; Sadun 2013) suggest that the current level of regulation may not be inviolable. This all suggests that this is an area of complexity. Baldock (2012b: 2) considers that 'Retail studies undertaken for local authorities are particularly demanding in a technical sense, because such studies form part of the evidence base for Local Plans, and so are potentially open to challenge', but he also notes that, unfortunately, 'there are no longer any specialized retail planning modules in UK planning degree courses'.

Finally, most of the above discussion has been framed in terms of expansion of space. However, at the time of writing, a greater concern is with the number of vacant shops[28] and the threat posed to the High Street and perhaps particularly to local shopping by adverse economic circumstances and, in the long run, by the Internet. This concern was reflected in a debate devoted to the future of the town centre in the House of Commons in January 2012, the tone of which was summed up by Gareth Johnson, MP for Dartford, when he said 'the future prosperity of British high streets is one of the biggest challenges the country faces. There is no simple solution to the problem' (Hansard, HC Deb, 17 January 2012, Vol. 538, c. 637). Similar views could be found in national[29] and local government

bodies throughout the UK and in commercial bodies with an interest in retail. This suggests that an equally important task for retail planning is how to manage decline and retrenchment, or how to develop policy frameworks that can facilitate or fuel regeneration.

Whilst no universally accepted strategy for the future of town centres has emerged, a number of themes are widely discussed. First, the idea that the 'business of town centres is more than just retail' (Experian 2012: foreword). All sorts of businesses are present in town centres and all can contribute to shaping their futures. Some uses, such as offices, face uncertainty like many retail uses[30] and this adds to the challenge of planning for future uses. This means that planning strategies for town centres need to be based on accommodating and encouraging a mix of uses, including community, leisure and residential, in addition to the traditional retail, commercial and employment uses. Second, longer-term trends and concerns, such as an ageing population and climate change, add to pressures to adapt approaches across a range of matters, such as structure and access. Third, whilst e-commerce might be seen as a threat to the traditional high street, some see m-commerce – the use of mobile technology – as a source of new opportunities for retailers and others through the development of a 'networked high street'. Many of these aspects will need to be combined with more readily tangible planning concerns, such as presenting an attractive environment, allied to other established approaches based on partnership working and management of the town centre 'asset'. This may mean that established approaches to the development of planning policy will need to be adapted to reflect changing business requirements and a changing mix of uses.

In a foreword to draft Supplementary Planning Guidance for London's centres (GLA 2013), Mayor Boris Johnson noted:

> town centres face considerable challenges, including changing consumer behaviour, the growth of internet spending, and competition from out of centre retail and leisure development. Yet they also possess considerable strengths and opportunities as the focus for business growth and development.

The task of planning policy is to capitalise on such strengths and opportunities to maintain the role of town centres as the 'hub of their communities' (DTCPTF 2013: 3), but this requires a focus which extends beyond what might be considered traditional planning policy competences.

CHALLENGES OF DEVELOPING PLANNING POLICIES

From the three broad but significant areas of policy considered here, it is possible to identify a number of important variables to be considered in developing planning policies. As with many areas of work, for planning at the local level an important foundation is national planning guidance. As we have seen, this guidance varies in nature over time. At points, it has been very detailed but recently it has become more synoptic, though whether the quest for shorter, more easily accessible guidance has been at the expense of certainty remains to be clarified through practice and experience. At various times it has attempted to 'roll back' the level of control exercised by planning to produce a more 'market-friendly' approach, although whether this results in too low a priority being given to some matters which the public thinks important is decided more by personal political ideology than evidence; however, the fact that there are impacts from adopting a more laissez-faire approach is clearly illustrated by the example of the Merry Hill shopping centre discussed above.

Government guidance extends to the methodological approach that LPAs should adopt in developing their policies. Sometimes this will emerge in the form of specific guidance, such as on how to assess local housing market conditions, or in other cases it will become embedded through the results of planning inquiries and professional guidance developing 'best practice' on how to approach a particular element of policy development. In many cases, such guidance will result in LPAs having to resort to outside experts to gather and interpret evidence as, for whatever reason, such expertise is not typically available within the authority itself. However, the LPA retains a responsibility for specifying the nature of the evidence required and applying it to the process of policy development.

At this point, it is perhaps important to recall the debate about the nature of evidence at the beginning of this chapter. Whilst the process of testing the soundness of local plans and their policy content has possibly shone a brighter light on the weight given to evidence underpinning policy development, it should not obscure the levels of uncertainty and the important assumptions that are an inherent part of gathering and interpreting this evidence. Whilst the issue of uncertainty is perhaps reflected in the continuing adherence to the 'plan, monitor, manage' mantra, this does not limit the need to retain an acute awareness of the impacts of uncertainty when developing policy. We also need to remember that policies are more than the product of evidence collection and interpretation. Planning has always espoused, to a greater or lesser extent, a 'visionary' element, wanting to improve and change rather than merely accommodate. However, blending supposedly objective evidence and these normative considerations is also problematic. A quest for a more pragmatic, 'realistic' approach to planning would prioritise the former over the latter, but whatever the relative priority, planning will need to test the realism and achievability of its visions against the established and possibly dominant trends.

Finally, in an era of 'spatial planning', with its renewed emphasis on implementation, the importance of developing policy which has a clear focus on deliverability has been enhanced. It might be argued that planning should always have developed policy which will make an impact because it can be delivered, but maybe in the period immediately after the Planning Act 1947, when the responsibility for much of the delivery of the construction guided by planning policy rested with the public sector, such a consideration was of less concern. However, this is no longer the case and a concern for implementation requires that policy be developed with an awareness of physical and market constraints and the processes and programmes available to address them. Such concerns are reflected in the methodological guidance given by government but they are an inherent component of developing a plan which will stand the best chance of achieving its objectives. However, whether this more pragmatic approach pushes planning closer to market objectives and away from its social and environmental objectives is a matter for debate.

Further reading

General

As this chapter is focused on making planning policy, the web-based *Practice Guidance* issued by DCLG in March 2014 is a valuable resource. It can be found at http://planningguidance.planningportal.gov.uk.

Evidence in planning

A good place to start is the special issue of *disP* of 2006 (42(165)) edited by Andreas Faludi and devoted to evidence-based planning. A more general text on the use of evidence in policy by Davies *et al.* (2000) *What Works? Evidence-based Policy and Practice in Public Services* provides a number of useful insights and allows a comparison to be made between planning and other activities in the public sector. For a more practice-focused approach, the DCLG publication (2007b) *Using Evidence in Spatial Planning* is useful.

Housing and population

A useful compendium of information is provided by *Focus on People and Migration* published by ONS in 2005. The most accessible discussions of household projections are given in Breheny and Hall (1996) *The People: Where Will They Go?* More technical is Bramley *et al.* (1997) *The Economic Determinants of Household Formation: A Literature Review,* and in Bramley *et al.* (2010) *Estimating Housing Need.* Holmans (2013) TCPA paper *New Estimates of Housing Demand and Need in England, 2011 to 2031* combines some up-to-date forecasts with explanations of their derivation. See also Allinson (1999) 'The 4.4 million households: do we really need them anyway?' A useful guide, linking to many other resources relevant to planning, is PAS (2013b) *Ten Key Principles for*

Owning your Housing Number: Finding your Objectively Assessed Needs.

Planning and affordable housing

The use of planning powers to require the provision of affordable housing has attracted much debate. See, for example, Kirkwood and Edwards (1993) 'Affordable housing policy: desirable but unlawful?'; Barlow *et al.* (1994) *Planning for Affordable Housing*; 'Planning mechanisms to secure affordable housing' in Joseph Rowntree Foundation (1994) *Inquiry into Planning for Housing*; Crook *et al.* (2006) 'Planning gain and the supply of new affordable housing in England'; and Crook and Whitehead (2002) 'Social housing and planning gain: is this an appropriate way of providing affordable housing?'. On the specific matter of affordable rural housing, Taylor's 2008 report *Living Working Countryside* places rural housing in its wider context; the final report of the Affordable Rural Housing Commission (2006) gives a very good overview of issues and initiatives, including the use of the planning system; whilst the National Housing Federation's 2009 report *A Place in the Country* offers a perspective from Northern England. On the strange 'exceptions policy' (the exceptional release of land for local needs housing), see Annex A to PPG 3 (or its replacement) and Circular 6/98 *Planning and Affordable Housing*; and Gallent and Bell (2000) 'Planning exceptions in rural England'. A more general perspective on affordability and social housing is offered by Hills (2007) *Ends and Means: The Future Roles of Social Housing in England,* and by the brief DCLG paper (2010d) *Affordability and Housing Market Areas,* which gives some insight into matters such as measuring affordability

New settlements

New settlements in the UK have a distinctive history, and reviewing this history is a valuable element in understanding the part that new settlements can play in accommodating housing growth. Two reviews of the New Towns programme are a good place to start: Ward's *New Town, Home Town: The Lessons of Experience* (1993) and the more recent text by Alexander (2009) *Britain's New Towns: Garden Cities to Sustainable Communities.*

More wide-ranging consideration of the contribution of new settlements is provided by the TCPA's publications such as *Best Practice in Urban Extensions and New Settlements* (2007) and *Re-imagining Garden Cities for the 21st Century* (2011c).

Economic development

Whilst there are many texts on regeneration and economic development, there are far fewer which focus specifically on the role of planning. However, many of the works on regeneration and economic development give valuable background and insights: a reasonable place to start is Jones and Evans (2008) *Urban Regeneration in the UK,* but there are many other books with a similar mission. For a particular take on planning and economic development, Raco (2007) *Building Sustainable Communities: Spatial Policy and Labour Mobility in Post-war Britain* is worth a read. For more specific practical aspects, various items of government guidance, such as the *Employment Land Reviews Guidance Note* (ODPM: 2004c), are essential.

Retail trends and policy

Guy's (2007) *Planning for Retail Development: A Critical Review of the British Experience* provides an excellent critical review of the development of policy, together with a useful brief review of retail trends. It is the most authoritative text on retail planning. For an interesting take on the problems of retailing in town centres, see *The Portas Review: An Independent Review into the Future of our High Streets* (2011). For the problems and policy approaches for small towns, Powe *et al.* (2007) *Market Towns: Roles, Challenges and Prospects* provides a useful overview. The technical issues of forecasting and measuring shopping needs are covered very effectively in England (2000) *Retail Impact Assessment: A Guide to Best Practice*, which marries academic and practical insights.

Notes

1 Whether such a policy approach actually reflects how people live or want to live is disputed by some: see, for example, Shorten (2004). However, such debates do not invalidate the objective of seeking a reduction in the need to travel.

2 More detailed discussion of settlement policies, particularly as they affect rural areas, can be found in Chapter 10.

3 It might be argued that moves towards quantification and quests for efficiency whose development can be associated with the Conservative governments from 1979 (Morrison and Pearce 2000) are all part of a similar management culture for the public sector which nurtured 'evidence-based planning'. Monitoring of matters such as housing land availability and efforts to measure the efficiency of planning processes clearly predate the Planning and Compulsory Purchase Act 2004.

4 Comparison of results from the 2001 and 2011 Census showed that the increase in single-person households levelled off after 2001.

5 The Use Classes Order was amended in 2013, but LPAs had the option of making use of Article 4 Directions to remove the new permitted development rights. Following concern that some LPAs may be 'misusing' these Directions, the planning minister, Nick Boles, felt obliged to state in February 2014, 'To ensure the permitted development rights are utilised fairly across England, my department will update our planning practice guidance to councils.'

6 Despite the uncertainties, the reports invaluably brought together a wide range of data, information and explanation – to make some telling points: for example, at the current new house replacement rates a home built now would have to last 1,200 years (Barker 2004: 47).

7 Hall *et al.*'s (1973) *The Containment of Urban England* noted that the planning system had contributed to rising land and property values through – effectively – rationing land supply, but a result was that regressive social redistribution had occurred, with the already affluent gaining most. In effect, along with other changes, such as the separation of home and workplace, the results of planning had been almost the opposite of what had been intended. Interestingly, this work remains perhaps the only serious attempt to evaluate the impact of planning.

8   Just 10 per cent of the population of the UK own 90 per cent of the land (UK Foresight 2010).

9   The days of council housing were similarly affected by concerns for countryside protection, which led to the production of tower blocks as an attempted technological fix for reconciling welfare aims with rural preservation (Dunleavy 1981).

10   And presumably that portion of the uplift not secured through obligations, and therefore arising as unearned profit for developers, would be 'paid for' in terms of an increased cost of new housing for consumers.

11   There were a series of – often technically detailed – publications issued under the title *Geography of Housing Market Areas*. The research 'sought to identify the optimal areas within which planning for housing should be carried out' and linked 'places where people live, work, and move home'. It therefore drew on a range of data, but after examining a range of alternative formulations plumped for an approach which worked on three levels, but only defined the first two. These first two were termed 'framework housing market areas', which, like journey-to-work areas, were defined by a high level of commuting closure (77.5 per cent self-containment); and 'local housing market areas', defined by migration patterns (50 per cent self-containment). The third was termed 'submarkets', which were defined by neighbourhood or house type. The result was a set of seventy-five framework housing market areas, with a tier of 280 local housing market areas nested wholly within them.

12   In 2013, the ONS reported that rates of home ownership had fallen for the first time since the 1950s. It cited high house prices (ownership rates are low in London), low wage growth and a period of tighter lending requirements (2013a) as possible explanations. However, in 2012 the Future Homes Commission reported that standards of new build homes – and particularly space standards – were a disincentive to buyers. A report by RIBA (2011) found that UK space standards were amongst the lowest in Western Europe – houses in Ireland were 15 per cent bigger, in Denmark 53 per cent bigger and in Germany 80 per cent bigger.

13   The 2011 Census showed a 70 per cent increase in 'concealed families', who made up 1.8 per cent of all families. In total, 1.3 million families lived in households with other families or other single people.

14   This is distinct from the phenomenon in some localities of an increase in numbers of 'houses in multiple occupation', associated with concentrations of, for example, numbers of recipients on housing benefit in seaside towns or of students in certain neighbourhoods, which are regarded as a problem by some local authorities. The impact of new regulations to control such concentrations introduced by the Labour administration was perhaps somewhat softened by the incoming Coalition government in the autumn of 2010, when it announced that 'changes of use from family houses to small HMOs will be able to happen freely without the need for planning applications. Where there is a local need to control the spread of HMOs local authorities will be able to use existing powers, in the form of article 4 directions, to require planning applications in their area.' However, landlords of HMOs need to obtain a licence for properties over a certain scale. (See Circular 08/10 *Changes to Planning Regulations for Dwellinghouses and Houses in Multiple Occupation*.)

15   Empty Homes is a charity campaigning to bring more empty homes back into use. Government announced in 2012 grants totalling £100 million administered by the HCA to bring empty homes back into use; it will also give the New Homes Bonus for long-term empty homes brought back into use, and there are also potential council tax changes – the levying of an 'empty homes premium'.

16   In a review of strategies to tackle low demand (Cole *et al*. 2003), the varied manifestations of low or falling housing demand were described in the following terms: 'high household turnover, empty properties, falling property values and small or non-existent waiting lists for rented dwellings. The wider market context, however, varies considerably. In some parts of the North and Midlands low demand is structural and endemic in nature, while in London and more buoyant markets the main challenge is the transience of the incoming population.' Housing Market Renewal Pathfinders represented an attempt to rebuild housing markets and communities in these areas.

17 The claimed benefits of living in compact cities vary greatly. Arguments in favour include Jacobs (1961), Elkin *et al.* (1991), Sherlock (1991), ECOTEC (1993) and various official publications on sustainable development. Arguments suggesting that the benefits are illusory, infeasible or overstated include Breheny (1997), Hall (1999b) and K. Williams (1999).

18 The Cambridge News of 29 January 2014 reported that the £30 million agreement includes £10 million for a primary school, £8 million towards a secondary school, £1.5 million for a community centre and £2.6 million on sports facilities.

19 As reported in the *New Statesman* of 6 March 2011.

20 The RTPI website styles this as 'Myth number 4: planning is a drag on economic growth'. The view of the RTPI president is 'Planners are not the enemies of enterprise. They are not the Town Hall bureaucrats who obstruct economic growth. On the contrary, they provide policies that are integrated across areas to promote both growth and regeneration. They provide land allocations to enable commercial and industrial uses to be developed where they are needed. They also provide a basis for co-ordinating delivery to make sure that things happen.' (Communities and Local Government Committee: Examination of Witnesses, 3 November 2011, Questions 105–28).

21 For example, in work for an inquiry into the LDF for York, estimates of land needed for employment – based on the same jobs forecast – varied between 90 and 263 hectares, depending on different assumptions about types of jobs, styles of development and vacancy rates.

22 In response to the Portas Review, government established some funded advice and support programmes – twenty-seven 'Portas Pilots', each receiving £100,000, and a number of 'town team partners', each receiving £10,000 – together with other measures in areas such as markets, parking, empty properties, and business rates. In 2013, DCLG announced in *The Future of High Streets: Progress since the Portas Review* a continuing range of support.

23 The Market Towns Initiative was announced in the *Rural White Paper* of 2000 and initially taken forward by the Countryside Agency and RDAs, but the responsibility was eventually moved entirely to the RDAs. There was a programme of health checks and associated action plans for over 200 towns, carried out and put into place with financial support from the initiative. In addition, there was support for a 'trade body', Action for Market Towns, which now operates as a membership organisation (www.towns.org.uk).

24 The guidance in PPS 6 was consolidated into an overarching document, *Prosperous Economies*, published in 2009 as PPS 4.

25 Further guidance on what is 'edge of centre' has been provided by DCLG in *Practice Guidance on Need, Impact and the Sequential Approach* (2009).

26 See also the series of reports on the employment impact of out-of-town superstores published by the National Retail Planning Forum.

27 This was based on fifty case studies of major application following the NPPF coming into force in March 2012. It found that the ratio of permitted new retail floorspace in major retail developments in out-of-centre locations to that in town centres and edge-of-centre locations combined, had been in excess of three to one.

28 The seriousness of the problem was described thus by the Distressed Town Centre Property Task Force in its report of 2013: 'Shop vacancy rates have grown nearly fourfold since the beginning of 2008 from c.4% to a peak of 14.6% in early/ mid 2012 and have now stabilised at 14.1%. This equates to over 22,000 empty shops in the top 650 town centres and if one casts the net to all GB premises (shops & leisure) then the number rises to over 53,000 units. This is the equivalent of over 53 Sheffield city centres' (DTCTPF 2013: 11).

29 For example, DCLG produced a report *The Future of High Streets* in 2013, as did the Scottish Government with its publication *Town Centre Action Plan*.

30 For example, studies for the GLA by Ramidus Consulting in 2012 found that there had been a fall in demand for office space in outer London, reflecting factors such as changing work styles, changing cost differentials, pressures on the public sector and the poor quality of some of the existing stock.

7 Environment, sustainability and climate change

There are two terms that have come to define the twenty first century. The one that we hear often is the 'urban age' and the one that we hear rarely is the 'age of man'. Both are, however, closely related. It was on the path to urbanisation that we paved the Earth, retrofitted nature to fit our purposes, and created the Anthropocene. What has made this process materially possible and ethically acceptable is the anthropocentric (human-centred) view of the world that has prevailed since the Enlightenment era.

(Davoudi 2014: 372)

Introduction

The term Anthropocene was coined by the Dutch atmospheric chemist Paul Crutzen to argue that we no longer live in the Holocene (i.e. a 10,000-year-old geological epoch of relative climate stability compared with the previous period, which was distinguished by regular shifts into and out of ice ages). He argued that we have entered a paradigmatically different epoch, called the 'age of man', in which, 'dam by dam, mine by mine, farm by farm and city by city' (*Economist* 2011: 3), humans have remade nature. For the first time in history, human activities have brought about planetary changes, the significance of which is on a par with geological forces. Compelling evidence of this is the reconfiguration of the planet's carbon cycle by the anthropogenic (human-made) release of quantities of fossil carbon over the past couple of centuries that took the planet hundreds of millions of years to store away, as will be discussed later in this chapter.

THE MEANINGS OF THE ENVIRONMENT IN PLANNING

Concerns about the environment have a long history in planning. Indeed, the planning movement in the late nineteenth century in the UK was motivated not just by social but also environmental concerns. Since then caring for the environment has been a key part of the planning system. However, the meanings given and the values attached to 'the environment' have fluctuated substantially over time. A number of attempts have been made to trace the evolving treatment of the environment in planning. Among the first was Newby (1990: 3), who argued that, in the 1980s, 'environmental concerns in the United Kingdom mark a shift in perspective from an amenity-led to an ecology-led approach to environment'. Later, Whatmore and Boucher (1993) explored the role of planning in the social construction of nature and its manifestation in the tensions between development and conservation.

They argued that until the 1980s the conservation narrative was dominant. Building on these analyses, Healey and Shaw (1994: 427) provided a more detailed history of the environmental discourses in planning and identified five strands, including welfarist utilitarianism (1940s/1950s), growth management (1960s), active environmental care (1970s), marketised utilitarianism (1980s), and sustainable development (1990s). A recent analysis of English planning documents by Davoudi (2012b) provides a more nuanced and updated account of the change and identifies eight distinct forms of environmental discourses in development plans. These are the environment as local amenity, as a heritage landscape, as a nature reserve, as a storehouse of resources, as a tradable commodity, as a problem, as sustainability, and as a risk.

The next sections of this chapter do not necessarily map on these categories but it is useful to start the chapter with an understanding of the importance of perceptions and value treatments of the environment in the formation of environmental and planning policies, because, as Botkin (1990) suggests, the basic assumptions that we make about nature influence the level of progress that we make with regard to environmental issues. The following account, which draws heavily on Davoudi (2012a and 2014), provides a brief summary of the different meanings of the environment in planning.

Environment as local amenity has a long history in planning and in fact predates the use of the term 'environment'. Many of the conservation objectives of plans are driven by intangible amenities such as aesthetic and recreational values. Their protection was also an influential driver of the planning movement and the emergence of two different attitudes towards the countryside in the early twentieth century. One sought to weave the amenities of the countryside into the urban fabric to improve the standard of city living (as in Ebenezer Howard's famous magnet and the garden city movement). The other sought to separate the perceived idyllic rural life from the encroachment of urban sprawl (as in the green belt policy). Although the garden city idea has remained as a source of inspiration, it was the dichotomous view of urban versus rural which became dominant in the planning system

and found powerful advocates, including Patrick Abercrombie, whose 1944 Greater London Plan was based on urban containment through the green belt. Throughout the 1940s and 1950s, seeing the environment as a local amenity and a backcloth to development remained the dominant perspective (Healey and Shaw 1994). This has continued to the present day. For example, the local plan for Merton (LBM 2010: 4) refers to 'the wide open spaces' as 'beauty spots' which are 'important . . . [for] everyone's well-being'.

Environment as heritage landscape refers to both the built and the natural environment but is particularly associated with landscape. Like amenity, it has a long history in planning dating back to the mid-nineteenth century. 'However, unlike amenity which began as a welfare-oriented discourse, the early approaches to heritage were highly elitist' (Davoudi 2012b: 53). As Newby (1990: 6) argues, 'Environmental artefacts were to be preserved "*for* 'the Nation', but *from* 'the public'", who were seen as "representing a threat to this 'national heritage'"'. During the inter-war period, such elitist views clashed with the demand for wider access to the countryside. Seeing the environment as heritage landscape has led to the assignment of hierarchical values to it ranging from national to local significance and resulting in a descending level of protection. At the top of the hierarchy are National Parks. Further down are the Areas of Outstanding Natural Beauty, and lower still are various sites of locally significant landscapes (see Chapter 10). Such hierarchical designations lead to different degrees of protection for them. Seeing the environment as heritage landscapes still occupies a central place in planning discourses. Every contemporary plan has a section with a list of designated sites.

Environment as a nature reserve may seem similar to 'the environment as heritage landscape' but it differs from it in a fundamental way, because it is the only conception of the environment which is not rooted in a human-centred view of nature. Instead, it is based on valuing nature for its own sake. It shows concerns about human beings as well as non-human species. An early indication of ecological considerations can be detected in Patrick Geddes's idea of the 'natural region', but the real influence on planning discourses began in the 1970s and through a series of international designations and protection of sites for their biodiversity and habitat values and not just for human enjoyment (see Chapter 10). The same is true for 'green corridors' and 'green networks' that aim to 'create safe species movement and havens for nature' (LBM 2010: 130). An important feature of these new concepts is their articulation of nature as relational and fluid rather than bounded and fixed. Although valuing and protecting nature for nature's sake have not been a prominent environmental discourse in planning, they have continued their somewhat marginal presence. The RSS for the East of England, for example, urges the planning authorities to 'seek to . . . protect, for their own sake, all important aspects of the countryside' (GOEE 2008: 50).

Environment as a storehouse of resources and functions that can be exploited at will has been a powerful and enduring conception of the environment in industrial societies and has had a major influence on environmental discourses in planning. 'Conquering nature' through technologies and exploiting environmental resources for economic growth is still considered a hallmark of progress. However, these extreme views are unfounded in the planning system, partly because planning itself emerged as a response to the excessive damage caused to the environment by urban industrialisation in the late nineteenth century. Nevertheless, a more subtle version of this extreme view has had a profound influence on planning. This is reflected 'in the utilitarian approaches to the environment and its treatment as a container of material resources, scientific repository and functional services, and also as a 'sink' for processing waste and pollution' including 'carbon sink' (Davoudi 2012b: 56). Examples of the functional utilitarian view of the environment can be found throughout the history of planning (Healey and Shaw 1994; Davoudi *et al.* 1996) and are still abundant in contemporary plans. For example, the local plan for Chelmsford (CBC 2008: 56) states that 'the importance of open spaces relates to their function and also amenity value . . . in providing a "green lung" and visual break in the built environment'. This view is now reinforced by seeing the environment as a container for 'ecosystem goods and services'. It should be noted that although the discourse of 'eco-system goods

and services' is ultimately a functional utilitarian one, it nevertheless puts the emphasis on making space for nature (e.g. freeing up land for flooding or coastal retreat) rather than on controlling nature (e.g. building flood and coastal defences).

Environment as a tradable commodity marked a major shift in the 1980s. Prior to this, exploitation of environmental resources had been largely justified in planning on the basis of being in the 'public interest'. In the 1980s, this welfare-oriented approach was sidelined by a market-led approach which considered the environment as a commodity. The change was partly driven by the 'presumption in favour of development' which became dominant at the time. As a result, 'a vision of planning as a bargaining process' and 'of nature as a social product valued through the market place' was constructed (Whatmore and Boucher 1993: 170).

> Environment was treated as a commodity which could be traded with other commodities through the bargaining processes of planning gains. The assumption was that the loss of one environmental 'parcel' can be offset by gaining another as long as the bargaining process did not jeopardise development projects.
>
> (Davoudi 2012b: 58).

The treatment of the environment as a tradable asset continued in the 1990s and underpinned the economically driven approaches to sustainable development and the notion of 'constant capital stock' in the *Blueprint for a Green economy* (Pearce *et al.* 1989). It continues to run through contemporary plans. For example, the local plan for Lancaster lists the city's 'natural and built "Environmental Capital"' and presents them as 'a major economic asset' (LCC 2008: 55). The RSS for the North West Region considers 'Access to greenspace' as having 'a central role to play in securing successful and sustainable economic regeneration' (GONW 2008: 89). In the London Plan, measures for reducing carbon emissions are framed in terms of their economic values: 'preventative and adaptive measures will generate long term savings . . . and should have positive impacts on property values' (GLA 2009: 94).

This approach is reflected in and reinforced by the notion of 'natural capital', which is defined in a government White Paper (DEFRA 2011a: 11) as

> the stock of our physical natural assets (such as soil, forests, water and biodiversity) which provide flows of services that benefit people (such as pollinating crops, natural hazard protection, climate regulation or the mental health benefits of a walk in the park). Natural capital is valuable to our economy. Some marketable products such as timber have a financial value that has been known for centuries. In other cases (e.g. the role of bees in pollinating crops), we are only just beginning to understand their financial value.

Environment as a problem made its 'dramatic leap to the top of the political agenda' in the 1960s (Dryzek 1997: 21), when 'attentions began to shift from what nature does to us and what we can get out of it, to what we do to nature and the problematic consequences of our actions' (Davoudi 2012b: 59). This triggered two different responses: the first one came from the Club of Rome's *Limits to Growth* (Meadows *et al.* 1972) and advocated radical actions to curb human demand; the second one promoted problem-solving approaches. It was the latter that influenced planning most, partly because it portrayed environmental problems as being 'tractable within the basic framework of the political economy of industrial society, as belonging in a well-defined box of their own' (Dryzek 1997: 61). In planning, the 'boxing' approach could be found in the plans of the early 1990s, where everything that was considered as environmental issues went into one chapter of the plan and was often colour-coded green (Davoudi and Layard 2001), a practice which was largely abandoned later in favour of a more integrated approach. Planning was given a major role in solving environmental 'problems' in the 1990s, mainly in response to the growing number of EU environmental directives (on waste, habitat, water, birds, etc. discussed below) and particularly the EU requirement to carry out environmental impact assessments for major development projects, as mentioned in Chapter 4. 'Planning documents have continued to present both *concerns* for the environmental problems and *reassurances* for the ability to solve them without having to change the existing socio-economic structures' (Davoudi 2012b: 59).

Environment as sustainability found salience after the publication of the Brundtland Report (WCED, 1987) and its famous definition of sustainable development. As discussed in the next section of this chapter, two fundamentally different interpretations of sustainability emerged, but a 'weaker' approach became dominant in planning.

Environment as a risk was introduced into planning discourses through the climate change agenda (see below for the discussion on climate change). While its mitigation discourse follows that of sustainability by emphasising energy efficiency and the use of renewable energies, its adaptation discourse marks a departure from sustainability in the sense that it considers 'the environment as a *risk to,* rather than as *an asset for,* human wellbeing' (Davoudi 2012b: 62). It should be noted that the emphasis here is on those discourses of climate change which portray apocalyptic imaginaries of the future. One implication of seeing nature as a risk is the growing concern with security. Because risk and security feed from each other, a growing number of social and environmental problems, including climate change, are now articulated as security problems. This may lead to a shift in traditional environmental concerns in planning away from development versus environment to a focus on which security should take precedence. 'Increasingly food security trumps biodiversity, energy security trumps renewable energy, and climate security trumps sustainability' (Davoudi 2014: 368). Therefore, contrary to the rhetoric of sustainable development, which imbued environmental discourses of plans with the optimism of win-win solutions (to be discussed in the next section), the risk-laden language of climate change discerns a subtle sense of pessimism. This in turn may lead to a 'shift from an environmental politics of cooperation and consensus', which underpinned sustainability, 'to a politics of securitization' (Blowers *et al.* 2009: 313).

While the above summary may give the impression that one meaning of the environment has replaced another, this is clearly not the case. By contrast, 'over time new discourses have been added to planners' repertoire to create what is now a multi-layered narrative of the environment in contemporary plans' (Davoudi 2012a: 66). Some discourses may complement each other, while others may be in conflict, but most are in competition for defining the policy agenda. Their creation and continuation are infused with power relations. The question of values and the question of power matter, even though they do not feature in the description of environmental legislation and policy that follows in this chapter, because of the scope and purpose of this volume.

EU ENVIRONMENTAL POLICY

By its very nature, the environment transcends political, legal and physical boundaries. That is why it is paramount that different countries across Europe, and indeed throughout the world, cooperate with each other to protect and enhance the environment and tackle challenges of climate change. This has been an important justification for the EU having competence in environmental policy:

> The underlying aim of EU environmental policy is to enhance natural capital, promote a resource-efficient economy and safeguard people's health. A coordinated environmental strategy across the Union ensures synergies and coherence between EU policies and, given the relevance of environmental legislation for many business sectors, will ensure a level playing field for their activities.
>
> (CEC 2013a:3)

Since the 1970s, the EU has agreed over 200 pieces of legislation to protect the environment (CEC 2013a). Indeed, much of the UK government's policies on the environment has been influenced and sometimes determined by EU directives and regulations (Milton 1991: 11; Wilkinson *et al.* 1998), with major implications for the planning system. However, both the EU and the UK have often faced the problem of

non-implementation as well as frequent prioritisation of economic growth over environmental sustainability.

Over the last four decades, the focus of EU policy on environment has shifted significantly. In the 1970s and 1980s, the emphasis was on traditional environmental themes, such as protecting species and improving the quality of the air and water by reducing emissions and pollutants. Today, the emphasis is on a more systematic approach that takes account of the interrelationship between various environmental and ecological issues and their global dimension. Thus, the EU has moved from mere remediation to include prevention of environmental degradation (CEC 2013a: 3).

As mentioned in Chapter 3, the Treaty of Rome imposed no environmental obligations on member states and the Community initially had no environmental competences. Indeed, Article 2 of the Treaty provided that sustained rather than sustainable growth was the aim: 'a continuous and balanced expansion'. The international scene changed in the late 1960s and early 1970s, with a significant influence being the UN Conference on the Human Environment which was held in Stockholm in 1972. In the same year, the European Community determined that economic expansion should not be 'an end in itself', and that 'special attention will be paid to protection of the environment' (Robins 1991: 7). In 1973, the first EC *Action Programme on the Environment* was agreed, covering the period 1973–6. The Single European Act 1987 gave added legitimacy by including environmental goals in the Treaty and crucially adding the important provision that 'environmental protection requirements shall be a component of the Community's other policies' (Haigh 1990: 11). Since then, the European Environment Agency (EEA) has been established, with headquarters in Copenhagen, providing a monitoring service for the European institutions.

The Action Programmes on the Environment have had an increasing impact on policy and practice in member states. They are 'forward planning' documents for emerging policies to be implemented by the EU and followed by national, regional and local governments. While they have no binding status, many of the proposals are followed through in EU directives and other actions. The *Fifth Action Programme*, published by the EC in 1992 and covering the period 1993–2000, brought a comprehensive and long-term approach. The overriding aim of the programme was to ensure that all EU policies have an explicit environmental dimension. It also stressed the importance attached to spatial planning instruments:

> The community will further encourage activities at local and regional level on issues vital to attain sustainable development, in particular to territorial approaches addressing the urban environment, the rural environment, coastal and island zones, cultural heritage and nature conservation areas. To this purpose, particular attention will be given to: further promoting the potential of spatial planning as an instrument to facilitate sustainable development.
>
> (CEC 1992: 70)

The EAP's objectives need to be transposed into agreed Community law and action to be implementable. The great majority of EU environmental laws are in the form of directives, which set objectives to be achieved but which give member states some freedom to choose the manner in which they are transposed into national law. It is unusual for directives to be transposed into national legislation by the due date – which is typically two months after adoption by the Council of Ministers. Nevertheless, they must be implemented 'in a way which fully meets the requirements of clarity and certainty in legal situations'. States cannot rely on administrative practices carried out under existing legislation (Wägenbaur 1991). Moreover, if a directive is not implemented by national law, it is possible for legal action to be taken by private parties to seek enforcement. The use of Community legislation has tended to give way in some areas to more general agreements and guidelines. (Chapters 3 and 4 consider the institutions and spatial planning actions of the EU.)

The 1987 Amsterdam Treaty incorporated sustainable development as a fundamental objective of the EU and since then there have been commitments to ensure environmental appraisal of all Community policies and actions. In 1999, the Commission undertook an evaluation of the Fifth Action Programme and reported in the *Global Assessment*, which recognised that despite some environmental improvements 'less progress has been made overall in changing economic and societal trends which are harmful to the environment'. The report notes that economic growth 'simply outweighs the improvements attained by stricter environmental controls' (CEC 1999b: 7).

A Sixth Environmental Action Programme was published in 2002: *Environment 2010: Our Future, Our Choice*. Its proposals range over such matters as environmental taxation, improving the implementation of existing initiatives, completing the European network of habitats through *Natura 2000*, and preventing urban sprawl, especially along coasts. Two of the proposals in the agenda are of particular interest here. The first is the commitment to encourage better land use planning and management decisions while ensuring that 'environmental issues are properly integrated into planning decisions'. Although the Commission was given the task of following this up through the preparation of a communication on the environment and planning, no further progress was made. Instead, more attention was given to the second main commitment in the Sixth EAP, the *Thematic Strategy on the Urban Environment*, one of seven thematic strategies. The Commission began a wide-ranging consultation on the Urban Environment Strategy in 2003 and published a communication, *Towards a Thematic Strategy for the Urban Environment* in 2004. This set out the overall aim of improving the 'environmental performance and quality of urban areas'. Four themes were identified: sustainable urban management, sustainable construction, sustainable urban design and sustainable transport.

The final version of the Thematic Strategy was published in 2006 and was issued as a joint decision of the European Council and Parliament. It could have been of great help to local authorities, but lack of resources among other factors inhibited delivery (Atkinson and Mills 2005: 107). However, the EU Green Capital Award has been set up to encourage environmental promotion and to 'showcase the environmental care and imagination that cities across Europe are displaying'

(CEC 2013a). Many UK cities have competed successfully for such recognition.

In 2010, the Commission published *Europe 2020: A Strategy for Smart, Sustainable and Inclusive Growth* (CEC 2010). This reflects rising awareness of the deficiencies in the conventional economic model, which does not account for environmental externalities in decisions on natural resource use and allocation. The Strategy explicitly acknowledges the need to create synergies between environmental and economic goals, and argues for a transition towards a 'green economy'. It stresses the use of environmental policy to promote the efficient use of natural resources and the strengthening of the resilience of ecosystems. EU Environmental policy is explicitly described as a 'quest for sustainability' in the way that it 'aims to strike a balance between our need to develop and use the planet's natural resources, and the obligation to leave a healthy legacy for future generations' (CEC 2013a: 8). The role of planning in delivering these aims is recognised. It is suggested that to be sustainable, sectors such as urban planning 'need to deliver the services we need without compromising the health of the natural world we all depend on' (ibid.).

As concerns over climate change have grown in the last decade, the EU has increasingly focused on policies related to climate change. Improving resource efficiency is seen as a cornerstone of this goal, and concrete targets are set in the 2011 *Roadmap* towards a resource-efficient Europe. The Roadmap includes a vision that 'by 2050 the EU's economy has grown in a way that respects resource constraints and planetary boundaries, thus contributing to a global economic transformation'. In 2011, a *European Resource Efficiency Platform* was also launched, followed by a Manifesto in 2012 and *Action for a Resource Efficient Europe* in 2013 (EREP 2013). The aim is to reduce the total material requirements of the EU economy by between 17 and 24 per cent by setting ambitious, credible targets as soon as possible. This fits with the EU 2020's objective of decoupling resource use and its environmental impacts from economic growth. Indicators to measure progress towards these targets relate to the four key resources of carbon, materials, water and land (EREP 2013: 2). Specific attention is to be given to valuing

ecosystems, identifying the opportunities arising from waste management and recycling and developing footprint indicators to account for EU imports.

The EU also plays an active role in international negotiations on sustainable development, biodiversity and climate change. A notable example is the UN Millennium Goal of long-term environmental sustainability. In January 2014, the Seventh Environment Action Programme to 2020: *Living Well, Within the Limits of our Planet* was published to guide European environment policy and set out a long-term vision:

> In 2050, we live well, within the planet's ecological limits. Our prosperity and healthy environment stem from an innovative, circular economy where nothing is wasted and where natural resources are managed sustainably, and biodiversity is protected, valued and restored in ways that enhance our society's resilience. Our low-carbon growth has long been decoupled from resource use, setting the pace for a safe and sustainable global society.
>
> (CEC 2014: 1)

As mentioned above, the Seventh EAP is not binding, so it is up to the EU institutions and the member states to ensure its implementation and the achievement of its priority objectives as summarised in Box 7.1.

European Environment Agency

An important EU institution as far as the environment is concerned is the European Environment Agency. Based in Copenhagen, the EEA began work in 1994, collecting national data to produce European datasets, developing and maintaining indicators and reporting on the state of the environment. Its mandate is to help the EU and its member states to make informed decisions about improving the environment, integrating environmental considerations into economic policies, and to coordinate the European environment and information network (CEC 2013a: 5).

The EEA produces annual reports of the state of the environment based on a set of 146 environmental indicators, grouped into twelve environmental themes.

BOX 7.1 THE EU SEVENTH ENVIRONMENT ACTION PROGRAMME (2014)

The EAP identifies three key objectives (emphasis in original):

- to protect, conserve and enhance the Union's *natural capital*
- to turn the Union into a *resource-efficient*, green, and competitive low-carbon *economy*
- to *safeguard* the Union's citizens from *environment-related pressures* and risks to health and well-being.

Four so called 'enablers' will help Europe deliver on these goals:

- better *implementation* of legislation
- better *information* by improving the knowledge base
- more and wiser *investment* for environment and climate policy
- full *integration* of environmental requirements and considerations into other policies.

Two additional horizontal priority objectives complete the programme:

- to make the Union's *cities more sustainable*
- to help the Union *address international environmental and climate challenges more effectively.*

Thirty-seven are designated 'core set indicators'. Most are explicitly designed to support environmental policies. They are organised into a five-point causal framework (Driver-Pressure-State-Impact-Response, or DPSIR) as follows:

> social and economic developments drive (D) changes that exert pressure (P) on the environment. As a consequence, changes occur in the state (S) of the environment, which lead to impacts (I) on society. Finally, societal and political responses (R) affect earlier parts of the system directly or indirectly.
>
> (EEA 2013: 28)

This framework helps to structure thinking about the interplay between the environment and socio-economic activities. Furthermore, in the absence of continued sustainability indicator monitoring in the UK (as they were stopped by the Coalition government), the EEA monitoring can be used as a source of information on sustainability at both European and UK level. The EEA's 2013 monitoring report notes that 'Spatial planning and land management emerge

as key approaches for framing governance strategies capable of increasing resource efficiency, maintaining environmental resilience and maximising human well-being' (EEA 2013: 12).

Despite its apparent disgruntlement with EU environmental regulations, the UK government has, by and large, been supportive of, and often a major contributor to, EU environment policy. This is reflected in the 2011 White Paper, which states that 'The Government's aim is to provide environmental leadership in the EU to put it on a path towards environmentally sustainable, low-carbon, resource-efficient growth, resilient to climate change' (DEFRA 2011a: 63). Government's priorities for influencing the EU include (ibid):

- achieving competitive agriculture, fisheries and food sectors which use and protect natural resources in a sustainable way and meet the needs of consumers;
- integrating the EU's objectives for environmental sustainability into all EU policies and spending;

- protecting, managing and using natural resources sustainably, so as to support economic growth; and
- implementing the stock of legislation properly across the EU, reviewing older directives and applying the principles of better regulation to new legislation.

Later in this chapter we will discuss how various EU regulations and directives have been incorporated into UK environmental regulations and how they have affected planning policies and practices.

ENVIRONMENT IN THE UK PLANNING SYSTEM

In one sense, all spatial planning is concerned with the environment, but the reverse is not true, and it is difficult to decide where to draw the boundaries. The difficulty is increased by the rate of organisational change over recent years, including the shifting of responsibilities from local government to ad hoc bodies, and by the flood of new legislation, prompted in part by the EU. The emphasis (at least in theory) on sustainable development as a political goal and concerns about climate change add to the complications of the links between planning and environmental policy in general and the division of responsibilities for environmental regulations between the planning system and other public bodies such as the Environment Agency (EA) and the newly established local nature partnerships (to be discussed below).

Notwithstanding these fragmentations, planning continues to play a significant role in delivering government's environmental policy. It is therefore, paramount that planning policies and decisions are based on up-to-date information about the natural environment and related areas. This is often provided in a growing number of management plans, such as river basin management plans, shoreline management plans, and waste management plans (discussed in relevant sections of this chapter). To ensure that planning policies and proposals are not detrimental to the environment (as well as social and economic considerations), local plans are subject to a sustainability appraisal which meets the requirements of the European Directive on Strategic Environmental Assessment (Directive 2001/42/EC). They may also be subject to a number of other environmental assessments, including those under the Habitats Directive (Directive 92/43/EEC), where there is a likely significant effect on a European wildlife site, strategic flood risk assessments and assessments of the physical constraints on land (as discussed in Chapter 4).

ENVIRONMENTAL POLITICS AND INSTITUTIONS

Environmental politics has become an energetic force on the British scene and this is reflected in the growth of environmentally related government units and agencies, advisory panels, interest groups and campaigners. Thus the environment has become part of the political coinage, and the parties vie with each other in producing convincing statements not only of their concern but also of their workable programmes of action.

Curiously, part of the growth of environmental consciousness was due initially to the lack of government concern. The environment was rarely the subject of political battles. Yet England has been a world pioneer on a number of environmental issues. The Alkali Inspectorate, which was established in 1863, was the world's first environmental agency. Some of the earliest voluntary organisations had their origins in England: for example, the Commons, Open Spaces and Footpaths Preservation Society in 1865, and the National Trust in 1895 – an organisation that (with over 2 million members) has grown to be the largest conservation organisation in Europe. The Town and Country Planning Act 1947 introduced a remarkably comprehensive land use planning system (even though, in the circumstances of the time, much of rural land use was purposely omitted). Legislation on clean air has a long

non-spatial, apolitical regulatory criteria' (Davoudi 2001: 90). Weak sustainability simply gives environmental capacities greater weight in the decision process (Owens 1994).

Another, more radical interpretation of sustainability is as 'strong' sustainability or 'reflexive ecological modernisation' (Hajer 1995). According to Davoudi (2001: 90), this approach (which she calls 'risk society' after Beck) considers the current mode of production as irreconcilable with maintaining the state of the environment and ecosystems. In this framework, planning would 'defend the environment against risks associated with economic processes, and focus on strategic and holistic approaches to place-making'. Owens (1994) explains that the strong definition places fixed and inviolable constraints on economic activity, and challenges whether it is right to continue to meet various demands and needs if this cannot be accomplished without reducing current levels and quality of environmental stock. Thus demand management of resource use should be the central policy response. In practice, Cowell and Owens (1998) have shown how the planning system mediates the questions of demand management and spatial location in a case study of aggregates planning – though the general argument can be applied more widely.

The strong sustainability approach is hardly found in the environmental narratives of planning. According to Blühdorn and Welsh (2007: 198), this seems to reflect the wider 'politics of unsustainability' which is due to two simultaneous trends: first, 'a general acceptance that achievement of sustainability requires radical change in the most basic principles of late-modern societies' and second, a general 'consensus about the non-negotiability of democratic consumer capitalism'. Both trends are happening 'irrespective of mounting evidence of unsustainability' (ibid.) and a growing awareness about climate change.

Those who advocate that sustainability is familiar in the history of planning (Hall *et al.* 1993) are in effect presenting the weak interpretation: the planning system's traditional role has been to deal with locational issues so as to reduce environmental damage and achieve some sort of 'balance', or, more correctly, 'trade-off' between new urban development and environmental protection. But strong (or even moderate) interpretations of sustainability raise questions about the capability of the planning system to deal with structural questions of the relationship between social justice – the distribution of costs and benefits – economic demands and environmental capacities. This is not to say that spatial or territorial questions are unimportant – they are – but that additional dimensions should also be considered, not least in demand management, for example, in relation to meeting housing land requirements (discussed in Chapter 6). They also reflect a growing consensus about the fundamental and very challenging principles which should govern public policy for sustainability.

In this respect, it should be noted that the UK approach has traditionally differed from that in other European countries, particularly Germany. An important difference in principle (differences in practice may be less marked) is that of 'anticipation', as distinct from reaction. Whereas the UK has taken the view that environmental problems should be defined in terms of their measurable impacts, other countries have gone beyond this, anticipating problems before the degree of environmental damage can be ascertained and adopting a *precautionary principle*. Shed of its more philosophical overtones, the issue is fundamentally 'whether to protect environmental systems before science can determine whether damage will result, or whether to apply controls only with respect to a known likelihood of environmental disturbance' (O'Riordan and Weale 1989: 290). The Labour government sanctioned the use of the precautionary principle in planning and suggested that it should be invoked when

- there is good reason to believe that harmful effects may occur to human, animal or plant health, or to the environment; and
- the level of scientific uncertainty about the consequences or likelihood of the risk is such that best available scientific advice cannot assess the risk with sufficient confidence to inform decision-making.

(PPS 23 *Planning and Pollution Control* (2004), para. 6)

Table 7.1 presents a framework of sustainability principles, developed specifically for spatial planning with the aim of assisting in transposing the very general notions of sustainability into planning and development practice and also for appraising existing planning policies and actions.

Assessment of the take-up of sustainability principles into aspects of spatial planning have grown

Table 7.1   Sustainability principles for spatial planning

Principles	Criteria
Overarching	
Futurity and intergenerational equity	Precautionary principle (no irreversible decisions) Include cumulative and long-term impacts in decision-making
Intersocietal equity	Commitment to equity at local, national and international levels Ensure commitment to equity so environmental impacts and the costs of protecting the environment do not unfairly burden any one geographic or socio-economic sector
Local and regional self-sufficiency	Reducing externality effects so that environmental impacts and costs do not unfairly burden any one geographic group or socio-economic sector Using close in preference to distant resources
Risk prevention and reduction	Natural disasters Human-made disasters
Environmental	
Maintain the capacity of natural systems	Absolute protection of critical natural capital Defence of improvement of soil quality and stability Defence and improvement of key habitats and biodiversity Respecting absorption and assimilation capacities of natural systems Efficient use of renewable resources
Minimise resource consumption	Minimum depletion of renewable resources Minimum depletion of non-renewable resources Energy efficiency Minimisation of waste, recycling and reuse
Environmental quality	Reduction of pollution emissions; protection of air and water quality and minimisation of noise Protection and enhancement of environmental amenity and aesthetics Protection of natural and cultural heritage
Economic and societal	
Protect and develop the economic system	Encourage and develop connections between environmental quality and economic vitality Satisfy and protect basic needs (shelter, food, clean water, etc.) Provide entrepreneurial and employment opportunities
Develop the human social system (education, democracy, human rights)	Protect basic human rights Ensure health and safety Improve local living conditions Satisfy the economic and living standards to which people aspire
Develop the capacity of the political system	Ensure transparent decision-making processes Develop open, inclusive and participatory governance Apply subsidiarity and ensure that competences are exercised at the most appropriate level

considerably. In sum, there has been only partial and fragmented conversion of the principles into planning policies and actions. Policies tend to follow well-worn formulae or 'checklists' and are seldom ambitious in addressing the strong definition of sustainability through, for example, demand management. Where there is a stronger position on sustainable development, the planning response tends to be understood in relatively narrow terms, predominantly the organisation of land uses and transport links, and because of institutional fragmentation there has been difficulty in coordinating impacts in fields such as energy, waste, air, noise and water. Policy compartmentalisation and departmentalism are strong barriers to effective coordinated approaches to sustainable development.

In the 1990s and early 2000s, the amount of advice to planning authorities on how to incorporate sustainability into their plans and decisions increased sharply, but aspirations outstripped achievements. Owens (1994) suggested there was a lot of 'sustainability rhetoric' but in practice 'business as usual'. Counsell (1998) reported that translation of sustainability principles into operational policies in structure plans was still 'proving difficult', though there was great variation in performance – perhaps as much related to local short-term self-interest as concerns about long-term intergenerational equity. Even the most ambitious experimental projects, such as the former government's *Millennium Villages*, did 'not . . . deliver the order of magnitude of improvement needed to demonstrate true sustainability' (Llewelyn Davies 2000: 3). The explanation is, of course, complex and the references noted here point to many factors, but planners often cited the contradictory and unhelpful nature of national policy and actions (especially outside the planning system) and the limited scope of planning. The *Sustainable Communities Plan* was notable in the controversy it engendered about just how sustainable it was (a question discussed in Chapter 6). One facet of the plan, the Millennium Communities Programme, was of particular interest, since it had the central task of providing demonstration projects to promote more sustainable development in mainstream housing development, and this is discussed in Chapter 10. Despite the shortcomings, the momentum on sustainable development and now climate change has continued (as can be seen in Box 7.2) albeit often at the level of policy talk rather than action.

BOX 7.2 MAIN EVENTS IN KEEPING THE MOMENTUM ON SUSTAINABLE DEVELOPMENT

(See Box 7.8 for climate change-related events.)

1962	Carson's *Silent Spring* was published, helping to launch the environmental movement
1972	Club of Rome's *Limits to Growth* was published, advocating radical actions to curb human demands on natural resources
1972	UN Conference on the Human Environment, Stockholm
1973	First EC Action Programme on the Environment
1985	First EC Directive on Environmental Assessment
1987	World Commission on Environment and Development: Brundtland Report, *Our Common Future*
1990	*This Common Inheritance: Britain's Environmental Strategy*

1992	UN Conference on Environment and Development (UNCED or the Earth Summit), Rio and creation of the UN Commission on Sustainable Development (CSD)
	Agenda 21
	Biodiversity Convention: international agreement to protect diversity of species and habitats
	Statement of Forest Principles for management, conservation and sustainable development of the world's forests
1994	*Sustainable Development: The UK Strategy*
1996	UN Habitat II Conference, Istanbul
	EU Expert Group on the Urban Environment Report on European Sustainable Cities
	The Aalborg Charter on Local Agenda 21 and the setting up of the European Sustainable Cities and Towns Campaign
1997	Earth Summit +5, five-year review and adoption of Programme for the Further Implementation of Agenda 21 by UN General Assembly
	EU Amsterdam Treaty incorporates sustainable development as a fundamental objective of the EU
1998	EU Communication on *Sustainable Urban Development: A Framework for Action*
1999	*A Better Quality of Life: A Strategy for Sustainable Development for the United Kingdom* (and additional special papers)
2000	EU Global Assessment of the Fifth Action Programme on the Environment
	The Convention on Persistent Organic Pollutants sets goals to eliminate a dozen of the world's most dangerous chemicals
	The establishment of the UK Sustainable Development Commission
2001	EU Sixth Framework Programme on the Environment
	OECD Analytical Paper on Sustainable Development
	UN Habitat III Conference
	Millennium Development Goals established
	EU leaders launch the first EU sustainable development strategy at the Gothenburg Summit (Gothenburg Strategy)
2002	World Summit on Sustainable Development, Johannesburg
2005	*Securing the Future: Delivering UK Sustainability Strategy* (DEFRA)
	Kyoto Protocol comes into force, limiting carbon emissions of signatory countries
	Millennium Ecosystems Assessment published
2010	The UK Coalition government announces its intentions of becoming the 'greenest government ever'
	UN Year of Biodiversity including the Tenth Conference of Parties for the International Convention on Biological Diversity, held in Nagoya
	Europe 2020: A Strategy for Smart, Sustainable and Inclusive Growth
	Making Space for Nature: A Review of England's Wildlife Sites and Ecological Network (report for DEFRA)
	The UK national ecosystem assessment (UK NEA)
2011	The UK's Green Investment Bank is formed to fund and drive investment in green infrastructure, including low-carbon projects in the UK
	The Natural Choice: Securing the Value of Nature (DEFRA)
	Mainstreaming Sustainable Development – the Government's Vision and What this Means in Practice (DEFRA)
2012	Rio+20 held, leading to the promised creation of sustainable development goals and strengthening of institutional structures
2013	The Ecosystems Markets Taskforce publishes its final report *Realising Nature's Value*
2014	EU Seventh Environment Action Programme to 2020: *Living Well, within the Limits of our Planet*

Agenda 21 in the UK

As mentioned above, the 1992 (Rio) Earth Summit gave a major impetus to the elaboration of 'sustainable' policies. Agreement was given to Agenda 21, a comprehensive worldwide programme for sustainable development in the twenty-first century. In formulating this programme, major emphasis was placed on a wide degree of participation. In the UK, this has been organised at central and local government levels. Two years after the Rio Summit, the government published *This Common Inheritance: Britain's Environmental Strategy*, which was followed by annual monitoring reports. In 1994, this was effectively replaced by *Sustainable Development: The UK Strategy*, which was the first national sustainability strategy arising from the Rio Declaration to be published. The 1997 Labour administration undertook to revise the strategy and published numerous consultation documents during 1998. By the mid-2000s, the following sustainability strategies were published:

- England – *A Better Quality of Life*, 1999
- Scotland – *Down to Earth*, 1999
- Wales – *Sustainability Strategy*, 2004
- Northern Ireland – *First Steps towards Sustainability*, 2005.

The 1999 Strategy for England promoted four main objectives: social progress, protection of the environment, prudent use of natural resources and maintenance of high levels of economic growth. It identified 147 sustainable development indicators, including fifteen headline indicators, and made an assessment of the baseline position and trends for each. The devolved administrations developed their own indicators. Levett (2000) described the list of indicators as 'a towering achievement', especially in their breadth, but noted that many were concerned with inputs as proxies for ends or measuring actual progress towards greater sustainability – as, for example, in measuring the existence of Agenda 21 strategies rather than their impacts. Such criticisms of indicators are well known. Selection is intensely political, because the indicators are in effect the definition of sustainability, and they may

reveal great shortcomings. Above all, as the strategy itself accepted, increasing eco-efficiency would not be able to keep pace with 'business as usual economic growth'. As Levett (2000) explained, 'eco-efficiency may have a useful contribution to make, but it is fanciful to the point of irresponsibility to expect it to be the main means of reconciling economic and environmental aims'. Thus ecological modernisation is not a long-term solution. Nevertheless, improvement in sustainability indicators was fast becoming an end in itself, while the political significance and impact of the strategy was questioned, especially in relation to public awareness. Consequently, considerable effort went into publicising the sustainable development goals of government, although the economic growth elements were often stressed. Indeed, Davoudi (2001) pointed out that the foreword to the 1999 *Sustainable Development Strategy* by Tony Blair barely mentioned the environment.

Reflecting on progress made on indicators, the Sustainable Development Commission concluded its 2004 review with *Shows Promise But Must Try Harder*, which also became the title of its report. The Commission challenged the government 'to create a new Strategy that is unified and much more strongly driven by a fundamental overarching commitment to sustainability at all levels and in all parts of Government; it should be a core part of the programme of all Departments, led from the centre' (p. 4). After a consultation which attracted only 900 written responses, a new Strategy was published in 2005, *Securing the Future: Delivering UK Sustainability Strategy*. It reported that government departments cherry-picked from the four principles set out in the 1999 version. Climate change figured much more prominently in the 2005 strategy and the planning system was identified as a 'key lever' for making necessary changes to help meet targets for slowing the growth of greenhouse gases and energy use. Crucially, the Strategy took account of the changed structure of government in the UK with devolution to Scotland, Wales and Northern Ireland and a greater emphasis on delivery at the then regional level. It, therefore, provided a set of shared UK principles for achieving sustainable development agreed by the UK government and

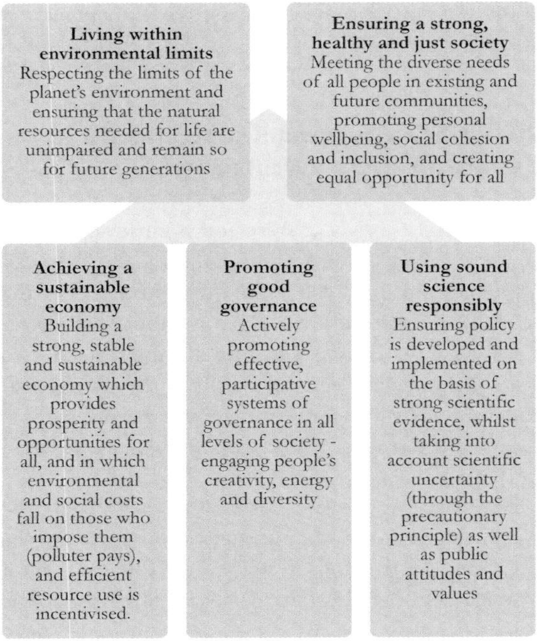

Living within environmental limits	Ensuring a strong, healthy and just society
Respecting the limits of the planet's environment and ensuring that the natural resources needed for life are unimpaired and remain so for future generations	Meeting the diverse needs of all people in existing and future communities, promoting personal wellbeing, social cohesion and inclusion, and creating equal opportunity for all

Achieving a sustainable economy	Promoting good governance	Using sound science responsibly
Building a strong, stable and sustainable economy which provides prosperity and opportunities for all, and in which environmental and social costs fall on those who impose them (polluter pays), and efficient resource use is incentivised.	Actively promoting effective, participative systems of governance in all levels of society - engaging people's creativity, energy and diversity	Ensuring policy is developed and implemented on the basis of strong scientific evidence, whilst taking into account scientific uncertainty (through the precautionary principle) as well as public attitudes and values

Figure 7.1 Guiding principles for sustainable development strategy in the UK

Source: HM Government 2011b: 17

devolved administrations. Its overarching approach and guiding principles (see Figure 7.1) were to be shared by the four separate strategies across the UK.

The contribution of planning to sustainable development is summarised in the Strategy on one whole page of the document (p. 116), which refers to policies already in place (in some cases for a considerable time), such as the brownfield land targets, sequential test and others discussed elsewhere in this book. The centrepiece of this explanation is section 39 of the Planning and Compulsory Purchase Act 2004, which (though rather distorted by the legal construction) had the effect of requiring those operating the planning system at both regional and local levels 'to exercise the function with the objective of contributing to the achievement of sustainable development'.

Although the 2005 Strategy has not been updated (as of 2014), the Coalition government published in 2011 its *Vision for Sustainable Development* as focusing

on 'Stimulating economic growth and tackling the deficit, maximising wellbeing in society and protecting our environment, without negatively impacting on the ability of future generations to do the same' (DEFRA 2011b: 2).

As can be seen, this moves the interpretation of sustainability further towards economic growth and away from environmental protection. This is despite the fact the Coalition government announced its intentions to become the 'greenest government ever'. As mentioned in Chapter 4, the government's initial poor definition of sustainable development in the National Planning Policy Framework (NPPF) created strong objections from many environmental and planning groups and forced it to be redrafted with a presumption in favour of sustainable development.

Local Agenda 21 in the UK

At the local level, Local Agenda 21 called for each local authority to prepare and adopt a local sustainable development strategy. These local efforts were aided by the work and publications of the Improvement and Development Agency (IDeA, formerly the Local Government Management Board). A major feature of the consultation programme at the local level was that it involved more than the term 'consultation' often means. Groups were established in local areas to debate the meaning of sustainability and to determine how progress towards it could be achieved and assessed, following the principle of 'you can only manage what you can measure'. These local endeavours were designed to produce policies and indicators which were locally appropriate. The research has underlined the importance of this local 'ownership'. There was a positive and a negative aspect to this. Positively, Agenda 21 was as much concerned with the *process* of sustainable development – participative, empowering, consensus-seeking, and democratic – as it was with *content*, and 'social processes of securing agreement on and commitment to sustainability aims were indispensable', even where the requirements for sustainability were determined externally (LGMB 1995a). Some argued that, although sustainable development strategies drew together many actors into an inclusive network, 'this, paradoxically, is potentially its greatest weakness, as excessive inclusivity may lead to a lack of clear purpose, direction and commitment' (Selman and Wragg 1999: 339).

In short, the changes in attitudes and behaviour which will be required by policies of sustainability will come about only if they are acceptable. The negative side to this is the widespread distrust of both local and central government which research has uncovered (Macnaghten *et al.* 1995). Agenda 21 emphasised equality and economic, social and political rights. Among the top concerns were poverty, unemployment and deterioration in the quality of life and the health of local communities. These were reflected to some extent in the local sustainability indicators chosen. But, as with practice at the national level, the indicators generally reflected the data that was routinely collected and readily available, and there was limited opportunity for comparison between one authority and another (Cartwright 2000). The process could 'easily become cosmetic and bogged down in group dynamics and inertia' (Scott 1999). In addition, although almost all local authorities prepared a Local Agenda 21 Strategy, their commitment varied considerably (Cartwright 1997). Local Agenda 21 certainly contributed to the growing awareness of environmental and sustainability issues in local politics, but the sum of evaluations (and a review of examples of strategies) suggests that they succeeded simply in presenting the agenda, with limited impact on mainstream policy. As concern for climate change moved to centre stage in politics, Local Agenda 21 quietly slipped down the agenda. In 2014, the term is rarely, if ever, mentioned.

As mentioned in 'The Meanings of the Environment in planning' section in this chapter, this reflects the wider shift of emphasis from sustainability to climate change and their associated discourses and practices (Hamdouch and Zuindeau 2010). Furthermore, a new concept appeared on the horizon called resilience, which began to replace sustainability in everyday discourses (Davoudi 2012c) in much the same way as the environment began to be 'subsumed in the hegemonic imperatives of climate change' (Whatmore

2008: 1777). This will be discussed in the next section of this chapter.

Rio+20 Conference: sustainable development's swansong?

In June 2012, heads of state and high-level officials of more than 190 nations met at the Rio+20 Conference to mark the twentieth anniversary of the 1992 Rio Earth Summit. The outcome document, *The Future We Want*, was the culmination of two years of intensive negotiations. It outlined practical and immediate actions to create 'a pathway for a sustainable century'. More specific outcomes are listed in Box 7.3.

In comparing the Rio+20 with its predecessor, the Rio Earth Summit, the foreword to a report by the Chartered Institution of Water and Environmental Management (CIWEM 2013) negatively sums up the conference as follows:

> In June 2012, Rio hosted the greatest failure of collective leadership since the First World War. The Earth's living systems are collapsing because politicians see the environment as a brake on growth and not an opportunity to grow sustainably. The outcome document, *The Future We Want*, might represent the future that most world leaders want, but it is not the future the world needs. You cannot have a document carrying this title without any mention of planetary boundaries, tipping points, planetary carrying capacity or hope for the future.

On a more positive note, there are increasing attempts to move away from an obsession with GDP as measure of progress to measures that include human and natural well-being (see Box 7.4). Examples include the EU's 'Beyond GDP' initiative (CEC 2013b), the OECD's 'Better Life' initiative (OECD 2013) and the Stiglitz-Sen-Fitoussi (2009) Commission on the Measurement of Economic Performance and Social Progress launched by the French government in 2008.

BOX 7.3 OUTCOMES FROM RIO+20 CONFERENCE, 2012

- the creation of a High Level Forum to act as a watchdog for sustainable development commitments
- a strategy to define Sustainable Development Goals
- a new strategic status for the protection of oceans
- efforts to outline a new inclusive metric to measure the wealth of countries beyond GDP that incorporates the three pillars of sustainable development
- a green light to a 'Green Economy' in the context of sustainable development and poverty eradication
- calls for greater effort to implement the Convention on Biological Diversity
- $323 billion devoted to achieving universal access to sustainable energy by 2030
- planting of 100 million trees
- empowering 5,000 women entrepreneurs in green economy businesses in Africa
- recycling 800,000 tons of PVC per year
- upgrading of the United Nations Environment Programme in key areas such as universal membership and improved financial resources
- recognition by all 192 governments that 'fundamental changes in the way societies consume and produce are indispensable for achieving global sustainable development' (para. 224).

BOX 7.4 WHAT IS 'WELL-BEING'?

Well-being is a positive physical, social and mental state; it is not just the absence of pain, discomfort and incapacity. It requires that basic needs are met, that individuals have a sense of purpose, that they feel able to achieve important personal goals and participate in society. It is enhanced by conditions that include supportive personal relationships, strong and inclusive communities, good health, financial and personal security, rewarding employment, and a healthy and attractive environment. Government's role is to enable people to have a fair access now and in the future to the social, economic and environmental resources needed to achieve well-being. An understanding of the effect of policies on the way people experience their lives is important for designing and prioritising them.

Source: DEFRA 2007: 7

Ecosystem services

Following the publication of the Millennium Ecosystem Assessment (MA) in 2005, the UK National Ecosystem Assessment (NEA) was published in 2010 and the UK became the first country to have undertaken a complete assessment of the benefits that nature provides, how they have changed over the past, the prospects for the future and their value to our society. The results of that assessment provided the rationale for many of the actions proposed in the White Paper *The Natural Choice: Securing the Value of Nature* (DEFRA 2011a).

The report was triggered by the Natural England report in 2008 stating that England had less natural diversity than 50 years ago and was under constant pressure from threats such as climate change, and similar patterns of loss have been seen across the UK. The need to ensure that UK ecosystems remain healthy and resilient required developing the baseline. As of 2014, DEFRA, the devolved administrations and partners are working towards implementing an ecosystem approach to conserving, managing and enhancing the natural environment of the UK. For this to happen decision-making needs to move away from sector-specific or habitat-specific approaches and towards an integrated approach based on whole ecosystems and ensuring the value of ecosystem services is fully reflected in decisions (UK NEA 2010). Box 7.5 provides the key definitions related to ecosystem services, while Box 7.6 outlines the different types of ecosystem services.

BOX 7.5 KEY DEFINITIONS RELATED TO ECOSYSTEM SERVICES

Ecosystem: a natural unit of living things (animals, including humans; plants; and micro-organisms) and their physical environment.

Ecosystem services: the benefits provided by ecosystems that contribute to making human life both possible and worth living.

Ecosystems approach: a strategy for the integrated management of land, water and living resources that promotes conservation and sustainable use in an equitable way.

Source: UK NEA 2010: 2

BOX 7.6 TYPES OF ECOSYSTEM SERVICES

The UK NEA classifies services into four types:

- *Provisioning services*: the products we obtain from ecosystems such as food, fibre and fresh water.
- *Regulating services*: the benefits we obtain from the regulation of ecosystem processes such as regulation of pollination, the climate, noise and water.
- *Cultural services*: the non-material benefits we obtain from ecosystems, for example through spiritual or religious enrichment, cultural heritage, recreation and tourism or aesthetic experience.
- *Supporting services*: ecosystem functions that are necessary for the production of all other ecosystem services such as soil formation and the cycling of nutrients and water.

Source: UK NEA 2010: 6

The results of the NEA provided the reason for many of the actions proposed in the White Paper *The Natural Choice: Securing the Value of Nature* (DEFRA 2011a). The White Paper used the term 'natural environment' in its broad meaning, covering 'living things in all their diversity: wildlife, rivers and streams, lakes and seas, urban green space and open countryside, forests and farmed land', and suggesting that 'it includes the fundamentals of human survival. . . . And it embraces our landscapes and our natural heritage, the many types of contact we have with nature in both town and country' (DEFRA 2011a: 7).

In response to Sir John Lawton's 2010 report, *Making Space for Nature*, the White Paper proposed the establishment of Nature Improvement Areas (NIAs). In 2012, England's first twelve NIAs (selected through a competition process) began to work, with funding provided by the government, Natural England and other sources. They extend from Morecambe Bay in the North West to the Wild Purbeck area in the South West. The aim is to make these areas better places for wildlife, creating more and better-connected habitats at a landscape scale and providing space for wildlife to thrive and adapt to climate change. A new institutional framework in the form of local nature partnerships (LNPs) has been established to strengthen local action. The White Paper also made a commitment that the government will

> take a strategic approach to planning for nature within and across local areas . . . guide development to the best locations, encourage greener design and enable development to enhance natural networks. . . . retain the protection and improvement of the natural environment as core objectives of the planning system.
>
> (DEFRA 2011a: 13)

RESILIENCE

If sustainability was the linchpin of planning in the 1990s, resilience is in the 2000s. The term appears everywhere and a growing number of reports are being produced that advocate resilient building. If in the 1990s planners were encouraged to plan for sustainable communities, in the late 2000s they are asked to plan for resilient communities. In an attempt to unpack resilience and trace its origin, Davoudi (2012c) refers to three fundamentally different definitions introduced by Holling (1973) and later in collaboration with his fellow ecologists in the Resilience Alliance Network based in Stockholm. The engineering definition of resilience put the emphasis on the ability of a system to return to equilibrium after a disturbance. So, the *resistance* to disturbance and the *speed* by which the system returns to equilibrium are the measure of resilience. The ecological definition of resilience put the emphasis on 'the *magnitude* of the disturbance that can be absorbed before the system changes its structure' (Holling 1996: 33). What is common to both perspectives is the belief in the existence of *equilibrium* in systems. The third definition is called socio-ecological or evolutionary resilience. This view challenges the idea of equilibrium and advocates that the very nature of systems may change over time with or without an external disturbance. Here resilience is not necessarily about returning to normality, but the ability of complex socio-ecological systems to change, adapt and, crucially, transform in response to disturbances and shocks (see Davoudi *et al.* 2013; Special Issue of *Planning Practice and Research* 2013).

Figure 7.2 provides a graphical representation of the adaptive cycles in evolutionary resilience, referring to four distinct phases of change in the structures and function of a system: growth or exploitation, conservation, release or creative destruction, and reorganisation (Holling and Gunderson 2002). The first loop of the cycle relates to the emergence, development and stabilisation of a system's structure and functions, while the second loop relates to their eventual rigidification and decline, but at the same time the opening up of new and unpredictable possibilities (Davoudi 2012c).

There are some promising parallels between evolutionary resilience and the relational understanding of spatiality (see Healey 2007; Davoudi and Strange 2009). From the evolutionary resilience perspective, places are not just neutral containers: they are complex, interconnected socio-spatial systems with extensive

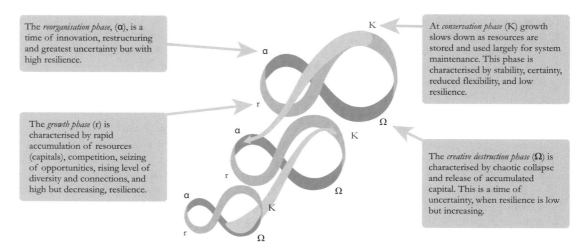

The *reorganisation phase*, (α), is a time of innovation, restructuring and greatest uncertainty but with high resilience.

The *growth phase* (r) is characterised by rapid accumulation of resources (capitals), competition, seizing of opportunities, rising level of diversity and connections, and high but decreasing, resilience.

At *conservation phase* (K) growth slows down as resources are stored and used largely for system maintenance. This phase is characterised by stability, certainty, reduced flexibility, and low resilience.

The *creative destruction phase* (Ω) is characterised by chaotic collapse and release of accumulated capital. This is a time of uncertainty, when resilience is low but increasing.

Figure 7.2 The adaptive cycle in evolutionary resilience
Source: Davoudi 2012c: 303 adapted from Holling and Gunderson 2002: 34–41 and Pendall *et al.* 2010: 76

and unpredictable feedback processes which operate at multiple scales and timeframes. However, planners should not uncritically translate concepts such as resilience that have been developed in natural sciences and engineering into the social context. This is because applying resilience to the social context raises a number of normative and political questions such as, what is the desired outcome of resilience and for whom, or 'resilience of what to what and who gets to decide' (Davoudi and Porter 2012: 331; Shaw 2012).

Resilience and the complexity theory which underpins it represent a new paradigm in systems thinking and challenge planners to seek transformative opportunities at a time of crisis. However, some of the criticism of the 1970s systems approach may well apply to this new wave of systems thinking. Further-more, like sustainability, resilience is in danger of becoming an empty signifier which is filled by various definitions depending on what is expected to be gained from them.

CLIMATE CHANGE

The climate change issue is part of the larger challenge of sustainable development. As a result,

climate change policies can be more effective when consistently embedded within broader strategies designed to make national and regional development paths more sustainable.

(IPCC 2001:4)

Over the past six years scientific opinion has moved decisively to an almost universal consensus that climate change is happening and is the result of human activity. That means we can move the debate from whether there is a problem to how to deal with it.

Yes, climate change represents a potentially catastrophic threat, but it is within our control to address it – and address it we must.

(The UK Government Sustainable Development Strategy 2005: 3)

Climate change is defined differently by the two major international organisations. The Intergovernmental Panel on Climate Change (IPCC) defines climate change as

a change in the state of the climate that can be identified (e.g., by using statistical tests) by changes in the mean and/or the variability of its properties, and that persists for an extended period, typically decades or longer. Climate change may be due to natural internal processes or external forcings such as modulations of the solar cycles, volcanic eruptions and persistent anthropogenic changes in the composition of the atmosphere or in land use.

(IPCC 2013: 1450)

The *United Nations Framework Convention on Climate Change* (UNFCCC) in its Article 1 defines climate change as a change of climate which is attributed directly or indirectly to human activity that alters the composition of the global atmosphere and is in addition to natural climate variability observed over comparable time periods (quoted in IPCC 2013: 1450).

However, in the basis of compelling scientific evidence, both organisations agree that first, the climate system is changing as a result of global warming. The latest estimates suggest that 'the globally averaged combined land and ocean surface temperature data as calculated by a linear trend, show a warming of 0.85[0.65 to 1.06] °C, over the period 1880 to 2012, when multiple independently produced datasets exist' (IPCC 2013: 3). Second and more importantly, human activity is responsible for a large proportion of observed changes. The latest estimates suggest that 'it is extremely likely that more than half of the observed increase in global average surface temperature from 1951 to 2010 was caused by the anthropogenic increase in greenhouse gas concentrations and other anthropogenic forcings together' (IPCC 2013: 17).

Global carbon emissions from fossil fuels have significantly increased since 1900. Emissions increased by over sixteen times between 1900 and 2008 and by about 1.5 times between 1990 and 2008 (see Figure 7.3). Emissions of non-CO_2 greenhouse gases have also increased significantly since 1900.

Observed changes in the climate system

The IPCC Fifth Assessment Report (IPCC 2013) is unequivocal about the *observed* changes in the climate

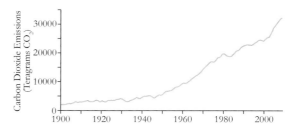

Figure 7.3   Global Carbon Dioxide (CO₂) emissions from fossil fuels 1900–2008

Source: US EPA website (www.epa.gov/climatechange/ghgemissions/global.html; accessed 4 April 2014. Based on data from Boden *et al.* 2010

system (including the atmosphere, the ocean, the cryosphere and the land surface), on the basis of direct measurements and remote sensing from satellites and other platforms. At the global scale such measurements became available for the period 1950 onwards (although measurement of temperature and other variables have been available since the mid-nineteenth century). On the basis of direct observations, the report makes the following headline statements (pp. 4–11):

- Warming of the climate system is unequivocal, and since the 1950s, many of the observed changes are unprecedented over decades to millennia. The atmosphere and ocean have warmed, the amounts of snow and ice have diminished, sea level has risen, and the concentrations of greenhouse gases have increased.

- Each of the last three decades has been successively warmer at the Earth's surface than any preceding decade since 1850. In the Northern Hemisphere, 1983–2012 was *likely* the warmest 30-year period of the last 1400 years.

- Ocean warming dominates the increase in energy stored in the climate system, accounting for more than 90% of the energy accumulated between 1971 and 2010. It is *virtually certain* that the upper ocean (0–700 m) warmed from 1971 to 2010, and it *likely* warmed between the 1870s and 1971.

- Over the last two decades, the Greenland and Antarctic ice sheets have been losing mass, glaciers have continued to shrink almost worldwide, and Arctic sea ice and Northern Hemisphere spring snow cover have continued to decrease in extent.

- The rate of sea level rise since the mid-19th century has been larger than the mean rate during the previous two millennia. Over the period 1901 to 2010, global mean sea level rose by 0.19 [0.17 to 0.21] m.

- The atmospheric concentrations of carbon dioxide, methane, and nitrous oxide have increased to levels unprecedented in at least the last 800,000 years. Carbon dioxide concentrations have increased by 40% since pre-industrial times, primarily from fossil fuel emissions and secondarily from net land use change emissions. The ocean has absorbed about 30% of the emitted anthropogenic carbon dioxide, causing ocean acidification.

Future climate change

In order to understand recent changes in the climate system, the IPCC uses the combined results from observations as well as studies of feedback processes and model simulations. These are used to develop a hierarchy of climate models ranging from simple to intermediate complexity, and to comprehensive climate models and Earth System Models. These models simulate changes based on a set of scenarios of anthropogenic forcings only. IPCC (2013) makes the following projections (pp. 20, 23):

- Global surface temperature change for the end of the 21st century is *likely* to exceed 1.5°C relative to 1850 to 1900 for all RCP scenarios except RCP2.6 (lowest forcing). It is *likely* to exceed 2°C for RCP6.0 and RCP8.5, and *more likely than not* to exceed 2°C for RCP4.5. Warming will continue beyond 2100 under all RCP scenarios except RCP2.6. Warming will continue to exhibit interannual-to-decadal variability and will not be regionally uniform.

- It is virtually certain that there will be more frequent hot and fewer cold temperature extremes over most land areas on daily and seasonal timescales

as global mean temperatures increase. It is very likely that heat waves will occur with a higher frequency and duration. Occasional cold winter extremes will continue to occur.

- Changes in the global water cycle in response to the warming over the 21st century will not be uniform. The contrast in precipitation between wet and dry regions and between wet and dry seasons will increase, although there may be regional exceptions.

- Globally, it is likely that the area encompassed by monsoon systems will increase over the 21st century. While monsoon winds are likely to weaken, monsoon precipitation is likely to intensify due to the increase in atmospheric moisture. [There will be a] lengthening of the monsoon season in many regions.

Drivers of climate change

Natural and anthropogenic (human-made) substances and processes that alter the Earth's energy budget are drivers of climate change. They lead to changes in the atmospheric concentration of greenhouse gases (GHG), which are shown in Figure 7.4, plus aerosols, solar radiation, and land surface properties. These in turn alter the energy balance of the climate system by exerting warming or cooling influences on global climate. The extent to which various gases contribute, to the greenhouse effect depends on their characteristics, abundance, and the indirect effects they may cause. Changes in energy balance of the climate system are expressed in terms of *radiative forcing*. Radiative forcing quantifies the change in energy fluxes caused by changes in human and natural drivers. The latest measurement confirms that 'total radiative forcing is positive, and has led to an uptake of energy by the climate system. The largest contribution to total radiative forcing is caused by the increase in the atmospheric concentration of CO_2 since 1750' (IPCC 2013: 13).

While both natural and human-made factors are responsible for climate change, the IPCC is unequivocal:

Human influence has been detected in warming of the atmosphere and the ocean, in changes in the

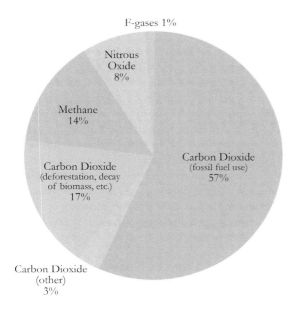

Figure 7.4 GHG emissions by types

Source: Adapted from IPCC 2007a, based on global emissions from 2004

most important anthropogenic GHG, increased by 80 per cent in that time (IPCC 2007a: 5). Global increases in CO_2 concentrations are due mainly to fossil fuel use and to a lesser extent land use change. Increases in methane (CH_4) concentrations are predominantly due to agriculture activities, waste management and fossil fuel use. Nitrous oxide (N_2O) emissions are primarily from agricultural activities, such as the use of fertilisers, and Fluorinated (F) gases are from industrial processes and refrigeration. The sectors which were most responsible for growth in GHG emissions are the energy supply sector, transport, industry, land use/forestry and agriculture, as shown in Figure 7.5.

Two important drivers of the rise in energy-related emissions are the global average growth of population and wealth. However, average figures mask national differences. The contribution of individual countries to global warming varies substantially. In 2008, the top CO_2 emitters were China, the United States, the European Union, India, the Russian Federation, Japan,

global water cycle, in reductions in snow and ice, in global mean sea level rise, and in changes in some climate extremes. This evidence for human influence has grown since AR4 (2007). It is *extremely likely* that human influence has been the dominant cause of the observed warming since the mid-20th century.

(IPCC 2013: 17)

The burning of fossil fuels and deforestation, large-scale cattle raising, the use of synthetic fertilisers and emissions from specific industrial activities have increased the atmospheric concentration of GHGs and aerosol particles. In 2008, the level of GHG in the atmosphere was about 430 parts per million (ppm) compared with 280ppm before the Industrial Revolution. This is estimated to reach 550ppm by 2050 at the current rate of increase, but given that levels are rising faster than expected, 550ppm could be reached as early as 2035 (HM Treasury 2006b). The emission of carbon dioxide (CO_2), which is the

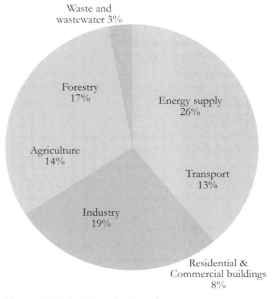

Figure 7.5 GHG emissions by sectors

Source: Adapted from IPCC 2007a, based on global emissions from 2004

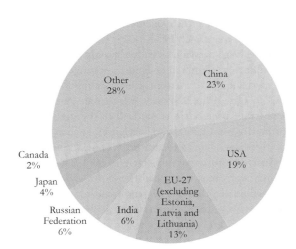

Figure 7.6 Global CO$_2$ emissions from fossil fuel combustion and some industrial processes, by regions

Source: US EPA website (www.epa.gov/climatechange/ghgemissions/global.html; accessed 4 April 2014). Based on National CO$_2$ Emissions from Fossil-Fuel Burning, Cement Manufacture, and Gas Flaring: 1751–2008

and Canada (see Figure 7.6). There are also differences between rich and poor countries. In 2004, for instance, high-income nations accounted for 20 per cent of world population, produced 57 per cent of gross domestic product (GDP) and generated 46 per cent of global GHG emissions (IPCC 2007b). Per capita emissions from developing countries in 2004 were one quarter of per capita emissions from developed countries. While a progressive decoupling of income growth from GHG emissions has taken place through measures such as reducing the energy intensity (i.e. energy used per unit of GDP), the level of improvement has not been sufficient to counteract the global rise in emissions.

In addition to national variations, there are also variations within countries and between cities. Here again, variation in wealth is an important factor. For example, total GHG emissions ranged from 44.3 million tonnes (mt) in London in 2006 to 64.8mt in Mexico City in 2000 and a mere 1.8mt in Dhaka in 1999. The per capita emissions were respectively 6.18, 3.6 and 1.7 tonnes, i.e. much higher in the

wealthy city of London than in Mexico City or Dhaka (Romero-Lankao 2007). Furthermore, it is not helpful to focus merely on settlement types (such as urban versus rural areas) in attributing GHG emissions, because, as Satterthwaite (2008: 547) stresses, 'the driver of most anthropogenic carbon emissions is the consumption patterns of middle- and upper-income groups, regardless of where they live, and the production systems that profit from their consumption'. However, this is not to suggest that the spatial relations and the configuration of the settlements are not important drivers of emissions. Indeed, as discussed later in this chapter, it is here that planning can make an important contribution in tackling climate change.

Impacts of climate change

The IPCC uses the term *impacts*

> primarily to refer to the effects on natural and human systems of extreme weather and climate events and of climate change. Impacts generally refer to effects on lives, livelihoods, health, ecosystems, economies, societies, cultures, services, and infrastructure due to the interaction of climate changes or hazardous climate events occurring within a specific time period and the vulnerability of an exposed society or system. Impacts are also referred to as *consequences* and *outcomes*. The impacts of climate change on geophysical systems, including floods, droughts, and sea-level rise, are a subset of impacts called physical impacts.
>
> (IPCC 2014: 4)

It is evident that in recent decades, changes in climate have caused impacts on natural and human systems on all continents and across the oceans, leading to the following global impacts observed with varying degrees of confidence (IPCC 2014: 6–8):

- In many regions, changing precipitation or melting snow and ice are altering hydrological systems, affecting the quantity and quality of water resources.

- Many terrestrial, freshwater, and marine species have shifted their geographic ranges, seasonal activities, migration patterns, abundances, and species interactions.
- Negative impacts of climate change on crop yields have been more common than positive impacts.
- At present the world-wide burden of human ill-health from climate change is relatively small compared with effects of other stressors and is not well quantified.
- Differences in vulnerability and exposure arise from non-climatic factors and from multidimensional inequalities often produced by uneven development processes which shape differential risks from climate change.
- Impacts from recent climate-related extremes, such as heat waves, droughts, floods, cyclones, and wildfires, reveal significant vulnerability and exposure of some ecosystems and many human systems to current climate variability.

- Climate-related hazards exacerbate other stressors, often with negative outcomes for livelihoods, especially for people living in poverty.
- Violent conflict increases vulnerability to climate change.

As shown in Figure 7.7, the IPCC considers the risk of climate-related impacts as the result of the interaction of climate-related hazards (including hazardous events and trends) with the vulnerability and exposure of human and natural systems. It shows that changes in both the climate system and socio-economic processes, including adaptation and mitigation, are drivers of hazards, exposure and vulnerability.

Different global regions are expected to experience different changes as a result of global warming. For example, while Europe is expected to experience an increase in inland flash floods, Africa will see a rise in arid and semi-arid land (IPCC 2007a). Some impacts may be slow to become apparent but likely to last for a long time. Beyond certain thresholds, some impacts

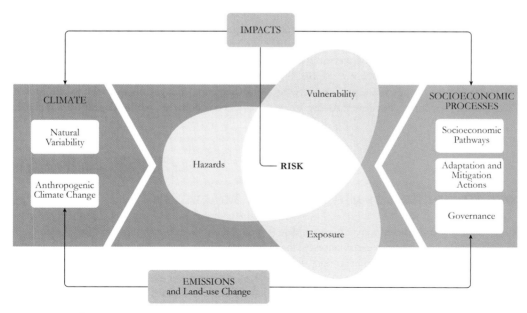

Figure 7.7 The interrelationship between hazard, exposure, vulnerability and impact
Source: IPCC 2014: 35, Figure SPM. 1

could be irreversible (Davoudi *et al.* 2009). For example, 'major melting of the ice sheets and fundamental changes in the ocean pattern could not be reversed over a period of many human generations' (IPCC 2001: 16–17). There is little doubt that projected changes are likely to transform the physical geography of the world with millions of people facing starvation, water shortages or homelessness.

The nature and intensity of impact will vary depending on the level of exposure and vulnerability of people and places. The IPCC (2014: 4) defines exposure as 'The presence of people, livelihoods, species or ecosystems, environmental functions, services, and resources, infrastructure, or economic, social, or cultural assets in places and settings that could be adversely affected', and vulnerability as 'The propensity or predisposition to be adversely affected. Vulnerability encompasses a variety of concepts and elements including sensitivity or susceptibility to harm and lack of capacity to cope and adapt.'

The level of vulnerability differs not only between places, but also between population groups. Differences in demographic and socio-economic profiles affect level of vulnerability considerably. Hence, children and the elderly are often the most vulnerable groups, as are those who already suffer from poor health or lack the capacity to reduce the impacts of climate change on their well-being (Davoudi *et al.* 2009). The latter are lower-income groups with little resources at their disposal, for example, to move to safer areas, insure their assets or gain access to adequate water, electricity, sanitation, sewage, and other basic utilities (Satterthwaite *et al.* 2007).

Mitigation and adaptation

> It is . . . no longer a question of whether to mitigate climate change or to adapt to it. Both adaptation and mitigation are now essential in reducing the expected impacts of climate change on humans and their environment.
>
> (IPCC 2007a: 748)

Prior to statements such as the one quoted above, adaptation and mitigation (see Box 7.7 for definitions) were dealt with separately in intergovernmental negotiations as well as in governmental climate policy.

The split is reflected in the structure of the IPCC Working Groups (discussed later in this chapter). During the initial climate change negotiations, adaptation was not only treated separately from mitigation, but also was given little attention on the assumption that it might distract attention from mitigation (Swart and Raes 2007). Another reason was larger uncertainties about adaptation measures and the limited role of international organisations for ensuring their implementation at the local level. Furthermore, mitigation was seen as the problem of developed countries, while adaptation was considered

BOX 7.7 DEFINITION OF MITIGATION AND ADAPTATION

The IPCC defines *mitigation* as 'a human intervention to reduce the sources or enhance the sinks of greenhouse gases' (IPCC 2013: 1458); and *adaptation* as 'the process of adjustment to actual or expected climate and its effects. In human systems, adaptation seeks to moderate or avoid harm or exploit beneficial opportunities. In some natural systems, human intervention may facilitate adjustment to expected climate and its effects' (IPCC 2013: 5). While mitigation measures aim to avoid the adverse impacts of climate change in the long term, adaptation measures are designed to reduce unavoidable impacts of climate change in the short and medium terms.

as the problem of developing countries. These assumptions have now lost their appeal and as Swart and Raes (2007: 301) put it, 'the question is not whether the climate has to be protected from humans or humans from climate, but how both mitigation and adaptation can be pursued in tandem'. They propose five ways for developing links between adaptation and mitigation measures, as follows:

- avoiding trade-offs between the two and in designing adaptation measures taking into account the consequences for mitigation strategies;
- identifying synergies between the two in response to options within specific policy sectors, notably through spatial planning and design;
- enhancing both adaptive and mitigative response capacity simultaneously and putting such capacity into action particularly in developed countries;
- building institutional links between two and bridging the communication gap between policy-makers;
- mainstreaming climate policies into overall sustainable development policies at all levels of governance.

The IPCC also confirms that 'making development more sustainable can enhance both mitigative and adaptive capacity, and reduce emissions and vulnerability to climate change' (IPCC 2007a: 22). However, there is an emerging concern about potential tensions between mitigation and adaptation measures. These can raise difficult conundrums for spatial planners and decision-makers for whom 'truly "win-win" situations may be few and far between' (McEvoy *et al.* 2006: 186). Some commentators (Pizarro 2009) have challenged the assumed synergies between the two, arguing that a development pattern that helps mitigate climate change may not be the best one for adapting that settlement to the negative effects of global warming in that geographical location. Others (Howard 2009) contend that the interrelationships between mitigation and adaptation have been largely neglected in planning literature and that 'adaptation turn' in planning may be at the cost of paying less attention to measures that reduce GHG emissions.

Researchers now advocate an integrated approach and urge planners to adhere to three key principles: (a) mitigation has priority, (b) mitigation is a primary form of adaptation, and (c) effective local adaptation requires a long-term global perspective. It is emphasised that the most desirable form of adaptation is adaptation that is not necessary (Howard 2009). The need for developing an integrated approach, particularly at the regional and local level, to avoid maladaptation and promote synergies and complementarities is widely acknowledged (EEA 2007). However, there remain a number of barriers and limitations, such as the mismatch in spatial and temporal scales, the balance of costs and benefits, and the different stakeholders and policy sectors involved (Hall 2009).

International climate policy

The development of international policy on climate change began a long time ago. Central to the events listed in Box 7.8 is the role of the United Nations (UN), which has shaped the global policy context in this field. One year after the publication of the Brundtland Report (see the section on sustainable development), in 1988, the UN Environment Programme along with the World Meteorological Organisation (WMO) established the IPCC. The initial role of the IPCC was to prepare a comprehensive review of the state of knowledge of the science of climate change, the social and economic impact of climate change, and possible response strategies and elements for inclusion in a possible future international convention on climate. Today the IPCC's role, as stated on their website (www.ipcc.ch), is:

to assess on a comprehensive, objective, open and transparent basis the scientific, technical and socio-economic information relevant to understanding the scientific basis of risk of human-induced climate change, its potential impacts and options for adaptation and mitigation. IPCC reports should be neutral with respect to policy, although they may need to deal objectively with scientific,

BOX 7.8 KEY EVENTS IN INTERNATIONAL POLICY DEVELOPMENT ON CLIMATE CHANGE

1979	First WMO World Climate Conference, 'continued expansion of man's activities on earth may cause significant extended regional and even global changes of climate'
1985	UNEP/WMO/ICSU conference on 'assessment of the role of carbon dioxide and of other GHG in climate variations and associated impacts'
	Ozone hole identified by British Antarctic Survey
1987	Montreal Protocol opened for signature
1988	WMO and UNEP establish the Intergovernmental Panel on Climate Change (IPCC)
1989	Montreal Protocol enters into force
1990	IPCC First Assessment Report
1991	First EC strategy to limit CO_2 emissions and improve energy efficiency
1992	UNFCCC adopted, IPCC Supplementary Reports
1994	UNFCCC came into force
	IPCC Special Report Radiative Forcing of CC
1995	First conference UNFCCC COP-1, laid foundations for Kyoto Protocol
	IPCC Second Assessment Report
1997	Adoption of Kyoto Protocol
2000	First European CC Programme (2000–4) launched
2001	UN Conference of Parties (COP) 7 Marrakech Accords, resolving remaining Kyoto issues and leading to faster ratification
	IPCC Third Assessment Report
2002	EC ratifies Kyoto Protocol
2004	COP-10, programme of work on adaptation and response measures
2005	Kyoto Protocol come into force with three mechanisms
	Second European CC Programme launched
2006	CDM became operational
	Al Gore's 'An Inconvenient Truth' released, bringing climate change and planetary limits to a broad audience
	The Stern Review on the Economics of Climate Change was published showing the effect of global warming on the world economy
2007	COP-13, Bali road map and action plan: enhanced action on adaptation
	IPCC Fourth Assessment Report
	Upgraded EU commitment: independent 20 per cent reduction, 30 per cent reduction if others also commit to higher reduction levels
2008	COP-14 Kyoto's adaptation fund, progress on issues affecting developing countries: adaptation, finance, technology, reducing emissions from deforestation and forest degradation (REDD) and disaster management.
	The UK passes the Climate Change Act, committing to an 80 per cent reduction in greenhouse gas emissions by 2050

2009	EU White paper *Adapting to Climate Change*
	COP-15 Copenhagen
	The G20 group resolves to phase out fossil fuel subsidies
2010	COP-16 Cancun
2011	COP-17 Durban
2012	COP-18 Qatar, end of first commitment period of Kyoto Protocol and year by which a new international framework had to be negotiated and ratified
2014	IPCC Fifth Assessment Report

technical and socio-economic factors relevant to the application of particular policies.

The IPCC is an intergovernmental body and is open to all member countries of both the UN and the WMO. As of 2014, 195 countries are members of the IPCC. It is hosted by the WMO, whose headquarter is in Geneva. The IPCC is organised in three working groups and a task force that are assisted by a technical support group. The focus of their work is as follows:

- Working Group I: The Physical Science Basis of Climate Change;
- Working Group II: Climate Change Impacts, Adaptation and Vulnerability;
- Working Group III: Mitigation of Climate Change;
- Task Force: National Greenhouse Gas Inventories.

Since its inception, the IPCC Working Groups have regularly delivered comprehensive scientific reports about climate change. These are called the assessment reports (AR) (see Box 7.9), of which five have been produced, with the first in 1999 and the latest in 2014.

The scientific evidence contained in the first IPCC AR of 1990 underlined the importance of climate change as a challenge requiring international cooperation to tackle its consequences. It therefore played a decisive role in leading to the creation of the UNFCCC, which was adopted in 1992 at the Earth Summit in Rio de Janeiro. The UNFCCC is the key international treaty to reduce global warming and cope with the consequences of climate change, and became the driving force behind the Kyoto Protocol, which was adopted in 1997 and came to force in 2005.

BOX 7.9 THE IPCC ASSESSMENT REPORTS

Assessment reports (AR) are published materials consisting of the full scientific and technical assessment of climate change, generally in three volumes, one for each of the working groups of the IPCC, plus a synthesis report. Each of the working group volumes is composed of individual chapters, an optional technical summary and a summary for policymakers. The synthesis report synthesises and integrates materials contained in the assessment reports and special reports, is written in a non-technical style suitable for policymakers and addresses a broad range of policy-relevant but policy-neutral questions. It consists of a longer report and a summary for policy-makers.

The Kyoto Protocol

The Kyoto Protocol is an international agreement linked to the UNFCCC. It commits its Parties/member countries (see Box 7.10 for a definition) by setting internationally binding emissions reduction targets. By 2008, 180 nations had ratified the Kyoto Protocol. The exact target for each member state varies depending on its historic emissions levels and capacity to change. Recognising that developed countries are principally responsible for the current high levels of GHG emissions in the atmosphere as a result of more than 150 years of industrial activity, the Protocol places a heavier burden on developed nations under the principle of 'common but differentiated responsibilities'. Indeed some countries were allowed to increase their emissions while others were required to make significant cuts. During the first commitment period (up to 2012), thirty-seven industrialised countries and the EU committed to reducing GHG emissions by an average of five per cent against 1990 levels. The UK committed to achieve a 12.5 per cent reduction but the largest per capita polluter in the world – the United States – did not sign up to any mandatory targets.

Under the Protocol, countries must meet their targets primarily through national measures. However, the Protocol also offers them three market-based mechanisms to do so:

- **International Emissions Trading:** if countries emit below their Kyoto target, they can trade (sell) their excess capacity to countries that have not met theirs in the international carbon market.
- **The Clean Development Mechanism:** countries can implement emissions reduction projects in

BOX 7.10 DEFINITION AND TYPES OF 'PARTIES' UNDER THE UNFCCC

The UNFCC Convention divides countries into three main groups according to differing commitments with respect to emission targets:

Annex I Parties (forty-two countries plus the EU) include the industrialised countries that were members of the Organisation for Economic Cooperation and Development (OECD) in 1992, plus countries with economies in transition (the EIT Parties), including the Russian Federation, the Baltic States, and several Central and Eastern European States.

Annex II Parties consist of the OECD members of Annex I, but not the EIT Parties. They are required to provide financial resources to enable developing countries to undertake emissions reduction activities under the Convention and to help them adapt to the adverse effects of climate change.

Non-Annex I Parties are mostly developing countries. Certain groups of developing countries are recognised by the Convention as being especially vulnerable to the adverse impacts of climate change, including countries with low-lying coastal areas and those prone to desertification and drought.

The forty-nine Parties classified as least developed countries (LDC) by the UN are given special consideration under the Convention on account of their limited capacity to respond to climate change and adapt to its adverse effects.

other (often developing) countries and use these to offset against their Kyoto target.

- **Joint Implementation:** countries can earn emissions reduction units (equivalent to one tonne of CO_2) from an emission reduction project in another country, and these can be counted towards meeting their Kyoto target. The idea is that the host country benefits from foreign investment and technology transfer.

Together, the UNFCCC and the Kyoto Protocol established a global policy framework for climate change which underlies an array of national policies. They also created an international carbon market and set up new institutional mechanisms to provide the foundation for climate policies. The Kyoto Agreement came to an end in 2012 but, following international negotiations, a second commitment period was agreed (although as of April 2014 this has not yet come into force) whereby Parties committed to reducing GHG emissions by at least 18 per cent below 1990 levels in the eight-year period from 2013 to 2020. However, the composition of Parties in the second commitment period is different from that of the first. Critics have argued that Kyoto has failed to reduce emissions significantly at the global level (Prins *et al.* 2010; Latin 2012). This is because, while the sum of emissions from nations with Kyoto targets has fallen significantly, emissions in the rest of the world (with no initial target) have increased sharply, especially in China and other emerging economies.

Several other international organisations (such as the World Bank) have also responded to the call for tackling climate change with policy measures, financial assistance and awareness-raising activities. While climate change is a global problem requiring coordinated global action, climate change responses are enacted and governed on multiple scales.

The European Union has fully supported the Kyoto Protocol, as is reflected in its 2005 European Climate Change Programme. This promised to cut the EU's emissions to 20 per cent, below 1990 levels by 2020, and by 30 per cent if other countries take part. To take action, it has set up the EU Emissions Trading System (EU ETS). The EU ETS is the largest multi-country, multi-sector GHG emissions trading system in the world. It is central to the EU's meeting its 20 per cent emissions reduction target by 2020. It covers around 11,000 energy-intensive industrial installations throughout Europe, including power stations, refineries and large manufacturing plants, with the aviation industry being added in 2012. Progress towards meeting Kyoto targets has varied across the EU, with the UK being able to cut its emissions to 21 per cent below 1990 levels, nearly double what was promised at Kyoto and just above its own target of 20 per cent.

National climate policy

Climate policy in the UK is underpinned by the Climate Change Act, 2008 – the world's first long-term legally binding framework for climate change mitigation and adaption – and the Climate Change (Scotland) Act 2009. The evidence base for government activities on projections of future climate change in the UK to 2100 is provided through the UK Climate Projections 2009 (UKCP09).

As regards mitigation, the 2008 Act introduced binding GHG emissions reduction targets, through action in the UK and abroad, of at least 80 per cent by 2050 against a 1990 baseline. The Act also made the UK the first country to set legally binding carbon budgets. These place a restriction on the total amount of GHG the UK can emit over a five-year period. The aim is to halve UK emissions relative to 1990 by 2027. Under a system of carbon budgets, every tonne of greenhouse gases emitted between now and 2050 will count. Where emissions rise in one sector, the UK will have to achieve corresponding falls in another. Therefore, the maximum emissions for the UK are as follows (HM Government 2011a):

- 3,018 million tonnes of carbon dioxide equivalent ($MtCO_2e$) over the first carbon budget period (2008–12);
- 2,782 $MtCO_2e$ over the second carbon budget period (2013–17);
- 2,544 $MtCO_2e$ over the third carbon budget period (2018–22);

The Act provided for the decisions on major infrastructure, including large renewable energy facilities (such as renewable electricity generating plants of over 50 megawatts onshore and 100 megawatts offshore in England and Wales), to be taken centrally (see Chapter 4). The Act also puts a statutory duty on regional and local planners to take action on climate change. Regions were expected to set targets in line with national targets or better. Similarly, local planning authorities were expected to go beyond encouraging the development of renewable energies to meet specific targets for new capacities. The 2011 reform of the planning system abolished the regional planning level and the obligation for meeting specific targets.

However, local planning authorities have been proactive in responding to climate change and have produced innovative planning responses in relation to smaller, on-site, renewable energy facilities. A notable example is the 'Merton Rule' devised in 2003 by planners in the London Borough of Merton. It required the incorporation of at least 10 per cent (of estimated energy requirement) in developments over 1,000 square metres. By 2007, the Rule was implemented by an estimated 100 local authorities (LGA 2007: 34). The Greater London Authority sought to incorporate into the 2007 amendments to the London Plan a policy for 20 per cent of energy to be met by on-site renewable sources. Although this policy built on the Merton Rule, it went wider in allowing the 20 per cent to be from on-site and/or decentralised sources. These local initiatives went beyond the PPS 22 policy, which required an undefined percentage of the energy to be used in new residential, commercial and industrial developments to come from on-site renewable sources, provided it is suitable and does not put 'undue burden on developers' (p. 8). In the late 2000s, national policy became widened through the PPS 1 Supplement and a number of Supplementary Planning Guidance documents which provided advice on climate change mitigation measures to planning applicants. For example, PPS 1 Supplement, published in 2007, recommended that decentralised, renewable and low-carbon energy supplies be incorporated into new development. Furthermore, review of the permitted development rights for households aimed to speed up the take-up of small-scale renewable installations.

Following the reform of the planning system in 2011, the NPPF (pp. 22–23) urges local planning authorities (LPAs) to help increase the use and supply of renewable and low-carbon energy by recognising 'the responsibility on all communities to contribute to energy generation from renewable or low carbon sources'. It therefore requires LPAs to:

- have a positive strategy to promote energy from renewable and low carbon sources;
- design their policies to maximise renewable and low carbon energy development while ensuring that adverse impacts are addressed satisfactorily, including cumulative landscape and visual impacts;
- consider identifying suitable areas for renewable and low carbon energy sources, and supporting infrastructure, where this would help secure the development of such sources;
- support community-led initiatives for renewable and low carbon energy, including developments outside such areas being taken forward through neighbourhood planning; and
- identify opportunities where development can draw its energy supply from decentralised, renewable or low carbon energy supply systems and for co-locating potential heat customers and suppliers.

Furthermore, the NPPF (p. 23) states that

When determining planning applications, local planning authorities should:

- not require applicants for energy development to demonstrate the overall need for renewable or low carbon energy and also recognise that even small-scale projects provide a valuable contribution to cutting greenhouse gas emissions; and
- approve the application if its impacts are (or can be made) acceptable. Once suitable areas for renewable and low carbon energy have been

identified in plans, local planning authorities should also expect subsequent applications for commercial scale projects outside these areas to demonstrate that the proposed location meets the criteria used in identifying suitable areas.

Reducing energy demand

Transforming the UK into a low-carbon economy requires policies and actions that are aimed at not only increasing the supply of low-carbon and renewable energy, but also substantially reducing energy demand. The action plan for transition to a low carbon economy emphasises that 'reducing our demand for energy from the energy system is fundamental to the Government's strategy, particularly because in many cases doing so saves money for households and businesses, whilst maintaining or improving our standards of living' (HM Government 2009a: 171). Managing energy demand through land use policies has been a major part of planning's sustainable development objective since the 1990s, as has been mentioned earlier. Two areas in particular have been at the centre of attention: the need to reduce car travel through policies on the location of new development and accessibility, and the need to increase the energy efficiency of the built environment through design policies and the layout of new developments. These will be discussed in turn.

Reducing car travel

Numerous studies have tried to establish the link between urban form, land use and travel patterns. While socio-economic variables often explain the variation in trip making more significantly than land use factors (Hickman and Banister 2005), evidence shows that at the regional and city levels three land use characteristics have major impacts on travel behaviour. These are density of development, settlement size, and access to facilities and services (Banister and Anable 2009), with density having a greater impact than settlement size in encouraging walking and cycling. The much cited research by Newman and Kenworthy (1999), which compared

eighty-four cities, has shown that density has an important impact on the distances travelled too.

Banister and Anable (2009) summarise the main conclusions about the impacts of land use factors on travel behaviour as follows:

(a) At the regional and city level, to reduce travel, the size of new development, especially housing, should be substantial (25–50,000 population) and located near to or within existing settlements, with the provision of local facilities and services phased so as to encourage local travel patterns.

(b) While average journey lengths by car are relatively constant (about 12 kilometres) at densities of fifteen persons per hectare, at lower densities it increases by up to 35 per cent. Similarly, as density increases the number of trips by car decreases from 72 per cent of all journeys to 51 per cent.

(c) Mixed uses reduce trip lengths and car dependence particularly with regard to proximity of jobs to houses.

(d) As settlement size increases, the trips are shorter with more trips taking place by public transport.

(e) Development which is near public transport interchanges and corridors (transit-oriented development) have a higher level of accessibility and are less car dependent.

(f) The availability of parking is a key determinant in the level of car use.

At the local neighbourhood level, the New Urbanism debate has highlighted a number of design factors which can reduce short-distance car travel and be incorporated in new development using SPGs and other planning regulatory interventions. These include direct routing for slow modes of travel, quieter and narrower streets, accessible neighbourhoods, street connectivity, thriving town centres and high streets, pedestrianisation, parking management, higher density of dwellings and using brownfield sites for infill.

In all these areas, spatial planning can use its proactive and regulatory interventions to make a difference. While planning may have a limited role in the short term, compared with fiscal measures for example, it certainly has a more significant role in the longer term by fostering sustainable location choices, facilitating other policy areas, and acting as a complementary policy for technologically driven and demand-management policies so that their benefits are 'locked in'. Furthermore, given the unequal distribution of GHG emissions from personal travel in the UK (Brand and Boardman 2008), the role of planning in providing for local services and access to them by sustainable modes of transport is pivotal to ensuring accessibility for lower-income groups. Overall, there is now compelling evidence which shows that the location of new housing and other developments in the UK has 'substantial implications for: the level of demand on transport systems, journey distances, and the use of different modes of transport over the next 20–30 years' (Banister and Anable 2009).

Increasing the energy efficiency of the built environment

The role of the planning system in the energy efficiency of the built environment can be achieved through first, appropriate location, layout, landscaping and site design of new development; second, appropriate design and materials for individual buildings; and third, appropriate environmental standards in larger new settlements, such as the new towns, eco-towns or garden cities. Here, we focus on the second area, buildings:

> In 2009, 37 per cent of UK emissions were produced from heating and powering homes and buildings. By 2050, all buildings will need to have an emissions footprint close to zero. Buildings will need to become better insulated, use more energy-efficient products and obtain their heating from low carbon sources.
>
> (HM Government 2011a: 5)

Planning provisions for increasing the efficiency of new buildings date back to the late 1990s, when pioneering local councils (such as Newcastle City Council) incorporated energy efficiency measures in their development plans (Owens 1994). Such practices became more widespread across the UK following publication of PPG 3 *Housing*, which suggested that planning authorities should promote the energy efficiency of new housing where possible. However, the scope for planning intervention in this area remained limited, as the standards of design in new buildings are controlled by the Building Regulations. While steps were taken to revise the Regulations to achieve more sustainable design and construction, until the mid-2000s progress was limited. This left a regulatory gap into which the planning system gradually stepped. It should be noted that a new version with more stringent energy efficiency measures in Part L took effect in 2006, which increased efficiency standards by 40 per cent over 2002 levels.

At the same time, the government introduced a package of measures labelled *Towards Greener Building*. These were aimed at achieving zero-carbon homes by 2016. Part of this package was the *Code for Sustainable Homes*, a government-endorsed rating system for new housing, with the sixth rating star awarded to zero-carbon development (DCLG 2006c). Although achieving a specific rating of the Code is voluntary, all new buildings have to be assessed against the Code as part of the planning permission process. Planning also plays a role in strategic coordination and 'in bringing together interested parties and facilitating the establishment of decentralized energy systems', as emphasised in *Building a Greener Future* (DCLG 2007c: 15). New developments with major planning inputs were also piloted to meet the highest environmental standards on a large scale, notably the eco-towns, the Thames Gateway eco-region and the London Olympic Park.

As well as with new development, there is a need for reducing the energy demand of existing buildings, because if we build all the houses we need, by 2050 there will still be a large number of existing stock in the region of about two-thirds of the total housing stock. Improving the energy efficiency of these is therefore paramount. Planning's regulatory intervention can be drawn upon to move this agenda forward. This is already taking place at the local level, using supplementary planning guidance to 'require cost-effective energy efficiency measures to be carried out for the existing building as a condition of planning consent for a home extension' (LGA 2007: 34). Some have suggested more drastic measures, arguing that meeting the national target for GHG emissions in the housing sector requires the demolition of 80,000 dwellings per year (Boardman 2007). While demolition was firmly on the agenda of the Housing Market Renewal programme, government is now putting more emphasis on refurbishment, as is reflected in the ambitious Green Deal initiative, which is aimed at improving the energy efficiency of homes (through loft insulation, wall insulation, new double- and triple-glazed windows, new energy-efficient boilers, draught-proofing and the installation of solar panels). The role of spatial planning in this area of retrofitting is not limited to regulatory measures deployed at the point of planning consents. It also extends to more strategic interventions within the framework of urban regeneration schemes. In fact, as Rydin (2009) suggests, there may be scope for returning to some of the 1970s ideas about housing improvement and bringing together housing and planning policy in new ways. Similar place-making endeavours can be sought in commercial areas in the context of town centre management and Energy Action Areas, where low-carbon technology has been piloted in, for example, New Wembley, Barking Town Centre, Merton and Southwark.

In conclusion, having outlined the great potential of planning to contribute to low-carbon agenda, we should note that 'planning is only one way to respond to climate change. In the UK a whole range of policy instruments and programmes are being used including: taxation, regulation of markets, subsidies and programmes' (ODPM 2004a: 27). Furthermore, the potential for spatial planning to reduce emissions or indeed achieve other sustainability objectives has been persistently undermined by an overriding expectation that the planning system should provide for the predicted demand for the growth of housing, economic activity, traffic volume, waste generation, construction activity, out-of-town shopping and so on. Such potential may be further hampered as a result of the current economic recession, as the emphasis is not just on providing for but also stimulating growth.

Adaptation to climate change

The UK is already experiencing the impacts of climate change. This includes, for example, extreme weather events, such as the 2007 summer floods, when thirteen people lost their lives and about 48,000 homes and 7,000 businesses were flooded (Pitt Review 2008), the 2004–6 drought and the 2003 heatwave (with temperatures in excess of 35°C in South East England). The Association of British Insurers estimated that claims for storm and flood damage in the UK doubled to over £6 billion over the period 1998–2003, with the prospect of a further tripling by 2050 (ABI 2004). These extreme events have had a major impact on the natural environment, households, businesses, infrastructure and the health of particularly vulnerable sections of society, such as low-income households or the elderly.

Developing resilience to the inevitable impacts of climate change is another area in which spatial planning has a significant role to play. Evidence on the extent to which planning has become engaged with adaptation is mixed. While some criticise planners for being fixated on mitigation to the near exclusion of adaptation (LGA 2007), others disapprove of them for not paying enough attention to mitigation policies (FoE 2005; Howard 2009). However, as mentioned earlier, the emerging consensus is that emphasis should be placed on integrating both measures and ensuring that adaptation policies do not jeopardise, in the long term, efforts to mitigate the causes of climate change (Pelling 2010). To this aim, integrated scenarios and models are being developed to assist complex decisions on the right course of action. There is now a clear governmental expectation of the planning system with regard to adaptation. It is expected that 'national policy statements on nationally significant infrastructure projects, regional strategies and local development documents must all take account of a changing climate . . . to deliver planning strategies that secure new development in ways that minimise vulnerability and provide resilience to climate change' (HM Government 2009a: 110). One of the key objectives of the National Adaptation Programme is 'to provide a clear local planning framework to enable all participants in the planning system to deliver sustainable new development, including infrastructure that minimises vulnerability and provides resilience to the impacts of climate change' (NAP 2013: 20).

The NPPF (p. 23) also stresses that

Local Plans should take account of climate change over the longer term, including factors such as flood risk, coastal change, water supply and changes to biodiversity and landscape. New development should be planned to avoid increased vulnerability to the range of impacts arising from climate change. When new development is brought forward in areas which are vulnerable, care should be taken to ensure that risks can be managed through suitable adaptation measures.

Four areas of climate risk have been at the centre of adaptation efforts. These are related to risks of flooding, coastal erosion, heatwaves, and drought (particularly in the south of England). The role of spatial planning has been mainly related to:

(a) the location of new development away from the areas of risk;
(b) the design and layout of buildings and urban areas which are resilient;
(c) the promotion of sustainable water management in new developments.

The TCPA guideline on *Climate Change Adaptation by Design* (Shaw *et al.* 2007) considers how adaptation options are influenced by geographical location and the scale of development, and considers the interrelated roles of the planning system, communities, other stakeholders and delivery bodies. In the following account, the focus will be on adaptation to flood risks and heatwaves, which are major concerns.

Flood risks

As NAP (2013: 16) suggests,

most of the highest order risks for the built environment highlighted in the CCRA are associated with the impacts of flooding, which is expected to become more common throughout the twenty-first century. The cost of expected annual damage to residential properties alone from tidal and river flooding in England and Wales is projected to increase from £640 million at present to over £1.1 billion by the 2020s under the CCRA mid-range climate change scenario. This does not account for potential population increase.

In England and Wales, planning policy on flood risk was first introduced in 1992. In the aftermath of major floods in 2000, the Select Committee on Environment, Transport and Regional Affairs undertook a review of *Development on or Affecting the Flood Plain* and highlighted the critical effect of increased run-off caused by new development, development in the flood-plain and particularly development in the functional floodplain or washlands that are used for storage during floods.

The revised government guidance in PPG 25 *Development and Flood Risk* issued by the DETR in 2001 made it clear that 'the susceptibility to flooding is a material planning consideration' and planners should 'consider how a changing climate is expected to affect the risk of flooding over the lifetime of developments' (p. 4). This was issued well before the Foresight Future Flooding study (DTI 2004), which led to a major reframing of government's long-term strategy for flood risks and coastal erosion. Instead of focusing only on building flood defences, attention was placed on recognising the need for *Making Space for Water* (DEFRA 2005) and protecting floodplains from development. Spatial planning decisions can influence both the probability of flooding and its consequences. As regards the former, PPS 25 *Development and Flood Risk* required planners to adopt a 'risk-based' approach 'to ensure that flood risk is taken into account at all stages in the planning process to avoid inappropriate development in areas at risk of flooding and to direct development away from areas at highest risk' (para. 8).

In 2009, the European Floods Directive (Directive 2007/60/EC) was issued and its requirements were implemented through the Flood Risk Regulations 2009 (SI 2009/3042) in England and Wales and the

Flood Risk Management (Scotland) Act 2009 in Scotland. These required the publication of preliminary flood risk assessments and hazard and risk maps by the end of 2013 and flood risk management plans by 2015. For areas at 'significant risk' the extent of flooding and its adverse impacts have to be mapped. Objectives and measures must then be developed to reduce the risk in flood risk management plans.

The requirements of the Flood Risk Regulations were incorporated into the 2010 revision of PPS 25, which continued to require planners to adopt a 'risk-based' approach. Local authorities have to conduct a sequential test to steer new development towards the lowest probability flood zones, identified in a strategic flood risk assessment (SFRA) based on the Environment Agency flood maps. Also, planning applications have to be supported by a flood risk assessment. By 2009, 85 per cent of local authorities had completed an SFRA and in over 96 per cent of cases where the EA has objected to planning applications on flood risk grounds, the final decision was in line with the EA advice

While some criticise local planning decisions for allowing development to go ahead on floodplains, others blame national planning policy for being too 'restrictive' and inflexible in 'areas that have limited land available for development' (DCLG 2006c: 14), particularly for the provision of much-needed housing. This clearly shows the context in which planning decisions have to be made. It also shows that planning can use not only its regulatory tools to protect 'at risk' areas, but also its collaborative practices to provide arenas for discussing different sides of the argument, and negotiating the terms upon which trade-offs need to be made.

Current national policy as presented in the NPPF retains aspects of PPS 25 but also introduces the Exception Test as a way of responding to the need for the above-mentioned trade-offs (see Box 7.11). It states that 'inappropriate development in areas at risk of flooding should be avoided by directing development away from areas at highest risk, but where development is necessary, making it safe without increasing flood risk elsewhere' (p. 23). What informs this process is the SFRA and advice from the Environment Agency

and other relevant flood risk management bodies, such as lead local flood authorities and internal drainage boards. The *Technical Guidance to the NPPF* (2012) provides further details about sequential and exceptional tests. It is also worth mentioning that the Guidance defines 'flood risk' as 'risk from all sources of flooding – including from rivers and the sea, directly from rainfall on the ground surface and rising groundwater, overwhelmed sewers and drainage systems, and from reservoirs, canals and lakes and other artificial sources' (p. 2). As regards surface water flooding, which is estimated to put over 2.8 million people at risk, the Flood and Water Management Act 2010 contains powers to change the law so that water and sewage companies will be responsible for new sewers.

Heatwaves

As regards the risk of heatwaves, the headline for spatial planning is the urban heat islands (UHI). This refers to air temperature in urban areas, which is several degrees warmer than in the countryside, partly because of the surface cover. The urban heat island effect in turn has a major impact on human health, energy use and biodiversity. It has been suggested that mitigating the effects of urban heat islands 'requires significant local powers in terms of planning and design' (RCEP 2007: 83). This echoes the recommendations of the *Urban Task Force* in 1999 that called for an integrated approach to planning, urban design, and management with a view to enhancing the potential amenity value of the public realm. Multifunctional green networks or 'green infrastructure' (Handley *et al.* 2007; Shaw *et al.* 2007) can provide cooler microclimates, reduce surface water run-offs, and help urban areas adapt better to climate change. The NPPF states that 'Local planning authorities should: set out a strategic approach in their Local Plans, planning positively for the creation, protection, enhancement and management of networks of biodiversity and green infrastructure' (NPPF 2012: 26).

Protecting local amenities, notably green and open spaces, has been an integral part of the planning system. However, the rationale for it has changed over time (see the section on 'The Meanings of the

BOX 7.11 DEVELOPMENT IN AREAS AT RISK OF FLOODING

A sequential approach should be used in areas known to be at risk from any form of flooding.

If, following application of the Sequential Test, it is not possible, consistent with wider sustainability objectives, for the development to be located in zones with a lower probability of flooding, the Exception Test can be applied if appropriate. For the Exception Test to be passed:

- it must be demonstrated that the development provides wider sustainability benefits to the community that outweigh flood risk, informed by a Strategic Flood Risk Assessment where one has been prepared; and
- a site-specific flood risk assessment must demonstrate that the development will be safe for its lifetime taking account of the vulnerability of its users, without increasing flood risk elsewhere, and, where possible, will reduce flood risk overall.

Both elements of the test will have to be passed for development to be allocated or permitted.

Source: NPPF 2012, pp. 23–25

Environment in Planning' above). The need to adapt to climate change has extended their functional values from aesthetics to biodiversity and the ecosystem. Green infrastructure resources (including private gardens and parks) need to be strategically planned, at both regional and local planning levels, and designed and managed to maximise their climate-related functionality (Gill *et al.* 2009). Planning clearly plays a role in achieving this. However, even here, some commentators have raised the potential for conflict between the environmental and the social dimensions of place making (Hebbert 2009).

Overall, the role of spatial planning in adapting to climate change is still at the developmental stage. Some even argue that it is taking place 'on the fringes' of the planning system (Bulkeley 2006). Institutionally, this is because the growing stakeholder-based *climate change partnerships* (Climate Local and UK Climate networks, for example) that have been set up across the UK to pursue local adaptation strategies are operating largely outside the formal arenas of the planning system. Some are drawing on alternative rationalities to encourage resilience. In London, for example, they are encouraging 'businesses to consider re-locating flood-sensitive IT equipment and archives out of

London to areas with negligible flood risks' (CLC 2007: ii).

Furthermore, in responding to climate change, planners are faced with a number of challenges which are arising from the inherent complexity of dealing with climate change issues, such as the interaction between energy, transport and settlement pattern, and between energy and building performance; transition from current state of the built environment to one which is less dependent on fossil fuel; timescale and dynamics of change (e.g. an extended, sometimes millennial, timescale of climate change and the traditional planning timescale of ten to twenty years); interactions of various spatial scales (e.g. mitigation of GHG emissions has aggregate effects at a global level but derives from cumulative actions on smaller spatial scales); the evolving policy context and the need for adaptive management; and potential conflicts between adaptation and mitigation measures. These complexities, coupled with climate change uncertainties, require a portfolio of policy responses and not just planning. However, there is no doubt that planning is seen as the key delivery mechanism for a series of government policies on adaptation. The need for adaptation to climate change is also raising important conceptual issues for planners. It highlights the need for understanding space and place in relational, rather than absolute terms, and for taking into account the ubiquity of change and uncertainty.

MARINE SPATIAL PLANNING

The division between the planning control system at sea and on land may be regarded as forming the root of many of the problems with current coastal protection and planning policies . . . Harmonising the planning systems of below and above the low water mark seems to us to be the basic requisite for an integrated approach to planning in the coastal zone.

(House of Commons Select Committee on
Coastal Zone Protection and Planning, 1992: 30)

Despite this early insight, it took ten years before *Safeguarding Our Seas* was published by the government (DEFRA 2002). It showed how coastal zones are affected by the interface between inland water catchment and high seas influences, and that coast is the interface between marine and terrestrial environment. However, coastal management activity in the UK continued to focus on the land and hence was subsumed under the framework of terrestrial spatial planning.

The sea is a complex ecosystem that cuts across national and international administrative borders. Even within the context of the British Isles alone, 'the marine geography of the seas . . . is both complex and diverse not only in physical terms, but also with respect to the human dimension' (Smith *et al*. 2012: 30). In the last two decades, coastal areas in the UK have been subject to a number of policy initiatives, which, as in other environmental areas, have been largely triggered by the EU. For example, in the 1990s the European Commission issued the Bathing Water and Urban Waste Water Directives and the Wild Birds and Habitats Directives, all relevant to the management of coastal areas.

A significant step towards environmental improvement in coastal areas was the adoption of the European Water Framework Directive (WFD) in 2000 (Directive 2000/60/EC) (see the section on 'Water quality' in this chapter) under whose terms coastal waters extend to all territorial waters for the achievement of 'good chemical status' but one nautical mile for achieving 'good ecological status'. In England and Wales, the coastal boundary is defined as only one nautical mile into territorial waters, while in Scotland the coastal boundary is three nautical miles from the baseline. The WFD required the first set of River Basin Management Plans to be in place by 2009.

During the 2000s, debate focused on 'integrated spatial approaches' to coastal development which in turn brought to the fore the role of spatial planning (Kidd and Ellis 2012). In 2002, the Commission's *Recommendation on Integrated Coastal Zone Management* (ICZM) raised concerns about increasing urbanisation in coastal areas and its environmental implications, particularly in places with a large number of tourist attractions. It highlighted the need to integrate both the marine and the terrestrial components of the coastal zones in ICZM. In 2006, the EU Green Paper *Towards a Future Maritime Policy for the EU* was adopted, followed by, first, the Blue Paper in 2007 – which proposed an Integrated Maritime Policy (IMP) for the EU and a detailed action plan – and second, the *Roadmap for Maritime Spatial Planning* in 2008. The Roadmap uses the term 'maritime-spatial' – instead of 'marine-spatial' – planning (MSP) in order 'to underline the holistic cross-sectoral approach of the process' (CEC 2008: 2). It suggests that 'MSP is a tool for improved decision-making. It provides a framework for arbitrating between competing human activities and managing their impact on the marine environment. Its objective is to balance sectoral interests and achieve sustainable use of marine resources' (CEC 2008: 2).

In 2008, the Marine Strategy Framework Directive (MSFD) (Directive 2008/56/EC) provided the environmental pillar of the IMP. It required

Member States to achieve good marine environmental status by 2020, to apply an ecosystem approach, and to ensure that pressure from human activities is compatible with good environmental status. Member States are required to cooperate where they share a marine region or sub-region and use existing regional structures for coordination proposes, including with third countries.

(CEC 2008: 7)

Although the MSFD does not directly regulate maritime activities, the impact of such activities must be taken into account for the determination of good environmental status. Some member states, including the UK, have declared that they would use MSP to implement the MSF Directive.

The MFSD came into legislative force through the Marine and Coastal Access Act 2009 in England and Wales, the Marine (Scotland) Act 2010 and the Marine Act (Northern Ireland) 2013. These form the basis of a new plan-led system for marine activities across the UK marine area, which comprises inshore and offshore regions for each of the four national jurisdictions.

The cornerstone of the marine planning system is the 2011 UK *Marine Policy Statement* (MPS), which has statutory weight across all the jurisdictions. This commits all four UK administrations to ensure that coastal areas and activities are managed in an integrated and holistic way in line with the principles of ICZM. Drawing on the five guiding principles of sustainable development (shown in Figure 7.1 above), MPS specifies a number of high-level marine objectives, such as achieving sustainability and living within environment limits. Therefore, the marine planning system has links to the establishment of an ecologically coherent network of marine protected areas (MPAs) in UK waters. The MPA network primarily consists of marine conservation zones (MCZs) designated under the Act and European marine sites designated under the EC Wild Birds and Habitats Directives (Natura 2000 sites).

The 2009 Act defined arrangements for a new system of marine management, including the introduction of marine planning, across the UK. The new arrangements provided for the creation of two new authorities in England: the Marine Management Organisation (MMO), based in Tyneside, and the Inshore Fisheries and Conservation Authority (IFCA). The MMO is responsible for delivering UK marine policy objectives for English waters through a series of statutory marine

plans (discussed below) and other measures. In the devolved administrations, similar responsibilities rest with the Welsh Assembly government, Marine Scotland and Northern Ireland Marine Task Force. The MMO is an executive non-departmental public body (NDPB) like the Environment Agency and is tasked with carrying out planning functions for English waters. To fulfil this function, the MMO has been delegated most of the Secretary of State's functions as the marine plan authority for the English inshore and English offshore regions. The decisions which remain with the Secretary of State are those to

• approve a new or revised statement of public participation (SPP) prior to publication;
• approve a consultation draft of a marine plan prior to publication;
• publish a new or revised marine plan;
• adopt or withdraw a marine plan.

Marine plans

Marine plans interpret and present MPS policies at a subnational level. They 'must be in conformity with the MPS unless relevant considerations indicate otherwise, thereby ensuring a strong link between national policy and local application' (DEFRA 2011c: 5). The general structure of the marine planning system, as proposed by the DEFRA (2011c) guideline, is presented in Figure 7.9.

In defining what 'a well-designed marine planning system' is, DEFRA (2011c: 9) refers to principles of strategic thinking which, ironically, are missing from the terrestrial (spatial) planning systems after the reform of the planning systems in 2011 (see Box 7.12).

The geographical scope of marine planning is defined by the 2009 Act, which divides UK waters into marine planning regions with an inshore region (0–12 nautical miles) and offshore region (12–c.200 nautical miles) under each of the four Administrations (England, Northern Ireland, Scotland and Wales). The Act refers to marine plan authorities which are responsible for planning in each region, with the exception of the Scottish and Northern Ireland inshore regions,

Policy-making Process

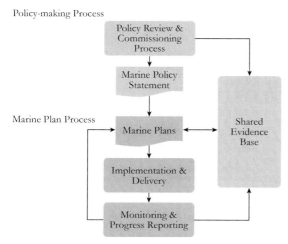

Marine Plan Process

Figure 7.9 Proposed structure of marine spatial planning

Source: DEFRA 2011c: 6, Figure 1

which are covered by separate legislation. In England's marine area, the Secretary of State has delegated the marine planning function for both the inshore and offshore regions to the MMO, as has been mentioned above. Another interesting aspect of marine planning is that marine plans have to

represent the three dimensional nature of the marine environment by addressing the seabed and the substrata below it, the whole of water column

and the area above it. They should also provide for a temporal dimension to cover seasonal, occasional, or time-limited activities, uses and designations.

(DEFRA 2011c: 18–19)

Following stakeholder consultation in 2010, eleven marine plan areas have been identified. As can be seen in Figure 7.10, the North West area is shown as a single colour divided by a dashed line to reflect the recommendation that here two plans should be prepared under a single process, because otherwise planning for a small area will entail two separate processes of consultation with the neighbouring areas of Scotland and Wales.

The order and timetable for developing plans in each marine plan area is decided by the MMO. The East Inshore and East Offshore areas are the first areas in England to be selected for marine planning. The East Inshore area includes a coastline that stretches from Flamborough Head to Felixstowe. The planning process officially started in April 2011 and by April 2014 (the time of writing) the UK's first ever marine plans for these areas were published in one document and approved by the Secretary of the State.

As regards the structure of a marine plan, the DEFRA (2011c) guideline proposes a structure which strongly resembles that of the pre-2011 local development framework (see Figure 7.11). It suggests that, 'most obviously, the Marine Plan Strategy Document and Policy Map bear similarities respectively

BOX 7.12 WHAT IS A WELL-DESIGNED MARINE PLANNING SYSTEM?

A well-designed marine planning system will determine the preferred scenario and future direction of any given Marine Plan area by ensuring that rather than reacting separately to each individual project/activity or looking in isolation at sectoral developments, as happens now, all decisions and future implications are made strategically, shaping the area according to a clearly set out vision, policies and objectives. In so doing, it should contribute to conserving and enhancing the value of the marine environment, including its biodiversity, its seascapes, the ecosystem services it provides and the heritage assets it contains.

Source: DEFRA 2011c: 9

The North West area is shown as a single colour divided by a dashed line to reflect the recommendation that the two plans here be prepared in a single process

Figure 7.10 Marine plan areas in England
Source: DEFRA 2011c: 23, Figure 2

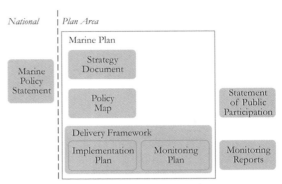

Figure 7.11 Proposed structure of a marine plan
Source: DEFRA 2011c: 33, Figure 4

which, as Jay (2012) demonstrates, is even more pertinent with regard to the relational complexities of the marine world.

PLANNING AND POLLUTION PREVENTION AND CONTROL

> The planning system should contribute to and enhance the natural and local environment by . . . preventing both new and existing development from contributing to or being put at unacceptable risk from, or being adversely affected by unacceptable levels of soil, air, water or noise pollution or land instability.
>
> (NPPF, para. 109)

Although local authorities are not environmental authorities, they have specific powers in relation to some environmental issues, such as certain aspects of the pollution of water, air, soil and noise pollution. Also, as planning authorities, they must pay attention to the pollution implications of development when deciding planning applications. Thus, in drawing up their local plans, they should ensure that a new development is appropriate for its location (NPPF, para. 120). This includes potentially polluting development and new developments (for example housing, schools

to the Core Strategy and Proposals Map in LDFs. One difference between Marine Plans and terrestrial planning documents is the geographic level at which Marine Plans operate' (DEFRA 2011c: 34). As in local plans, marine plans are subject to sustainability appraisal (SA). The SA process is based on that of the Strategic Environmental Assessment Directive and Regulations, but it covers social and economic effects of plans alongside the predominantly environmental topics in the Directive. They should also engage in a public participation process.

Marine planning is a new area of planning and a lot can be learned from the experience of a century of terrestrial planning in the UK (Kidd and Ellis 2012). One important lesson is the move away from a physically deterministic towards a relational and socially constructed understanding of spatial relations,

or hospitals) that may be affected by existing sources of pollution. They should also consider site-specific policies for the location of potentially polluting activities and criteria against which development management decisions will be made (NPPF, para. 109).

Pollution control framework

The planning and pollution control systems are separate but complementary. Planning has a major role in determining where development goes and whether its location may give rise to pollution directly or indirectly (through, for example, traffic generation). However, as was stressed in the former PPS 23 (para. 10),

> the planning system should focus on whether the development itself is an acceptable use of the land, and the impacts of those uses, rather than the control of processes or emissions themselves. Planning authorities should work on the assumption that the relevant pollution control regime will be properly applied and enforced. They should act to complement but not seek to duplicate it.

The NPPF follows a similar line and stresses that local planning authorities should not attempt to control processes or emissions which are subject to approval under other pollution control regimes (NPPF, para. 122). They should assume these regimes will operate effectively. They should, however, consult with the relevant pollution control authority before determining planning applications where pollution could be an issue.

One of the cornerstones of the pollution control system since the Alkali Act 1874 has been the concept of *best practicable means* (BPM), which focuses on achieving means which meet desirable standards for dealing with pollution without restraining the polluter's resources too far. This informal and relatively secretive system, which was traditionally based on voluntary compliance, avoiding confrontation and legalistic procedures, was a characteristic feature of practices such as waste management. It predominantly took place within a narrow and closed policy community dominated by the regulators and the industry, with environmental groups playing a marginal reactionary role. As McCormick (1991: 12) states, the Control of Pollution Act 1974, for example, 'was essentially shaped by industry and local government'. The subsequent implementation of the Act was heavily criticised by the House of Commons Environment Committee in 1989, which pointed to extremely low standards in many waste disposal authorities, which had 'encouraged contractors who had no regards for the potential dangers to the environment'. They concluded that 'the cheapest tolerable option was too often deployed instead of the best practicable environmental option' (Environmental Protection Act 1990, Part II).

By that time the BPM had been replaced by the concept of the *Best Practicable Environmental Option* (BPEO), which retains the element of negotiation but involves a wider consideration of environmental factors and an openness which was foreign to its predecessor (RCEP 1988: para. 1.3). Central to this principle is the recognition of the need for a coordinated approach to pollution control, taking into account the danger of the transfer of pollutants from one medium to another, as well as the need for prevention. Section 7 of the Environmental Protection Act 1990 (later updated by the Environment Act 1995) introduced a requirement for the regulating authority to ensure that the *best available techniques not entailing excessive cost* (BATNEEC) are being used for

- preventing the release of prescribed substances into an environmental medium, or, where that is not practicable, for reducing the release to a minimum; and
- rendering harmless any other substance which could cause harm if released into any environmental medium.

BATNEEC is a concept favoured by the industry and introduced in EU Directives which were adopted in UK environmental law. It is the responsibility of the operator to demonstrate that the requirements of BATNEEC are met and also to demonstrate their competence and experience, and that effective environmental management controls are in place. Additionally, certain statutory environmental standards ('quality objectives'), specified emissions limits or national quotas have to be met. The 1990 Act provided that where it is alleged that BATNEEC has not been used in a prescribed operation, 'it shall be for the accused to prove that there was no better available technique not entailing excessive cost than was in fact used' (section 25). This makes the offence one of 'strict liability', in contrast to the traditional one of 'fault-based liability'. Where a process involves the release of harmful substances into more than one medium, BPEO must also be demonstrated – thus there may be trade-offs between the effects in one environmental medium and another. In order to judge the effects of different emissions in different media, an integrated permit process has been adopted. The current framework has within its foundations the concept of 'best available techniques' (BAT) for dealing with potential pollution. Under this, the conditions in each installation's permit have to be based upon the application of BAT relevant to the industry sector concerned.

Environmental regulation is progressively based on an integrated approach which recognises that pollution does not abide by the boundaries of air, land or water, that pollution is mobile and is a 'cross-media' problem. This is the foundation of the current *integrated pollution prevention and control* (IPPC) regime, which regulates pollution from larger industrial installations and implements the 2008 EU Directive on Integrated Pollution Prevention and Control (Directive 2008/1/EC). IPCC applies to about 4,000 industrial installations in the UK (about 45,000 in the EU), ranging from refineries to breweries and from intensive pig farms to cement works. The IPPC Directive is implemented in England and Wales by the Environmental Permitting Regulations 2007 (SI 2007/3538), which came into effect on 6 April 2008. They brought waste management licensing, pollution prevention and control permitting and a number of Directives into a single regulatory regime. From 6 April 2010, the scope of the Regulations was expanded to include the permissions previously called water discharge consents, groundwater authorisations, radioactive substances authorisations and registrations. Permits are required for activities or operations comprising

- installations and mobile plant that carry out certain industrial, waste and intensive agriculture activities;
- waste operations;
- water discharge and groundwater activities;
- radioactive substances activities;
- mining waste operations.

In 2010, the Directive on Industrial Emissions (IPCC) (Recast) (Directive 2010/75/EU) was published and required transposition into UK law by January 2013. This represents a combining of seven directives, including the Waste Incineration Directive (Directive 2000/76/EC), into one piece of legislation. It is expected to be implemented gradually, starting in 2013, for various sectors and to be completed by 2016, when Industrial Emissions Directive will replace the IPPC Directive.

The first national Planning Policy Guidance, PPG 23 *Planning and Pollution Control*, was issued in 1994 by the then Department of the Environment. In 2004, its revised version, in the form of PPS 23, provided a review of who does what in pollution control and is especially helpful in explaining where and how the planning system should tackle pollution issues. It should be remembered that many of the 'polluting problems' that planners have to deal with do not come under the pollution regulation regime. The planning system considers a wider range of developments whose polluting activities may or may not be relevant to the pollution control regimes but constitute a *statutory nuisance* or result in a loss of amenity. In many cases, close cooperation is required between the planning authority and the Environment Agency, which took over the regulatory functions of Her Majesty's Inspectorate of Pollution (HMIP) in 1996.

Following the publication of the NPPF, the Environment Agency produced a 'Quick Guide' to set out 'the key messages and planning policy hooks on preventing and managing environmental risks in the NPPF' (EA 2012: 1). The Guide confirms that the planning system has a key role to play in protecting and improving the environment by directing potentially polluting developments to appropriate locations and facilitating the redevelopment of polluted sites. Again, the role of planning is seen as complementary to pollution control decisions in the attempt to deliver sustainable development (EA 2012: 2).

Air quality

Concern about air pollution is not new: it was as early as 1273 that action in Britain was taken to protect the environment from polluted air. A royal proclamation of that year prohibited the use of coal in London and one man was sent to the scaffold in 1306 for burning coal instead of charcoal. Those who pollute the air are no longer sent to the gallows, but, though gentler methods are now preferred, it was not until the disastrous London smog of 1952 (resulting in 4,000 deaths) that really effective action was taken. The Clean Air Acts 1956 and 1968 introduced regulation of emissions of dark smoke, grit and dust from furnaces, chimney heights and domestic smoke. Local authorities were empowered to establish *smoke control areas*, which were very effective (coupled with the switch from coal fires to central heating).

Air quality has improved considerably since the early 1960s: smoke emissions have fallen by 85 per cent since 1960, the notorious big-city smogs are a thing of the past, and hours of winter sunshine in central London have increased by 70 per cent. Over the past twenty years, the EU has successfully reduced the number of pollutants. Lead emissions, for example, have fallen by some 90 per cent. But air pollution is still a major cause of health and well-being concerns. While, air pollution from industrial activities has decreased and improvements have been made in respect of domestic sources of air pollution, pollution from increased traffic sources has increased considerably (Banister 1999; Stead and Nadin 2000), particularly in 'hotspots' such as congested urban centres.

Action to manage and improve air quality is largely driven by EU legislation and supported by UK legislation. Indeed, subsequent UK sustainable development strategies in 1994, 1999 and 2005 have given prominence to improving air quality. A major step forward came through the EU Air Quality Framework Directive in 1996 (Directive 96/62/EC), which set target values for twelve air pollutants which are elaborated and revised under related directives. This was followed by the UK *National Air Quality Strategy* (1997), which provided national standards and targets, with monitoring taking place through a comprehensive network of air quality monitoring stations and longitudinal data.

The 2008 Ambient Air Quality Directive (Directive 2008/50/EC) sets legally binding limits for concentrations in outdoor air of major air pollutants that impact public health such as particulate matter (PM_{10} and $PM_{2.5}$) and nitrogen dioxide (NO_2). These pollutants not only have direct effects, but also can combine in the atmosphere to form ozone, a harmful air pollutant (and potent GHG) which can be transported great distances by weather systems. An annual national assessment of air quality (based on modelling and monitoring) is undertaken by DEFRA to ensure compliance with EU limit values. The role of planning is to ensure that the potential impact of new development on air quality is taken into account where the national assessment indicates that relevant limits have been exceeded or are near the limit.

Under local air quality management (LAQM), local authorities are responsible for reviewing and assessing ambient air quality. If there is a risk that pollutant levels will exceed prescribed objectives, then local authorities are required to designate *air quality management areas* (AQMA) and write action plans to meet the objectives. These plans seek to reduce emissions through addressing the sources and distribution, especially traffic (discussed in Chapter 12). They may, in principle, designate areas which should be closed to traffic or be restricted to low-emission vehicles – although care will be needed to avoid displacement effects and some 'local' pollution will have a non-local source. In practice, they are not proving to be so radical (Miller 2000). Other developments could also affect existing air quality or create new exposure to poor air quality. Thus, local plans and decisions on planning applications should help achieve EU air quality limit values and local air quality action plans (NPPF, para. 124).

The European Commission conducted a wide-ranging review of the EU's air quality policies, taking into account the latest science and cost-effective measures. On the basis of this, in December 2013, it adopted a new Clean Air Quality Package, which can be viewed online. Also, the UK Public Health Outcomes Framework, published in 2012, includes a number of sections on air pollution (DoH 2013: 12) and noise. These show that the question of significance is obvious in air quality management areas, where the local authority is seeking to lead improvement in air quality, but it may also be necessary to consider the 'cumulative impacts of a number of smaller developments on air quality, and the impact of development proposals in rural areas with low levels of background air pollution' (PPS 23, Appendix A). Although restricted to questions of the use and amenity of land (as opposed to health), there is a clear signal to authorities to use planning powers to generally improve air quality conditions. This is important not only in relation to human health and well-being, but also environmental health, because poor air quality affects biodiversity and may therefore impact on the UK's international obligations under the Habitats Directive. For these reasons, air quality is a consideration in strategic environmental assessment, and sustainability appraisal can be used to shape an appropriate strategy, for example, through establishing a 'baseline', appropriate objectives for the assessment of impact and proposed monitoring. Odour and dust can also be a planning concern because of their impact on local amenity.

Water quality

Concerns about the broader water environment came to the fore with the publication of the WFD in 2000, which set out a programme of assessing, protecting and improving the aquatic environment of both freshwater and coastal waters so that 'good ecological status' is achieved by 2015. The directive uses river basin districts as the units of management and is based on a six-year cycle of planning.

The WFD was transposed into UK legislation by the Water Environment (Water Framework Directive) (England and Wales) Regulations 2003 (SI 2003/3242), the Water Environment and Water Services (Scotland) Act 2003 and Water Environment (Water Framework Directive) Regulations (Northern Ireland) 2003 (SI 2003/544). River basin management plans are the responsibility of the Environment Agency in England, Wales and Northern Ireland, and the Scottish Environmental Protection Agency.

The environment agencies hold the main regulatory powers over the water environment, although they have no operational responsibilities (these are carried out by the water service companies or in some cases local authorities). The agencies have statutory functions in relation to water resources and the control of pollution in inland, underground and coastal waters. In England, the Environment Agency is responsible for monitoring and reporting on the objectives of the WFD on behalf of government. It works with government, Ofwat, local government, non-governmental organisations (NGOs) and a wide range of other stakeholders, including local businesses, water companies, industry and farmers, to manage water. The EA produces a river basin management plan for each river basin district in England to address the main issues in protecting and improving our water environment. A river basin is the area of land that runs or drains down into a river. It published its first set of river basin management plans in 2009, along with an impact assessment. These are being reviewed, and updated plans will be published in 2015.

Environment agencies have wide-ranging powers, but there are three critical issues in relation to planning: water quality and pollution, the maintenance of water supplies, and flooding. With regard to water quality, the agencies can take preventive action to stop pollution, take remedial steps where pollution has already occurred and recover the reasonable costs of doing so from a polluter. The agencies have inherited and continued to develop a sophisticated and relatively public regulatory system, which involves the setting of water quality objectives and a requirement that consent is obtained for discharges of trade and sewage effluent into controlled waters. Extensive monitoring programmes include surveys of the quality of rivers, estuaries and coastal waters. These show that river quality is improving steadily. In 2003, only 4 per cent of rivers monitored (across 7,000 sites) were considered to be of poor quality and 1 per cent were bad, compared with 10 per cent in these categories in 1990. Although the majority of rivers still have high levels of phosphate (53 per cent) and nitrate (27 per cent), this was down on 1990 rates (64 per cent and 30 per cent).

Bathing water in the UK is protected by EU Directive 76/160/EEC, now replaced by Directive 2006/7/EC. The new Directive has a stricter definition of good water standards. The implementation of this Directive as well as the EU Urban Waste Water Directive (Directive 91/271/EEC) and investment in sewage treatment has resulted in improved bathing water quality. In 2000, 44 per cent of the 471 beaches tested in England and Wales and in 2004, 80 per cent of 491 beaches met the standards of the EU Bathing Waters Directive, and 95 per cent passed the mandatory tests. Despite all this, there is still a need to improve the quality of open waters, also known as 'water-bodies', which include rivers, streams, lakes, estuaries, coastal waters and groundwater. The WFD says that every EU member must reach 'good water-body status' by 2015, and cannot allow water-body standards to drop. Only 27 per cent of water-bodies in England are, as of 2014, classified as being of 'good status' under these standards, which have been set by the WFD.

The 2006 Groundwater Daughter Directive (Directive 2006/118/EC) of the WFD also requires action to prevent hazardous substances entering groundwater (underground water), which is an important resource, both for drinking water and for providing water for rivers. Developments, such as minerals activities, might be rejected or receive conditional approval on the basis of their impact on groundwater quality.

As the UK is a country surrounded by water and has a relatively high level of annual rainfall, one might expect that there would be no question of an adequate supply of water. However, rain falls unevenly over both area and time. In the mountainous areas of the Lake District, Scotland and Wales, average annual rainfall exceeds 2,400 millimetres, and for most of the country there is a significant margin between effective rainfall and abstraction. But in the Thames estuary rainfall is less than 500 millimetres and for much of the Thames and Anglian regions licensed abstractions are more than two-thirds of the effective annual average rainfall. This is of great concern, even given the high level of reuse, because these are also the regions with the highest demands for new development. The drought of 1988–92 and long, hot, dry summer of 1995 raised awareness about the impact of demand, with unacceptably low levels in some rivers and supply constraints. As a result, there has been a stream of official reports and consultation papers, together with the development of academic studies in this area, which had received remarkably little attention previously (at least in the UK).

The need for a major programme of new investment is widely recognised, not only to replace outworn facilities, but also to meet new demands for water, for environmental protection, and for sustainability. At the same time, increased concerns about water supply have come from developers and the public. The result is a renewed awareness of the importance of the relationship between water and land use planning (Slater *et al.* 1994: 376). In addition, government has made a requirement for twenty-five-year resource plans from water companies and targets for reduction of leakage of 25 per cent over three years (in 1997 about 25 per cent was lost through leakage). Thus, demand management is certainly coming to the fore in relation to water supplies. Most water companies are planning to publish the final version of their latest plans (covering 2015–40) later in 2014 and begin consulting on their next plans in 2018.

Noise

'Quiet costs money . . . a machine manufacturer will try to make a quieter product only if he is forced to, either by legislation or because customers want quiet machines and will choose a rival product for a lower noise level. So stated the Wilson Committee in 1963. This, in one sense, is the crux of the problem of noise. More, and more powerful, cars, aircraft, portable radios and the like must receive strong public opprobrium before manufacturers – and users – will be concerned with their noise level. Similarly, legislative measures and their implementation require public support before effective action can be taken.

As with other aspects of environmental quality, attitudes to noise and its control have changed in recent years, partly as a result of the advent of new sources of noise such as portable music centres, personal stereos, and electric DIY and garden equipment, as well as greatly increased traffic. The increased concern about noise is reflected in a succession of inquiries and planning policy (PPG 24 *Planning and Noise*). More substantively, two Acts have been passed to provide stronger measures for dealing with the problems. The Noise and Statutory Nuisance Act, which was passed in 1993, strengthened local authority powers to deal with burglar alarms, noisy vehicles and equipment, and various other noise nuisances. Second, the Noise Act 1996 provided a summary procedure for dealing with noise at night (11 p.m. to 7 a.m.). This includes powers for local authorities to serve a warning notice and to seize equipment which is the source of offending noise. The 1996 Act does not require local authorities to use its provisions, but the situation is to be reviewed in the light of experience.

There are three ways in which noise is regulated: by setting limits to noise at source (as with aircraft, motor-cycles and lawnmowers), separating noise from people (as with subsidised double glazing in houses affected by serious noise from aircraft or from new roads), and exercising controls over noise nuisance. Where intoler-able noise cannot be reduced and reduces property values, an action can be pursued at common law or, in the case of certain public works, compensation can be obtained under the Land Compensation Act 1973.

Noise from neighbours is the most common source of noise nuisance and complaints. This is a difficult problem to deal with, and official encouragement is being given to various types of neighbourhood action, such as 'quiet neighbourhood', 'neighbourhood noise watch', noise mediation and similar schemes (Oliver and Waite 1989). There is provision under the Control of Pollution Act 1974 for the designation by local authorities of *noise abatement zones*, though the statutory procedures for these are cumbersome and, in any case, they are not well suited to dealing with neighbourhood

noise in residential areas (though they are useful for regulating industrial and commercial areas).

Traffic noise takes many forms and is being tackled in various ways (as summarised in Chapter 4 of the Royal Commission on the Environment 1994 report on *Transport and the Environment*). Road traffic noise is the most serious in the sense that it affects the most people. Here, emphasis is being put on the development of quieter road surfaces and vehicles. Aircraft noise has long been subject to controls both nationally and (with the UK in the lead) internationally. The principal London airports are required by statute to provide sound insulation to homes seriously affected by aircraft noise and similar non-statutory schemes apply to major airports in the provinces.

Noise is a material consideration in planning decisions and development plans may contain policies on noise particularly where there are major noise generators such as airports (although the reproduction of detailed noise contours in plans is not recommended). PPG 24 set out four noise exposure categories (NECs), and in the worst case (category D), permission should normally be refused. The definition of boundaries between categories is difficult for non-experts, but they are clearly insufficient to prevent the building of houses adjacent to motorways, which continues regardless. Such decisions aside, local authorities are taking more interest in noise and one – Birmingham City Council – with the support of central government (and building on practice in other European countries) produced a noise map of the whole of the city, including the impact of road, rail and air traffic and ambient noise levels during both the day and night. CPRE has also produced a map of tranquil areas for the whole of England comparing the 1960s with the 1990s, which demonstrates the extensive intrusion of noise. CPRE then joined with the Environment Agency, the former Countryside Agency and the Countryside Council for Wales to designate *tranquil areas*. They estimated that England had lost 21 per cent of its tranquil areas since the 1960s.

The EU Environmental Noise Directive (Directive 2002/49/EC) covers noise from roads, rail, aviation and industry. It requires member states to carry out noise mapping of environmental noise sources every five years. These maps are used to produce noise action plans. The first noise maps were completed in 2007 and updated in 2012. The updated action plans take into account the 2012 noise mapping and views from the consultation held in 2013. Noise action plans provide a framework to manage environmental noise and its effects. They also aim to protect quiet areas in agglomerations (large urban areas) where the noise quality is good, and generally provide information on noise levels to the public.

In 2010, DEFRA published the Noise Policy Statement (NPS) for England, which sets out the long-term vision of government noise policy, to promote good health and a good quality of life through the management of noise. Its first aim is to 'avoid significant adverse impacts on health and quality of life from environmental, neighbour and neighbourhood noise within the context of Government policy on sustainable development' (DEFRA 2010: para. 2.22).

NPPF is clear that noise should be considered when new developments may create additional noise and when new developments would be sensitive to the prevailing acoustic environment. When local or neighbourhood plans are being prepared or decisions taken about new development, there may also be opportunities to consider improvements to the acoustic environment. NPPF (para. 123) states that planning policies and decisions should aim to:

- avoid noise from giving rise to significant adverse impacts on health and quality of life as a result of new development;
- mitigate and reduce to a minimum other adverse impacts on health and quality of life arising from noise from new development, including through the use of conditions;
- recognise that development will often create some noise and existing businesses wanting to develop in continuance of their business should not have unreasonable restrictions put on them because of changes in nearby land uses since they were established; and
- identify and protect areas of tranquillity which have remained relatively undisturbed by noise

and are prized for their recreational and amenity value for this reason.

In line with the Explanatory Note of the NPS for England (DEFRA 2010), the above entails identifying whether the overall effect of the noise exposure (including the impact during the construction phase, wherever applicable) is, or would be, above or below the 'significant observed adverse effect level' and the lowest observed adverse effect level for the given situation. As noise is a complex technical issue, it may be appropriate to seek experienced specialist assistance when applying this policy.

Light pollution

Artificial light provides valuable benefits to society, for example, through extending opportunities for sport and recreation, and can be essential to a new development. Equally, artificial light is not always necessary, and has the potential to become what is termed 'light pollution' or 'obtrusive light', and not all modern lighting is suitable in all locations. It can be a source of annoyance to people, harmful to wildlife, undermine enjoyment of the countryside or detract from enjoyment of the night sky. For maximum benefit, the best use of artificial light is about getting the right light in the right place and providing light at the right time.

Lighting schemes can be costly and difficult to change, so getting the design right and setting appropriate conditions at the planning stage are important. In particular, some types of premises (including prisons, airports and transport depots, where high levels of light may be required for safety and security reasons) are exempt from the statutory nuisance regime for artificial light, so it is even more important to get the lighting design for these premises right at the outset. The courts have ruled that lighting itself is not 'development'. However, on the basis of the Clean Neighbourhoods and Environment Act 2005, planning permission is required for lighting if it alters the material appearance of a building. Since 1997, local authorities have been able to consider lighting as part of the planning process for new residential and commercial buildings. Therefore, they can decide to regulate lighting under planning permission and set planning obligations for lighting to prevent light pollution (DEFRA 2006b: 10).

Waste planning

> Waste, is a deceptively simple term. Common usage defines waste as readily identifiable materials which are no longer useful and hence can be thrown away. However, while it is possible to recognise the ontology of waste (that waste exists), it is difficult to determine the boundaries between what *is* and what *is not* waste. In other words, it is difficult to determine exactly when a material ceases to be a resource (with social, economic or environmental values) and *becomes* a waste.
>
> (Davoudi 2009b: 131)

The question of defining waste is not just a technical or legal matter, important as this is. It also matters to how societies handle their waste problem. What constitutes waste is socially and culturally constructed and varies between places, societies and times. The popular saying that 'somebody's waste is another person's gold mine' is a simple reflection of the complexities of defining waste. As Worpole (1999: 24) suggests, 'a newspaper on the café table is a highly esteemed cultural artefact; blowing around the street an hour later, it becomes a threat to our very sense of meaning and belonging. Ten newspapers scattered on the pavement and there goes our neighbourhood.' Thus 'waste is matter in the wrong place' (Mary Douglas quoted in Worpole 1999:24) and at the wrong time.

As far as technical/legal definitions are concerned, the UK definition follows that in the EU Waste Framework Directive (Directive 2008/98/EC), which describes waste as 'any substance or object . . . which the holder discards or intends or is required to discard'. Within this definition, waste streams are employed to categorise particular types of waste which may be produced by individuals or organisations. Primarily these are (DEFRA 2013a: 7):

1. Municipal waste – household waste and commercial waste similar to household waste;
2. Industrial (including agricultural) and commercial waste;
3. Construction and demolition waste;
4. Hazardous waste.

On the basis of this definition, the EU's economy uses 16 tonnes of materials per person per year, of which 6 tonnes become waste, half of it going to landfill, despite the fact that some member states (such as Germany) have already achieved recycling rates of over 80 per cent and have virtually eliminated landfill. Others (such as the UK) still have some way to go (CEC 2013a). The reason for this lies in decades of policy neglect during which waste was not seen as a problem; neither was it considered as something that had to be managed rather than merely disposed of in voids created by minerals extraction (Davoudi 2006b). As a result, landfill remained (and still is) the main disposal option, because it was considered cheap, convenient and even beneficial for 'filling the holes in the ground'. Little attempt was made to introduce more sustainable waste management practices. Prior to the turn of the century, the vast majority of waste produced in the UK had been landfilled and even by 1997–8, only 7 per cent of household waste was recycled in England. Political apathy and over-reliance on landfill led to Britain's being labelled as the 'dirty man of Europe' in the 1990s and having difficulties in catching up with the emerging EU waste regulations.

This policy neglect was even more prominent within the planning system. Until 1994, when the first national Planning Policy Guidance Note was issued, government legislation and guidelines had concentrated solely on waste licensing and pollution regulation rather than spatial planning dimensions. Yet, at the local level this lack of national planning policy was not considered problematic because of the ample supply of quarries which provided convenient tipping sites (Davoudi 1999b). This is reflected in the following statement from a district planner:

> At the time, landfill seemed quite an attractive way of getting these holes in the ground dealt with. Fill

them up, put grass and trees on the top and hey presto! But then you have tricky things like landfill gas to deal with.

(interview, 1996, quoted in
Davoudi 2009c: 144)

Hence, for a long time the role of the planning system was reduced to that of reiterating a series of standard site-specific regulatory criteria. Planning policies for waste, often only a few policies tacked on at the back of development plans, were devoid of a long-term strategic approach to waste. This, however, changed in 1991, when local planning authorities were charged with the preparation of a complete waste local plan for their entire jurisdiction.

However, with stricter controls, EU policy, and rising development pressures to be accommodated in a plan-led process, together with more demand for waste sorting and bulking depots and recovery facilities, planning matters became more complex. Furthermore, decisions on the location of waste management facilities tend to generate considerable public concern. Thus, planning began to play a more central role in the waste management process in the late 1990s.

The main tenets of the EU policy on waste go back to 1989, when the *Community Strategy for Waste Management* was adopted. While this was later reviewed, in 1996, it led to significant pieces of legislation which together provided the foundation for the EU policy on waste. This general legal framework was supplemented by several other directives dealing with specific waste streams and setting technical standards for waste treatment and disposal facilities, notably for energy from waste (EfW) incineration and landfill. As will be discussed later, landfill became the subject of another influential EU Directive with far-reaching consequences for UK waste policy. It was against this backdrop that the UK waste strategy *Making Waste Work* was published in 1995. This was the government's first major policy document since the publication of the 1974 Green Paper *The War on Waste*.

The central plank of the strategy, reproduced from the EU Waste Framework Directive, was the concept of the waste hierarchy (Figure 7.12), an influential and enduring concept in terms of its impact on

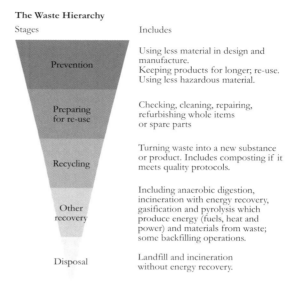

The Waste Hierarchy

Stages	Includes
Prevention	Using less material in design and manufacture. Keeping products for longer; re-use. Using less hazardous material.
Preparing for re-use	Checking, cleaning, repairing, refurbishing whole items or spare parts
Recycling	Turning waste into a new substance or product. Includes composting if it meets quality protocols.
Other recovery	Including anaerobic digestion, incineration with energy recovery, gasification and pyrolysis which produce energy (fuels, heat and power) and materials from waste; some backfilling operations.
Disposal	Landfill and incineration without energy recovery.

Figure 7.12 The waste hierarchy
Source: DEFRA 2013a

emerging waste policies and discourses in the UK (Davoudi 2000a). The waste hierarchy is both a guide to sustainable waste management and a legal requirement, enshrined in law through the Waste (England and Wales) Regulations 2011 (SI 2011/988). The waste hierarchy put the prevention, reduction and reuse of waste at the top of management priorities, followed by recovery of material and energy, with disposal (e.g. landfill) being considered as the least environmentally favourable option.

In 1999, the introduction of a series of stringent targets for the reduction in landfilling of biodegradable waste by the EU Landfill Directive (Directive 99/31/EC) in 1999 (see Box 7.13) provided further incentives for change in policy and practice, because non-compliance with the targets could lead to EU fines of the order of several million pounds. This, plus the rising cost of landfill and the shortage of sites, began to shift the policy towards the options at the top of the hierarchy and an emphasis on management rather than mere disposal. Waste management is defined by the revised Waste Framework Directive (Directive

BOX 7.13 EU LANDFILL DIRECTIVE AND ITS TRANSPOSITION INTO UK STATUTE

The *EU Landfill Directive* refers to Council Directive 1999/31/EC of 26 April 1999, [1999] OJ L 182/1 on the landfilling of waste. The main elements of the Directive are:

- targets to reduce the amount of biodegradable municipal waste sent to landfill;
- banning co-disposal of hazardous and non-hazardous waste, and a requirement for separate landfills for hazardous, non-hazardous and inert wastes;
- banning landfill of tyres;
- banning landfill of liquid wastes, infectious clinical wastes and certain types of hazardous waste;
- provisions on the control, monitoring, reporting and closure of sites.

The most significant of these is the first. With account taken of a four-year derogation offered to those countries heavily reliant on landfill, the targets for the UK are to reduce biodegradable municipal waste landfilled to:

- 75 per cent of that produced in 1995 by 2010;
- 50 per cent of that produced in 1995 by 2013;
- 35 per cent of that produced in 1995 by 2020.

EU member states had until 16 July 2001 to transpose the Directive into national law. The UK government started this process with the publication of a consultation paper, *Limiting Landfill*, in October 1999. However, the transposition deadline was missed and a second, delayed, consultation was issued by DEFRA in August 2001. This paper included draft regulations for the first time. Legislation to transpose the technical provisions of the Landfill Directive into law was finally laid before Parliament in May 2002, and took effect in England and Wales on 15 June 2002, only one month before implementation of the Directive commenced on 16 July 2002. In 2007, the UK government set new, higher recycling/composting and recovery targets (40 per cent and 53 per cent by 2010 respectively) for municipal waste in order to meet one of its key objectives, which is 'to meet and exceed the Landfill Directive diversion targets for biodegradable municipal waste in 2010' (DEFRA 2007: 11). In 2013, the government stated that 'as part of monitoring progress towards meeting EU Landfill Directive targets – we estimate that we will have sufficient residual waste treatment infrastructure, on reasonable assumptions, to meet our Directive obligations' (DEFRA 2013a: 31).

Source: Updated from Davoudi 2009c: 140, Box 1

2008/98/EC) as 'the collection, transport, recovery and disposal of waste, including the supervision of such operations and the after-care of disposal sites, and including actions taken as a dealer or broker' (Article 3(9)).

In the last two decades, waste management in England (and the UK) has undergone a sea change. Waste production is gradually declining, with total waste generation in 2010 in England being estimated

at 177Mt, down from 325.3Mt in 2004. Per head of population, it produces 526 tonnes of municipal waste per year (2009), down from a peak of 603 tonnes in 2004. Thus, the UK can now boasts that compared with the EU15 countries, the UK produces the sixth-lowest quantity of waste per capita. While this might be partly due to the economic crisis of the late 2000s, it is also due to a shift in policy and practice with regard to, for example, recycling. Recycling of household waste in

England increased from 7 per cent in 1997–8 to 36.3 per cent in 2007–8. By 2010, recycling and composting of household waste have increased to 43 per cent and business recycling rates have increased to 52 per cent.

Similar achievements have been made in commercial and industrial waste (C&I), of which 47.9 million tonnes were generated in England in 2009, compared with 67.9 million tonnes in 2002–3, A total of 25 million tonnes (52 per cent) of C&I waste was recycled or reused in England in 2009, compared with 42 per cent in 2002–3, and a total of 11.3 million tonnes (24 per cent) of it were sent to landfill in 2009, compared with 41 per cent in 2002–3 (DEFRA 2013a) (see Figure 7.13 and Box 7.14).

Local authorities, which cover all household waste and some commercial and industrial waste, have reduced the amount of waste they send to landfill by about 60 per cent since 2000.

As was mentioned above, the change has been driven by a range of EU and UK policies. The landfill tax escalator has created a strong incentive to divert waste from landfill. Additional funding for local authorities, for example, through the private finance initiative, has led to the development of new waste treatment facilities. National planning policy seeks to enable local authorities to put planning strategies in place through their local plans which shape the type of waste facilities in their areas and where they should go. All of these measures are helping to drive waste to

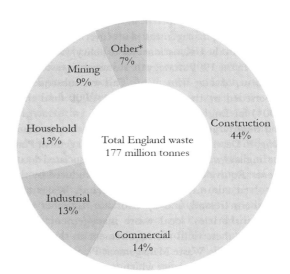

*Agriculture, forestry, fisheries and dredging spoils

Figure 7.13 The distribution of waste arising by key sectors, based on 2010 data

Source: DEFRA 2013a: 18

be managed further up the waste hierarchy to the extent that, from lagging well behind, the UK has now reached a comparable level of performance with many countries in the EU (DEFRA 2013a: 1).

Article 28 of the revised Waste Framework Directive (Directive 2008/98/EC) requires that member states

BOX 7.14 WASTE GENERATION IN THREE MAIN WASTE STREAMS, 2010, ENGLAND

- Household waste accounts for 24Mt (13 per cent) of waste generated in England. 1.0Mt of hazardous waste is produced from the household sector mainly from discarded vehicles and equipment.
- Commercial and industrial waste accounts for 48Mt (27 per cent) of waste generated in England. 4.6Mt of hazardous waste is produced from the C&I sector
- Construction and demolition waste (including mineral waste) accounts for 77Mt (44 per cent) of waste generated in England. 0.8Mt of hazardous waste is produced from the C&D sector.

Source: DEFRA 2013a

the Strategic Environmental Assessment Directive. The main stages in the SEA process are as follows (p. 24):

- Stage A: setting the context and objectives, establishing the baseline and deciding on the scope;
- Stage B: developing and refining alternatives and assessing effects;
- Stage C: preparing the environmental report;
- Stage D: consulting on the draft plan or programme and the environmental report;
- Stage E: monitoring the significant effects of implementing the plan or programme on the environment.

Figure 7.14 illustrates the relationships between the main stages. However, the SEA should be treated as a flexible process, tailored to the needs of the different types of plans and programmes to which the directive applies, while at the same time ensuring that the requirements of the directive are met. In using the method suggested by the guidance, it should be noted that in practice, the responsible authority may find it appropriate to vary its approach, for example, in combining qualitative and quantitative assessment. A detailed account of the SEA using several case examples is provided by Therivel (2010) and the OECD (2008 and 2010). While some have suggested that the SEA can act as instrument of 'good governance' (e.g. Ahmed 2008), others have called for a more radical and effective approach to environmental or ecological compensation schemes (Wilding and Raemaekers 2000), both of which are already employed in Germany.

Sustainability appraisal

Section 19 of the Planning and Compulsory Purchase Act 2004 requires a local planning authority to carry out a sustainability appraisal of each of the proposals in a local plan during its preparation. More generally, section 39 of the Act requires that the authority preparing a local plan must do so 'with the objective of contributing to the achievement of sustainable development'. Sustainability appraisals incorporate the requirements of the Environmental Assessment of Plans and Programmes Regulations 2004, which implement the requirements of the European SEA Directive. Sustainability appraisal ensures that potential environmental effects are given full consideration alongside social and economic issues.

SEA considers only the environmental effects of a plan, whereas sustainability appraisal considers the plan's wider economic and social effects in addition to its potential environmental impacts. Sustainability appraisal should meet all of the requirements of the Environmental Assessment of Plans and Programmes Regulations 2004, so a separate strategic environmental assessment should not be required.

A sustainability appraisal is a systematic process that must be carried out during the preparation of a local plan. Its role is to promote sustainable development by assessing the extent to which the emerging plan, when judged against reasonable alternatives, will help to achieve relevant environmental, economic and social objectives. The process is an opportunity to consider ways by which the plan can contribute to improvements in environmental, social and economic conditions, as well as a means of identifying and mitigating any potential adverse effects that the plan might otherwise have. By doing so, it can help ensure that the proposals in the plan are the most appropriate, given the reasonable alternatives. It can be used to test the evidence underpinning the plan and help to demonstrate how the tests of soundness have been met. Sustainability appraisal should be applied as an iterative process informing the development of the local plan (see Figure 7.15). The local planning authority must carry out an appraisal of the sustainability of their proposals to ascertain how the plan will contribute to the achievement of sustainable development.

Sustainability appraisal applies to any of the documents that can form part of a local plan, including core strategies, site allocation documents and area action plans. Neighbourhood plans, supplementary planning documents, the statement of community involvement, the local development scheme or the authority monitoring report are excluded from this requirement.

In 2004, government published specific guidance, *Sustainability Appraisal of Regional Spatial Strategies and*

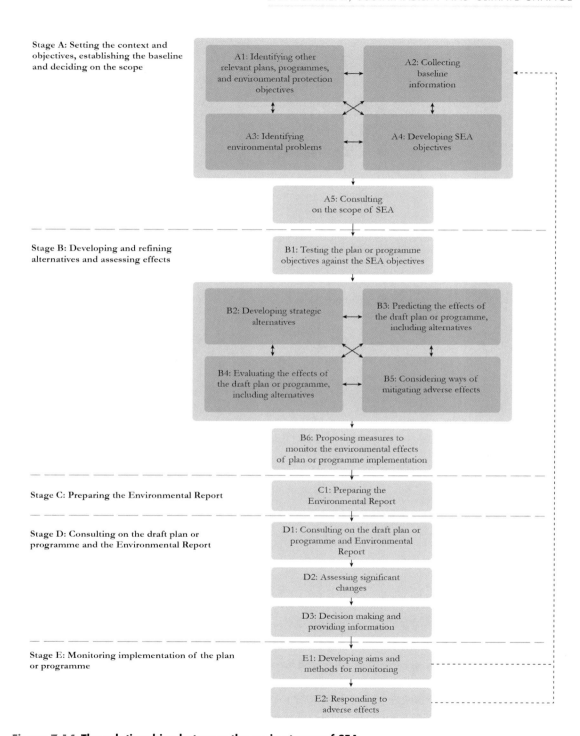

Figure 7.14 The relationships between the main stages of SEA

Source: ODPM 2005

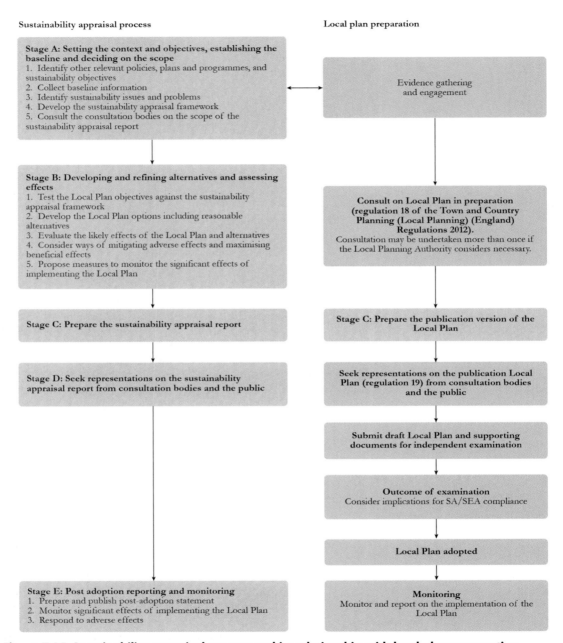

Sustainability appraisal process

Stage A: Setting the context and objectives, establishing the baseline and deciding on the scope
1. Identify other relevant policies, plans and programmes, and sustainability objectives
2. Collect baseline information
3. Identify sustainability issues and problems
4. Develop the sustainability appraisal framework
5. Consult the consultation bodies on the scope of the sustainability appraisal report

Stage B: Developing and refining alternatives and assessing effects
1. Test the Local Plan objectives against the sustainability appraisal framework
2. Develop the Local Plan options including reasonable alternatives
3. Evaluate the likely effects of the Local Plan and alternatives
4. Consider ways of mitigating adverse effects and maximising beneficial effects
5. Propose measures to monitor the significant effects of implementing the Local Plan

Stage C: Prepare the sustainability appraisal report

Stage D: Seek representations on the sustainability appraisal report from consultation bodies and the public

Stage E: Post adoption reporting and monitoring
1. Prepare and publish post-adoption statement
2. Monitor significant effects of implementing the Local Plan
3. Respond to adverse effects

Local plan preparation

Evidence gathering and engagement

Consult on Local Plan in preparation (regulation 18 of the Town and Country Planning (Local Planning) (England) Regulations 2012).
Consultation may be undertaken more than once if the Local Planning Authority considers necessary.

Stage C: Prepare the publication version of the Local Plan

Seek representations on the publication Local Plan (regulation 19) from consultation bodies and the public

Submit draft Local Plan and supporting documents for independent examination

Outcome of examination
Consider implications for SA/SEA compliance

Local Plan adopted

Monitoring
Monitor and report on the implementation of the Local Plan

Figure 7.15 Sustainability appraisal process and its relationship with local plan preparation

Source: Planning Practice Guidance website (http://planningguidance.planningportal.gov.uk/blog/guidance/strategic-environmental-assessment-and-sustainability-appraisal/sustainability-appraisal-requirements-for-local-plans/; accessed 7 July 2014)

Local Development Frameworks, which defined sustainability appraisal as 'an iterative process that identifies and reports on the likely significant effects of the plan and the extent to which implementation of the plan will achieve the social, environmental and economic objectives by which sustainable development can be defined' (para. 1.2.2).

Prior to implementation of the SEA Directive, it was argued that the appraisal process had not contributed to policymaking; that it was highly subjective; and that it led to inconsistent conclusions (Russell 2000). However, it is now widely applied and there is a strong recognition of its importance in the delivery of sustainable development (Short *et al.* 2004). The NPPF (para. 165) clearly states that 'A sustainability appraisal which meets the requirements of the European Directive on strategic environmental assessment should be an integral part of the plan preparation process, and should consider all the likely significant effects on the environment, economic and social factors.'

There remain some questions regarding the availability of expertise to conduct appraisals, especially in view of the demanding timescales for preparing plans. 'The bottom line is that if the government is committed to making the plan-making system more efficient, it needs to consider how to accommodate increasingly complex appraisals' (Holstein 2002: 219). Government's response is that

> Assessments should be proportionate, and should not repeat policy assessment that has already been undertaken. Wherever possible the local planning authority should consider how the preparation of any assessment will contribute to the plan's evidence base. The process should be started early in the plan-making process and key stakeholders should be consulted in identifying the issues that the assessment must cover.
>
> (NPPF, para. 167)

Further reading

For the changing environmental discourses, see Dryzek's *The Politics of the Earth* (1997). A more specific analysis of how the environment has been treated and interpreted in the planning system is offered in the following four articles: Newby (1990) 'Ecology, amenity and society: social science and environmental change'; Healey and Shaw (1994) 'Changing meanings of "environment" in the British planning system'; Whatmore and Boucher (1993) 'Bargaining with nature: the discourse and practice of "environmental planning gain"'; and Davoudi (2012b) 'Climate risk and security: new meanings of "the environment" in the English planning system'. On the notion of ecological modernisation, see Hajer (1995) *The Politics of Environmental Discourse: Ecological Modernisation and the Policy Process* and Jacobs (1999) *Environmental Modernisation*.

For a detailed stocktaking on sustainability, see Jordan (2008) 'The governance of sustainable development: taking stock and looking forwards' and (2009) 'Revisiting the governance of sustainable development: taking stock and looking forwards'. Among earlier reflections on the topic, the following offer useful critical insights: Owens (1994) 'Land, limits and sustainability'; Khan (1995) 'Sustainable development'; Real World Coalition's *From Here to Sustainability*, edited by Christie and Warburton (2001); and Church and McHarry (1999) *One Small Step: A Guide to Action on Sustainable Development in the UK*.

Publications that specifically address planning's contribution to sustainability are Layard *et al.* (2001) *Planning for a Sustainable Future*; Blowers (1993) *Planning for a Sustainable Environment*; Breheny (1992) *Sustainable Development and Urban Form*; Williams *et al.* (2000) *Achieving Sustainable Urban Form*; Buckingham-Hatfield and Evans (1996) *Environmental Planning and Sustainability*; Kenny and Meadowcroft (1999) *Planning Sustainability*; Hales (2000) 'Land use development planning and the notion of sustainable development'; and Wheeler (2013) *Planning for Sustainability: Creating Liveable, Equitable and Ecological Communities. The Earthscan Reader in Sustainable Cities* (1999), edited by David Satterthwaite and reprinted in 2001, compiles contributions from twenty-two authors on the issues of urban development and sustainable cities.

For up-to-date and reliable information about the science of climate change, mitigation and adaptation, see the latest IPCC assessment reports. Regarding the relationship between planning and climate change, two books provide details and examples: Wilson and Piper (2010) *Spatial Planning and Climate Change* and Davoudi *et al.* (2009) *Planning for Climate Change, Strategies for Mitigation and Adaptation for Spatial Planners*. For a specific account of the English planning system and its role in climate change, see Davoudi (2013) 'Climate change and the role of spatial planning in England'. Pelling (2010) *Adaptation to Climate Change* provides a good discussion of various aspects of climate adaptations. Jordan *et al.*'s (2010) *Climate Change Policy in the European Union: Confronting the Dilemmas of Mitigation and Adaptation?* addresses the issue at the EU level.

A detailed and useful introduction to environmental policy in the EU is provided by Jordan and Adelle's (2012) *Environmental Policy in the EU: Actors, Institutions and Processes*. On the politics of environmentalism, see Castree's double articles 'Neoliberalising nature: the logics of deregulation and reregulation'.

On marine spatial planning, see Qui and Jones (2011) *The Emerging Policy Landscape for Marine Spatial Planning in Europe*.

For a history of pollution control and much else on the origins of environmental policy, see Ashby and Anderson (1981) *The Politics of Clean Air*. A legal sourcebook is *Encyclopaedia of Environmental Law*, edited by Thornton *et al.* (looseleaf; updated regularly). Wood (1999)

'Environmental planning' also traces the history of the relationship. A detailed review of the evolution of waste planning in England is provided in Davoudi (2009) 'Planning for Waste Management: The Changing Discourses and Institutional Relationships', *Progress in Planning,* 53(3), pp. 165–216; and its update in Davoudi (2009c) 'Scalar tensions in the governance of waste: the resilience of state spatial Keynesianism'.

The principal texts on environmental impact assessment are Glasson *et al.* (1998) *Introduction to Environmental Impact Assessment*; Elvin and Robinson (2000) 'Environmental impact assessment'; Jones *et al.* (1998) 'Environmental assessment in the UK planning process'; Weston (2000) 'Reviewing environmental statements'; and Wood (2000) 'Ten years on: an empirical assessment of UK environmental statement submissions'.

Two sources provide a full account of SEA. One is Ahmed (2008) *Strategic Environmental Assessment for Policies: An Instrument for Good Governance*; the other is Therivel (2010) *Strategic Environmental Assessment in Action*. See also Short *et al.* (2004) 'Current practice in the strategic environmental assessment of development plans in England' and Russell (2000) 'Environmental appraisal of development plans'.

The best sources of up-to-date information about environmental regulations are the official websites of environment agencies for the EU, England and the devolved administrations in the UK.

8 Conservation of the historic environment

[preservation is a] subject on the edge of land Planning proper

(Keeble 1964: 315)

The Government believes that the historic environment is an asset of enormous cultural, social, economic and environmental value. It makes a very real contribution to our quality of life and the quality of our places.

(HM Government 2010b: 3)

Introduction

This chapter describes the conservation-planning system and how it operates. Like the planning system, heritage protection has evolved over more than a century. Initially heritage protection was quite separate from planning but since the great post-war planning acts of the 1940s there has been substantial convergence. However, whilst conservation planning has come to be firmly embedded within the town planning system, ideas of heritage and conservation also derive from wider conceptions of cultural heritage and these are briefly reviewed at the beginning of this chapter.

This is followed by a brief historical account of the evolution of the conservation-planning system and how it has moved from a being a peripheral concern of planning, as suggested by the Keeble quotation at the start of this chapter, to become a central (and not unproblematic) objective in many places. The focus then shifts towards the contemporary conservation-planning system – its structure, financing and principal components in terms of the protection of archaeology, historic buildings and areas as well as some of the trends and pressures that the historic

environment is facing today. As with other elements of the planning system, whilst the fundamental principles of the conservation-planning system are mostly the same across the UK, there are some significant differences of detail in the component nations. The main emphasis in this chapter is upon England, with some significant differences in the other nations highlighted.

Conservation values and the evolution of conservation planning

The conservation of the historic environment sits within a broader family of conservation activity, with some important shared values; we can regard conservation planning as part of a family of practice concerned with cultural, material heritage that encompasses, for example, seeking to sustain in some way a historic building or place (the theme of this chapter) or a painting or museum object. Muñoz-Viñas (2005) traces a lineage of conservation that emphasises shared values, philosophies and principles, and unity of practice, including central concepts, such as the idea of 'authenticity'. At the same time, different conservation activities have evolved in different contexts and have developed their own values and practices.

Conservation planning has strong roots in the older field of architectural conservation. The origins of modern architectural conservation in Britain rest with claims on two nineteenth-century polymaths, famous across a range of artistic and social domains, John Ruskin and William Morris. These figures represent an idea of conservation as critical opposition to practices of transformation and change. In particular, they were famous for opposing the restoration of ecclesiastical buildings through radical interventions in building fabric that also claimed to be conservation. Rather simplistically, this came to be represented as 'conservative repair', as advocated by Ruskin and Morris, versus 'stylistic restoration', which has become particularly associated with the French architect Eugène Emmanuel Viollet-le-Duc (Jokilehto 1999). Their views about *how* buildings should be treated were underpinned by a view of *why* such buildings should be valued, particularly centred on ideas of inter-generational stewardship.

A critical precept for Ruskin and Morris was that the value of the building, its authenticity, is closely associated with its material fabric. The goal of the architect or conservator should be to make as little physical alteration to the historic building as possible. Ruskin's ideas and protests were mobilised and codified by the *Manifesto* drafted by Morris in 1877 at the formation of the Society of the Protection of Ancient Buildings (SPAB) (Morris 1877: 2). Morris wrote:

> It is for all these buildings, therefore, of all times and styles, that we plead, and call upon those who have to deal with them to put Protection in the place of Restoration, to stave off decay by daily care, to prop a perilous wall or mend a leaky roof by such means as are obviously meant for support or covering, and show no pretence of other art, and otherwise to resist all tampering with either the fabric or the ornament of the building as it stands . . . thus, and thus only can we protect our ancient buildings and hand them down instructive and venerable to those that come after us.

Within this quote are some important principles about repairing buildings rather than 'restoring' them, undertaking the minimum work necessary, that new work that is necessary should not be concealed and the notion of custodianship of the heritage. This remains a touchstone document in the field of architectural conservation and the SPAB remains a significant amenity body. Thus Ruskin and Morris articulated principles of conservation action about *how* buildings should be treated, such as the idea of minimum intervention, that are still current.

Ruskin and Morris are particularly remembered for their battles over the ecclesiastical heritage. The material content, or *what*, of architectural conservation has undergone a subsequent extraordinary transformation. Definitions of buildings, places, and environments to which heritage value is ascribed have extended in ways one imagines they would have found difficult to contemplate, including some of the constructions that were the object of their fury. Thus, we have seen protection extended to a huge diversity of buildings and objects in terms of architectural style and temporal period. In addition, we have also seen conservation protection extend beyond architectural conservation into place management, into a new field of conservation planning.

This broadening has often come about through a continuation of the tactics of Ruskin, Morris and the SPAB of critical opposition, often represented, and proselytised for, by temporal period. Thus we have the formation in chronological succession of the Georgian Group (as a breakaway from last the SPAB), the Victorian Society and the Thirties Society, with the last in turn reconstituting itself as the Twentieth Century Society, each campaigning for the value of a particular historical period. Campaigns over the course of the twentieth century became increasingly focused on influencing the planning system through, for example, lobbying for buildings to be listed and against any subsequent moves to gain consent for demolition or unsympathetic alteration. Throughout the period that conservation planning has developed, its advocates have been involved in development struggles and in open conflict with other social interests (with, for example, 'philistine developers'). Moreover, the mission developed in other ways – most notably the increasing focus from the mid-twentieth century on looking beyond the monumental to historic *places*. This was perhaps best represented by the formation of the now defunct Civic Trust, alongside a myriad of local place-based groups, and the subsequent creation of the conservation area system. In recent decades, we can see the development of a professional infrastructure with representative professional bodies, such as the Institute of Historic Building Conservation.

In the process, key principles of intervention have endured, albeit within an evolving framework. Therefore, for example, this has included new concepts of approach to traditional architectural conservation problems. There is now greater tolerance of interventions if these can be considered reversible. There has also developed a greater emphasis on aesthetic considerations. The extension of the mission of conservation from object to place, and the management of place, reinforced this compositional element, for example through the influence of the townscape movement (Pendlebury 2009). Thus, as architectural conservation as an activity has been extended over a much more significant quantity of buildings and places, it has been increasingly linked with systems and processes of town planning, becoming something we can label conservation planning.

English Heritage (EH) produced an important statement on conservation values in 2008, *Conservation Principles: Policies and Guidance for the Sustainable Management of the Historic Environment* (English Heritage 2008a). It is intended first and foremost as an EH document – to ensure a consistent and reasoned approach from EH – but is also intended to have a wider influence. Drawing heavily upon the Australian Burra Charter and foregrounding the concept of significance (see below), *Conservation Principles* sets out four key sets of heritage value: evidential value, historical value, aesthetic value and communal value. Evidential value and aesthetic value represent a historical continuity of focus on fabric and aesthetics and historical value picks up the once subordinate area of historical association and the story to be told about the place, although this is often linked back to material evidence. More novel is the inclusion of communal value, although as Waterton (2010) has argued, this

is generally subsidiary and assumed to be dependent on the other, primary, values, in the planning process at least.

Valuable though these documents are, their primary focus still tends to be with protected historic buildings or sites and not the wider issue of the management of historic environments that planners often deal with. How far are principles evolved for our major monuments applicable and practical when dealing with more ordinary buildings and places? How far can they be applied to conservation areas in their complexity, diversity and dynamic nature? In many ways conservation thought has not kept pace with changing circumstances. So, for example, there is not as clearly articulated a tradition for the management of historic areas as there is for buildings. The principal approach has been that of *townscape*, especially linked with Cullen's seminal text of this name (Cullen 1961), but evident in a range of planners of the post-war period. This approach places great stress on aesthetics and picturesque composition. Less well known is the urban morphological tradition, focusing on the processes of evolution of land use in historic places (Larkham 1996; Pendlebury 1999). In practice, much of the debate about conservation areas has come to focus on concepts of 'character'.

Finally in this section it is worth briefly mentioning the proliferation of international and country-specific charters, many produced by the International Council on Monuments and Sites (ICOMOS). One particular charter that has been influential in the UK was produced specifically for an Australian context – the Australia ICOMOS Burra Charter, originally adopted in 1981 and revised a number of times subsequently (Australia ICOMOS 2013). The Charter is based on seven key principles for undertaking conservation projects:

1   The place itself is important.
2   Understand the significance of the place.
3   Understand the fabric.
4   Significance should guide decisions.
5   Do as much as necessary, as little as possible.
6   Keep records.
7   Do everything in a logical order.

Several of these principles are familiar from the SPAB tradition. However, there are some interesting differences from the SPAB. The Burra Charter was evolved in a specific context. 'Understanding the significance of the place' involves understanding the wider cultural significance of a place. As well as the significance of the fabric, this includes, for example, peoples' memory and association with the place. Another key contribution of the Burra Charter is in codifying systematic processes. Linked to this has been the wide-scale adoption of conservation (management) plans as a means of understanding historic sites and informing decision-making (Worthing and Bond 2008). Within this there are some critical concepts useful for an intelligent approach to managing historic assets. The emphasis placed upon defining significance, that is, exactly what it is that is important about a historic site, before deciding what can be done with it, is of particular value. Significance now forms the basis of policy advice for regulatory decision-making in town planning, albeit, importantly, not (as yet) in the primary legislation for heritage protection.

The UK is a signatory to two Council of Europe conventions concerned with cultural heritage: the Malta Convention on the Protection of the Archaeological Heritage and the Granada Convention for the Protection of the Architectural Heritage of Europe. In practice, these have next to no impact on the day-to-day management of the cultural heritage. At an international level, of rather more significance in practical terms is the United Nations World Heritage Convention. We return to this below.

A brief history of the heritage protection system

One issue that emerges from the preceding discussion is the importance that individuals and organisations outside the state have played, and continue to play, in conservation endeavours in the UK. Since the nineteenth century, national and local amenity bodies have been vigorously promoting and extending the conservation cause both in terms of what is considered to be heritage and the need for a developed system of policy and law to protect it and in terms of the fate of individual buildings and places up and down the country.

In practice, Britain was slow to legislate on heritage protection and was somewhat behind many other European countries. The first tentative legislation that was introduced with the aim of heritage protection was the Ancient Monuments Act 1882. This was promoted by the MP Sir John Lubbock and only became law after a number of failed bills over the previous decade. The legislation merely allowed the government to accept as a gift or to purchase a schedule of sixty-eight largely prehistoric monuments, with the provision that more 'of a like character' could be added. In the words of Thurley (2013: 41), 'Thus the mountain of ten years of debate produced the molehill of an act.' More effective legislation began to be introduced in the twentieth century. In particular the Ancient Monuments Act 1913 is seen as a landmark and the centenary worthy of celebration (Thurley 2013). This act expanded the definition of what could be classified as an ancient monument, introduced the possibility of Preservation Orders (a term still used colloquially, though no longer in the legislation) and allowed for much larger numbers of monuments to be 'scheduled', which would require owners to give a month's notice before undertaking any work on a monument.

Tentative but ultimately largely inconsequential beginnings were made in the inter-war period to extend heritage protection to buildings. The Housing Act 1923 authorised Ministers to make a town planning scheme to preserve the existing character and to protect the existing features of a locality where the 'special architectural, historic or artistic interest' warranted it (cited in Delafons 1997: 38). The origins of the clause in the Bill have been linked with concerns about the preservation of Oxford and it was apparently inserted at the initiative of Oxford graduate Civil Service lawyers (Sheail 1981; Cocks 1998) but in practice this clause was never used.

Concerns in Oxford were not unique. Lobbying for action also came from the newly formed Stratford-upon-Avon Preservation Committee, set up in response to a potentially damaging factory proposal.

It commissioned a report from Patrick Abercrombie on the future planning of Stratford to present to the Corporation (Abercrombie and Abercrombie 1923). The Stratford report was an early example of concerted action by a local group on preservation issues. Whilst the national focus was on the countryside and Georgian London, pressure for development and change was occurring in historic cities up and down the country. Although local amenity groups had existed since the nineteenth century, this period saw groups arise in some of the most significant historic cities. Preservation groups were formed in Oxford in 1925, in Cambridge in 1928 and in Bath in 1934 (in this last case growing out of an earlier society formed in 1909). A number of historic cities also gained local acts of legislation giving a degree of control over such matters as the aesthetics of new buildings and demolition. Bath had the Bath Corporation Acts 1925 and 1937, the latter of which brought in some controls over the façades of 1,251 buildings. Many college buildings in Oxford were scheduled as ancient monuments. York had an architectural panel reviewing the design quality of proposals. The Town and Country Planning Act 1932 contained further enabling provisions on historic buildings. However, again this led to an almost total lack of action. Thus, prior to the Second World War, though conservation planning legislation existed, it did not lead to systematic national action. However, at local level a variety of preservation groups and local authorities used what means they could to influence development in their towns and cities.

The stimulus to start what became the post-war 'listing' process developed from twin efforts: first, identification of important buildings to be kept, if possible, amongst the bomb-damaged buildings and second, the foundation of the National Buildings Record to record threatened buildings. Subsequently the listed building system was created in the Town and Country Planning Acts 1944 and 1947. A key purpose for post-war listing was the identification of potential constraints to be included in, or worked around, in post-war reconstruction. Post-war development was to be led by the public sector. Thus lists of buildings were not a means of resisting rapacious developers, but guides to inform rational

decision-making by municipal planners. At the time, it was generally accepted that many buildings so listed would be sacrificed in the higher interests of planning (Pendlebury 2009). During the latter war years and in its immediate aftermath, but mostly before the landmark 1947 legislation, many localities up and down the country produced, or commissioned from consultants, advisory plans, now generally referred to as reconstruction plans, including many smaller towns and cities, such as cathedral-type cities, pre-industrial resort towns and ancient university towns (Larkham 2003; Pendlebury 2003). Typically, plans for such settlements acknowledged their historic character whilst seeking to achieve a 'balanced approach' to development: becoming more functional and rationally organised twentieth-century towns whilst respecting place history. In practice, they were highly variable in how they engaged with historic character. Thomas Sharp's plans were perhaps the most sensitive, whilst also displaying a breathtaking confidence in the interventions proposed. For example, his Oxford plan proposed leaving much of the historic city untouched but included a radical inner 'substitute road' across Christ Church Meadow (Sharp 1948).

In practice, the economic deprivations of post-war austerity meant that relatively little development occurred in the decade or so after the 1947 Act, with the major exception of the public housing programme. However, towards the end of the 1950s and in the 1960s the momentum in favour of redevelopment of central areas grew rapidly. As the consequences of redevelopment in city centres became apparent, so opposition to the demolition of buildings grew. The exemplars of this in the early 1960s were the Euston Arch and the Coal Exchange, both in London. As public concerns about the transformation of urban areas developed, so did an official concern for area conservation. Richard Crossman, Secretary of State from 1964, and supported by junior minister Lord Kennet from 1966, was instrumental in putting the conservation of historic areas more firmly on the agenda, despite civil service resistance. A significant milestone in the development of national official thinking about the planning of historic cities was the four demonstration studies jointly commissioned by

particular controls are now much stricter and the policy presumption in favour of retaining listed buildings stronger than in the early days of the system. Decisions on what to list are made by the DCMS, advised by EH. This can be done as part of a comprehensive survey or following a request to look at a particular building (often referred to as 'spot listing'). Emphasis has recently shifted from comprehensive area surveys to thematic surveys, most controversially with modern buildings. An owner or developer concerned about the impact of listing can apply at any time for a 'certificate of immunity' from listing, which, assuming the Secretary of State decides not to list the building, lasts for five years.

The decision to list a building is undertaken on the basis of defined criteria. In summary these are:

- age and rarity, with the criteria becoming tighter closer in time (see Box 8.1);
- architectural interest (this started with a focus on major buildings by major architects, but is now much broader);
- historic interest (which might be, for example, social, economic or cultural significance);
- close historical association (for example, with a notable person or historical event);
- group value (a recognition – prior to the introduction of conservation areas – of the significance of groups of buildings).

There are approximately 375,000 list entries in England. However, older list entries can encompass a whole street of buildings, so, in practical terms, the number of individual buildings, actually listed is considerably higher and estimates suggest this is over 500,000. More recent list entries follow a new and more comprehensive format. However, it will be a very long time before these new provisions can be applied to all existing listings. It is important to note that a building is listed in its totality, including its interior as well as the exterior, (apart from a very small number of cases where interiors are specifically excluded), regardless of what is mentioned in the list description.

Buildings are graded as:

- I (only 2.5 per cent of listed buildings);
- II* (5.5 per cent) ('two star');
- II (92 percent).

Though the grading indicates the relative importance of the building, the statutory controls that apply are the same for each grade. The grade can be a material factor in making planning decisions and there are technical differences over which applications require consultations and referrals to other bodies. By far the biggest type of listed buildings is dwellings. Other major categories of protection include, for example, churches, but listing does extend to a range of structures that are not habitable buildings as such, for

BOX 8.1 CRITERIA FOR LISTED BUILDINGS IN ENGLAND

- Before 1700: all buildings which survive in anything like their original condition are listed.
- 1700–1840: most buildings are listed, though selection is necessary.
- After 1840: progressively greater selection is used.
- Particularly careful selection is used for buildings after 1945.
- Buildings under thirty years old are normally listed only if they are of outstanding quality and under threat.

In Scotland, similar criteria are applied to the periods prior to 1840, 1840–1914, 1914–1945, and post-1945.

Plate 8.1 Listed buildings in Edinburgh (photograph courtesy of Tom Parnell)

example milestones and telephone boxes, as well as curiosities such as a racing pigeon loft in Sunderland.

One of the changes introduced by the Enterprise and Regulatory Reform Act (ERRA) 2013 was to enable greater clarity to be given to the practical definition of listed buildings. The legal definition of a listed building is as follows:

> "listed building" means a building which is for the time being included in a list compiled or approved by the Secretary of State . . . and . . .
>
> (a) any object or structure fixed to the building;
> (b) any object or structure within the curtilage of the building which, although not fixed to the building, forms part of the land and has done so since before 1st July 1948, shall be treated as part of the building.
>
> (Planning (Listed Buildings and Conservation Areas) Act 1990, section 1(5))

This introduces us to some important notions about listings, i.e. that what is listed may encompass 'objects and structures' fixed to the building and the wider 'curtilage' of the building. In both cases, the precise legal definition has been subject to case law, none of which can ultimately be considered that definitive. Though legal debates about whether objects and structures count as part of the fixtures of a listed building might seem very dry, because of the value many such fixtures might have (for example, fireplaces), this is a very important issue. The legal tests revolve around such concepts as the relationship of the object or structure to the original design. Curtilage is a potentially even more wide-reaching issue, as it can potentially extend listed building control to large sites, far beyond the primary building. Early lists confused this issue by often only listing a main building and not many of the other structures or buildings which might have importance. More recent lists try to explicitly list these features separately. The ERRA 2013 allows EH to state definitively whether attached or curtilage structures are protected and also that a part or feature of a listed building is *not* of special interest, for the purposes of listed building consent. Whilst greater precision over the nature of special interest is to be welcomed, there is also the potential for future problems, as understandings of a place and its significance changes over time.

Works that would affect the character of a listed building as a building of special architectural or

**Plate 8.3 Modern infill: York – a scheme featured in *Building in Context*
(English Heritage and CABE 2001)**

approaches, albeit critically informed by historical precedent. However, the 'traditional' architect Robert Adam argued that buildings are *inevitably* of their time, but that what is implicit in this phrase is the self-conscious Modern Movement fracture from historical continuity. This modernism, he argued, is deeply embedded in conservation culture and instruments such as conservation charters. It stems from academic archaeological and historical values and a stress on authenticity, which is quite at odds with the experience of place that most people have. He argued instead for a literate continuity of architecture.

Following the theme of a wider public response to architecture, the chapter by the Planning Officer for Canterbury charted the way unpopular post-war buildings have been replaced by others that seek to reproduce historic styles. A major justification for this is their public popularity (Jagger 1998) (see Plate 8.4).

Though no great evidence was presented to demonstrate this popularity, few, including opponents of this architectural approach, would deny that it is probably the case. Hobson (2004), in his case studies, argued that while conservation professionals advocate contemporary design, planning officers look for conservative, visually acceptable schemes, and councillors seek 'safe' approaches. Thus, historic reproduction and weak interpretations of historic form have perhaps been the dominant form of new construction in practice in historic centres over the last thirty years. Equally, however, the 2000s brought a new popularity for building tall. This is most marked in London, with a whole series of self-consciously 'iconic' skyscrapers springing up in the historic heart of the city, but it is a trend also evident in most of the major provincial cities across the country (Short 2012). Whilst these tall buildings are generally not to be found directly in

Plate 8.4 Neo-traditional infill: Canterbury

the historic core, they clearly have a significant impact on the wider urban character and have caused significant problems in relation to World Heritage Sites (discussed below).

As the number of demolitions of listed buildings and other historic buildings in conservation areas declined, debate shifted from total demolition to the degree of permissible intervention. In commercial areas, a frequently agreed 'comprise' solution to the tension between the demands of the market and the retention of listed buildings, was 'façadism', the retention of a historic façade or façades as the public face of what were essentially brand new constructions.

At the heart of the orthodox 'conservative repair' tradition of conservation practice is a concern for the historic fabric and 'authenticity' of the cultural object. From this conventional conservation position the practice of façadism is a destruction of cultural value and a despicable deceit. Façadism equally has attracted the opprobrium of conservation critics. Martin Pawley, in his polemic *Terminal Architecture* (1998), linked façadism with what he termed 'stealth architecture'. In describing the façading of a Royal Mail Sorting office, transformed from a five-storey building to ten with four storeys underground, he stated that this is 'a veritable "stealth bomber" of a building' and 'post-Modern architecture (can) preserve old buildings, while at the same time utterly destroying their identity' (Pawley 1998: 134).

Thus orthodox conservationists and those who consider that conservation is now too extensive an activity concur in their dislike of façadism. As a practice, it arises through the reality of negotiating for development and the compromises that are offered. For developers, the additional cost of façade retention is presumably still preferable to having to work within the constraints of a retained building or achieving an acceptably high quality of architecture for an entire replacement building. For the public-sector decision-makers, a retained façade presumably is regarded as at least a partial success for the conservation cause, and perhaps as something that will prove more acceptable to wider public opinion.

In defending façadism, Richards (1994) argued the façade to be generally capable of differentiation from the building as a whole and to be usually more important both in architectural terms and in terms of its contribution to the public domain. Furthermore, he argued that users may want the image and perceived prestige of a historic façade combined with modern functionality. However, this is very much a minority view in the conservation and architecture worlds. Façadism is the most radical transformation a building can endure short of total demolition and, for better or worse, is explicitly concerned with creating an artificial historic backdrop to urban space.

The wider historic environment: other designated heritage assets and characterisation

The other designated heritage assets in England are:

- World Heritage Sites, designated by the UNESCO World Heritage Committee for their outstanding universal value
- protected wreck sites, designated by order under the Protected Wrecks Act 1973 for their historical, architectural or artistic importance
- registered parks and gardens, designated by English Heritage under the Historic Buildings and Ancient Monuments Act 1953 for their special historic interest
- registered battlefields, designated by English Heritage on a non-statutory basis.

World Heritage Sites

In 1972, the UNESCO General Conference adopted the Convention Concerning the Protection of the World's Cultural and Natural Heritage, otherwise known as the World Heritage Convention. The rationale of the convention was that there were places of 'outstanding universal value', that these were part of the heritage of all humankind and that their protection was therefore a shared responsibility. The best-known outcome of this was the identification of cultural and natural properties and their listing as World Heritage Sites (WHS). These are considered on the basis of nominations put forward by national governments.

The first twelve WHS were designated in seven countries in 1978. By 2013, the total had reached 981 sites (759 cultural, 193 natural and twenty-nine 'mixed') in 160 states. By the 1990s, various issues were emerging within UNESCO about the nature and status of the list. One outcome of this has been the development of 'mixed' sites, or 'cultural landscapes', where, in the interaction of their influence, man and nature are deemed to be inextricably linked. This allows for symbolic understandings of cultural significance, rather than relying necessarily on material fabric. Other issues of concern have included the large number of applications coming forward, the predominance of certain types of site and the over-representation of western countries already well represented on the list. In order to address and manage these issues, countries are now expected to produce tentative lists of possible future submissions for inclusion, looking ahead over a five- to ten-year period and most states are now only allowed to make one submission per year. There has also been a focus on site types; for example, an effort has been made to add more sites related to industrialisation, which has had implications for the UK.

The UK ratified the World Heritage Convention in 1984 and between 1986 and 1988 the first thirteen sites within the UK were designated. Eleven of these sites were in the cultural category and displayed a strongly ecclesiastical and archaeological emphasis. Coincident with Britain's withdrawal from UNESCO between the mid-1980s and 1997 was hiatus of activity in nominating new sites, though Edinburgh Old and New Towns were listed in 1995 and Maritime Greenwich just as the UK rejoined UNESCO under a new Labour government in 1997. Reflecting the emphasis UNESCO was putting on the theme of industrialisation, six subsequent new sites have a principally industrial significance (Blaenavon Industrial Landscape, Derwent Valley Mills, New Lanark, Saltaire, Cornwall and West Devon Mining Landscape and Pontcysyllte Aqueduct and Canal), whereas only one of the earlier sites fell into this category (Ironbridge Gorge). For now there are no more historic towns to be nominated, such as York or Oxford, nor Royal Palaces such as Hampton Court, nor more great Gothic cathedrals such as Lincoln (Whitbourn 2002). Thus, by 2013, in the UK there were twenty-two cultural sites, two natural and one mixed site, and the UK had responsibility for a further three sites in dependent territories (see Box 8.2).

BOX 8.2 WORLD HERITAGE SITES UNDER THE JURISDICTION OF THE UK, WITH DATE OF DESIGNATION

1986	Giant's Causeway and Causeway Coast
	Ironbridge Gorge
	Stonehenge, Avebury and Associated Sites
	Durham Castle and Cathedral
	Studley Royal Park, including the Ruins of Fountains Abbey
	Castles and Town Walls of King Edward in Gwynedd
	St Kilda
1987	Blenheim Palace
	City of Bath
	Frontiers of the Roman Empire (including Hadrian's Wall)
	Palace of Westminster and Westminster Abbey, including Saint Margaret's Church
1988	Canterbury Cathedral, St Augustine's Abbey and St Martin's Church
	Tower of London
	Henderson Island
1995	Old and New Towns of Edinburgh
	Gough and Inaccessible Islands
1997	Maritime Greenwich
1999	Heart of Neolithic Orkney
2000	Blaenavon Industrial Landscape
	Historic Town of St George and Related Fortifications, Bermuda
2001	Derwent Valley Mills
	New Lanark
	Saltaire
	Dorset and East Devon Coast
2003	Royal Botanic Gardens, Kew
2004	Liverpool – Maritime Mercantile City
2006	Cornwall and West Devon Mining Landscape
2009	Pontcysyllte Aqueduct and Canal

Management issues have been a concern with UK World Heritage Sites. Some of these have been extremely contentious. Improvements to the surroundings of Stonehenge, probably the most iconic of all WHS, were on the agenda throughout the twentieth century, with a new visitor centre finally opening in 2013. Though not as contentious, management issues have been a major focus for the Hadrian's Wall WHS, which in 2005 became absorbed into the transnational Frontiers of the Roman Empire site, spread also across Scotland, Germany and Ireland. The sheer size and complexity of the Roman remains, combined with the complexities of ownership and use of the site, including high visitor pressure in some locations, make management a difficult task. Stakeholders include English Heritage, the National Trust, twelve local authorities, five central government departments, a variety of tourist bodies, the National Farmers' Union, the Countryside Agency, Natural England and many, many more. Hadrian's Wall was the first UK WHS to produce a management plan in 1996, as is now required for all sites by UNESCO, with the latest version covering the 2008–14 period (Hadrian's Wall Management Plan Committee 2008).

The management of WHS in constantly changing and evolving urban areas brings another layer of complication. The primary means by which WHS are managed in the UK is through the planning system: through planning policies, through the integration of management plans with planning policies and through decision-making about individual proposals. On the face of it, provisions on WHS in UK planning law are not especially strong, with no specific legislative provision or additional statutory controls, although government guidance (in England) has made the existence of a site a key material consideration in planning decisions for some time and there has been a minor tightening of both regulatory controls and the scrutiny process more recently (Department for Communities and Local Government and Department for Culture Media and Sport 2009; English Heritage 2009). WHS management is also subject to the scrutiny of international conservation bodies and this has been important in these sites, in an era when British governments have taken a fairly flexible view of development and conservation in line with economic priorities. These issues have raised concerns in the Bath, Edinburgh and central London WHS, which has resulted in reactive monitoring visits by UNESCO and its advisors the International Council on Monuments and Sites (ICOMOS). The problems are most severe, however, with the Liverpool Maritime Mercantile City WHS. Liverpool became the first UK WHS to be officially placed on the List of World Heritage in Danger in 2012. Concerns over the management of the WHS had been evident since its initial designation (Pendlebury et al. 2009), reaching a head with the approval of the planning application for Liverpool Waters, an enormous proposed redevelopment, partly high-rise, partly in the WHS and partly in the adjacent buffer zone.

Historic parks and gardens

Historic parks and gardens have a very significant place in the design history of the UK. Yet they have rarely had the attention that has been given to the architectural heritage. Even the amenity movement body formed to focus attention on garden heritage, the Garden History Society, is a relative newcomer, formed in 1965. By way of contrast, some other countries have protected important parks and gardens through their general heritage legislation since the nineteenth century. The Register of Historic Parks and Gardens in England was developed after the formation of English Heritage in 1984, and whilst a statutory document, it carries no additional statutory controls. Parks and gardens are graded like listed buildings, but with a larger proportion of sites in the higher-grade categories.

Initially, the Register was asserted to be 'overwhelmingly, a record of the aristocratic country estates of the south' (Roberts 1995: 47), although subsequent iterations have sought to address this through thematic reviews of, for example, cemeteries, hospitals and other institutional grounds. Of the 1,600 or so sites now on the Register, 300 are public parks and 100 municipal cemeteries. In Scotland there is an Inventory of Gardens and Designed Landscapes with different selection criteria from England, whereby more weight is attributed to horticultural importance. In Wales, the register was prepared in conjunction with ICOMOS UK and in Northern Ireland there is a Register of Parks, Gardens and Demesnes of Special Historic Interest.

Though parks and gardens have gradually assumed greater weight in the planning system, problems over their status remain. For example:

- Many agricultural activities can occur as permitted development, even within registered sites. Thus major changes to the landscape can take place without the need for planning permission.
- 'Enabling development' has been a major concern. A typical scenario would be that a major listed building, such as country house, is uneconomic to repair. To facilitate its reuse, developers will argue for a package that includes ancillary development that normally would not be allowed in order to create value to put into the listed building. The problem with this has been that some schemes have allowed such enabling development that is either directly damaging to the heritage asset that it aims to save, or it is damaging to some other heritage asset. Very commonly, this might be a park and garden sitting around the building heritage asset. EH has a guidance note on enabling development (English Heritage 2008b) and other nations have also addressed this issue; for example, Northern Ireland has a Planning Policy Statement on the topic.

- Lack of weight in planning decisions. Overall there can still be a feeling that parks and gardens issues are not taken as seriously in the planning system as issues relating to, for example, listed buildings.

Numbers of historic parks and gardens, along with the range of other 'heritage assets', are summarised in Table 8.3.

Characterisation

English Heritage has an interest in the historic environment that extends beyond legally protected or identified heritage assets. The principal way in which it seeks to engage with this wider historic environment has, since the 1990s, been through the process of characterisation. Initially, there was a rural emphasis to this work, through the Historic Landscape Characterisation (HLC) programme, which was effectively a form of landscape archaeology, focusing on the historical development of landscapes. Subsequently, characterisation ideas have fed into, amongst other things, conservation area character appraisals (see above) and the controversial Housing Market Renewal (HMR) programme in the 2000s. The HMR programme initially promised the large-scale demolition of housing stock on a scale unparalleled since the slum clearance heyday of the 1950s and 1960s. This raised the spectre of past mistakes of large-scale slum clearance and, amongst other issues, raised concerns of the potential impact on the historic character of northern

Table 8.3 Numbers of listed buildings, scheduled monuments, conservation areas and World Heritage Sites in the UK

	Listed buildings	Scheduled monuments	Conservation areas	World Heritage Sites	Historic parks and gardens	Battlefields
England	375,725	19,792	9,820	17	1,626	43
Scotland	47,649	8,000	645	5	394	39
Wales	30,000	4,000	500	3	400	(under preparation)
Northern Ireland	9,000	1,800	60	1	154	n/a

Note: The numbers of heritage assets are in some cases estimates. For buildings they tend to be register entries, which gives a lower figure than the actual number of buildings, as, for example, one entry can include a terrace of buildings.

towns and cities. One of the ways EH engaged with HMR was through promoting characterisation work to help guide decision-making. It worked in the Anfield and Breckfield areas of Liverpool, using techniques it termed 'whirlwind surveys' and rapid area assessment.

Conservation and regeneration

As the protection and use of the heritage has developed in scale and become more central to the planning of settlements of all sorts across the UK, so its connections to other policy areas have developed. Heritage is now often explicitly seen as one means of achieving a range of desirable policy goals, rather than being important just for its own sake. This is perhaps most marked in the relationship with efforts at physical regeneration.

Conservation and regeneration have in the last two decades often come to be considered as essentially complementary processes. For example, historic buildings have been helpful in many flagship property regeneration projects in adding quality and place distinctiveness to schemes, which has been of increasing importance during a period when 'culture-led regeneration' has become more and more fashionable. Conservation bodies in turn have generally embraced these relatively new economic and social roles for the historic environment, which are quite different from the rather narrow cultural role conservation once assumed, and indeed it is an agenda often promoted by English Heritage and other conservation organisations.

In Britain until the 1970s, most state efforts at improving urban economic conditions were based on renewal and physical redevelopment. This decade saw a significant shift in direction as the developing

Plate 8.5 Heritage saved and commodified: Covent Garden, London

field of urban policy saw a growing emphasis on the recycling and rehabilitation of existing building stock, initially in housing areas but gradually extending to other contexts. Towards the end of the 1970s, conservation organisations directly linked to this emergent urban policy of regeneration and sought to demonstrate the economic sense of a conservation approach (for example, SAVE Britain's Heritage 1978). Examples of this process were becoming evident. The Covent Garden area, slated for demolition by the Greater London Council a few years before, became a festival marketplace-type shopping arena. Saved from wholesale clearance by community activism, the area was steadily gentrified, as investors realised its potential (see Plate 8.5). Perhaps the exemplar of heritage being positioned at the front of regeneration programmes in the 1980s was the way in

which the restoration and reuse of Albert Dock, a large complex of Grade I listed warehouses in Liverpool, became the regeneration flagship of the Merseyside Development Corporation. This combined the physical regeneration of superb-quality industrial buildings with a focus on culture; Albert Dock hosts both a maritime museum and an outpost of the Tate gallery and, more recently, an International Slavery Museum (see Plate 8.6).

By the early 1990s, significant parts of the conservation sector in the UK had fully embraced these more economically instrumental relationships. A key body in promoting this agenda has been English Heritage. An EH publication, *The Heritage Dividend* (English Heritage 1999), was part of a continuing process of more thorough documentation of the economic impact of heritage spending and the ability to present this in

Plate 8.6 Albert Dock, Liverpool. Regenerated in the 1980s, World Heritage Site 2004, 'in danger' 2012

terms of the performance measures and indicators that might be recognised by evaluators of mainstream regeneration funding schemes. It was followed up with a second *Heritage Dividend Report* (English Heritage 2002), and there subsequently have been a number of other reports seeking to demonstrate the economic impact of heritage spending, for example, on waterways (ECOTEC 2003), traditional farm buildings (English Heritage and DEFRA 2005) and seaside towns (English Heritage 2007), as well as on the economic contribution of cathedrals (ECOTEC 2004). In reflecting on the history of the Heritage Dividend initiative, a 2005 report noted:

> The brand has played a key role in the promotion and repositioning of English Heritage as a proactive, enabling organisation, fully engaged in regenerating some of the UK's most economically deprived and physically run-down communities.
>
> (English Heritage 2005: 3)

It also pointed out that the measures adopted have focused on positive by-products rather than the core purpose of the schemes. That is to say, what has been measured has been related to mainstream regeneration indices rather than to the success of schemes in terms of conservation measures and values. The ability of EH to embed conservation in mainstream public investment is more important, as its own ability to invest through conservation area and other capital funding continues to be diminished (see above).

Constructive conservation and modern heritage

English Heritage frequently produces glossy, image-laden case study-based documents generally with the purpose of illustrating good practice on an overall theme. Five of these have appeared in the 'Constructive Conservation' series: *Capital Solutions* (2004), *Shared Interest: Celebrating Investment in the Historic Environment* (2006), *Constructive Conservation in Practice* (2008), *Valuing Places: Good Practice in Conservation Areas* (2011) and *Sustainable Growth for Historic Places*

(2013). These are significant documents not only in terms of describing practice, but also in terms of EH benchmarking what it considers *good* practice.

The case studies in this series have sometimes extended the 'market friendliness' of EH to an extraordinary degree. *Capital Solutions,* a series of case studies from London, showcased a number of substantive alterations to historic buildings, often through dramatic contemporary design. The first line in the first case study, Sadlers Wells Theatre, is 'Sometimes demolition is the best solution', a strange jumping-off point in a document prepared by the national conservation agency. Perhaps the most extreme highlighted case in *Shared Interest* was the Free Trade Hall in Manchester. Here English Heritage supported the removal of the major part of the concert hall and its replacement with a fifteen-storey hotel – all that was retained of the historic building was the Italian palazzo façade. As the document notes 'our public support for the proposal was, initially, also controversial' (p. 43).

Constructive Conservation in Practice features Park Hill, Sheffield on the cover. This substantial, pioneering 1950s deck-access council-housing scheme sits in a prominent location overlooking Sheffield city centre. Listed in 1998, this was always a controversial and problematic listing, given the general unpopularity of much post-war modernist council housing and the poor physical condition of the estate, combined with major social problems. The scale of the physical and social problems, together with a lack of available finance to invest in social housing, ultimately led to Park Hill being handed over to the niche developers Urban Splash, the physical fabric was stripped back to the concrete frame and new flats created, principally for sale. *Constructive Conservation* asserts that the heritage values lay

> not only in the site's history but in the scale and vision of the original council housing scheme, in the expressed reinforced concrete frame and the relationship of the building to the landscape in which it sits. Substantial changes to the internal layout and the infill panels within the frame could therefore be introduced without damaging its historic significance.
>
> (English Heritage 2008c: 14)

Plate 8.7 Park Hill, Sheffield, showing an original block to the left and a block regenerated by Urban Splash to the right

This is very different from a traditional emphasis on the authenticity of fabric. On the one hand, EH is very anxious to be positively perceived by developers and its political paymasters; to demonstrate conservation as a socially beneficial agent of change and, conversely, not as an impediment to development. On the other hand, it maybe that different conservation approaches are needed for modernist mega-structures such as Park Hill than those which were evolved in the nineteenth century for ecclesiastical buildings (see Plate 8.7). This is particularly the case where short-life materials, such as plastic, must be replaced in order to protect the design they are used to create.

Conserving the modern

The listing of post-war buildings was made possible in 1987, when eligibility for listing was extended to any building at least thirty years old. This 'thirty-year rule' allowed for constant extension, and included additional provision to list buildings over ten years old if they were considered to be both outstanding and threatened. The case that triggered the bringing forward of the legislative change allowing post-war listing was a 1950s neoclassical office building, Bracken House; but in accepting the principle of post-war listing the contentious possibility of protection for modernist buildings of the welfare state was opened up. Considering modernism as heritage introduced new technical and functional challenges and costs associated with the dominant building forms and materials of the 1950s and 1960s, whether 'the seemingly intractable problems of rust-stained dirty concrete, falling tesserae and leaky flat roofs, finding a new use for inflexible 1960s megastructures, or adapting 1950 and 1960s buildings to

changing urban design and mobility principles'
(While 2007: 649).

However, even more significant was the political
issue of the perceived deep unpopularity of the built
environments of the period and the pervasive collective
narratives of 'modern movement failure' (Gold 1997).
This was seen as a period of architecture and planning
that had produced environments that were not just
temporarily unpopular and out-of-step with contemporary design principles, but were seen as contributing
to urban decline and social breakdown (Cunningham
1998). Most controversial of all was the listing of
public housing.

Given this context, the government proceeded
cautiously, accepting only eighteen of English
Heritage's first inventory of seventy buildings proposed
for listing, with an emphasis on relatively uncontentious
buildings such as Basil Spence's Coventry Cathedral
or the Royal Festival Hall on London's South Bank.
Subsequently, post-war listing has proceeded in fits
and starts, often influenced by the views and tastes of
individual ministers. Whilst post-war listing is very
selective compared with that of earlier periods, by
September 2013 there were approximately 700 post-
war listings. However, the story of post-war Britain is
as much about its planning as its architecture and, as
While (2006) has argued in relation to Coventry and
Plymouth, the heritage movement has struggled to
make the case for the protection of wider post-war
planned environments, with, for example, conservation
areas focused on post-war development few and far
between. Even Byker, the 1970s housing estate in
Newcastle upon Tyne, where all 1,200 or so buildings
are listed grade II* and which relies for its special
interest on a public realm fully integrated with the
housing, is not designated as a conservation area.

Summary and prospects

Today the British system of heritage protection is both
extraordinarily extensive and relatively flexible, both of
which can be argued as being a strength or a weakness.
Conservation has enjoyed remarkably stable political
support over the last forty years and seems to remain a
popular cause. Flexibility is a characteristic of the discretionary British planning system. The scope this gives
for negotiation can allow for sensible and creative solutions to conservation problems, but the absence of fixed
rules can create confusion and poor compromises.

Since the rise of the conservation movement, there
have been constant anxieties about sustaining hard-
won gains. Changes in government, all of which tend
to have a pro-development rhetoric, are particular
moments of concern. This was evident with the
election of a self-consciously modernising New Labour
government in 1997 and with a neo-liberal Coalition
government in 2010. As discussed, the emphasis on
'sustainable development' above everything else and
the compression of planning policy into the NPPF
certainly rang alarm bells in the heritage sector, but
whether these represent a major long-term problem
for heritage protection remains unclear.

What is certainly an issue is the loss of resources
and skills consequent on sustained austerity measures
and a reduction in the scale of the state. This can be
seen in the loss of capital spend on the historic
environment from EH and the public sector generally,
although this is mitigated by HLF, which continues
to bring substantial monies into a variety of heritage
projects, albeit with a focus on engagement and
intellectual access. It is more profoundly felt in the
loss of personnel capacity in local authorities. Research
has shown that conservation specialist staff numbers
have declined by 33 per cent in England between 2006
and 2013 (English Heritage 2013c) and it is estimated
there was a decline of the order of 15 per cent in
Scotland between 2011 and 2013 (IHBC 2013). These
figures are of major concern in a system that struggled
to effectively manage the extent of the protected
historic environment even in better times. The
catastrophic impact of skills shortages was highlighted
in a report by a perhaps unlikely source, the Country
Land and Business Association (2011), which made
the point that the uncertainty created by a lack of
specialist staff was not only bad for heritage protection
but could deter investment and sensible solutions for
the reuse of historic buildings.

It might be argued that localism is one potential
means of plugging this gap. Localism is discussed

more fully in other chapters, but in the heritage sector there has been something of an ambivalence to the localism agenda. On the one hand, this resonates well with the long tradition of voluntarism and local action in conservation matters, but on the other there were concerns about sustaining or bypassing national systems of protection and also the role of the national amenity bodies, i.e. civil society is not only found at the neighbourhood level.

The extension of planning powers to neighbourhoods through the Localism Act 2011 could potentially lead to more inclusive, more diverse, more democratic readings of heritage and how these might be planned for. Neighbourhood plans could, from a heritage perspective, develop more holistic conceptions of place and place value, with less emphasis upon locally applied expert-led judgements of national criteria. Equally, however, we might see subversions of the process: the status of locally valued places put under challenge by commercial or other local interests directly through the micro plan-making process. Indeed, we may see both.

Further reading

There is no single up-to-date textbook on British conservation planning. The best general books on the subject include Pendlebury (2009) *Conservation in the Age of Consensus*, which explores the historical development of conservation and seeks to put it in the context of wider social and political forces, and Rodwell (2007) *Conservation and Sustainability in Historic Cities*, which considers the management of historic cities through a sustainability framework. Larkham (1996) *Conservation and the City* still makes an interesting read, for example, on the relation between urban morphology and conservation, and Hobson (2004) *Conservation and Planning: Changing Values in Policy and Practice* contains some quite revealing material on the attitudes of conservation experts and the way that conservation priorities and practice can vary between local authorities. Books of interest with a more specialist focus include Worthing and Bond (2008) *Managing Built Heritage* for discussions of the concept of significance, Short (2012) *Planning for Tall Buildings* and

Macdonald *et al.* (2007) *Conservation of Modern Architecture*. Introductions to the subject of architectural conservation include a series of books by Forsyth (e.g. 2013) and Orbasli (2008) *Architectural Conservation*.

The definitive text on conservation law is currently probably Harwood (2012) *Historic Environment Law: Planning, Listed Buildings, Monuments, Conservation Areas and Objects*. Whilst this supplants the now somewhat outdated Mynors (2006) *Listed Buildings, Conservation Areas and Monuments*, the latter book remains useful for its more detailed treatment of some subjects. In terms of debates over the evolution of the conservation system, English Heritage (2000) *Power of Place: The Future of the Historic Environment* remains a worthwhile read, if it is perhaps primarily of historical interest now, whilst the report by the Country Land and Business Association (2011) *Averting Crisis in Heritage: CLA Report on Reforming a Crumbling System* is interesting and provocative. English Heritage (2008c) *Constructive Conservation in Practice* is important for developing an understanding of conservation principles, as is its inspiration, the Burra Charter (Australia ICOMOS 2013).

Recent years have seen the development of a large and varied multidisciplinary academic field of heritage studies. Graham *et al.* (2000) *A Geography of Heritage* is an accessible introduction to geographical perspectives, whereas the contributors in Gibson and Pendlebury (2009) *Valuing Historic Environments* discuss historic environment from the perspective of values. A key critical perspective is presented by Smith (2006) *The Uses of Heritage* and her notion of the 'Authorised Heritage Discourse' has had a significant impact on academic literature at least. One book that is interesting for relating these ideas to English Heritage is Waterton (2010) *Politics, Policy and the Discourses of Heritage in Britain*.

Delafons (1997) *Politics and Preservation* narrates a lively history of British heritage policy, with an insider's view as a former senior civil servant, whereas Stamp (2007) *Britain's Lost Cities* gives a nostalgic and impassioned survey of buildings lost. Major historical overviews of architectural conservation extending across the world are given by Jokilheto (1999) *A History of Architectural Conservation* and Glendinning (2013) *The Conservation Movement: A History of Architectural Preservation*. Bandarin and van Oers (2012) *The Historic Urban Landscape: Managing Heritage in an Urban Century* is an important text on the emergent international concept of historic urban landscapes. Contemporary writings can make interesting and valuable reading. So 1970s polemics such as Fergusson (1973) *The Sack of Bath*, Aldous (1975) *Goodbye Britain?* and Amery and Cruikshank (1975) *The Rape of Britain* give a vivid sense of the battles of the 1970s. If we look further back, Cullen (1961) *Townscape*, Sharp (1968) *Town and Townscape* and especially Worskett (1969) *The Character of Towns: An Approach to Conservation* are all interesting on the relationship between townscape, urban design and conservation.

The English Heritage and the Historic Environment Local Management websites both have extensive useful material, including downloadable pdfs of the reports featured in this chapter as well as much, much more. The annual 'Heritage Counts' reports are a particularly useful resource. Similarly the websites of Historic Scotland, CADW and the Northern Ireland Environment Agency have much valuable material. The websites of the national amenity societies are full of interesting reading and cases and the publicly accessible archive of the journal 'Context' on the website of the IHBC is a great resource (www.ihbc.org.uk/context_rw/). The archive of English Heritage's professional journal 'Conservation Bulletin' is also available online.

9 Design and the planning system

Introduction

The emphasis on how much the planning system should be concerned with design issues has fluctuated over the post-war period. Furthermore, what is denoted by the term 'design' has also been at times unclear and the associated terminology has changed and evolved. The purpose of this chapter is to outline where design has been seen to fit within the planning system from the national policy perspective, by briefly reviewing how such policy has developed over the post-war period. However, the chapter also attempts to resolve the confusion between the terms 'design' and 'urban design' and briefly introduces some key elements of academic thinking where appropriate.

When design has been talked about in relation to the planning system during the post-war period, what has mostly been referred to is some kind of design 'control' that is, attempts to regulate in some way issues such as the height and massing of new buildings. Design is often wrongly assumed to involve only *aesthetics* – a confusion which, as outlined below, has not been helped by central government advice and guidance in the recent past. In part this is because 'design' is quite an ambiguous word in English – it encompasses multifarious meanings. While aesthetics *are* a major part of design concerns, design encompasses much wider matters. As much as anything, in relation to the built environment it is about the process of delivering successful buildings and places that function effectively, as well as appearing pleasing to the senses. More recently, and particularly through the work and advice issued by the Commission for Architecture and the Built Environment (CABE),[1]

emphasis has shifted from mere design 'control' to greater consideration of design at all stages of the development process.

PLANNING AND DESIGN IN THE POST-WAR PERIOD

Immediate post-war

In the immediate post-war period, there was a clear divide among experts in the planning field between those who regarded planning as an economic and social activity, whose prime concern was the right use of land (with design seen as secondary to this), and those who considered the creation of physical design – including both built form and land use – as the key purpose of planning. The latter groups considered disciplines such as economics and sociology as playing a supporting role to the designer, who retained supremacy (Cherry and Penny 1986). A key figure in the debates at this time was William Holford, who was chief advisor to the Ministry for Town and Country Planning 1943–7. Holford's perspective is illustrated by the Ministry's first advisory handbook, an important element of which was a simple tool called the 'Floor Space Index', which set a maximum extent of development for a given plot of land – in effect controlling building mass in core urban areas (MTCP 1947). However, the guide also introduced 'modernist'[2]-inspired planning ideas, including separating out land uses and emphasising vehicular circulation. Holford was resistant to the idea of ministerial design guides, although, largely through local authority pressure, the Ministry did produce *Design in Town and Village*. This guide advocated that designers should consider how local, regional, or metropolitan characteristics of place could be intensified by town planning, through new buildings, landscape, street furniture, floorscape, and public art (Sharp *et al.* 1953). It thereby set in print a theme which is recurrent throughout post-war debates on design in the built environment, that is, how new development should respond to the place in which it is built. What might be called 'contextualisation', the approach of the key authors of the guide (Sharp, Holford and Gibberd), is unquestionably a modernist approach[3] – albeit a constrained one.

But whatever the noble aspirations of the period, with a few notable exceptions, much of the development itself was largely disappointing. Bland office buildings in London, neo-Georgian council housing, shopping precincts in blitzed city centres and the 'prairie' developments of private suburbia were all 'extensively criticised' (Punter and Carmona 1997: 20). Ian Nairn was a notable critic of the time, attacking planning and what he termed 'subtopia', which represented the blending of the most basic design contexts 'urban' and 'rural' (Nairn 1955).[4] While Nairn was something of a maverick, it is likely he expressed views shared by not only many built environment professionals, but also by the wider public. Dissatisfaction with the emerging built environment was reflected in the establishment of the Civic Trust in 1957 and through mass membership of civic societies concerned with the local environment at this time.

During the 1950s and 1960s, there were a number of ambitious city centre redevelopments. Bomb-damaged Coventry, for example, gained a pedestrianised precinct in a contemporary style. However, the highlight of this scheme was the development of Basil Spence's Cathedral, built at right angles to what had survived of the mediaeval one and providing a vista from the new centre. Plymouth's reconstruction to plans by Abercrombie (one of England's most influential planners of the mid-twentieth century) failed to live up to their ambitious aspirations. The quality of the commercial development that came to line the streets simply did not match up to the grand beaux

arts-style planning approach. Cherry also suggests that plans to redevelop Piccadilly Circus in the late 1950s provide an illustration of 'the frailty of the planning system' in dealing with major design proposals originating with the private sector (1988: 169). A deal between London County Council (which acquired land for road widening) and a property developer who obtained the green light for a redevelopment scheme started to founder when the brash, commercial development (featuring a 100-foot-high soft drinks commercial) was revealed at a press conference. The scheme was publicly vilified and subsequently rejected by a public inquiry.

The design of public housing schemes was in many cases worse than the redevelopment of commercial cores. High-density 'high-rise' and 'slab-block' estates, poorly designed, often cutting essential corners in terms of construction and community provision, and not well managed, quickly became stigmatised and a byword for social ills. Park Hill, Sheffield, built 1957–1961, is an interesting example and survival. The redevelopment started out positively. The deck-access flats were light and airy compared with the cramped back-to-back housing they replaced. They were inspired by ideas from Le Corbusier and British architects Alison and Peter Smithson, in particular the concept of 'streets in the sky' which allowed free pedestrian movement (as well as electric vehicles, such as milk floats) and were meant to provide space for socialisation and community interaction. As with other such schemes, in the long term they did not live up to these lofty aspirations. A combination of factors – rehousing policies, maintenance issues, as well as the decline of traditional industries and sources of employment for many residents – meant that by the late 1970s the reputation of the estate was in tatters and the estate flats were let only to the most desperate tenants, while many stood empty and boarded up. Many estates resembling Park Hill, such as the physically similar Kelvin Flats in Sheffield, were demolished. However, in recognition of its pioneering architecture, Park Hill was listed by English Heritage in 1998. From 2003, a Manchester-based developer, Urban Splash, has been involved in the regeneration of the building. The structure was stripped back to its concrete skeleton and then effectively rebuilt, provoking debate about whether this was an appropriate approach for a 'listed building', and about the resultant space standards of the flats, much reduced from their original form. However, while the transformation is some way from completion, the first phase, realised in 2013, was nominated for the RIBA Stirling Award in that year.

Parker Morris

No overview of post-war housing would be complete without reference to the Parker Morris standards. Internal space standards have never been an integral part of the planning system. However, Sir Parker Morris was appointed by the Ministry of Local Housing and Government in 1961 to consider standards of residential design (both public and private). The resultant report was an extremely thoughtful document which carefully explored the everyday needs of families. Though it stopped short of outright criticism of small privately built homes – accepting that there might be some need for them – it did warn against building too many, stating that planning needed to consider the balance of need within neighbourhoods (MHLG 1961). Parker Morris set standards for floorspace which were applied to all public housing from the late 1960s and have remained a frequently cited benchmark. Though undoubtedly progressive, one problem was that these quickly became *the* standards that local authorities built to, while the intention had been to set *minimum* standards. The Parker Morris standards remained in place until 1979.

Today, Parker Morris standards look positively generous in comparison with modern housing specifications and there have been calls to return to a similar approach. However, the Parker Morris standards were based on the ways families lived in the 1950s. Today, it is likely that families have quite different and possibly more diverse ways of living, although there is actually little robust research on which to base space assumptions. However the Greater London Authority (GLA) commissioned a report in 2006 which drew together the evidence available from the UK and internationally at that time (Drury *et al.* 2006), and this

was later subsumed into standards for the London Plan (see below).

The 1970s

During the 1970s, more commonplace development and, in particular, privately developed mass housing proved notably problematic in terms of design quality. During this period the development of 'identikit' housing developments really took hold. Towns and cities were expanded by rings of suburban development, remarkably similar in appearance from one part of the country to the next. In 1973, the Essex Design Guide was the first real attempt to address this issue. Drawing on Townscape principles (see the section on 'Design' or 'Urban Design' below) and related to the strengthening of conservation in planning – the guide was produced by Essex Council's conservation team – its principles of spatial organisation represented a progressive stance against the tyranny of the rigid geometry of highways standards in particular. However the neo-vernacular architecture it promoted, while appreciated by the general public, drew criticism from architects, developers and the DoE alike (Punter and Carmona 1997: 23). The Essex Guide format was also followed by a number of other authorities – although generally less successfully. An unfortunate consequence was that although housing developers were initially sceptical about Essex Design Guide housing, it proved remarkably popular – too popular in some ways – and 'Essex'-style housing began to appear all over England.

The problems of creating good-quality new housing, with a real sense of place, continue to this day (see the CABE housing reviews 2005, 2007), a theme we return to later. The Essex Design Guide itself has proved itself remarkably resilient. It was comprehensively overhauled in 1997 and again in 2005. However, many of the original principles, such as the emphasis on informal spatial arrangement and a rejection of suburban 'norms', have remained unchanged (Essex County Council 1997, 2005) (see Plate 9.1).

In the 1970s, economic contingencies brought to an end most large-scale redevelopment schemes, both for city centres and for public housing. However, one

Variety of houses mainly wide frontage shallow plan, mainly joined together, some without on-plot parking. Most houses front back edge of footway without front gardens. This is a practical and flexible format for the typical residential layout at urban densities (8 dwellings per acre, 20 dwellings per hectare and above).

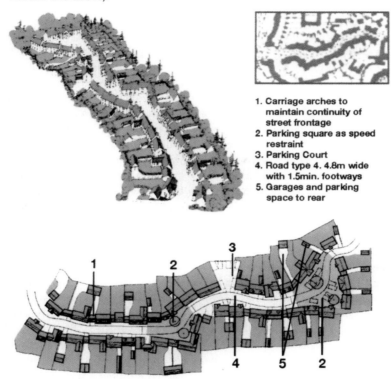

1. Carriage arches to maintain continuity of street frontage
2. Parking square as speed restraint
3. Parking Court
4. Road type 4. 4.8m wide with 1.5min. footways
5. Garages and parking space to rear

Plate 9.1 Essex Design Guide

new trend did gain emerging prominence and that was the introduction of pedestrianisation. Most cities and towns of any size introduced some kind of pedestrianisation scheme at this time. Traffic would be rerouted, which would leave the main shopping streets largely vehicle free; new, level paving that stretched the width of the street would be introduced alongside seats, lighting columns, bins and other items of 'street furniture'. At a time when the term 'urban design' was becoming more commonly used in relation to the built environment (see below), it is perhaps not surprising that the two ideas were sometimes incorrectly confused. As discussed below, though important, the

concept of 'urban design' encompasses far more than physical interventions such as pedestrianisation; yet 'urban design' is still often entirely equated with such public-realm schemes.

The 1970s also saw central government increasingly move away from design control. Circular 142/73 *Streamlining the Planning Machine* infamously implicated design control as the cause of delays in the planning system, while the interim report of the Dobry Commission advocated dropping design control outside of conservation areas and relaxing controls over architect-designed buildings – although this was reversed in the final report (Dobry 1975). In 1979, the newly elected Conservative government, firmly committed to the deregulation of planning and to freedom of private enterprise, adopted a firm anti-design control stance.

New towns

The design of 'new towns' in the post-war period deserves separate mention, as they were widely considered vanguards of contemporary design and planning during that period. There were several generations of new towns. The early new towns were designated between 1946 and 1950. Eight designated around London were designed to relieve housing pressure in the city; elsewhere in the UK, such as Corby and East Kilbride, new towns were linked to anticipated industrial expansion. There were many commonalities in overall approach between the different settlements. Each town was designed to provide its own employment and industrial areas and residential neighbourhood 'units' (each with a population of 3,000 to 12,000 people) were physically separated and located around a town centre. The housing was generally low rise and relatively low density (around 35 homes per hectare), with private gardens, set within communal landscaping. Much of the housing did not differ radically from the typical local authority estates built around towns and cities across the UK during this period. 'Radburn' design was introduced, which meant vehicular access to the rear of the property, while the front was accessed by pedestrian pathways. This was popular with architects and planners, though not necessarily with those who lived in the houses, since increasing reliance on motorised transport often made front entrances effectively redundant. The other design feature that was common to all and became increasingly replicated in other locations was the pedestrianised shopping precinct that formed the town centre.

The next generation of new towns was designated in the 1960s. They were mostly located in areas of anticipated population growth, such as Runcorn in Merseyside and Redditch in the West Midlands. Unlike the early new towns, this group was more varied in its built form and design and often included

design innovations. Not all innovation was successful, however, and some of the more experimental housing has subsequently undergone demolition. For example, Castlefields, a system-built estate in Runcorn, proved very unpopular and the large, deck-access flats have now been replaced with more conventional housing.

The best-known town of the later generation is probably Milton Keynes, effectively a 'greenfield' city. Milton Keynes was innovative in a number of respects, not least in being entirely designed around the requirements of the private car. The overall plan of the town is a 1 kilometre irregular road grid design to be 'draped like a mesh over the landscape' (Macrae 2007), around two-thirds of which allows dual lanes of traffic in either direction – a form unique in the UK. Aesthetic control in Milton Keynes was also unusual, in that it was divided into zones with varying levels of control, for example high control in the city centre and intermediate control in most housing areas, which allowed freedom for housing developers and home owners to come up with their own designs (see Plate 9.2). It is true to say that for a long period Milton Keynes suffered from an image problem, related to both the road network and housing design. The addition of a sculpture of a group of half life-sized concrete cows in 1978 (replicas of which are visible from the main rail line – the originals are in the shopping centre) seemed to accentuate its artificial nature. Milton Keynes was built with significant off-road walking and cycling facilities, but despite this it has high car dependence in an era of increased costs and concerns the over the sustainability of this mode of transport. However, it has proved to be a remarkably successful and popular place to live.

Road design

Road design during the post-war period also deserves separate consideration, since it has been one of the most strictly controlled elements in the urban environment. It has had a dramatic impact over the layout and appearance of our towns and cities, yet lies outside the usual control of planners and planning. The first manual on *Design and Layout of Roads in Built-up Areas* was produced by the Ministry of Transport in 1946

Plate 9.2 Housing, Bradwell Common, Milton Keynes: Architect E. Cullinan, 1979

and subsequent revisions ran until 1968. It encouraged the development hierarchies, for example 'inner ring roads', distributor and local roads. The needs of non-vehicular traffic, that is, pedestrians and cyclists, were largely ignored. Design Bulletin 32, on *Residential Roads and Footpaths*, was first published in 1977 and became the new 'roads bible' with a lifespan in various releases lasting twenty years, but was largely similar to its predecessor in prioritising vehicular movement. We return below to road design with consideration of more recent government guidance.

'Design', or 'urban design'?

If the word 'design' has at times been ambiguous in relation to the planning system, the term 'urban design' has been unequivocally misused and misunderstood – for example in the 1970s, it was often equated with pedestrianisation schemes discussed above. While 'urban design' is widely used, there is no internationally, nor even nationally, agreed definition of the term. However, in a UK context most writers would agree that it implies the multidisciplinary activity of creating, managing and improving the urban environment. It brings together, and attempts to balance, aesthetic concerns with socio-economic and environmental ones and while it has a focus on public spaces, is not solely concerned with them. In simple terms, urban design aims to create the types of places people enjoy and where they want to live, work and spend their leisure time. It is interested in both the process of producing such places and the product itself. However, 'urban design' has also come to imply a rejection of the certainties of modernism and the unsatisfactory urban conditions created by modernist-inspired city redevelopment and renewal in the 1960s.

Building on a narrative first developed by Jarvis,[5] Carmona states that urban design is the bringing together of two broad traditions in design appreciation – 'visual artistic' and 'social usage' – into the synthesis of 'place-making' (see Carmona et al. 2003: 6). To begin with the visual artistic tradition, the most significant contribution to this from a UK perspective was provided by Gordon Cullen. Cullen's notion of 'Townscape' emphasises the relationship between the dynamic interplay of buildings from different periods and their landscape setting that creates visual interest and sense of place (1961).[6] While Cullen's work is seductive, he is open to criticism for underplaying the role of 'people' in place making. In contrast, his US contemporary Kevin Lynch prioritised the perceptions and experience of ordinary people in his analysis of the built environment (1960).[7] The other key US writer of the time, Jane Jacobs, attacked modernist planning for its dehumanising impact on cities. She also focused on the lives of ordinary people and how, in combination, these everyday lives generated the 'organised complexity' which leads to the vitality of cities (1961: 567). These two US writers, therefore, emphasise the social usage tradition of understanding places.

It is difficult to identify the exact moment when these two traditions came together to form the basis of 'urban design' as we know and commonly understand it – which, as Carmona et al. suggest, is dominated by a 'place-making' ethos in the UK (2003: 7). However, it is clearly expressed in Responsive Environments, a 1985 work by Bentley et al. at Oxford Polytechnic (now Oxford Brookes University). The work drew together the ideas of Lynch, Jacobs and Cullen (along with other key writers) into seven key principles, which were: permeability, legibility, variety, robustness, visual appropriateness, richness and personalisation. Though certain limitations to the principles were subsequently recognised – for example Bentley later suggested the inclusion of some issues of sustainability (1990) – the text was extremely influential, not only for academic writing, but also in subsequent policy guidance, such as in By Design, discussed below.

Finally a key concept in urban design from the early 1990s has been that of New Urbanism. New Urbanism is a movement associated with a group of American architects, including Andrés Duany and Elizabeth Plater-Zyberk. Frustrated with the seemingly unending automobile-orientated urban sprawl associated with late twentieth-century North American cities, their aspiration was to encourage the (re)creation of compact 'walkable' communities. These would feature vibrant mixed-use cores, public parks and squares, integrated into a well-connected street network – in essence, therefore, taking inspiration from traditional settlements of the pre-motorised traffic era (Leccese and McCormick 2000). A number of complete settlements have been created to New Urbanist principles in the US, though undoubtedly the best known is Seaside, Florida.[8] In the UK, the principles were closely aligned with the concepts of 'urban villages' (Aldous 1992) – a key example of which is Poundbury, Dorset. The principles are also clearly articulated through By Design.

The 1980s and 1990s

While, in theory, concepts of urban design in the early 1980s were becoming more sophisticated (see above) – this was not initially reflected in official government policy. The return of a Conservative government in 1979 had a major impact on the planning system. In terms of design control, the most important document was Circular 22/80 Development Control: Policy and Practice, which applied in England and Wales. This circular contained four paragraphs under the title 'aesthetic control' – although the word 'design' was also used in the text – which clearly suggested that the Secretary of State at the time regarded the two terms as synonymous. Even if the terminology was confused, the thrust of these four paragraphs was clear: design control was only considered necessary in special interest areas, for example, conservation areas, and while design guidance 'may' be useful, it was only to be used as 'guidance and not detailed rules' (p. 21). The advice was largely reiterated in Circular 31/85 Aesthetic Control.

While the government position was that planning had little to do with design, issues of 'planning' and 'design' nevertheless hit the media headlines on several

principles; while 24 per cent were 'poor', which meant they should not have been built at all. CABE did not allocate blame in their audits. However, the fact that speculative house builders displayed little interest in the long-term quality of their developments was clearly a major structural barrier to improving volume house building (CABE 2005a, 2007). This situation remains largely unchanged today.

By design – urban design in the planning system: towards better practice

In terms of day-to-day working practice, *By Design: Urban Design in the Planning System* (DETR and CABE 2000) is CABE's first significant publication to bear the organisational name (though in fact it had been in production some time before CABE's existence). Drawing on contemporary urban design thinking, it sets out the key objectives of urban design as promoting character, continuity and enclosure, quality of the public realm, ease of movement, legibility, adaptability and diversity. It also set out a 'toolkit' of planning policies, guidance and development control. Most significantly the text suggested that urban design was not a blueprint for creating successful places, but should be considered an approach to a better understanding of how people felt about the places they lived and worked in, and, moreover, that collaboration with the general public was an essential part of the urban design process.

By Design remains available and is still frequently quoted. It is seen as the closest document to a national urban design policy for England ever produced. However, in 2012 it was specifically named in the Taylor Review of planning documents as one that should be deleted and in March 2014 was superseded by the *National Planning Practice Guidance* (NPPG). These are both discussed further at the end of this chapter.

Post-2011

From March 2011, CABE cease to exist as an independent organisation and was merged with the Design Council. The work of CABE has continued, but with 85 per cent fewer staff and a severe reduction in other resources, it is necessarily operating at a much lower level than previously. At the time of writing, it is still somewhat early to predict the outcome of CABE's altered circumstances in relation to its work and the resultant quality of the built environment. Moreover, since the UK development industry has been effectively in recession since late 2008, many major schemes (at least outside London) have effectively stalled. The official rhetoric from Design Council CABE is generally upbeat about its role as a focus for policy debate and best-practice dissemination. In particular there is a commitment to continuing the Design Review process (see below). However, the organisation also raises concerns about the erosion of capacity at local authority level (Design Council CABE 2011). In essence, public spending cuts have meant that many skilled design teams in local authorities have been decimated, as the authorities scale back to core planning requirements.

It should be noted that Design Council CABE operates in England only. Design advocacy in Scotland is provided by Architecture and Design Scotland, in Wales by the Design Commission for Wales, and in Northern Ireland by a ministerial advisory group. Although each has its own identity, they fulfil roles similar to that of Design Council CABE; but unlike CABE, they continue to receive public funding directly through their devolved governments.

Planning and Compulsory Purchase Act 2004

The Planning and Compulsory Purchase Act 2004 ushered in a new system for England – replacing 'land use' planning with the concept of 'spatial' planning. Although the significance of this change goes beyond debates about design, it was widely welcomed by those who regard design as core to mainstream planning. The emphasis on planning as an activity which is visionary, integrative, inclusive and action-orientated was regarded as 'intimately linked to a concern for and an engagement with design' (CABE 2005b: 5). In

2005, PPG 1 was replaced by PPS 1 *Delivering Sustainable Development*. The design principles outlined therein resembled those of its predecessor. However, not only did PPS 1 call for 'high quality and inclusive design' over merely the 'good' design aspired to in PPG 1, but it stated unequivocally that good design was 'indivisible from good planning', a mantra that has been repeated many times since. Furthermore PPS 1 emphasised that design policies should be based on the understanding of an area's characteristics and, potentially even more importantly, based on its *needs* – a fundamental principle drawn from contemporary urban design concepts.

Design and the local development framework and local plans (from 2012)

Spatial planning was defined by the new PPS 1 as 'integrating the use of land with other policies and programmes which influence the nature of places and how they function' (p. 30): in short, proactive place making. The main thrust of PPS 1 reinforced the importance of design in the planning system. An opportunity to embed design concerns at all levels was seen to be provided by the new system of the local development framework (LDF), which saw the introduction of a suite of planning documents which would outline the local authority's planning policies. The core strategy of the LDF was to contain design priorities and principles and non-development control design policies. Optional documents – area action plans (AAPs) – should set out detailed design policies for those areas designated as needing intervention, whereas more generic guidance should be included in supplementary planning documents (SPDs). From 2012, LDFs have been referred to as 'local plans'

Design in area action plans

The purpose of AAPs is to set out three-dimensional visions for allocated areas where significant intervention, or conservation, is needed. In an AAP, design issues sit alongside a range of key topics, such

as employment, housing and transportation. A key feature of AAPs is a focus on the implementation of policies. An example is provided by Northampton Borough Council's Central Area Action Plan (adopted in January 2013). This AAP includes a number of design documents, such as a character area assessment. The assessment subdivides the central area into ten 'character areas', each having its own identity in relation to movement, key buildings, topography, heritage, built form and so on (Northampton Borough Council 2013). On the basis of this analysis, each area has a set of priorities, defined as 'dos and don'ts', which act as proactive and reactive tools for guiding policy and development management. Plate 9.3 illustrates a plan of movement and Lynchian[13] analysis from the plan. In this character area, for example, the analysis suggests that two gasometers due for demolition which currently act as local landmarks should be replaced by structures which adopt that landmark role, in order to strengthen the sense of arrival in the town centre. In contrast, a restrictive measure discourages more surface car parking in the area.

Supplementary planning document

SPDs – previously referred to as supplementary planning guidance (SPG) – are part of the suite of documents comprising the LDF. Because they are optional, they hold less weight in relation to planning decisions than compulsory components, such as the core strategy. However, they are still material considerations, that is, they have to be taken into consideration when determining a planning application and in defending decisions challenged at appeal.[14] SPDs can take a number of different forms. For example development briefs set out development parameters for specific sites; and issue-based documents, many of which are in the form of 'design guides', cover specific issues – for example guidance for building extensions and alterations in conservation areas, and shopfront design guides. These guides are often relatively brief pamphlets dealing with aspects of physical design. However, some are more substantial and ambitious in their scope, such as the latest edition of the Essex Design Guide (see Plate 9.1).

Character Area 01 - Brewery District

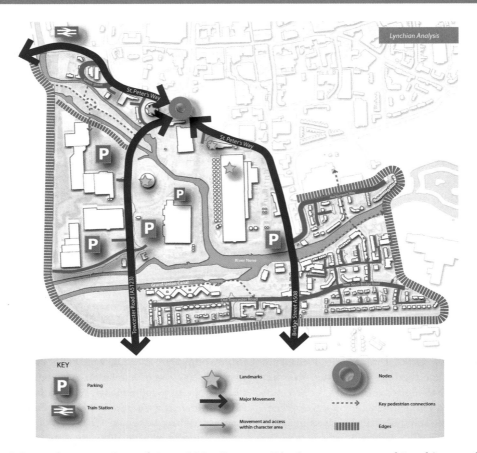

Plate 9.3 Northampton Central Area AAP – Brewery District movement and Lynchian analysis

Design and access statements

The Planning and Compulsory Purchase Act 2004 brought in a requirement for design and access statements. These must accompany all major applications that require planning permission or listed building consent. The statements outline the design thinking behind the application, for example, how the proposed development responds to the site and how the applicant has addressed the needs of all users. They are currently covered by the Town and Country Planning (Development Management Procedure) (England) (Amendment) Order 2013 (SI 2013/1238). They are not generally required for householder applications, except in specifically designated areas. The level of detail in the statements must be proportionate to the complexity of the application and although they should not be long, they should address the following issues:

- **Use:** what buildings and spaces will be used for and how this relates to local need;
- **Amount:** how much development is being applied for and why this amount is appropriate for the site;
- **Layout:** why the layout has been chosen and how it relates to its surroundings;
- **Scale:** why the size and massing of the buildings are correct for the site and how they relate to adjacent development;
- **Landscaping:** the principles that have been used to create landscape details and why these are appropriate;
- **Appearance:** the rationale that underpins the aesthetics of the development.

In addressing access, the statement should also focus on a range of issues, including how the development links to existing road and path networks; how people access and move through the site and buildings – for example, how level changes affect circulation and how signage will be used; how design will ensure safety of users; that the design is inclusive for all – including, for example, those with mobility problems; and how the design ensures safety during an emergency. For all elements, appropriate consultation should be carried out and the statement needs to outline how that consultation informed the design. CABE produced a guide on design and access statements which is now part of their publication archive, but is still useful (CABE 2006).

Manual for Streets

In 2007, the newly founded Department for Communities and Local Government (DCLG) published *Manual for Streets* in conjunction with the Department for Transport (2007). This was in response to an earlier ODPM report entitled *Better Streets, Better Places* (ODPM 2003b) – which identified highways standards as a key barrier to place making. The resultant document is a significant departure from its predecessors. It was underpinned by evidence developed by the Transport Research Laboratory,[15] to deliver streets that are pleasant, cost-effective and safe, and significantly foster ideas of 'community'. In particular, it prioritises the needs of people – pedestrians – in residential areas, over the requirements of vehicular traffic. This new priority is also followed in the structure of the document, where topics such as the context of streets and the quality of places are addressed, before it turns to more pragmatic issues, such as those of street geometry. Emphasis is also placed on issues such as reducing traffic speeds, integrating parking to increase safety for pedestrians and the introduction of 'home zones'. Home zones are areas where pedestrian/vehicular emphasis is rebalanced through elements such as shared surfaces.[16] However, the guide makes it clear that home zones should only be implemented after thorough consultation with local communities, emphasising the need for inclusive processes for successful delivery. In scope and tenor, *Manual for Streets* is one of the most significant publications that has been published by central government in relation to design issues.

The National Planning Policy Framework

The planning system underwent further major reform in 2012 with the introduction of the National Planning Policy Framework (NPPF) (see Chapter 4). This reduced central government guidance to just sixty-five pages of script. In relation to design, a section – paragraphs 56–68 – is encouragingly devoted to 'requiring good design' (p. 14). If the wording is compared with that of PPS 1, which it replaces, the nuances in language are interesting. The key statement 'good design is indivisible from good planning' is directly transplanted from the former document, and is, moreover supplemented by the statement that design should 'contribute positively to making places better for people' (ibid.). This reinforces previous guidance and academic writing on the importance of 'place making', as opposed to the tradition of design being limited to a concern with individual buildings. The guidance goes on to mention buildings, public and private spaces and the wider development and creation of attractive and comfortable places to live, work in and visit.

All of the above seems positive and reassuring and appears to take a broad, holistic view of the scope of design, which accords with contemporary urban design thinking. There is, however, one jarring element: paragraph 59 states that design polices should 'avoid unnecessary prescription or detail' and 'should concentrate on guiding the overall scale, massing, height, landscape, layout, materials and access of new development'. Of these only 'access' might be thought of as a broad concept – the rest are very much focused on the physical/material nature of places. This seems to directly contradict the idea of 'place making' and the wider social, environmental and economic concerns that this concept encapsulates, returning instead to a much narrower focus on design as relating only to physical concerns. Thus the document returns to previous ambiguities in the meaning of the term 'design' in central government guidance.

The other main issue with the NPPF is that there is precious little on how design quality will be delivered – two approaches are specifically mentioned, *design reviews* and *design coding*, as explained below. However, these will only ever be suitable for certain types of development and in respect of codes, these generally only apply beyond a certain scale. How design quality is delivered in relation to the myriad of day-to-day development management decisions, for example, which can so critically impact on the places people live, work and spend their leisure time, is far from clear. Moreover this directly relates to the vastly reduced capacities of CABE (see above). In essence the NPPF makes (some) positive statements in relation to design, but there is ambiguity and little of real substance.

Design review

Design review has been part of the planning system for some time. The Royal Fine Art Commission began reviewing building proposals of national interest in the 1920s, a system that operated up to the advent of CABE. Furthermore, at local level many larger local authorities have operated local (single authority, or in conjunction with neighbouring authorities) review panels, particularly for sensitive developments, such as those in conservation areas, for many decades – although there has always been patchy coverage. In 1999, responsibility transferred to CABE for London schemes and those considered of national importance. From 2002, regional panels for the review of significant schemes were set up. It should be noted that 'significance' is open to interpretation here and does not imply 'scale' – a single house in a rural setting for example may be of immense significance to a local community.

From 2013 and the integration of CABE into the Design Council, effectively a new system has come into operation. Design Council CABE undertakes reviews for London. There is also a network of eight regional design review boards – which provide national coverage with common protocols. The key principles for the operation of these boards are as follows: they are an independent, multidisciplinary team of experts; their role is advisory, and that advice must be proportionate, timely, objective and accountable (that is, in the public's best interest); and their processes must be transparent and accessible (understandable by all) (Design Council *et al.* 2013). Where projects are of national significance (for example, involving major infrastructure) CABE will add expertise to the regional board. Specialist panels may also be set up for specific programmes of activity; for example, a panel was set up for London's Crossrail project, specifically including infrastructure specialists. As part of its role, this has reviewed the designs of station buildings, ensuring that they not only relate to their surroundings and are of high quality, but that they serve passenger's needs in relation to issues such as clarity, accessibility and inclusiveness.

An example of a regional board undertaking a review that also included input from Design Council CABE – because of the project's national significance – is provided by a recent scheme, Freeman's Reach in Durham. The site is located in the centre of Durham City, on the eastern bank of the River Wear and adjacent to the World Heritage Site which includes Durham Cathedral, Castle and adjacent buildings. To the north, the Penny Ferry footbridge crosses the river, linking east and west banks, and to the south is Milburngate Bridge, which carries road traffic across the river. At a higher level is the Millennium Walkergate complex, which includes the City's public library and a theatre. The proposed development for a new office/mixed-use scheme was, therefore, in a highly sensitive location. The scheme, designed by architects Faulkner Browns (a well-known practice in North East England, based in Newcastle), was taken through design review by the North East Design Review and Enabling Service (NEDRES) twice (May and October) in 2012. Although the initial scheme was commended by the panel, particularly with regard to its composition, massing and site organisation, the review process led to a number of revisions – particularly, for example, at roof level, with the inclusion of roof terraces and a revised roof line, to address the highly visible nature of the scheme from adjacent higher vantage points (see Plates 9.4a and 9.4b).

Design codes

In addition to *design review*, the NPPF also encourages the use of design codes (para. 59) – although, again, as with review, this is not a new idea. The introduction of design codes was initially championed in the UK by the ODPM in 2004, which, working in partnership with CABE and English Partnerships, instituted an action research programme to monitor and evaluate the roll-out of nineteen design codes across England (DCLG 2006e). Design codes are a particular form of design guidance. They offer detailed three-dimensional direction for development, while not actually prescribing the outcome, that is, they provide a detailed compendium of design instructions and advice, usually heavily illustrated with visual examples.

The notion of design coding is not essentially new and there are many historical precedents. However in recent times they have been most closely associated with the New Urbanism movement. Many schemes with design codes in the UK have resulted in – and been subject to criticism for – somewhat traditional 'neo-vernacular' architectural design, though this is not necessarily the intent of coding.

Design codes provide certainty for developers, by providing clear instructions as to what is acceptable and reassurance to local communities that development will meet certain quality standards. On large sites, that involve many developers, codes can foster the delivery of a coherent vision. They are most effective when they build upon a physical design contained in a master plan or design framework. This should set out broad parameters, for example, establishing a road hierarchy, although some flexibility is generally desirable within the master plan, to allow the coding process to refine what is delivered on the ground. Coding is resource-intensive and can lead to higher construction costs for developers, as they are often unable to deliver their favoured standardised units. However there is evidence that it delivers enhanced land and sale values and helps new housing compete more effectively with the existing housing market. This allows for the offset of increased upfront resources (Adams *et al.* 2011).

In England, the best-known use of a design code is probably in Poundbury, Dorset, in the estate built on land owned by the Duchy of Cornwall – which belongs to Prince Charles – and the resultant development very much reflects the Prince's rejection of modern architecture in its styling. The scheme has received much criticism from the architectural profession, which has viewed it as boring and a pastiche. But its network of interconnected streets and pathways does create a highly walkable environment, and it has proved popular with the general public. A slightly more recent example of design coding is Upton, Northampton. An urban extension of around 1,300 homes has been set around a core with a school and shops, of which 22 per cent is affordable housing, discreetly distributed through the site and indistinguishable from private homes. Upton demonstrates that adherence to a design code can deliver a high-quality public realm, and the hierarchical street network is particularly noteworthy and works well at the scale of the development. The integration of flood attenuation measures, through sustainable urban drainage (SUD), demonstrates that high environmental performance can be achieved by dwellings which are not unusual in appearance (see Plate 9.5).

Plate 9.4a Freemans Reach Scheme, Durham: Faulkner Browns Architects, as first take to review

Plate 9.4b Freemans Reach Scheme, Durham: Faulkner Browns Architects, final proposal

Plate 9.5 Upton housing

There is no doubt that design coding has the potential to increase the quality of residential developments in England, not least by creating more cohesive new developments on larger sites, involving multiple developers and forcing developers to think more carefully about design quality. The most recent research in this area suggests, however, that many speculative house builders have yet to be convinced of the advantages of design codes, and substantial transformation of the market has yet to be realised (Adams *et al.* 2011).

CONTEMPORARY ISSUES

The following section reviews a number of issues of concern, which do not necessarily fit neatly into the trajectory of national guidance.

Housing: space standards and density

As outlined above, the standard of volume housing has been a focus of concern in the UK in the recent past. Concerns have addressed both poor design in residential layouts and inadequate internal space standards. In relation to the latter, there has been no national-level attempt to tackle internal space standards since Parker Morris (see above). However, in 2005 the Greater London Authority (GLA) commissioned a number of reports as background reviews for the London Plan; one was on residential space standards (Drury *et al.* 2006). This report concluded that there was an increasing mismatch between the types of dwelling built in the capital (increased provision of one- and two-bedroom flats) and the needs and preferences of householders. Furthermore, cramped and overcrowded conditions carried with them the danger of adverse impacts to the health and well-being of occupants. The report concluded that under the then Planning and Compulsory Purchase Act 2004, space standards were both a material consideration and a component of sustainable development, and therefore they were a legitimate concern of the planning system. The planning system should, therefore, be a vehicle for allowing such requirements 'to be set, implemented and enforced' (p. 73). Subsequently, space standards have been incorporated as a component of the London Plan (albeit with slight variation from the original recommendations). Policy 3.5 of the London Plan (GLA 2011) set outs minimum floorspace requirements, for example 37 square metres for the smallest one-person dwelling and 50 square metres for a two-person, one-bedroom, dwelling (p. 87): Details are shown in Table 9.1.

The London Plan also includes guidelines on housing density, with ranges dependent on location and built form (defined as central, urban and suburban) and public transport accessibility. The lowest ranges, deemed suitable for suburban areas with poor public transport are 35–55 units/hectare (u/ha). The highest densities for central areas with good public transport accessibility are 215–405 u/ha: Table 9.2 shows the range of recommendations. However, as set out in the Housing SPG, the density figures are not to be applied mechanistically, but proposals should also be assessed in terms of the capacity of 'local amenities, infrastructure and services' to support the development (GLA 2012a: 40). While the standards are intended for London only, in the absence of national standards, it is very likely they will be adopted (at least informally) by urban authorities elsewhere.

Table 9.1 GLA minimum internal space standards

	Dwelling type (bedroom/persons)	Essential GIA (m²)
Flats	1p	37
	1b2p	50
	2b3p	61
	2b4p	70
	3b4p	74
	3b5p	86
	3b6p	95
	4b5p	90
	4b6p	99
2-storey houses	2b4p	83
	3b4p	87
	3b5p	96
	4b5p	100
	4b6p	107
3-storey houses	3b5p	102
	4b5p	106
	4b6p	113

Table 9.2 Density matrix (habitable rooms and dwellings per hectare)

Setting	Public transport accessibility level (PTAL)		
	0 to 1	2 to 3	4 to 6
Suburban	150–200 hr/ha	150–250 hr/ha	200–350 hr/ha
3.8–4.6 hr/unit	35–55 u/ha	35–65 u/ha	45–90 u/ha
3.1–3.7 hr/unit	40–65 u/ha	40–80 u/ha	55–115 u/ha
2.7–3.0 hr/unit	50–75 u/ha	50–95 u/ha	70–130 u/ha
Urban	150–250 hr/ha	200–450 hr/ha	200–700 hr/ha
3.8–4.6 hr/unit	35–65 u/ha	45–120 u/ha	45–185 u/ha
3.1–3.7 hr/unit	40–80 u/ha	55–145 u/ha	55–225 u/ha
2.7–3.0 hr/unit	50–95 u/ha	70–170 u/ha	70–260 u/ha
Central	150–300 hr/ha	300–650 hr/ha	650–1100 hr/ha
3.8–4.6 hr/unit	35–80 u/ha	65–170 u/ha	140–290 u/ha
3.1–3.7 hr/unit	40–100 u/ha	80–210 u/ha	175–355 u/ha
2.7–3.0 hr/unit	50–110 u/ha	100–240 u/ha	215–405 u/ha

Source: GLA 2012a: Table 3.2

Building for Life

The Building for Life concept was initially launched as a collaboration between CABE, the Civic Trust and the House Builders Federation (HBF) in 2001, to identify best practice in housing and neighbourhood design – and as an attempt to raise housing standards. This was followed by a CABE guide published in 2005 which set twenty key criteria in the form of questions to explain the Building for Life Criteria. While the aims of Building for Life were admirable, its initial impact in improving housing quality was limited – see reviews undertaken by CABE above. However, despite continuing problems in the housing market, there is some evidence that more quality housing is being developed and the number of housing developments reaching Building for Life criteria has been slowly increasing.

Most recently Building for Life has been effectively simplified and relaunched as Building for Life 12 (BfL 12) – on the basis of common problems identified in housing schemes across the country. The guidance is reformatted into twelve questions based on topics of connections; facilities and services; public transport; meeting local housing requirements; character; contextualisation; creating well-defined streets and spaces; legibility; inclusiveness; car parking; public and private spaces; and external storage and amenity space. The main aim of the redesign of the scheme is to emphasise facilitating discussion to raise design quality, rather than assessment (Building for Life Partnership 2012). At the time of writing, the effectiveness of the changes to Building for Life and whether they will accelerate improvements has yet to be evaluated.

Tall buildings

Tall buildings are worthy of specific consideration, since their impact is felt over very wide areas and they introduce unique issues in relation to design, such as overshadowing of neighbouring areas and impacts on historic skylines. Debates about tall buildings in England date back to the 1950s, when new fire regulations allowed for taller buildings to be built. The first 'tall' building of the period was the Shell Centre (twenty-seven storeys), which was built on part of the cleared Festival of Britain site in London. In the interim period, building heights have increased dramatically. There is no universally agreed definition of a tall building. However, a CABE/English Heritage report (2007) described a tall building as one which is significantly taller than those structures around it and which breaks the skyline, and this remains a useful description. Skylines, often a familiar view of a town or city, can capture the essence of place and may be a defining element in the *genius loci* or sense of place – for example, on the basis of its skyline, Oxford is known as the 'city of dreaming spires'. Conversely, inappropriately located tall buildings can ruin familiar and cherished views and, moreover, can lead to impoverished environments in their immediate surroundings – where they meet the ground – unless designed to the highest standards. And, furthermore, attempts to mitigate this, such as glass façades, create their own problems – both aesthetic and functional. Many towns and cities have therefore developed tall buildings strategies, often looking at key viewpoints from movement corridors and areas of open space.

The London View Management Framework (GLA 2012b) is the most comprehensive SPG in England to address the issue of tall buildings. The policy aims to guide the management of twenty-seven views worthy of protection. These are divided into four groups – panoramas, linear views, river prospects and townscape views – across the capital. The guidance states that new development should make 'a positive contribution' to views and landmark elements, and that the architecture of individual projects should be outstanding. However, there has been much debate about the effectiveness of the policy.

The Shard, for example, London's first super-tall structure, designed by Renzo Piano and currently the tallest structure in Europe at eighty-seven storeys, looms over views of St. Paul's from Hampstead Heath – a view the policy states should be protected (Panorama – Parliament Hill to Central London). However, there has been even more vociferous criticism of a number of other schemes. Stata SE1, for example, winner of Building Design's 'Carbuncle Cup' in 2010 (for worst new building) greatly impinges on the view of the Houses of Parliament from the Serpentine Bridge in Hyde Park, another view that is supposed to be protected by the management framework (Townscape View – Bridge over the Serpentine, Hyde Park to Westminster).

The impact of these towers at ground level is also often overlooked. Many public spaces, even at the base of otherwise good examples, are windswept, unpleasant and largely unusable. Given the large number of tall building proposals already in place for London and other major UK cities, it is likely that the debate and controversy will intensify in the future.

Crime and design

Theories about the influence of the design of the physical environment on rates of crime, vulnerability to crime and fear of crime have a relatively long trajectory in urban design writing. A key text was Newman's *Defensible Space: Crime Prevention Through Urban Design (CPTUD)* (1973). Newman studied two housing estates, with residents of similar socio-economic profiles, but with different physical designs and different crime rates. He concluded that the estate that displayed higher crime rates had a number of design features that encouraged criminality to flourish, whereas the estate with lower crime rates had design features which encouraged residents to be agents of their own security. An example was how overlooked entrances to the blocks were from nearby apartments – overlooking was seen to be beneficial, as it promoted natural surveillance, that is, people were likely to watch people coming and going, and criminals would feel more open to observation. A key notion of Newman's work was also that of 'territoriality', that is, residents needed to feel ownership of space. This was invoked by clearly defined private space and limiting the numbers of people who shared communal spaces – such as common entrance areas. CPTUD had much in common with Crime Prevention Through Environmental Design (CPTED), a contemporaneous theory developing among US criminologists, which also emphasised the impact of physical environment on the propensity to commit crime.

In the 1980s, CPTUD/CPTED was translated to an English context through the work of the geographer Alice Coleman. Coleman looked at the design of a number of local authority housing estates which displayed high rates of crime and antisocial behaviour, and found many of the features that US theorists had identified as problematic. Coleman's work took place at a time when the Thatcher government was dismantling the public housing framework, and the notion that the very design of some public housing schemes was responsible for the problematic behaviour of residents suited the government's position very well. Though Coleman herself may have been well intentioned, the work greatly underplayed the complex root causes of social conditions in the estates she studied.

In Britain, the key components of increasing territoriality, maximising surveillance (including CCTV, etc.), improved physical security (target hardening) and minimising access and escape routes have become associated with the term 'designing out crime' and have been promoted through 'architectural liaison officers' in the UK police force. Aspects of CTPED, for example, avoiding alleyways running along blank gable ends of dwellings, have been found to reduce the burglary rate when used judiciously. However, if CPTED is used as an inflexible 'rule book', as proponents often suggest it should be, the resultant quality of the environment is often questionable. Indeed, taking the approach to its logical conclusion might imply that the entire residential stock should be (re)built as disconnected, bungalowed cul-de-sacs[17] – a situation that even the most ardent advocate of CPTED would surely see as undesirable.

There is an alternative theory to CPTED, which in essence traces its origins to the writings of Jane Jacobs and, more latterly, to New Urbanism. This is about putting 'eyes on the street' to discourage criminality by encouraging pedestrian movement, that is, creating well-connected streets, high-quality public spaces and mixed uses – in essence making it easy and pleasant to walk and providing people with a reason to do so. More generally, this reflects a position promoted by much urban design writing in the past few decades, or, put another way, it represents the antithesis of the CPTED approach.

The most comprehensive assessment of which perspective (CPTED or New Urbanist) might be correct has been carried out by Hillier and Sahbaz (2012) in what they term high-resolution analysis. This is using a case study where the spatial factors are well defined and as many as possible of the aspects that influence crime rates are taken into consideration. Space prevents detailed exploration of their research, but their overall conclusion was that in reducing criminal activity a key concept was 'safety in numbers' (p. 135). This means that higher residential populations associated with street segments are associated not only with lower burglary rates, but also lower street crime. This appears to reverse the CPTED 'small is beautiful' logic, that is, that smaller, more separated spaces enable communities to inhibit crime. In contrast it suggests that co-presence is more important than interaction. However, it also suggested that longer streets with more dwellings co-present were better than shorter streets with lots of interconnection (highly permeable) – often a feature of New Urbanism. In other words, though they largely supported the New Urbanism 'eyes on the street' perspective, they disagreed as to the best way in which this could be achieved.

The *National Planning Practice Guidance* published in 2014 and reviewed below states the need for well-connected, safe streets. However, it also has a separate section on crime prevention that appears to reinforce the 'designing out crime' (CPTED) position. Although Hillier's work has yet to be validated by other case studies – the empirical research covered a single London Borough – the most likely conclusion is that the issues are complex and interrelated. The main value of CPTED may be in increasing avoidance of crime-friendly design features, rather than the hope of entirely designing out crime.

influence over landscapes in some areas, such as National Parks or other areas seen as requiring some form of protection, has steadily grown. Second, the reference to tourism presages a growing appreciation that agriculture is not the sole economic driver in the countryside, and planning must take account of the needs of other forms of business when preparing plans or making decisions on applications. Third, the countryside is a place where one in five of the population lives and it is important that new development contributes to meeting the needs of this population. At the time of the 1947 Act, a main concern was the movement of population *from* rural areas, whereas today planning is more likely to be seeking to deal with the impacts of more people wanting to move *to* the countryside; but even if the direction of movement of population has changed, a concern remains to seek to provide adequate access to services and housing for all rural residents. Finally, as was the case in 1947, agricultural policy remains an important influence over the rural economy and society (even though numbers directly employed in agriculture have been significantly reduced) and the nature of the rural landscape and environment. Whilst between 1947 and 1973 the principle influence was the national policy of deficiency payments to farmers, since 1973 it has been the Common Agricultural Policy of the EU.

That the relationship between agricultural and planning policy domains can do much to shape a number of important outcomes for rural communities is illustrated by the two 1947 Acts. However, the post-war years were littered with numerous examples of agricultural and planning policies pulling in different directions – for example, whilst agricultural policies to enhance levels of production were resulting in the diffusion of more intensive methods of farming and consequent losses of important habitats, planning was interested in protecting and enhancing biodiversity. Debates over how such inconsistencies could be avoided sometimes led to the pursuit of organisational 'fixes', such as the invention of DEFRA, ostensibly linking the three key policy domains of environment, food and rural affairs (see Chapter 3). The need to improve coordination was perhaps first given some formal recognition when John Major's government

published its White Paper *Rural England: A Nation Committed to a Living Countryside*, which was jointly prepared by the ministries responsible for planning and for agriculture (DoE/MAFF 1995), an approach which was also taken by the successor Labour government with its own Rural White Paper *Our Countryside: The Future* (MAFF/DETR 2000).

The content of these Rural White Papers and a number of subsequent reports – most recently, the Taylor Review *Living Working Countryside*, published in 2008 – reflects a recognition of the need to develop a policy agenda which has a focus on factors affecting quality of life for rural residents. Such debates had been happening since the 1960s, with perhaps the most consistently high-profile issue being that of access to affordable housing for rural residents, although there are others. Many of these matters relate closely to planning policy concerns, although whether planning is the most suitable implement for tackling some of the issues is a matter for debate. In the view of Taylor:

> The planning system has a crucial role to promote and deliver sustainable communities – ensuring development occurs in the right place at the right time and makes a positive contribution to people's lives – providing homes, jobs, opportunity and enhancing quality of life. It must simultaneously protect and enhance the natural and historic environment, and conserve the countryside and open spaces that are important to everyone.
>
> (p. 7)

The National Planning Policy Framework (NPPF) has only a limited dedicated rural content, but it does offer guidance on supporting a prosperous rural economy, providing affordable housing, as well as some guidance on the protection of important environments, perhaps reflecting that planning has to have regard to an agenda made up of economic, social and environmental matters, rather than a narrower focus on agriculture and the protection of land from development, which was felt to be the emphasis of early post-war policy.

To help establish the context in which policy has emerged, this chapter begins with a review of some of the key trends affecting the countryside, including the impacts of changing agricultural policy. This is also the context within which planning has to seek to make an impact. It then moves on to a review of the evolution of planning policy for development in the countryside, before giving some detail of key policies designed to protect rural environments.

The changing rural context

For much of the first fifty years of the twentieth century, Britain's rural areas were areas of population loss or very slow growth, which can be related both to declining demand for labour in agriculture and growing demand for labour in other industries, most of which were based in urban areas. In the 1930s, some slow growth could be correlated with the development of suburbanisation. However, from the 1960s the population of rural areas began to grow, which reflected what Champion (2003: 19) describes as 'the strong and enduring attraction' of the deep countryside. Bolton and Chalkley (1990) pointed out that growth in rural populations was the result of in-migration, and identified three drivers behind what is sometimes know as 'counter-urbanisation': about half came for a mix of work and environment, about a third were motivated solely by the economic reasons provided by new employment, whilst a fifth were mainly 'lifestyle' migrants, often for retirement. Migration flows are a product of moves to and from rural areas, and one significant factor is that many of those leaving are young, which contributes to rural areas having a population which is older than the average (Champion and Shepherd 2006). Whilst net in-migration can be seen as an indicator of prosperity, it can also be a mechanism whereby a number of problems are generated, including those related to service provision, housing markets and potential environmental conflicts. It is the role of planning to seek to manage the process of accommodating this growth, through decisions about where growth should occur and at what rate.

As has been noted above, growth in employment has been a driver for population growth in rural areas. Whist not all parts of the country have benefited to

the same extent, in recent years rural areas have outperformed urban areas in their rate of employment growth, with the main type of new employment being in the service sector. One explanation for this was what was known as the 'urban-rural shift', which explains growth in terms of both push and pull factors. In the 1980s, manufacturing industry in urban areas was often burdened with constricted sites with little scope for expansion, tight planning controls and high labour turnover, whilst rural areas offered better site availability, better environments and favourable land prices (Fothergill *et al.* 1985). Movement was facilitated by improvements in transport and more recently by the growing influence of information technology, although there is a continuing struggle to achieve an equal quality of broadband service in all rural communities. Some quantification of these trends, along with other aspects of change, is given in Box 10.1.

Whilst this is a generally positive picture, a number of studies have established that rural communities do experience economic and social problems – as with urban communities, not everyone can share equally in prosperity. Since the 1980s, it has been pointed out that about a quarter of rural households experience poverty (McLaughlin 1986; Cloke *et al.* 1994; CRC 2010). This is a product of a number of factors, including low rates of pay, more seasonal and part-time work, significant levels of underemployment (that is, people working below their capacity in some way) and a higher cost of living stemming from higher housing and transport costs. A combination of high cost and lower incomes means that in rural areas the cheapest house is six times the average income of the lowest-income households compared with five times in urban areas (CRC 2010). Transport costs reflect the low density of public transport networks and the need to own private transport to access work, services and leisure. So, a pressing issue for rural planning is, as Silkin put it, to make life more convenient in rural areas. A principal concern has been to address the issue of a shortage of affordable housing, but there is also the general matter of trying to develop a pattern of growth which will help address issues of access to services, work and leisure. The different ways

BOX 10.1 CHANGING COUNTRYSIDE IN ENGLAND

Rural areas are in constant change. The main features over recent years are as follows:

- *population growth:* net migration of 60,000 people per year into wholly or predominantly rural districts between 1991 and 2002;
- *an ageing population:* the number of people aged 65 or over in wholly or predominantly rural districts increased by 161,000 (12 per cent) between 1991 and 2002, while the number aged 16–29 decreased by 237,000 (18 per cent);
- *relative prosperity, especially in more accessible areas:* higher income per head than the national average – but with a disadvantaged minority amidst prevailing affluence;
- *economic weaknesses, with associated social deprivation, in a minority of 'lagging' rural areas:* characteristically in areas adjusting to a decline in mining, agriculture and fishing, and tending to be in more peripheral areas;
- *convergence between the urban and rural economies:* though agriculture is still at the core of the rural economy and society, employment in agriculture has decreased by 30 per cent (151,000) since the early 1980s, and employees in rural businesses are now more likely to be in manufacturing (25 per cent), tourism (9 per cent) or retailing (7 per cent) than in agriculture (6 per cent);
- *increased mobility through the car:* bringing benefits for many but reducing the customer base for public transport and thus creating difficulties for those without access to a car; half a million (14 per cent) rural households do not have a car, and many people in households which do have a car do not have access to it when they need to travel;
- *pressures on the countryside – especially through demand for housing and transport:* rural areas remain a rich resource, valued by both residents and visitors for fine landscapes, biodiversity and open space; these contribute to enjoyment and general well-being as well as enhancement for the benefit of all.

Source: Adapted from DEFRA 2004: 8–9

---

that planning has approached this challenge are considered later.

Changes in agriculture

Whilst British agriculture existed under the system of deficiency payments introduced by the 1947 Act for twenty-five years, for the last fifty years it has been the Common Agricultural Policy (CAP) of the EU that has been the key influence. Whilst this is not the place to discuss the intricacies of the CAP, it is worth highlighting some of the changes in that policy and

the influences these have had on agriculture and rural areas in Britain, particularly as these impact on issues of the rural economy, society and environment which are of central interest and concern to planning.

Like the 1947 Act, the CAP originally had the objectives of ensuring the security of food supply and limiting dependence on outside sources. It was successful in its pursuit of this objective such that by the mid-1980s the EU had become self-sufficient in all but a few staple products. The mechanism by which this was achieved was principally an intervention system[2] – at times of plenty, products would be bought in and stored, to maintain prices for suppliers and in the longer term to provide security of supply for consumers. The 'success' of these measures led to production growing faster than consumption, the build-up of stocks and overall EU expenditure on agriculture growing. Because the measures led to an intensification of production methods, there were a number of adverse environmental impacts. Further mechanisation led to impacts on the character of landscapes and biodiversity; land improvements and especially land drainage led to the loss of habitats; increased use of fertilisers also affected biodiversity and in some cases contaminated water supplies; and intensive livestock production methods led to problems with waste generation and disposal as well as concerns for animal welfare. The financial impact was also regressive in that 80 per cent of support was concentrated on 20 per cent of the largest farms and it also acted as a prompt to farm rationalisation. Generally, there was little favourable impact on the incomes of the majority of farmers and employment in agriculture continued to fall. Whilst farm incomes have always fluctuated for climatic reasons, policy changes became a significant factor in influencing income changes, which could be substantial: for example, between 1990 and 1995 UK farm incomes doubled, but between 1995 and 1999 they halved, and such changes affect rural communities as a whole, not just farmers. So, the social, economic and environmental consequences of the policy were not necessarily seen as favourable for rural areas.

These problems led, in the late 1980s, to the EU reconsidering the nature of its agricultural policy – some might say moving away from an agricultural policy to a rural policy. The then Agriculture Commissioner, Ray MacSharry, identified some of the imperatives for change in a speech[3] in 1990. These included the continuing pressure for change in agriculture and the need to help communities adapt to these changes; the need to develop an agricultural policy that was more environmentally sensitive; and the need to give a greater focus to the wider economic and social needs of rural communities, not just the needs of agriculture. Proposals for reform, which became effective in 1992, involved four main changes: changes in price support to reduce surpluses; new measures to promote environmentally friendly methods of production; promotion of forestry; and schemes to encourage the retirement of elderly farmers to promote restructuring and open up opportunities for younger farmers.[4] Subsequent reforms in 2000, 2003 and 2008 continued the changed emphasis on 'extensification', greater environmental awareness and a focus on rural development in a wider sense. Reforms to be introduced after 2013 include switching the emphasis of price support to farmers' incomes rather than product prices; making 30 per cent of payments available for promoting environmentally friendly agricultural practices; and a range of measures to increase competitiveness and growth in rural areas, but in ways which are environmentally friendly. These changes in the CAP are summarised in Figure 10.1.

Rural development measures were to be a significant part of the CAP when it was first developed. In addition to the funding of price guarantees, there was a 'guidance' element of the fund which was there to promote development in rural areas. This element was intended to be one-third of spending, but in 1976, about fifteen years after the launch of the CAP, it amounted to 1.5 per cent and it has never approached the intended target. This element is now known as the European Agricultural Fund for Rural Development (EAFRD) and it now operates in a similar fashion to the other EU structural funds in that each member state has to prepare a strategy outlining how the funds allocated to them will be spent within the general guidelines for the operation of the fund. There are separate plans/strategies for England and each of the devolved administrations. England's strategy, the Rural Development Programme for England (RDPE) for the period 2007–13 had a value in excess of 5 billion and had four main foci for action: improving the competitiveness of farming and forestry, for example, through better marketing or processing of agricultural products; improving the environment and the countryside through various forms of 'agri-environmental' action (see Box 10.3); measures to improve the rural quality of life and diversification of the rural economy, such as the development of new tourism initiatives; and a 'bottom-up' approach to stimulating rural development through local partnerships, LEADER (Liaison entre actions de développement de l'économie rurale).

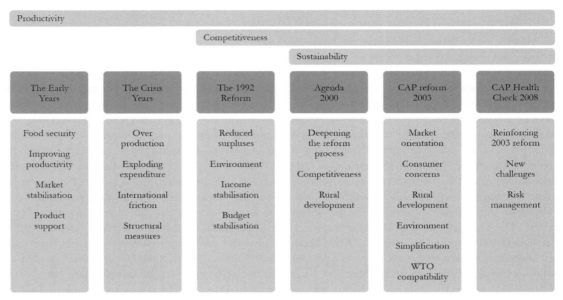

Figure 10.1 Development of the CAP

Source: European Commission (http://ec.europa.eu/agriculture/cap-history/cap-history-large_en.png; accessed 7 July 2014)

So, fifty years of the CAP has seen significant changes in agriculture – to some extent policy-induced – and a widening of interest in, and assistance for, a more extensive agenda of rural development and change. As well as helping focus interest and policy action on the protection of important and endangered environments, these changes have helped focus attention on a changing and more diverse rural economy. Whilst direct employment in agriculture has diminished, it is important not to ignore the wider agricultural economy, upstream and downstream of the farm. A prosperous agriculture can sustain a range of supply and support industries, whilst the processing and distribution of agricultural products are often an important source of local employment and income. In a move to maintain or improve incomes, many farmers have sought to diversify on-farm activities, sometimes to include processing and sale of products – for example, producing cheese or ice cream from milk – or sometimes by moving into other areas, such as farm-based tourism or using farm buildings for other forms

of enterprise. Whilst income from diversification might, in aggregate, represent only 6 per cent of farm income,[5] for some farms it can be important, and planning guidance and policy have developed to reflect this growing importance. In terms of rural areas generally, it is now firmly a service industry-based economy, rather than relying on agricultural and primary products, and this change is something which is supported and maybe partly driven by policy. Planning policy has evolved to attempt to realise the potential and meet the challenges which accompany these trends.

Rural planning policy

Whilst, from the days of Silkin onwards, there have been plenty of government statements and guidance on the objectives of planning for rural areas, in some ways these are more sharply revealed in the policy that was developed in local plans. The idea of 'key settlements' or 'rural service centres' is long established at the heart of rural local plans – in the view of Cloke (1979), it was the dominant approach to rural planning in Britain and had been so since the early 1950s. It can also be seen as an essential component of current policies for 'sustainable rural development'. Cloke (1979) suggested the emergence of key settlement policies can be traced to the Scott Report of 1942 and the Town and Country Planning Act 1947. These favoured the concentration of new development in larger settlements possessing an endowment of basic services. At that time, the (imputed) objectives of the planning system of countryside protection and the presumption against building on good agricultural land (Hall *et al.* 1973) coincided with a continuing decline in the rural population. So, the objectives of settlement policy included the maintenance of the status and position of rural communities – or, in some cases, their managed decline[6] – and achieving an efficient use of resources, particularly in the sense that the economic provision of services could only be brought about by the selection of certain villages for expansion.

Key settlement policies began to emerge in county development plans from the 1950s and grew into the cornerstone of rural planning policy. They aimed to confine the growth of housing, services and employment to a small number of settlements. These settlements were selected by reference to their role and function within the rural area under consideration, as well as by reference to a range of broader strategic objectives for the rural area as a whole. The rationale for the policy was by no means uniformly articulated in development plans, but reasons given for its adoption included 'restraining the pressure for development on agricultural land, reducing the costs of services, maintaining the quality of the environment and improving the quality of rural life' (Martin and Voorhees Associates 1980: 1). However, the assumptions underpinning these policies were by no means unchallenged. The economic arguments have been contested (Gilder 1984),[7] as has the idea that it is possible to achieve both lower costs and high-quality accessible services (Moseley 2003).

The advantages and disadvantages of key settlement policies were reviewed in 1980 (Martin and Voorhees

Associates 1980). The study found that such policies were more effective in dealing with physical development issues than with the wider range of factors of concern to rural communities, and that they had little impact in diverting the path of market forces. They were generally unsuccessful in promoting rural services, in encouraging economic development, or in addressing social problems, principally because the planning policy was – at the time – not accompanied by the powers necessary to ensure the implementation of the broader policy objectives. The study concluded by arguing that 'the uniqueness of rural areas defies the adoption of a single type of settlement policy for all rural areas . . . rural planning should always be a problem-solving exercise' (p. 223). Empirical study by Shorten (2004) suggested that the hierarchical assumptions underpinning these policies bore little relation to how rural residents lived their lives and he also concluded that 'the need for tailored policies to fit local circumstances is imperative' (p. 191).

The 1990s saw a somewhat reinvigorated debate on policy for rural areas. In a formal arena, this could be seen as starting with the first Rural White Paper, published in 1995. In his foreword to the White Paper, John Major characterised rural England as an area of continuous change and saw the purpose of the White Paper as being to examine this change, look for ways of responding that assured sustainable development of jobs, services, and the environment, and launch an 'open debate about rural England and its future'. This judgement was echoed by the next study to be published, *Rural Economies*, produced in 1999, which stated: 'rural life is changing, as rural areas are subject to the impact of big social and economic forces. Government cannot stop these forces; but it can influence some of the changes and help rural areas to adjust' (PIU 1999: 7). The questions and options suggested were to be part of a discussion of the revised Rural White Paper, but amongst other matters, the report addressed the role of the planning system. Among the ideas put forward for discussion were the extension of planning controls to agricultural development, the relaxation of development control in rural areas to allow 'sympathetic and appropriate economic activity, and a change in the presumption against development of the best and most versatile agricultural land' (p. 75) (but with appropriate protection for areas of high environmental value).

The White Paper of 2000, *Our Countryside: The Future – A Fair Deal for Rural England*, called for a more flexible and positive attitude to development in the countryside, especially where this supports the provision of services and affordable homes and the diversification of the rural economy. These views seem to be based on the (largely erroneous) assumption that planning is the problem. Studies have shown that the planning system has not generally blocked rural diversification (Shorten *et. al.* 2001), although there is room for improvement in positive action. The problems of providing for affordable housing have long been recognised in the planning system, but a wider agenda was set, focusing on improving access to services, supporting economic diversification, improving access to a protected countryside and enhancing the capacity of local communities to help shape their future. Interestingly, it also established a system of 'rural proofing', a mechanism to assess the impact on rural areas of policies developed across government and, where necessary, 'to make positive adjustments that take into account the needs of rural communities and businesses' (DEFRA 2013b: foreword). The process by which rural proofing is undertaken is set out in Figure 10.2.[8]

Policy with a sharper focus on planning had been in place since 1992, when PPG 7 *The Countryside and the Rural Economy* was issued. More recent guidance – PPS 7 *Sustainable Development in Rural Areas* – was issued in 2004, though much of this was closely based on its predecessor, though with some development of policies for community services, tourism, leisure, and countryside designations. Both the 2000 White Paper and PPS 7 closely adhered to the hierarchical settlement policy paradigm, stating that 'planning authorities should focus most new development in or near to local service centres, where employment, housing (including affordable housing), services and other facilities can be provided close together.' (p. 6).

Both the White Paper and PPS 7 developed policy within the overriding objective of the achievement of sustainable development and sustainable rural

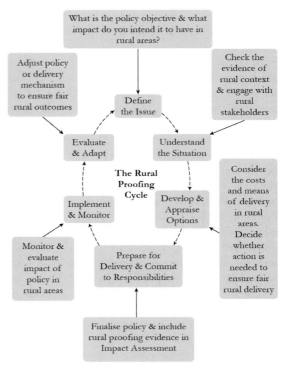

Figure 10.2 The process of rural proofing
Source: DEFRA 2013b

communities. Guidance issued on the nature of sustainable communities (ODPM 2003a) highlighted economic, social, environmental and political features that are key requirements. Some emphasis is given to issues of scale – 'sufficient size, scale and density . . . to support basic amenities' (ODPM 2003a: 4) – perhaps implicitly supporting a quest for more self-contained rural communities with functionally linked surrounding areas and generally reducing the need to travel: 'it is an implicit assumption of national policy that massing of new development in larger settlements with better accessibility and a fuller range of local services will produce more sustainable outcomes, chiefly reducing travel by private car' (Shorten *et al.* 2001: 39).

The White Paper also included a 'new commitment to market towns', with market towns rather loosely defined as towns with a population of between 2,000 and 30,000.[9] This followed a lead set in the *Rural Economies* report and included a number of objectives: to help market towns manage the process of change affecting them; strengthen the role of service provision to surrounding rural areas; help regenerate the most deprived areas; and ensure that central and local government recognises the role of market towns in their strategies. This was a regeneration initiative soon absorbed into the programmes of regional development agencies and assisting around 250 towns (more than one in five of the towns in the population range), but the role identified for such towns as 'rural service centres' and their role in acting as a focus for meeting the needs of a wider rural area re-emphasises a commonality with a central element of hierarchical settlement policy.

Whilst sustainable development is a well-established component of planning policy, moving from principles to developing practice and a policy framework can be challenging. This dilemma is well summed up by Owens:

> Because land-use is so closely bound up with environmental change, land-use planning demands the translation of abstract principles of sustainability into operational policies and decisions. Paradoxically, this process is likely to expose the very conflicts that 'sustainable development' was meant to reconcile . . . The planning system is likely to remain a focus of attention because it is frequently the forum in which these conflicts are first exposed.
>
> (Owens 1995: 8)

Concerns about the interpretation of sustainable development in the context of rural communities and planning have been articulated in other quarters. For example, the (former) Rural Development Commission[10] claimed that the planning system took sustainable development

> to mean environmental protection and reducing travel needs by concentrating development into larger settlements. Those strands of sustainable development relating to economic development and

social equity tend to be overlooked . . . Emphasis should be given to the social and economic implications of not providing for development.
>
> (RDC 1998)

Similarly, the Countryside Agency, in its *Planning for Quality of Life in Rural England* (1999), stressed 'the essential interdependence between a thriving rural economy, sustainable communities and the proper care and enjoyment of the countryside'. It argued that, in judging development proposals, the new philosophy should be 'Is it good enough to approve?', not 'Is it bad enough to refuse?' This calls for a proactive approach by local planning authorities: their policies need to be 'criteria based' and clearly state 'the qualities the plan wishes to pursue'.

Most recently, the Taylor Review (2008) identified a 'sustainability challenge' in pursuing its objective of a living, working countryside:

> Planning must not determine the future development of rural communities against a narrow tick-box approach to sustainable development, assessing communities as they are now and not what they could be. In too many places this approach writes off rural communities in a 'sustainability trap' where development can only occur in places already considered to be in narrow terms 'sustainable'. The question planners must address is 'how will development add to or diminish the sustainability of this community?' taking a better balance of social, economic, and environmental factors together to form a long term vision for all scales of communities. A mix of housing and employment opportunities are essential for the sustainability of rural communities.
>
> (p. 26)

The Coalition government elected in 2010 made a number of changes in both organisation and emphasis of policy. Its 2012 *Rural Statement* continued the commitment to 'living, working countryside', with a significant emphasis on facilitating the rural contribution to economic growth (DEFRA 2012b). This reflects the then current planning guidance in the NPPF, where the clearly dedicated rural section states 'Planning policies should support economic growth in rural areas in order to create jobs and prosperity by taking a positive approach to sustainable new development' (p. 9). To this end, neighbourhood and local plans should support business development, including agricultural diversification and tourism development, while at the same promoting the retention and development of local services.

So, planning for the development of rural areas – as opposed to planning policy mainly focused on countryside protection – has a long history developed from a hierarchical view of the role and function of settlements. The development of a mission for planning focused on the achievement of sustainable development seemed to do little to move policy away from this hierarchical model, and the guidance in the NPPF or other government rural policy documents did not signal fundamental change from the model. However, the quest for greater self-containment for larger rural settlements – a greater proportion of the population can live, work and access services in the same settlement – whilst an attractive proposition in the context of planning for sustainable development, does not sit comfortably with the evidence on how people in rural areas live their lives. Of course, this is not to say that either through changing context or the influence of policy such patterns of living cannot be changed – planning, after all, has always embraced a normative element – but it suggests that change will not be easily achieved. Hierarchical policies also need to have regard to Taylor's warning of the effects of the 'sustainability trap': tightly constraining the amount of development which can take place in smaller settlements can mean that the development they need to accommodate change will not be able to take place, and in effect they will be condemned to a slow decay and decline. So, in both times of decline and growth of population in rural areas, planning has an important role to play in managing and guiding new development, as part of the continuing mission to ensure that life in the countryside is, in the words of Lewis Silkin, 'more convenient, and its attractiveness maintained'.

Similarities and differences in Scotland, Wales and Northern Ireland

There are a number of similarities in the profiles of the four countries – the pattern of in-migration of people to rural areas, the ageing of the rural population, a lack of affordable housing and a challenging environment for the provision of a good standard of accessible services – but there are basic differences in a number of areas such as geography – for example, according to standards set by OECD (2011a), 75 per cent of Scotland is 'predominantly rural' whereas none of England fits into that category. So, whilst the similar profiles of the countries set similar policy agendas, there are some differences in the approaches taken.

In Scotland, policies for rural development generally are to be found in *Scottish Planning Policy*, issued in 2010. This identified the 'overarching aim [as] supporting diversification and growth of the rural economy' (p. 19), but this is set within a range of objectives supporting the rural quality of life. In a reflection of the dispersed and sometimes sparse nature of the rural population, outside more accessible areas, it is noted that

> small scale housing and other development which supports diversification and other opportunities for sustainable economic growth whilst respecting and protecting the natural and cultural heritage should be supported in a range of locations. In these areas, new housing outwith existing settlements may have a part to play in economic regeneration and environmental renewal.
>
> (ibid.)

An OECD study of Scotland's rural policy published in 2009 characterised it as more sectoral than place-based, but the new policy guidance does stress that rural development policies should reflect specific local circumstances. However, there remain some distinct sectoral threads to policy which in themselves can be seen to reflect national characteristics, such as the importance of industries like fish farming or the distinctive patterns of land ownership and tenure. It is also interesting to note that current policy guidance includes a presumption against development on prime-quality agricultural land. Such a stance would have been a prominent feature of English policy in earlier years but it is not so to the same extent now.

In Wales, as in Scotland, there is a greater degree, than in England, of rurality and more communities which do not have ready access to the facilities of larger urban centres. These 'deep rural' areas generally have poor access to services, and there is evidence of concern about the sustainability of communities in these areas (WRO 2009). This greater degree of rurality can perhaps be seen as reflected in a rural policy which pays greater attention to agriculture (WRO 2004), in spite of a commitment to integrated rural development. Indeed, *Planning Policy Wales* (2012) states that in pursuing policy priorities for rural areas 'it will be essential that social, economic and environmental policies are fully integrated' (p. 51). Together with *Planning for Sustainable Rural Communities* (Technical Advice Note 6, 2010), *Planning Policy Wales* provides policy guidance for local planning authorities. It contains a presumption against development away from existing settlements and favours development in the countryside being located 'within and adjoining those settlements where it can besic best be [*sic*] accommodated'. One of the priorities for rural areas is identified as 'a thriving and diverse local economy where agriculture-related activities are complemented by sustainable tourism and other forms of employment in a working countryside' (p. 51). The importance attached to developing the rural economy is underlined by guidance in TAN 6 for 'rural enterprise dwellings'. This refers to applications for isolated rural dwellings which would otherwise be resisted but which are necessary to support the development of land-based tourism or leisure businesses. There is scope for local plans to extend this set of qualifying rural enterprises. This focus on developing rural enterprise represents a development in current policy as compared with earlier national guidance; other developments include a greater attention to the need to provide affordable rural housing and for new development to have regard to impacts on climate change.

In Northern Ireland, agriculture is a more significant industry than in other parts of the UK, in terms of employment and in GVA. In addition, farms are smaller and are almost all family farms, although there are some signs of convergence with broader UK patterns of employment, farm size and ownership. Rural Northern Ireland is experiencing some of the same counter-urbanisation pressures as the rest of the UK, although with perhaps a greater emphasis on movement for commuting to main urban employment centres or for second homes. This has impacts on rural society and environment, and these pressures along with a number of important contextual changes provide the basis for the rural element of the *Regional Development Strategy*. As well as support for agriculture and the agri-food sector, there is a strong emphasis on economic diversification across a range of land-based and other sectors. This is coupled with a desire to maintain vibrant rural communities with access to services and housing. There is something of an emphasis on focusing development in larger centres with policy focused on developing 'a network of strong main towns as the major locations providing employment, services and a range of cultural and leisure amenities for both townspeople and rural communities' (p. 100). In addition to the RDS, policy guidance is provided by PPS 21 *Sustainable Development in the Countryside*. This gives more detailed guidance and policy, within the objectives of economic diversification, supporting rural communities and environmental protection, including the importance of new development respecting the character of the countryside. It also helps identify some distinctive features of rural development, particularly the large numbers of dispersed rural communities and individual dwellings in the countryside. Policy attempts to address the impacts that such factors may have on sustainability.

As well as individual planning policy frameworks, each country has its own rural development programme for the use of EU resources. Whilst all programmes have to comply with overall EU priorities – each has measures for farming and food, environment and countryside, rural communities, and local action within the LEADER framework – there are distinctive features within each programme reflecting local economic differences and priorities.

Planning and managing rural development

The population of rural areas has been growing, which has given added pressure for development. The nature of this population growth and the fact that it is taking place in what planning has in some sense traditionally regarded as an area where development should be limited gives some particular issues to be dealt with. The nature of the migrant population – largely older and predominantly people who are making a 'lifestyle' choice, choosing to live in a rural area for environmental or social reasons – means that proposals for future development can be particularly contentious, as it can be perceived as threatening the very qualities that migrants sought in moving to the area. Rural communities have often been characterised as having a strong community spirit and institutions, but these structures will often espouse a more forward-looking, less NIMBYist approach than is prototypically associated with in-migrants. Community institutions, such as parish councils or other local partnerships, have frequently taken a lead in seeking to address issues which are firmly on the planning agenda. Matters such as access to services and the availability of affordable housing have frequently been brought to prominence by community groups and organisations. Such bodies also come close to or engage with planning processes through taking a leading role in the production of community level appraisals or strategies,[11] such as *parish plans* or *village design statements*. Rural communities have been among the pathfinders in the neighbourhood planning process which emerged as part of the Coalition government's localism agenda.

Local plans thus need to address issues of significant development pressure taking place in the context of economic and social change. Current planning guidance in the NPPF to assist the development and diversification of the rural economy in many ways merely reiterates imperatives that have long been present. Pressures posed by changing patterns of provision of private and public services and how rural quality of life can be maintained in the face of these changes provide a context for the development of planning policies to guide future development. Most

development plans for rural areas rarely cover just collections of small villages in the countryside: for the most part district-wide local plans cover a range of rural communities – villages and medium-sized towns, often in the area of influence of much larger urban communities. It is in this context that the guiding planning principle of sustainable development has to be operationalised and where the possibly conflicting imperatives of avoiding Taylor's 'sustainability trap' and seeking to promote a pattern of development which will produce more 'self-contained' communities, by concentrating growth, have to be rationalised. However, the prominence now given to these essentially socio-economic questions does not mean that the importance of countryside protection, which has been a part of the planning agenda since 1947, is fundamentally diminished. Rather, these issues have to be addressed whilst maintaining a focus on the stewardship of the countryside. The remainder of this chapter reviews the mechanisms available to aid in this objective.

The national parks

Perhaps the most notable long-term policy relating to the countryside has been the establishment of the national parks, areas of outstanding natural beauty and other areas that were designated for protection. The national parks were a response to a very long-term public demand. This stretched from the early nineteenth-century fight against enclosures, James Bryce's abortive 1884 Access to Mountains Bill and the attenuated Access to Mountains Act 1939, to the promise of the National Parks and Access to the Countryside Act 1949, an Act which, among other things, poetically provided powers for 'preserving and enhancing natural beauty'. Many battles have been fought by voluntary bodies such as the Commons, Open Spaces and Footpaths Preservation Society and the Campaign to Protect Rural England, but they worked largely in a legislative vacuum until the Second World War. The mood engendered by the war augured a better reception for the Scott Committee's emphatic statement that 'the establishment of national parks is long overdue' (1942: para. 178). The Committee had very wide terms of reference, and for the first time an overall view was taken of questions of public rights of access to the open country and the establishment of national parks and nature reserves within the context of a national policy for the preservation and planning of the countryside.

Government acceptance of the necessity for establishing national parks was announced in the series of debates on post-war reconstruction which took place during 1941 and 1943, and the White Paper on *The Control of Land Use* referred to the establishment of national parks as part of a comprehensive programme of post-war reconstruction and land use planning. Not only was the principle accepted but, probably of equal importance, there was now a central government department with clear responsibility for such matters as national parks. There followed the Dower (1945) and Hobhouse (1947) reports on national parks, nature conservation, footpaths and access to the countryside, and in 1949, the National Parks and Access to the Countryside Act, which established the National Parks Commission and gave the main responsibility for the parks to local planning authorities.

The administration of the national parks has been a matter of controversy throughout their history. Dower (1945) had envisaged that there would be ad hoc committees with members appointed in equal numbers by the Commission and the relevant local authorities. Local representation was necessary since the well-being of the local people was to be the first consideration, but the parks were also to be *national*, and thus wider representation was essential. The lengthy arguments on this issue were eventually resolved by the 1949 Act in favour of a local authority majority, with only one-third of the members being appointed by the Secretary of State. (In line with his conception of truly national parks, Dower had proposed that the whole cost of administering them should be met by the Exchequer – an idea which was never accepted.) Increasing pressures on the countryside have led to a succession of policy and legislative changes. In 1968, the Countryside Act replaced the National Parks Commission with a more powerful Countryside Commission. A Countryside Commission

for Scotland was established under the Countryside (Scotland) Act 1967. The Wildlife and Countryside Act 1981 strengthened the provisions for management agreements and introduced compensation for farmers whose rights were restricted (a major change in principle). Later there were major structural changes in the organisation of agencies responsible for countryside matters, including the establishment of a separate Countryside Council for Wales and the merging of the countryside Commission for Scotland with the Nature Conservancy for Scotland as Scottish Natural Heritage. The Environment Act 1995 established independent national park authorities, which took over the responsibilities previously exercised by local government. Circular 12/96 *Environment Act 1995, Part III: National Parks* set out advice to the newly established national park authorities (NPAs) on how they should discharge their responsibilities. NPAs are now the sole local planning authority for a national park area. In addition to the normal plans, a national park authority is required to prepare a *national park management plan* (Countryside Commission, 1997). This goes further than the scope of development plans: in addition to establishing policies, it is intended to spell out how the park is to be managed.

Since Circular 12/96 was published, there has been much new legislation enacted which changed the way in which the authorities set up to manage the parks need to operate and engage with local authorities and other key delivery partners. Consequently, a new circular was published in 2010 which updated the vision for national parks, suggested priorities and actions for the parks and codified the many changes which had taken place in the intervening fourteen years. The contents of this circular were subsequently endorsed by the incoming Coalition government. There are many diverse matters where change occurred, ranging from countryside access to consultation arrangements, from mineral workings (many of which are present in national parks) to conservation of biodiversity. Readers wishing to gain a comprehensive overview are directed to Chapter 5 of the 2010 circular. Little specific mention is made of national parks in the NPPF, other than to reinforce the importance of their role in landscape conservation, although clearly its wider provisions also apply to national parks. National Parks in England have claimed[12] that they are well placed to deliver the localism agenda, in part because they are continually seeking to balance the national purposes for parks with local needs and aspirations. This is to be achieved through their role as a planning authority, with a higher than national approval rate for planning applications at 89 per cent (for 2010) being cited as evidence of successful adoption of 'localism'.

From their inception, the national parks have had two purposes: 'the preservation and enhancement of natural beauty' and 'encouraging the provision or improvement, for persons resorting to national parks, of facilities for the enjoyment thereof and for the enjoyment of the opportunities for open air recreation and the study of nature afforded thereby' (National Parks and Access to the Countryside Act 1949, section 1). There is inevitably some conflict between these twin purposes, and the National Parks Review Panel (Edwards 1991), set up by the Countryside Commission, recommended that they be reformulated to give added weight to conservation – as did the earlier Sandford Report (1974). This argument, which continues, was a major issue in the debates on the sections of the Environment Act 1995 which established independent national park authorities. Controversy centred on the need 'to promote the *quiet* enjoyment and understanding in national parks'. Unlike in some other countries, most of the land in the national parks is in private ownership (74 per cent), and only 2 per cent is owned by the park authorities. In addition it is important to remember that there are many residents and businesses present in national parks who have similar needs to residents and businesses in other rural areas. National park authorities need to balance these local needs against the national purposes for which the parks were established. The current 'vision' for the parks, taken from the 2010 circular, is set out in Box 10.2.

There are now thirteen national parks in England and Wales, none in Northern Ireland, and two in Scotland, which are discussed separately later. The location of these national parks is shown in Figure 10.3. The Broads were for many years viewed somewhat differently from other parks. They were also proposed as a national park by the Dower Report (1945), but at that time were rejected because of their deteriorated state and the anticipated cost of management (Cherry 1975: 54). From 1968, they were managed by the Norfolk and Suffolk Broads Authority, which was initially a voluntary consortium formed by the relevant public authorities (with powers and financial resources under the provisions of the Local Government Act 1972, and with 75 per cent Exchequer funding). Discussions continued over several years among the large number of interested bodies and, in 1984, the Countryside Commission reviewed the problems of the area and the progress that had been made by the Broads Authority. Its conclusion was that, despite some achievements, the authority had not made significant improvements in water quality. Moreover, an effective framework for the integrated management of water-based and land-based recreation had not been established and the loss of traditional grazing marsh was continuing. The outcome was the designation of the area as a body of equivalent status to a national park, but with a constitution, powers and funding designed to be appropriate to the local circumstances. A new Broads Authority (with the same name as its predecessor) was established by the Norfolk and Suffolk Broads Act 1988. The duties of the authority are extensive, with very similar duties to those of the national parks and the same level of protection. It is the local planning authority and the principal unit of local government for the area. It has strong environmental responsibilities, and is required by the Act to produce a plan which has a wider remit than those required under the Planning Acts: it is more akin to a national park management plan. The separate legislation, it is argued, has advantages in dealing with the special circumstances of the Broads – the very extensive navigation of the waterways. However, the 2010 circular felt able to state that 'Whilst the National Parks and the Broads are established under two separate Acts of Parliament, the similarities between them are such that this circular has been produced to apply equally to them all' (p. 3).

BOX 10.2 VISION FOR THE ENGLISH NATIONAL PARKS AND THE BROADS

By 2030 English National Parks and the Broads will be places where:

- There are thriving, living, working landscapes notable for their natural beauty and cultural heritage. They inspire visitors and local communities to live within environmental limits and to tackle climate change. The wide range of services they provide (from clean water to sustainable food) are in good condition and valued by society.
- Sustainable development can be seen in action. The communities of the Parks take an active part in decisions about their future. They are known for having been pivotal in the transformation to a low carbon society and sustainable living. Renewable energy, sustainable agriculture, low carbon transport and travel and healthy, prosperous communities have long been the norm.
- Wildlife flourishes and habitats are maintained, restored and expanded and linked effectively to other ecological networks. Woodland cover has increased and all woodlands are sustainably managed, with the right trees in the right places. Landscapes and habitats are managed to create resilience and enable adaptation.
- Everyone can discover the rich variety of England's natural and historic environment, and have the chance to value them as places for escape, adventure, enjoyment, inspiration and reflection, and a source of national pride and identity. They will be recognised as fundamental to our prosperity and well-being.

Source: English National Parks and the Broads: UK Government Vision and Circular 2010, DEFRA, March 2010, p. 5

Landscape and countryside designations

Both the Dower and Hobhouse Reports proposed that, in addition to national parks, certain areas of high landscape quality, scientific interest and recreational value should be subject to special protection (see Table 10.1). These areas were not considered, at that time, to require the positive management which it was assumed would characterise national parks, but 'their contribution to the wider enjoyment of the countryside is so important that special measures should be taken to preserve their natural beauty and interest' (Hobhouse 1947). The Hobhouse Committee proposed that such areas should be the responsibility of local planning authorities, but would receive expert assistance and financial aid from the National Parks Commission. In total, fifty-two areas, covering some 26,000 square

kilometres, were recommended, including the Breckland and much of central Wales, long stretches of the coast, the Cotswolds, most of the Downland, the Chilterns and Bodmin Moor (Cherry 1975: 55). The 1949 Act did not contain any special provisions for the care of such areas, the powers under the Planning Acts being considered adequate for the purpose. It did, however, give the Commission power to designate *areas of outstanding natural beauty* and provided for Exchequer grants on the same basis as for national parks. Thirty-nine areas have been designated in England and Wales, covering almost 20,000 square kilometres (roughly 15 per cent of the area of England, and 4 per cent of Wales). In Northern Ireland, there are eight areas of outstanding natural beauty (AONBs), covering 3,400 square kilometres. The Scottish equivalent of AONBs number forty and cover 10,000 square kilometres. Areas of outstanding natural beauty

Table 10.1 National parks, areas of outstanding natural beauty and national scenic areas

	Number	Area (km²)	Percentage of total land area
England			
National parks	10	12,126	9
Areas of outstanding natural beauty*	33	18,741	15
Northern Ireland			
Areas of outstanding natural beauty	8	3,414	25
Scotland			
National parks	2	5,680	7
National scenic areas	40	10,018	13
Wales			
National parks	3	4,129	20
Areas of outstanding natural beauty*	5	727	4

Note: * The Wye Valley AONB straddles England and Wales; it has an area of 326 square kilometres and is not included in the above figures

are, with some notable exceptions, generally smaller than national parks.

AONBs are designated solely for their landscape qualities, for the purpose of conserving and enhancing their natural beauty. Whilst they do not have a duty to promote recreation, they should be used to meet the demands for recreation insofar as this is compatible with their other objectives. In pursuing the primary objective of designation, account needs to be taken of the need to safeguard agriculture, forestry, other rural industries, and of the economic and social needs of local communities. They are the responsibility of local planning authorities, which have powers for the 'preservation and enhancement of natural beauty' similar to those of park planning authorities. The Countryside and Rights of Way Act 2000 brought in new measures to help protect AONBs further. The role of local authorities was clarified and this now includes the preparation of management plans to set out how they will care for their AONBs. The Act also introduced the option to create conservation boards to manage AONBs. There are currently two of these boards, in the Cotswolds and the Chilterns. There are local AONB partnerships in all AONBs led by local

authorities and including a wide range of key organisations, dedi-cated to the conservation of these areas. Staff teams funded mainly by the local authorities and Natural England – the body responsible for designating AONBS and advising government on how they should be protected and managed – are based locally to coordinate and deliver action on the ground.

In addition to AONBs, there are many local authority designations designed to assist in safeguarding areas of the countryside from inappropriate development; some of these have been given additional status through inclusion in development plans. Although these are like AONBs in that they involve the application of special criteria for control in sensitive areas, they do not imply any special procedures for development control. PPS 7 took a strong line against local designation which may unduly restrict sustainable development and economic activity without identifying the particular features of the local countryside which need to be respected or enhanced. It stated:

Local landscape designations should only be maintained or, exceptionally, extended where it can

National Parks Family

1. Cairngorms
2. Loch Lomond
   and the Trossachs
3. Northumberland
4. Lake District
5. Yorkshire Dales
6. North York Moors
7. Peak District

8. Snowdonia
9. Broads
10. Pembrokeshire Coast
11. Brecon Beacons
12. Exmoor
13. South Downs
14. New Forest
15. Dartmoor

Areas of Outstanding Natural Beauty

16. Northumberland Coast
17. Solway Coast
18. North Pennines
19. Arnside and Silverdale
20. Forest of Bowland
21. Nidderdale
22. Howardian Hills
23. Anglesey
24. Clwydian Range
25. Lincolnshire Wolds
26. Lleyn
27. Shropshire Hills
28. Cannock Chase
29. Norfolk Coast
30. Gower
31. Wye Valley
32. Malvern Hills
33. Cotswolds
34. Chilterns
35. Dedham Vale
36. Suffolk Coast and Heaths
37. Mendip Hills
38. North Wessex Downs
39. Surrey Hills

40. Kent Downs
41. Isles of Scilly
42. Cornwall
43. North Devon
44. Tamar Valley
45. South Devon
46. Quantock Hills
47. Blackdown Hills
48. East Devon
49. Dorset
50. Cranborne Chase and
    West Wiltshire Downs
51. Isle of Wight
52. Chichester Harbour
53. High Weald
54. Sperrins
55. Binevenagh
56. Causeway Coast
57. Antrim Coast and Glens
58. Lagan Valley
59. Strangford Lough
60. Lecale Coast
61. Mourne
62. Ring of Gullion

National Parks Family

Areas of Outstanding
Natural Beauty

Figure 10.3 National parks

Source: Association of National Park Authorities

be clearly shown that criteria-based planning policies cannot provide the necessary protection.

(para. 25)

This enthusiasm for criteria-based policies is maintained in the NPPF but it does not have the overt objections to local designations. Local designations need to be viewed as part of a hierarchy of designations, in which they are placed behind those of international and national importance, but there is tacit recognition of their value in protecting biodiversity through functions such as the protection of wildlife corridors.

Hedgerows

A significant feature of the countryside landscape is the hedgerows. Mainly because of changing agricultural practices, there has been a dramatic loss of hedgerows: since 1947 over half of these have disappeared. Although hedges are protected (under section 97 of the Environment Act 1995 and the Hedgerow Regulations 1997 (SI 1997/1160), made under the provisions of this Act), this protection is limited. The Regulations offer protection to 'important' hedgerows, but defining an important hedgerow can be complex: for example, amongst a number of criteria are that it 'marks a pre-1850 parish or township boundary, marks the boundary of, or is associated with, a pre-1600 estate or manor or contains certain species of birds, animals or plants'.[13] There are problems for LPAs in having information available to make judgements – for example, at certain times of the year it will be difficult to judge what species are present in a hedgerow. In recognition of the nature of these potential difficulties, the then minister, Michael Meacher, set up a Hedgerow Review Group in the same year as the Regulations were enacted. It recommended in 1998 that, in the long term, the statutory provisions should be amended, but changes have yet to be made. A survey of the effectiveness of the Regulations carried out by CPRE in 2010 found 'that hedgerow protection has increased since 1998' (p. 2) but that there were significant variations in the operation of the Regulations across the country. Perhaps for this reason 'local authorities are keen for improvements

to be made to the Regulations' (ibid.). One of CPRE's conclusions was that 'the lack of a landscape criterion means that locally distinctive hedgerows are not being protected' (ibid.). No reference is made to hedgerows in the NPPF and no further guidance has yet been issued, but indications are that the pace of loss of hedgerows has been slowed.

Scottish designations

Scotland contains large areas of beautiful unspoiled countryside and wild landscape. It has the majority of Britain's highest mountains, with nearly 300 peaks of over 900 metres; it has the great majority of the UK islands and its coast is over 10,000 kilometres in length. Despite expectations to the contrary, there were, until recently, no national parks in Scotland. Though a Scottish committee (the Ramsay Committee) recommended, in 1945, the establishment of five Scottish national parks, no action followed. The reasons for this inaction were partly political and partly pragmatic (Cherry 1975: Chapter 8). A major factor was that (with the exception of the area around Clydeside and, in particular, Loch Lomond) the pressures which were so apparent south of the Border were absent.

Nevertheless, the Secretary of State used the powers of the Planning Act 1947 to issue *National Parks Direction Orders*. These required the relevant local planning authorities to submit to the Secretary of State all planning applications in the designated areas (which included Loch Lomond/Trossachs, the Cairngorms and Ben Nevis/Glencoe). In effect, therefore, in an almost Gilbertian manner, while Scotland at this time did not have any national parks, it had an administrative system which enabled controls to be operated as if it did. But, of course, this approach was inherently negative, and it was not until the Countryside (Scotland) Act 1967 that positive measures could be taken on a significant scale. This Act provided for the establishment of the Countryside Commission for Scotland – later combined with the Nature Conservancy Council for Scotland to form Scottish Natural Heritage. It also enabled the establishment of regional parks and country parks. A policy framework for these was set out in the Commission's 1974 report *A Park System for Scotland*. The report also recommended the designation of national parks in Scotland, though the term *special park* was used. Until recently, this has not been accepted, though objectives similar to those of national parks have been achieved under other designations. Despite the reluctance to establish national parks in Scotland, increased pressure for them mounted (Rice 1998) and in 1997 the Secretary of State for Scotland announced:

> National parks would be the correct way forward for Loch Lomond and the Trossachs, quite probably in the Cairngorms, and possibly in a few other areas as well. I see national parks in Scotland as integrating economic development with proper protection of the natural heritage. Scottish Natural Heritage was asked to provide advice for action by the Scottish Parliament.

This it did by undertaking wide consultations, inviting views, and commissioning further work. In 1998, Scottish Natural Heritage produced a consultation paper *National Parks for Scotland* outlining its initial ideas. These emphasised the importance of an integration of social and economic purposes along with the protection and enhancement of the natural and cultural heritage, and the enjoyment, understanding and sustainable use of natural resources. Also stressed were local community involvement in national parks, and a strong park plan 'prepared through consensus with a zoning system to help reconcile differing needs'. These and other proposals make this document an outstanding statement on participatory planning.

In 2000, the National Parks (Scotland) Act was passed, and two national parks were designated: Loch Lomond and the Trossachs (2002) and the Cairngorms (2003). Areas proposed for national parks are large in area and small in number: the Cairngorms Park is 3,800 square kilometres and the largest national park in the UK. Scottish national parks have wider powers than those south of the Border, including statutory responsibilities for the economy and rural communities. They are central government bodies and wholly funded by the Scottish Executive: 20 per cent of the membership of the two parks is directly elected and the other 80 per cent is chosen by the Secretary of State, half of whom are nominated by the constituent local authorities.

There are numerous other designations in Scotland, including forty national scenic areas, three regional parks and thirty-six country parks. The *national scenic areas* are of similar status to areas of outstanding natural beauty in England. They extend over an area of more than 1 million hectares, and include such sites as Ben Nevis and Glencoe, Loch Lomond and the World Heritage Site of the islands of St Kilda. Development control in these areas is the responsibility of the local planning authorities, which are required to consult with Scottish Natural Heritage for certain categories of development. As in England and Wales, there is an increasing concern for 'positive action to improve planning and land use management' in these areas, and for dealing with the erosion of footpaths. There is also a similar complaint about the lack of resources. A *regional park* is statutorily defined simply as 'an extensive area of land, part of which is devoted to the recreational needs of the public'. The three parks are Clyde-Muirshiel, Lomond Hills, and the Pentland Hills. The regional parks are primarily recreational areas within easy reach of Scotland's main urban areas, and each has a local plan which sets out management policies. Emphasis is laid on *integrated land management* schemes to ensure that public access is in harmony with other land uses. In this, they give effect to Abercrombie's green belt philosophy, articulated in the Clyde Valley Regional Plan. He conceived these *outer scenic areas* not only as recreational areas but also as a means of protecting the rural setting of the conurbations (Smith and Wannop 1985). Since the passing of the 1967 Act, Scottish local authorities have provided thirty-six *country parks* spread across the central belt and north-east. The parks are 'registered' with Scottish Natural Heritage, which makes grants for capital development expenditure and also towards the cost of a ranger service. Country parks are not only of direct benefit to their 11 million annual visitors, but also have a conservation objective of 'drawing off areas that are sensitive due to productive land uses and fragile wildlife habitats.'

Northern Ireland designations

Northern Ireland boasts some of the finest countryside in the UK, such of which has special value as wildlife habitat. Some of the factors contributing to these qualities have been noted earlier. Progress with planning for landscape and nature conservation has, at points, been slower than in the rest of the UK. Legislation has been seen as far less developed, and there has been much criticism about the delays in designating areas needing protection and management (Dodd and Pritchard 1993). Criticisms of the backwardness of countryside and nature conservation led to a review on behalf of the Secretary of State by Dr Jean Balfour, whose 1983 report, *A New Look at the Northern Ireland Countryside*, confirmed the low priority given to conservation and its lack of status in the work of rebuilding as part of the 'Rebuilding of Northern Ireland' (Northern Ireland Department of the Environment). The legislative basis for the Department's actions was established by the Nature Conservation and Amenity Lands (Northern Ireland) Order 1985 (SI 1985/170) (NCALO). This provided that, as well as designating the finest landscape areas as either AONBs or national parks, the Department could take steps to manage them for the purposes of both conservation and recreation. An amendment to this Order in 1989 established an advisory Council for Nature Conservation and the Countryside and extended the Department's powers, in areas such as the ability to designate nature reserves.

However, designations remained in the view of some 'pitifully slow'. In 2003, a statement on protected landscapes was issued, *Shared Horizons*, with the aim of generating 'an enhanced interest in the subject' (foreword). This set out 'the Department's intention to complete the designation of Areas of Outstanding Natural Beauty and to establish management structures in these areas that will help deliver measures to conserve and enhance their special features and to promote their enjoyment' (ibid.). It also had the stated intention of progressing the designation of National Parks in Northern Ireland. A White Paper was published in 2011 on the enabling legislation for establishing the parks, but the majority of consultation responses to the White Paper were negative, either because they were opposed to national parks in any circumstances, or because the proposals were not felt to go far enough. In 2012, the then Environment Minister, Alex Attwood, announced three potential areas where parks might be established – the Mournes, the Causeway Coast and Antrim Glens, and the Fermanagh Lakelands – but in the face of some concerted opposition from the localities the minister declared that he would 'take stock'. Progress remains uncertain. However, there are numerous AONBs (there are eight, covering 341,000 hectares), nature reserves, and *areas of special scientific interest* (the equivalent of sites of special scientific interest (SSSI)). Whilst there is a desire to promote sustainable development in the countryside, interest in countryside protection is perhaps particularly evident through concern for the problem of dispersed development in the countryside – about half of a 2013 review into the operation of PPS 21 was devoted to this topic.

The coast

A few figures underline the particular significance of the coast, and therefore of coastal planning: nowhere in the UK is more than 135 kilometres from the sea; the coastline is around 19,000 kilometres in length and the territorial waters extend over about a third of a million square kilometres. About a third of the coast of England and Wales is included in national parks and AONBs, and large areas of the coast are owned or protected by the National Trust. Following the *Enterprise Neptune* fundraising appeal (now known simply as 'Neptune' and entering its fiftieth year in 2015), the Trust protects 1,200 kilometres of the coastline.

In spite of all this protection, the pressures on the coastline have proved difficult to cope with. Indeed the complexity of responsibilities, statutes and policy covering the coast is part of the problem. Between a quarter and a third of the coastline of England and Wales is developed and this has been increasing.[14] Growing numbers of people are attracted to the coast for holidays, for recreation and for retirement. There are also economic pressures for major industrial development in certain parts, particularly on some estuaries (which have international importance for nature conservation). The problem is a difficult one which cannot be satisfactorily met simply by restrictive measures: it requires a positive policy of planning for leisure provision. This has long been accepted, and the *heritage coast* designation, introduced in 1972, implies recreational provision as well as conservation. The then Countryside Commission (now Natural England) urged that every heritage coast should have a management plan. It established a *Heritage Coast Forum* as 'a national body to promote the heritage coast concept and to act as a focus and liaison point for all heritage coast organisations'.

However, core funding for this body ceased in the mid-1990s and its demise has led to 'little ongoing sharing of best practice between the AONBs and other bodies working in the Heritage Coasts' (Land Use Consultants 2006). There is a large overlap between heritage coasts and AONBs and, to a lesser extent, national parks. Overall, 8 per cent of the total heritage coast area in England lies within these designations. So, from 1995, when the priority attached to heritage coasts by the Countryside Agency declined, the delivery of heritage coast purposes was effectively superseded by the purposes of the AONB and national park designations. There are strong similarities between these purposes (particularly the primary purpose of the conservation of natural beauty), and the focus remains on natural beauty, enjoyment and understanding, and the economic and social well-being of communities. However, heritage coast makes up around a third of the coastline of England and Wales and its Scottish equivalent comprises around three-quarters of Scotland's mainland and islands coastlines.

The Environment Committee, in its 1992 report, *Coastal Zone Protection and Planning*, complained of the lack of coordination between the host of bodies concerned with coastal protection, planning and management. In England, there were over eighty Acts which deal with the regulation of activities in the coastal zone, and as many as 240 government departments and public agencies involved in some way. In Scotland, 'in 1995 there were 79 Acts of Parliament relating to the Scottish coastal zone and marine environment

(Cleator and Irvine, 1995); and this has increased in the intervening years especially following devolution' (Firn and McGlashan 2001: 14). Not surprisingly, there have been many suggestions that action is required to simplify, rationalise, coordinate or consolidate matters.

Integration has been a theme of action by the EU, initially through the *Demonstration Programme on Integrated Coastal Zone Management* (ICZM) which ran from 1996–9. The programme involved thirty-five projects throughout the EU, including seven in the United Kingdom. The experiences and outputs of this programme led to a recommendation[15] for the pursuit of an approach that involved the coordinated application of the different policies affecting coastal zones. Town and country planning was seen as being one of a number of policy areas. A Commission document of 2007 (COM (2007) 308 Final) described the essential elements of ICZM:

> Integration across sectors and levels of governance, as well as a participatory and knowledge-based approach, are hallmarks of ICZM. Based on these principles, the EU ICZM Recommendation invites coastal Member States to develop national strategies to implement ICZM. Given the cross-border nature of many coastal processes, coordination and cooperation with neighbouring countries and in a regional sea context are also needed.
>
> (p. 4)

In 2013, the Commission issued proposals to more closely integrate maritime planning and coastal management to improve the interaction between land- and sea-based policies, particularly in the area of environmental policy. A statement by the relevant English minister, Richard Benyon, indicated a broad agreement with the proposed approach.

The approach to ICZM outlined by the EU has been followed in the UK. A joint stocktake of the frameworks for managing coastal activity was undertaken and subsequently all four administrations have produced Integrated Coastal Zone Management Strategies.[16] Such strategies have to reconcile a number of potentially conflicting activities and interests, including facilitating economic development; meeting the demands of tourism and recreation; protecting areas of scenic, geological or ecological importance; protecting vulnerable communities against the effects of erosion and flooding. There is a clear and important link between coastal management, flood prevention and planning. In the foreword to *Understanding the Risks, Empowering Communities, Building Resilience*, a strategy document on flood prevention issued by DEFRA and the Environment Agency in 2011, the objective was articulated as to 'improve the environment through managing flood and coastal erosion risk as well as avoiding or off-setting damage to protected habitats'.

Current policy guidance documents underline the value of integrated approaches. So the NPPF (para. 105) advises that local planning authorities should 'apply Integrated Coastal Zone Management across local authority and land/sea boundaries, ensuring integration of the terrestrial and marine planning regimes'. Scottish Planning Policy (para. 103) clearly outlines the role of development plans in coastal areas:

> Development plans should protect the coastal environment, indicate priority locations for enhancement and regeneration, identify areas at risk from coastal erosion and flooding, and promote public access to and along the coast wherever possible. Where relevant, development plans should also identify areas where managed realignment of the coast may be appropriate, setting out the potential benefits such as habitat creation and new recreation opportunities. Planning authorities should take the likely effect of proposed development on the marine environment into account when preparing development plans and making decisions on planning applications.

Waterways

The programme of building canals began about 250 years ago and at its peak there were about 8,000 kilometres of inland waterways, carrying 30 million tonnes of freight. However, the growth of the railways and, later, road transport sounded their death knell as a freight transport system and by the 1950s many canals had fallen into disrepair and there was a danger that many could be lost. They had been nationalised after the war and had been in the care of the British Transport Commission (BTC), but it was the work of voluntary societies such as the Inland Waterways Association that saved them from destruction. Gradually people began to see the leisure possibilities in the network of canals that remained.

In 1962, British Waterways was created and the care of the canals passed to them from the BTC. At its first meeting it approved the closure of canals in St Helens, Birmingham and Macclesfield, and there was a significant level of animosity between British Waterways and voluntary groups striving to keep waterways open and develop their recreational potential. In 2012, the waterways of England and Wales passed into the care of a charitable trust, the Canal and River Trust (in Scotland, the 220 kilometres of canals are still managed in the public sector by Scottish Canals). In its final annual report, British Waterways pointed out that its approach had changed over fifty years: 'Officials denying access and managing closure have been replaced by a team dedicated to restoring our waterways and improving their condition for future generations' (p. 4). Partly as a result, canals have become a major recreational resource, receiving over 300 million visits in 2011.

For some time, it has been government policy to maximise the economic, environmental and social benefits offered by waterways. So, for example, in 2000 a DETR document *Waterways for Tomorrow* identified a number of roles they might perform. In addition to encouraging tourism, leisure and recreation, they were seen as having an important part to play in acting as a catalyst for urban and rural regeneration, maintaining an important built and natural environment, and developing new roles in passenger and freight transport. The development of these roles needed to be supported through the planning system and *Waterways for Tomorrow* contained a commitment to 'continue to review each PPG when it is revised with the aim of developing the potential of the inland waterways through the planning system' (p. 16).

British Waterways had published in 1992 *The Waterway Environment and Development Plans* and in 2003 it published a revision, *Waterways and Development Plans*, seeking to ensure that the new plans about to be launched 'not only protect waterways and related waterspaces from inappropriate development, but also encourage their use and unlock their potential' (p. 1). The potential is seen as existing in a number of areas, all of which closely interact with planning. For some time, canalside development has played a significant role in urban regeneration, but canals and their corridors are seen as having a part to play in rural diversification and strengthening the role of small and medium-sized towns. Part of this diversification can relate to further developing the tourist potential of canals. They are also seen as acting as 'important and valuable linear habitats, which act as "green lungs" and host rare and different species' (p. 5). Clearly, local plans can not only use this potential in support of economic, social and environmental policies, but also develop policies to protect and enhance canals as an asset.

The Canal and River Trust came into being on 2 July 2012. It took over all British Waterways' statutory functions, assets, rights and liabilities in England and Wales. This included its considerable commercial property portfolio, which is intended to provide income to help finance its responsibilities. Other funding will come from a fifteen-year funding agreement with government, trading activity – such as boat licences – and donations and sponsorship. As a charitable trust charged with the stewardship of its assets 'for the public good', the CRT is developing its focus on 'helping to meet the needs of local communities'. This can be seen as continuing the wide agenda of its predecessor, much of which will relate to planning priorities in the countryside.

Public rights of way

The origin of a large number of public rights of way is obscure. As a result, innumerable disputes have arisen over them. Before the 1949 Act, these could be settled only on a case-by-case basis, often with the evidence of 'eldest inhabitants' playing a leading role. The unsatisfactory nature of the situation was underlined by the Scott (1942), Dower (1945) and Hobhouse (1947) Reports, as well as by the 1948 report of the Special Committee on Footpaths and Access to the Countryside. All were agreed that a complete survey of rights of way was essential, together with the introduction of a simple procedure for resolving the legal status of rights of way which were in dispute. The 1949 Act attempted to provide for both. This Act has been amended several times. Under the current provisions, county councils have the responsibility for surveying rights of way (footpaths, bridleways and 'byways open to all traffic') and preparing and keeping up to date what is misleadingly called a *definitive map*. The maps are supposedly conclusive evidence of the existence of rights of way but, in fact, they are not necessarily either complete or conclusive. They are incomplete because inadequate resources have been devoted to undertaking the necessary surveys, and they can be inconclusive because a map may wrongly identify a right of way. The latter is a legal matter which is not discussed here (see Chesman 1991), but the former is a continuing problem of planning policy and administration. The Countryside Agency pointed out that

> the showing of a way as a footpath does not prove that there are not, for example, additional unrecorded rights for horse riders to use the way. Nor is the fact that a way is omitted from the definitive map proof that the public has no rights over it.
>
> (Countryside Agency 2003: 11)

The definitive maps show some 225,000 kilometres of rights of way in England and Wales. There are four categories: footpaths, which can be used on foot; bridleways, for use on foot, horse or bike; restricted byways, for all traffic except motorised vehicles; and byways open to all traffic. Previously there was a category known as 'roads used as public paths' (RUPPs): these were highways mainly used by the public for similar purposes to footpaths or bridleways, but which might or might not carry vehicular rights. In some parts of the country, RUPPs were reclassified

for the public sector and for landowners. Increasing access gives rise to a range of management issues and Natural England issued a *Countryside Code* in 2012, offering advice to the public and to land managers on making visits to the countryside, whilst Scottish Natural Heritage publicised the *Scottish Outdoor Access Code* in 2010, perhaps an essential corollary to efforts to expand access.

Provision for recreation and country parks

In the early post-war years, national recreation policy was largely concerned with national parks (and their Scottish equivalents), areas of outstanding natural beauty and the coast. Increasingly, however, there developed a concern for positive policy in relation to metropolitan, regional and country parks. The 1966 White Paper *Leisure in the Countryside* recognised the growing demand for leisure activities 'in the open' and saw country parks and the like as a way of meeting that demand without greatly increasing the need for travel, while at the same time easing pressure on 'more remote and solitary places'. This formed the basis for the Countryside Act 1968, which gave additional powers to the then Countryside Commission for 'the provision and improvement of facilities for the enjoyment of the countryside' (section 2), including experimental schemes to promote countryside enjoyment. At the same time, local authorities were empowered to provide *country parks*, including facilities for sailing, boating, bathing and fishing. These country parks are not for those who are seeking the solitude and grandeur of the mountains, but for the large urban populations who are 'looking for a change of environment within easy reach'. There is now a wide range of country parks, picnic sites, visitor-interpretative sites, recreation paths, interpretative trails, cycleways and similar facilities provided by local authorities and the Countryside Agency. About two-thirds of country parks are located in the urban fringe, many are either sites of historic and/or conservation interest and nearly 80 per cent have at least one landscape planning designation. A report produced in 2003 for the Countryside Agency, *Towards a Country Park Renaissance*, voiced some concerns about the poor state of a minority of country parks and suggested an expanded agenda for their future, including roles in the promotion of health and social inclusion, and establishing good practice in sustainability and biodiversity protec-tion. The report encouraged the Agency to reverse its slackening of interest in country parks which occurred in the 1990s, such that its 2003–6 Corporate Plan identified them as 'a recreational infrastructure which can easily be enjoyed by everyone' and 'a vibrant and diverse urban fringe providing a better quality of life'.[18]

The natural environment, including what we might think of as 'countryside' but also a range of other green open spaces, is a significant recreational resource and it has been estimated (Natural England 2013) that around two-fifths of the population make regular visits to the outdoors. However, the social distribution of visitors is significantly skewed, with people living in the areas of greatest deprivation least likely to have visited the natural environment in the previous seven days, whilst those living in the least deprived areas were significantly more likely to have made a visit. Findings such as these feed into a broader public health agenda which is increasingly being seen as an important focus for planning (Barton 2009). The NPPF notes that 'the planning system can play an important role in facilitating social interaction and creating healthy, inclusive communities' (para. 69). However, this slightly negative note should not overshadow the fact that the positive economic impact of visitors to the countryside is substantial: recreation and tourism make a significant contribution to the rural economy and total spending by all visitors was £9 billion in 1994 and £9.7 billion in 2003. Total employment supported by visitor activity has been estimated at a third of a million jobs (RDC 1997). Tourism employment is estimated to account for 15 per cent of employment in rural areas such as the far south-west of England and the west of Scotland (ONS 2012b).

For the most part, the impact of visitors on the environment is of manageable proportions, though there are conflicts in specific areas. As the Environment

Committee pointed out in its 1995 report *The Environmental Impact of Leisure Activities*, there are difficult problems of overcrowding, overuse and conflict of activities in certain areas. The favourite answer is 'good management', and there is no doubt that this can help in preventing visitors 'loving to death' the beauty spots they wish to visit (to use the apt phrase adapted as the title of a report on sustainable tourism in Europe by the Federation of Nature and National Parks). So can 'countryside codes', 'visitor awareness' campaigns and such like. However, traffic problems certainly can considerably reduce the 'quality of the recreational experience'; people with access to a car are twice as likely to visit the countryside. Measures to manage the impact of the car are important in sensitive environments. The Peak District, which is within easy access of about 20 per cent of England's population, can come under significant pressure and has developed a range of traffic management processes, including promoting the use of more sustainable forms of transport, such as cycling and the use of public transport, small-scale park-and-ride schemes, and in the Upper Derwent Valley, road closures accompanied by subsidised bus services at the busiest periods.

It has been suggested that the growth in the provision of country parks, as part of a growth in informal countryside recreation, led to, or at least affected, the decline of urban parks.[19] A report by Comedia/Demos in 1995, *Park Life: Urban Parks and Social Renewal*, noted that 'the declining quality of Britain's urban parks and open spaces is now a matter of extensive public concern, and is part of a wider fear that we can no longer manage safety and well-being in public spaces' (Greenhalgh and Worpole 1995: 3). There was something of a flurry of activity after this report and the ODPM issued *Living Places: Cleaner, Safer, Greener* in 2002, which put parks into the context of the public realm as a whole. There was a significant growth of awareness of the importance of parks in planning terms. National agencies such as English Heritage added nationally important public parks and open spaces to the Register of Parks and Gardens of Special Historic Interest. Local authorities included policies for the protection of parks and gardens in their development plans. PPG 17 *Planning for Open Space, Sport and Recreation* required audits of open space in terms of quality and quantity and for local authorities to plan strategically for their provision and management. This encouragement is continued in the NPPF, in which a specific link is made between access to open spaces and the health and well-being of communities. Recently, there has been a growth in good practice in terms of repairing public parks. This was stimulated by the Heritage Lottery Fund, which began funding repairs to public parks in 1996. Through initiatives such as Heritage Lottery Funding and listing on English Heritage's register of parks and gardens of historic interest, historic designs are increasingly being re-evaluated and appreciated. The Heritage Lottery Fund helped many local authorities restore their historic parks. This restoration has been the catalyst for the original, often Victorian, designs to be dusted off, reconsidered and brought back to their former glory. In *Park Life Revisited* published in 2013, Ken Worpole, one of the authors of the 1995 report, felt able to point to the contribution of public parks in 'raising the quality of the everyday townscape and sense of well-being' (p. 7) and 'the invisible but crucial environmental benefits which green space provides in terms of flood relief, heat-moderation, supporting bio-diversity and urban connectivity' (p. 8).

One particularly interesting initiative which started in the early 1980s and is now well established is the work of the *Groundwork Trusts*. Conceived as an *additional* resource for converting wasteland to productive uses, particularly in urban fringe areas, it facilitates cooperative efforts by voluntary organisations and business, as well as public authorities. The enterprise is wide-ranging and covers land reclamation, landscaping and environmental appreciation, as well as provision for recreation and many other activities seen as desirable and worthwhile in local communities. Groundwork works on 'thousands of projects across the UK each year' and these include major projects such as national grant schemes and programmes of community work funded by corporate partners, as well as smaller and locally important schemes. Examples include the development of the Taff Trail, which is a long-distance footpath and cycleway linking Cardiff and Brecon; recreating wildlife sanctuaries and access

Plate 10.1 Roberts Park, Saltaire – restored with National Lottery funding (photos courtesy of Hilary Taylor of HTLA Historic Landscape Consultants)

around mining villages in east Durham; the development of the Middleton Riverside Park on a totally derelict site a few miles from Manchester; and a programme which encourages owners of industrial and commercial premises 'to stand back and take a look at the external image of their premises and then to make practical landscape improvements' (Jones 1999).

There have been many forces driving the increased demand and supply of rural recreation, ranging from increased leisure time and wealth to imperatives for agricultural diversification. As has been illustrated above, leisure and tourism can bring substantial employment benefits to rural localities. Tourism also brings more indirect benefits to rural areas, such as the survival of bus services and village shops. There is, of course, a cost to be borne for these advantages – in terms of changed character and conflicts between the interests of visitors and residents (particularly in areas which have attracted new residents). Such conflicts can be reduced by 'good management', but they are inherent in the dynamics of social and economic change. Such conflicts and efforts to manage them are often a central issue to be considered when designing planning policies for a locality.

Green belts

The NPPF states that the 'fundamental aim of Green Belt policy is to prevent urban sprawl by keeping land permanently open' (para. 79). In this it echoes the purpose set for green belts in earlier guidance. However, green belts cover over 10 per cent of land in the UK – up to 16 per cent in Northern Ireland – so their role in protecting the countryside is significant, as is their role in providing accessible green space for recreational purposes. Additionally, a report by Natural England and CPRE (2010) *Green Belts: A Greener Future* emphasised the role of green belts in nature conservation and in the development of green infrastructure. Perhaps for reasons of their simplicity, green belts are among the most widely understood and enthusiastically supported of planning policies.

Green belt policy formally emerged in 1955, although the idea first gained currency in the 1920s and designations were made in 1938 around London, where the process was assisted by Act of Parliament, and Sheffield, where it was designated by local government. Unusually, the policy can be identified with a particular minister – Duncan Sandys (who later made another contribution to planning with the promotion of the Civic Trust and the Civic Amenities Act). Sandys' personal commitment involved disagreement with his senior civil servants, who advised that it would arouse opposition from urban local authorities and private developers, who would be forced to seek sites beyond the green belt. Experience with the Town Development Act 1952 (which provided for negotiated schemes of 'overspill' from congested urban areas to towns wishing to expand) did not suggest that it would be easy to find sufficient sites. Sandys, however, was adamant, and a circular was issued asking local planning authorities to consider the formal designation of clearly defined green belts wherever this was desirable in order to check the physical growth of a large built-up area, to prevent neighbouring towns from merging into one another, or to preserve the special character of a town.

The policy had widespread (and long-lasting) appeal to county councils, which now had another weapon in their armoury to fight expansionist urban authorities, but also more widely. One planning officer commented that 'probably no planning circular and all that it implies has ever been so popular with the public. The idea has caught on and is supported by people of all shades of interest' (quoted by Elson 1986: 269). Another noted that

> the very expression *green belt* sounds like something an ordinary man may find it worthwhile to be interested in who may find no appeal whatever in 'the distribution of industrial population' or 'decentralisation' . . . Green belt has a natural faculty for engendering support.
>
> (ibid.)

However, initially, its biggest support came from the planning profession, which in those days still saw planning in terms of tidy spatial ordering of land uses. Desmond Heap, in his 1955 presidential address to the (then) Town Planning Institute, went so far as to declare that the preservation of green belts was 'the very *raison d'être* of town and country planning'. Their popularity, however, has not made it any easier to reconcile conservation and development. The green belt policy commands even wider support today than it did in the 1950s. Elson concluded his 1986 study with a discussion of why this is so:

> It acts to foster rather than hinder the material and non-material interests of most groups involved in the planning process, although it may be to the short term tactical advantage of some not to recognise the fact. To *central government* it assists in the essential tasks on interest mediation and compromise which planning policy-making represents . . . To *local government* it delivers a desirable mix of policy control with discretion. To *local residents* of the outer city it remains their best form of protection against rapid change. To the *inner city local authority* it offers at least the promise of retaining some economic activities that would otherwise leave the area; and to the *inner city resident* it offers the prospect, as well as often the reality, of countryside recreation and relaxation. To the *agriculturist* it offers a basic form of protection against urban influences, and for the *minerals industry* it retains accessible, cheap, and exploitable natural resources. *Industrial developers* and *house builders* complain bitterly about the rate at which land is fed into the development pipeline, yet at the same time are dependent on planning to provide a degree of certainty and support for profitable investment. Planning may be an attempt to reconcile the irreconcilable, but green belt is one of the most successful all-purpose tools invented with which to try.
>
> (Elson 1986: 264)

The NPPF reiterates the purposes for green belts set out in earlier policy statements, namely that they should check the unrestricted sprawl of large built-up areas prevent neighbouring towns merging into one another; assist in safeguarding the countryside from encroachment; preserve the setting and special

character of historic towns; and assist in urban regeneration, by encouraging the recycling of derelict and other urban land. It emphasises that 'inappropriate development' should be resisted in the green belt. What could be considered appropriate development is set out in paragraphs 89 and 90. A list of around a dozen items includes outdoor facilities (sport, recreation, cemeteries) which do not impact on the open nature of the green belt; limited development of various sorts (extensions, infilling in villages); redevelopment or reuse of previously developed sites; mining, engineering operations and certain infrastructure; and development brought forward under a Community Right to Build Order.

The impact of the policy has been the subject of study, particularly in the early years. Elson *et al.*'s (1993) study concluded that green belts had been successful in checking unrestricted sprawl and in preventing towns from merging. In this context, the more recent study by Natural England and CPRE referred to above noted (2010: 32):

> In terms of development, the evidence suggests that designation does have a significant impact on the rate of built development which is much lower in the Green Belt than for urban areas and a third less than the Comparator Areas. Without the Green Belt designation it is likely that the rate of development would have been much higher leading to a loss of undeveloped land and the openness protected by Green Belt policy.

The relationship between the policy and the preservation of the special character of historic towns proved much more difficult to evaluate. Though the idea had 'a well-established pedigree', and though the green belt boundaries were particularly tight, there was little evidence to connect policy and outcomes. It was difficult also to assess how far green belts had assisted in urban regeneration. Though the policy did 'focus development interest on sites in urban areas', local authorities tended to regard the creation of jobs as more important than any land development objective per se. Indeed, urban regeneration was often seen as requiring the selective release of employment sites in the green belt. The supply of adequate sites within urban areas was not felt to be sufficient for development needs, and refusal to allow development on the periphery of an urban area could lead to leapfrogging beyond the green belt, or development by the intensification of uses in towns located within the green belt.

Data for England also suggests that the policy has been successful. In the fifteen-year period since 1997, less than 0.1 per cent of the 1.6 million hectares of land within the green belt was lost to development or changes in boundaries introduced in local plans. The local planning system has been supported by the appeal system, which has strongly upheld policy. However, this has not meant that there has not been criticism of the policy. A report issued by the Regional Studies Association (Herington 1990) identified a number of issues, including that green belts had restricted economic development, forced too much growth on towns and villages beyond the green belt and done little to improve the appearance or access to the open countryside. It was also the case that, since their inception, other policy measures had emerged to help manage development in the countryside. More recently, criticism has focused on the economic impacts of the green belt, in relation to their effect on housing supply. So, in its 2011 survey of the UK economy, the OECD stated: 'Green Belts constitute a major obstacle to development around cities, where housing is often needed. Replacing Green Belts by land-use restrictions that better reflect environmental designations would free up land for housing, while preserving the environment' (2011b: para. 19). Similarly, also in 2011, the Institute of Directors voiced the opinion that 'greater land release [in the Green Belt] could also lead to lower land and house prices and greater affordability' (IOD 2011: section 5). Perhaps surprisingly, the RTPI has expressed some doubts about the status of Green Belts. In a statement issued in 2007, it stated:

Green belt policy is now tired. It came in as a vital and innovative planning tool sixty years ago limiting urban sprawl but it now needs to be sharpened up. It is the planning equivalent of a rusting lawnmower, its creaking blades no longer up to the job.

It goes on to advocate that the policy should be reviewed and that we 'should be building on the green belt where it makes environmental sense to do so and still protecting it where this makes more environmental sense'.

However, any such attempts to weaken green belt designation tend to attract strong public resistance. In the debate over the introduction of the NPPF, one of the matters that excited concern was the possible impact on the green belt of the 'presumption in favour of sustainable development'. Similarly, the attempt by the Conservative government in the late 1980s to relax green belt rules to allow a ring of new villages around London proposed by a consortium of housing developers famously backfired when a network of commuter-belt residents mobilised to apply political pressure and changed the government's mind (Thornley 1993).

There is something of an uneven pattern of green belt designation in the devolved administrations. Wales has but one area, between Newport and Cardiff, and this is relatively recent in origin, but there has been pressure for more designations. Northern Ireland, as has been mentioned, has about thirty areas of green belt covering around 16 per cent of its land area. In Scotland, green belts have been established around Aberdeen, Greater Glasgow, Ayr and Prestwick, Edinburgh, Falkirk and Grangemouth, and Stirling and in Clackmannanshire, East Lothian and Midlothian. Interestingly, the Dundee green belt was replaced by a general countryside policy in 1989 (Herington 1990: 22) but recent local plans for Dunfermline, Perth and St Andrews proposed new areas of green belt. The consultation on the revision of Scottish Planning Policy (2013) sets out a somewhat different perspective on green belts, which are seen as supporting spatial strategy by directing development to the most appropriate locations and supporting regeneration, protecting and enhancing the character, landscape setting and identity of the settlement, and protecting and providing access to open space. It also notes that 'For most settlements a green belt is not necessary as other policies can provide an appropriate basis for directing development to the right locations' (para. 52), somewhat echoing the view taken in Dundee in the 1980s.

Recently, we have seen the emergence of the term 'green infrastructure' (GI). The NPPF defines green infrastructure as a 'network of multi-functional green space, urban and rural, which is capable of delivering a wide range of environmental and quality of life benefits for local communities' (Annex 2: Glossary) and it features in guidance in the context of climate change. Whilst the NE/CPRE document discusses the role of green belts in action for climate change generally, as Thomas and Littlewood (2010: 218) point out, there is little engagement between the two concepts and 'There is a notable avoidance of formal documentary engagement with the green belt in the recent GI strategy statements.' However, in England and Wales, the debate on green belts has largely ignored the issue of managing the countryside within green belts, but this may partly reflect the fact that this does not figure significantly in the public debate, where the overwhelming concern is with preventing development.

Figure 10.4 Green belts in the UK

Countryside grant programmes

Changes in policies relating to farming have had a more tangible effect than the heated arguments of those for and against freer countryside access. The lower priority for food production led to attempts to broaden the role of landowners as 'managers' of the countryside. Within this changing framework, 'access becomes a means of diversifying the agricultural economy'. Successive measures have reflected the changing priorities, and there has been increased emphasis on the role of farmers as 'stewards of the countryside', which has led to a greater concentration of funds on environmental schemes.[20]

The Countryside Agency took the lead in promoting conservation and recreation as explicit objectives of agricultural policy. Its 1989 policy statement *Incentives for a New Direction in Farming* argued that the diminishing need for agricultural production provided an opportunity for 'environmentally friendly' farming. It presented a menu of incentives for farmers and landowners to provide environmental and recreational benefits. These ideas were translated into the *countryside premium*: an experimental scheme which gave incentives for land to be set aside for recreation. It was followed by the *countryside stewardship*, which provides incentives for the protection and enhancement of valued and threatened landscaped. This scheme proved to be a successful one: in its first four years, some 5,000 agreements were concluded, covering 91,000 hectares. By the time applications closed in 2004, when countryside stewardship was replaced by the *Environmental Stewardship* scheme, there were about 3,700 agreements, managing around 100,000 hectares. Parallel schemes operated in Wales and Scotland.

The Agriculture Act 1986 made provision for *environmentally sensitive areas* (ESAs), where annual grants were given by MAFF to enable farmers to follow farming practices which would achieve conservation objectives. The introduction of ESAs marked a fundamental policy change (Bishop and Phillips 1993: 325). They provide financial support for practices which result in environmental benefits, in contrast to earlier schemes which gave compensation for forgone profits. ESAs developed into the main plank of the ministry's countryside protection policy. As part of changes to the range of schemes, ESAs now operate as part of the wider Environmental Stewardship scheme, but by the time applications closed in 2004 there were twenty-two ESAs covering 10 per cent of agricultural land in England, or 3,600 agreements covering about 269,000 hectares. The government has also operated grant schemes for farm diversification and for farm woodlands. The objectives were to enhance the farmed landscape and environment and to encourage a productive land use alternative to agriculture.

During the 1990s, new schemes were introduced at a bewildering rate: Box 10.3 gives an overview of the schemes designed to encourage environmentally friendly farming and public enjoyment of the country-side. It is apparent that a determined effort has been made to offset some of the effects of the stimulation to excess production and degradation of the countryside. The effectiveness of these various schemes is a matter of controversy (Winter 1996; Adams 1996) and at times the range of schemes in place was complex. The 2004 *Rural Strategy* promised to cut down the one hundred or so rural funding schemes to three major programmes – rural regeneration, agriculture and food industry regeneration, and natural resource protection – and some progress has certainly been made on ration-alisation. The agri-environmental element of funding was addressed by the 2002 *Strategy for Sustainable Food and Farming: Facing the Future*. The Strategy not only noted the progress made on slowing the deterioration in the quality of the rural environment, but also recog-nised that despite the introduction of legislation pro-tecting habitats, stronger planning policies and funding to encourage more environmentally friendly practices,

> significant problems remain, including continued attrition of the historic environment, serious overgrazing in some upland areas, declines in the population of widespread species and the loss of biodiversity within some surviving habitats.
>
> (p. 27)

Much of this activity has been prompted and funded by EU regulations and programmes. The EU supports a diverse range of agri-environmental measures, all of which have two broad objectives: reducing environmental risks associated with modern farming on the one hand, and preserving nature and cultivated landscapes on the other hand. Payments are given to farmers in return for the service of managing the environment, but this has to be in a way which goes beyond normal good farming practice. The idea was adopted by the European Community in 1985 but remained optional for member states. In 1992, it was introduced for all member states as an 'accompanying measure' to the CAP reform. It became the subject of a dedicated Regulation (EU 2078/92), and member states were required to introduce agri-environment measures 'throughout their territory'. In 1999, the provisions of the Agri-environment Regulation were incorporated into the Rural Development Regulation (EC 1257/1999) as part of the Agenda 2000 CAP reform. The area of agri-environment has developed into a key part of EU rural development policy, and it is now the only compulsory measure for member states in their rural development plans. These rural development measures are also integrated with other EU environmental measures. An example here would be the 1991 Nitrates Directive (Directive 91/676/EEC), which aims to protect water quality by preventing nitrates from agricultural sources polluting ground and surface waters and by promoting the use of good farming practices. If farmers fail to comply with the rules in *nitrate vulnerable zones*, they may be prosecuted and fined and may not be entitled to a full subsidy payment under the Single Payment Scheme (SPS) and other direct payments such as the Environmental Stewardship schemes. Programmes such as these are key means of moving towards central environmental objectives and as such are part of the context for spatial planning.

Nature conservation

The concept of wildlife sanctuaries or nature reserves is one of long standing and, indeed, it antedates the modern idea of national parks. In other countries, some national parks are in fact primarily sanctuaries for the preservation of big game and other wildlife, as well as

BOX 10.3 SOME ENVIRONMENTAL SCHEMES IN ENGLAND

(A similar range of schemes exists in other jurisdictions)

Environmental Stewardship

This scheme, funded through the Rural Development Programme and operated by Natural England, has replaced/absorbed a number of other designations, such as *Countryside Access, Countryside Stewardship, environmentally sensitive areas* and *the Moorland Scheme*, but some of these have agreements in place which have yet to be completed (many run for ten years). Its primary objectives are to conserve wildlife (biodiversity); maintain and enhance the landscape; protect the historic environment; promote public access and understanding of the countryside; protect natural resources; prevent soil erosion and water pollution; and support environmental management of uplands areas. There are three elements:

- Entry Level Stewardship, open to all farmers and landowners, promoting effective land management to maintain land in good agricultural and environmental condition: payments can be £30 per hectare of qualifying land but with more in some upland areas.
- Organic Entry Level Stewardship, open to all farmers registered with an organic control body: payments are higher than for basic ELS.
- Higher Level Stewardship involves more complex types of management aimed at meeting specific local targets: payments can range from £25 to £590 per hectare, depending on what is being delivered.

By the end of 2012 there were about 45,000 Environmental Stewardship agreements covering about 6 million hectares.

English Woodland Grant schemes

Grants are available from the Forestry Commission for woodland planting, woodland creation, woodland improvement, woodland regeneration, woodland management and the sustainable production of wood fuel.

Nitrate vulnerable zones

These are designated on all land draining into, and contributing to the nitrate pollution in, "polluted" waters. For farmers in the designated zones to qualify for full payments under the Single Payment Scheme of the CAP they must comply with rules for managing nitrogen use and managing land generally. Requirements cover areas such as risk assessment, planning use of manure and nitrogen fertiliser, providing adequate storage for manure and keeping records of nitrogen use.

Catchment-sensitive farming

A joint project between the Environment Agency and Natural England covering eighty priority catchments, this covers the entire range of agricultural pollutants and their impact on sensitive aquatic habitats. There are capital grants to support the improvement or installation of facilities that would benefit water quality by reducing diffuse pollution from agriculture, to meet the standards of the EU Water Framework Directive.

for the protection of outstanding physical features and areas of outstanding geological interest. British national parks were somewhat different in origin, with an emphasis on the preservation of amenity and providing facilities for public access and enjoyment (though, as was noted earlier, the trend has been to give increasing priority to conservation). The concept of nature conservation is primarily a scientific one, concerned particularly with the management of natural sites and of vegetation and animal populations. The Huxley Committee argued in 1947 that there was no fundamental conflict between these two areas of interest:

> their special requirements may differ, and the case for each may be presented with too limited a vision: but since both have the same fundamental idea of conserving the rich variety of our countryside and sea-coasts and of increasing the general enjoyment and understanding of nature, their ultimate objectives are not divergent, still less antagonistic.

However, ensuring that recreational, economic and scientific interests are all fairly met presents some difficulties. Several reports dealing with the various problems were published shortly after the war (Dower 1945; Huxley 1947; Hobhouse 1947). The outcome was the establishment of the Nature Conservancy, which was later replaced by English Nature, Scottish Natural Heritage and the Countryside Council for Wales. In Northern Ireland, Scotland and Wales, central responsibility for nature conservation and access to the countryside rests with a single body.

Legislation often emerges as a response to new perceptions of problems; but sometimes legislation itself fosters such perceptions. So it was with the Wildlife and Countryside Act 1981. Introduced as a mild alternative to the Labour government's aborted Countryside Bill (and stimulated by the need to take action on several international conservation agreements), the Conservative government expected no serious trouble over the Bill. It was very much mistaken: the Bill acted as a lightning rod for a host of countryside concerns that had been building up over the previous decade or so – moorland reclamation, afforestation and 'new

agricultural landscapes', loss of hedgerows, damage to SSSIs, and such like. The Bill had a stormy passage through Parliament, with an incredible 2,300 proposed amendments. Though most of these failed, the Bill was considerably changed in the process. The major focus of argument (with the strong National Farmers' Union and the Country Landowners' Association holding the line against a large but diffuse environmental lobby) was the extent to which voluntary management agreements could be sufficient to resolve conflicts of interest in the countryside. The government steadfastly maintained that neither positive inducements nor negative controls were necessary. Indeed, it was held that controls would be counterproductive in that they would arouse intense opposition from country landowners.

Of particular concern was the rate at which SSSIs were being seriously damaged, the speed at which moorland in national parks was being converted to agricultural use or afforestation, and the adverse impact of agricultural capital grants schemes both on landscape and on the social and economic well-being of upland communities. On the first issue, the government finally made a concession and provided for a system of 'reciprocal notification'. This required the then Nature Conservancy Council (NCC) to notify all landowners, the local planning authority and the Secretary of State of any land which, in their opinion, 'is of special interest by reason for any of its flora, fauna, or geological or physiological features', and 'any operations appearing to the [NCC] to be likely to damage the flora or fauna or those features'. Landowners were required to give three months' notice of intentions to carry out any operation listed in the SSSI notification. This was intended to provide the NCC with an opportunity 'to discuss modifications or the possibility of entering into a management agreement'. This much vaunted voluntary principle did not work: sites were damaged while consultations were under way. Amending legislation, passed in 1985, was designed to prevent this. Subsequent legislation – the Countryside and Rights of Way Act 2000 and the Natural Environment and Rural Communities Act 2006 – has further strengthened the role of Natural England in enforcement: if prosecution is deemed to be the route to be followed, unlimited fines can be imposed in the Crown Court for the most serious cases (see Box 10.4 for further information on SSSIs).

Another issue of contention arose when, during the debates, attempts to extend grants from 'agricultural business' to countryside conservation were defeated, and a host of amendments divided the Opposition and confused the issues. As passed, the amendments did little more than exhort the Minister of Agriculture, when considering grants in areas of special scientific interest, to provide advice on 'the conservation and enhancement of natural beauty and amenities of the countryside' and suchlike 'free of charge'. There is, however, power to refuse an application for an agricultural grant on various 'countryside' grounds, but such a refusal rendered the objecting authority (the county planning authority in national parks and the NCC in SSSIs) liable to pay compensation. This was a return to the pre-1947 planning system (even though it applied to only a small part of the country) and it did, not surprisingly, give rise to a considerable amount of debate. William Waldegrave (then Minister for the Environment, Countryside and Local Government) argued:

> I think that the moral and logical position of the farmer who finds that his particular bit of flora or fauna is now rare is such that there should be no hesitation in saying that he deserves public money if he is asked to do better than those who have been allowed to extinguish their bits, and if it is expensive for him to do so. I believe such flows of money, from taxpayer to land-user, for conservation expenses, are thoroughly justified and should become a useful and permanent adjunct to farm incomes for quite a considerable number of farmers, often in the rather more marginal farming areas where the inherent difficulty of farming has prevented our predecessors from extirpating species which may have gone for good elsewhere.
>
> (Waldegrave *et al.* 1986: 10–11)

The issue here goes much further than appears at first sight, since it raises questions about the ownership of development rights. While post-war planning legislation nationalised rights of development of land, it effectively excluded agriculture and forestry. The owner of a listed building receives no compensation for the restrictions which are imposed, and may even be charged for repairs deemed necessary by the local authority in default. The farmer, on the other hand, expected – and obtained – payment for 'profits forgone' in 'desisting from socially undesirable activity, or merely for departing from what is conventionally regarded as good agricultural practice'. Thus, to quote Hodge (1999), there was 'the irony of one UK government agency being obliged to buy out the subsidies being offered by another'. Moreover, there was concern that some landowners threatened changes merely as a means of extracting compensation. A few very large payments had much publicity. Not surprisingly, the Labour government objected in principle to this system and proposed that payments should be made to landowners only where this was in furtherance of conservation management.

> Public support for the proper management of SSSIs is essential and appropriate, but it should be given where positive management prescriptions are required. The Government is not prepared, in future, to pay out public money simply to dissuade operations which could destroy or damage these national assets.
>
> (DETR (1999) *Sites of Special Scientific Interest: Better Protection and Management*, p. 4)[21]

The case of the moorland of Exmoor National Park perhaps exemplifies the 'policy journey' that was made. Encouraged by grant aid and a policy drive to increase production, moorland improvement and ploughing on Exmoor during the 1960s and 1970s came to be characterised as a conflict between farmers and conservationists. It appeared that the characteristic defining feature of Exmoor – its moorland – was under threat. The Porchester Inquiry (which reported in 1977) calculated that 4,900 hectares of moorland had been lost between 1947 and 1976 and that another 1,380 hectares were at risk in the near future. The combination of a change of government in 1979 and the willingness of a few Exmoor farmers to pioneer a

BOX 10.4 SITES OF SPECIAL SCIENTIFIC INTEREST*

The nature of SSSIs

SSSIs represent some of the very best wildlife and/or geological sites and are often also designated as special areas of conservation (SACs), special protection areas (SPAs) or Ramsar sites. Many are also national nature reserves (NNRs) or local nature reserves (LNRs). SSSIs contain some of the rarest and most threatened geology and plants and animals, some of which find it more difficult to survive in the wider countryside. The habitats are often the product of management and active management to maintain their conservation interest.

Designation of SSSIs

The first SSSIs were identified in 1949 by the then Nature Conservancy. Natural England now has responsibility for identifying and protecting SSSIs in England under the Wildlife and Countryside Act 1981. Similar bodies perform this role in other administrations. The designation of SSSIs includes a two-stage process: notification of all owners and occupiers of land making up potential sites, together with a range of public agencies; and confirmation, following a period of consultation, (in the case of England) by the board of Natural England.

Protection of SSSIs

In England, SSSIs are protected under the Wildlife and Countryside Act 1981 and subsequent amending acts. Whilst there are enforcement mechanisms, Natural England sees its role as 'Maintaining goodwill and building upon the enthusiasm, knowledge and interest of owners to protect and manage the sites'. It claims to work with 26,000 owners and land managers in carrying out its role.

Examples of SSSIs

Peat bogs, felt to be unique in world terms, both in themselves and in the birds, invertebrates and plants which they support. Bolton Fell Moss in Cumbria is mostly designated as an SSSI and an SAC.

Maritime heathlands include heathers, mosses, and a range of invertebrates. The Dorset Heaths include thiry-seven SSSIs.

Limestone pavements notable for their 'paving blocks', known as clints, and crevices, known as grikes, together with some unusual vegetation. Scar Close at Ingleborough in North Yorkshire is an SSSI and also a SAC

* known as areas of special scientific interest (ASSIs) in Northern Ireland

new approach to conservation management on farmland meant that Porchester's recommended compulsory moorland conservation orders were never implemented. Instead, for a relatively short period, the Exmoor approach to moorland management agreements provided a blueprint for management agreements nationally. These voluntary agreements included requirements for maintenance of existing vegetation characteristics, adherence to the rules of good husbandry and constraints on improvement. A key aspect of the Exmoor scheme, and one which would influence national policy, was the linking of financial compensation to the potential 'profit foregone' by the farm business. In other words, in recognition that farmers who provided better environmental outputs would suffer a financial penalty (by comparison with their position were they to carry out the proposed 'improvement'), the scheme attempted to redress the balance. These calculations perhaps helped to demonstrate just how expensive conservation could be. However, in less than twenty years there was a movement from a situation where farmers were offered grant aid to destroy important environmental assets to one where they were increasingly paid for supporting the environment and penalised for damaging it. These principles have become embedded in both European and national programmes and legislation.

Biodiversity

Concern for conservation of natural resources has risen on the political agenda with growing worldwide concern for biodiversity. Official biodiversity policy was originally set out in *Biodiversity: The UK Action Plan* and *Sustainable Development: The UK Strategy*, both published in 1994. The action plan was the UK Government's response to the Convention on Biological Diversity (CBD), which the UK signed in 1992 in Rio de Janeiro. It described the biological resources of the UK and provided detailed plans for conservation of these resources. Action plans for the most threatened species and habitats were set out to aid recovery. Following devolution, the four countries of the UK developed their own strategies, allowing conservation

approaches to differ according to the different environments and priorities within the countries.[22] In 2007, however, a shared vision for UK biodiversity conservation was adopted by the devolved administrations and the UK government: *Conserving Biodiversity – the UK Approach*. From the outset, the task was seen as a partnership activity between a variety of professional and community interest groups. The role for planning has evolved since this date and it is currently outlined in the NPPF, which gives about twenty mentions to 'biodiversity'. Local plans have to be based on an understanding of local ecological networks and need to take a strategic approach for the creation, protection, enhancement and management of networks of biodiversity. All development should minimise impacts on biodiversity and provide net gains in biodiversity where possible (paras 109, 114 and 117).

Whilst PPG 7 and PPG 9 *Nature Conservation* (1994) gave guidance on the importance of the conservation of habitats and the role of planning, PPS 9 *Biodiversity and Geological Conservation* (2005) offered more sharply focused guidance on the role of planning in the protection and enhancement of biodiversity: the guidance given could have been a template for that offered in the NPPF. PPS 9 was followed in August 2005 by Circular 06/05 *Biodiversity and Geological Conservation – Statutory Obligations and Their Impact Within the Planning System*, which 'provides administrative guidance on the application of the law relating to planning and nature conservation as it applies in England'. The four main sections dealt with planning controls in relation to sites of international conservation importance; covered planning and development control in relation to SSSIs; reminded local planning authorities and developers that important species and habitats (including those included in the UK Biodiversity Action Plan and Local Biodiversity Action Plans) occur outside of protected sites and can be a material consideration in planning; and established the importance of protected species and their treatment in the plan-making and development control processes. Whilst the detail has become somewhat outdated[23] by amendments to the Habitats Regulations in 2007 and the passing of the Natural Environment and Rural Communities Act 2006, it remains a useful framework for identifying the relationship between planning and biodiversity.

The United Nations Rio convention on biodiversity was updated in Nagoya in 2010 and, following this, the 1994 UK Biodiversity Action Plan was also updated. However, the *UK Post-2010 Biodiversity Framework*, published in 2012, recognises the devolution of powers which had taken place in the intervening period, and that most work which was previously carried out under the UK Biodiversity Action Plan (UK BAP) is now focused on the four individual countries of the United Kingdom and Northern Ireland, and delivered through the countries' own strategies. However, coordination between the countries remains important. The five strategic goals of the Nagoya convention[24] cover areas such as addressing causes of biodiversity loss and safeguarding better policy implementation. The achievement of many of these can be assisted by focused planning policies and actions, such as the mapping and monitoring of key environmental assets and managing land use to protect these key assets.

The 1994 UK Biodiversity Action Plan recognised that biodiversity is ultimately lost or conserved at the local level and would require a partnership approach. There are about 120 local partnerships working on local biodiversity action plans (LBAP)[25] which identify local priorities for biodiversity conservation and work to deliver agreed actions and targets for priority habitats and species and locally important wildlife and sites. The work of these local partnerships and plans closely relates to local planning policy frameworks.

There are some aspects of conservation policy which are worth emphasising. The first, as was made clear by Circular 06/05, is that it is not restricted to designated sites (despite their large number): in addition to these nationally (and internationally) important sites, there are many more which are of local importance. The now superseded PPG 7 drew attention to the importance of these sites for local communities, since they often afforded people the only opportunity of direct contact with nature, especially in urban areas. A second, related, point is that, since wildlife does not respect human-made boundaries, it is important to safeguard 'wildlife corridors, links or stepping stones from one habitat to another'. The Habitats Directive specifically requires member states of the EU

> to encourage the management of features of the landscape which . . . by virtue of their linear and continuous structure (such as rivers with their banks or the traditional systems for marking field boundaries) or their function as stepping stones (such as ponds or small woods), are essential for the migration, dispersal and genetic exchange of wild species.
>
> (Directive 92/43/EEC, Article 10)

What is particularly significant about this approach (which is only briefly illustrated here) is that it is mandatory on local authorities, and that, in addition to land use designations, actual *management* is involved. (The mandate comes from a combination of two legal instruments: the requirement for plans to include policies in respect of 'the conservation of natural beauty and amenity of the land' and the provision in the Conservation (Natural Habitats, etc.) Regulations 1994 (SI 1994/2716) that these policies 'shall be taken to include policies encouraging the management of features of the landscape which are of major importance for wild flora and fauna' (reg. 37).)

Forestry

In 2012, forest and woodland covered about 3 million hectares in the UK or 12 per cent of the land area, though this is much lower than the EU average of 37 per cent. The woodland is made up of about 52 per cent conifer and 48 per cent broadleaved, though this ratio has changed, as the majority of new planting in recent years has been broadleaved – for example, in 2004 comparable figures would have been 59 per cent and 41 per cent. There has been a steady increase in the forest area as shown in Table 10.2: during the 1980s the increase was of some 300,000 hectares and from 1990 to 2004 another 400,000 hectares was added; by 2012 there had been a 42 per cent increase since 1980. It should be remembered here that forestry is not generally within planning control, though, as

PPS 7 points out, it can prevent the conversion of woodland to urban and other uses. However, as part of the role of planning in protecting the environment and biodiversity, the NPPF notes that 'planning permission should be refused for development resulting in the loss or deterioration of irreplaceable habitats, including ancient woodland' (para. 118).

About 28 per cent of woodland is managed by Forest Enterprise, the development and management arm of the Forestry Commission set up in 1996 (although the detail of this arrangement has been changed since devolution[26]). A substantial reorientation of forestry policy has emerged since the late 1970s. In particular, there has been a major move away from a preoccupation with production to a more balanced approach which places importance on amenity and environmental factors. This has come about after a lengthy period of debate and scrutiny. There was, for example, much argument, both before and after the Wildlife and Countryside Act 1981, which centred on the effects of hill farming and forest policies. A succession of reports concluded that these policies, far from sustaining the economies and landscapes of the uplands of England and Wales, were major factors in their decline (MacEwen and MacEwen 1982; Sinclair 1992).

The predominant concern for production was badly affecting the vitality of rural communities and the conservation of the countryside. It was argued that much more employment could be created by coordinated policies sensitively directed to the problems of the uplands as a whole, rather than to

separate aspects of them. This was essentially a call for 'integrated' policies, which began to emerge later. The immediate result, however, was a provision in the Wildlife and Countryside (Amendment) Act 1985 requiring the Forestry Commission to attempt a reasonable balance between the interests of forestry and of conservation and enhancement of the countryside. Forestry policy in the UK has been increasingly influenced by international commitments such as the 1992 Rio Earth Summit, which led to a statement on forestry principles, the Rio Declaration, and Agenda 21, which provided an agenda for sustainable development. A review of forestry policy in 1994 (*Our Forests: The Way Ahead*) proclaimed that 'UK forestry policy is based on the fundamental tenet that forest resources and forest lands should be sustainably managed to meet the social, economic, ecological, cultural and spiritual human needs of present and future generations'. So now, the Forestry Commission lists among its functions that of 'conservationist', protecting habitats and conserving species, whilst one of the roles of Forest Research, the Forestry Commission's research agency, is to 'promote high standards of sustainable forest management'.

In addition to the broader economic implications of forestry policy for rural areas, there are several issues which are of particular relevance to countryside policy: amenity, wildlife, access and recreation. Problems arise because, though forest production is essentially a very long-term enterprise, there is a need, in the words of the National Audit Office, *Review of Forestry Commission Objectives and Achievements* (1986), 'to have regard to a

Table 10.2 Areas of woodland in the UK 1924–2012

	1924	1947	1965	1980	1999	2003	2012
England	660	755	886	948	1,097	1,110	1,295
Scotland	435	513	656	920	1,282	1,327	1,392
Wales	103	128	201	241	287	285	304
N Ireland	13	23	42	67	81	85	105
	1,211	1,419	1,785	2,176	2,747	2,807	3,097

Sources: DEFRA (2004) *eDigest of Environmental Statistics*; Forestry Commission (2012) *Facts and Figures*

number of broadly drawn secondary objectives [which] can produce conflicts with and constraints upon the Commission's primary aim of increasing the supply of timber'. One of these secondary objectives is the preservation (and enhancement) of the landscape and of the wildlife it sustains. Concern for such wider issues has increased as environmental awareness has grown. This concern is highlighted in the Minister's foreword to the 2013 *Government Forestry and Woodlands Policy Statement*:

> In line with our commitments in the Natural Environment White Paper, we want to do more than just maintain our woodland assets. We want to improve them. This will help everyone derive the greatest possible economic, social and environmental benefit from them, while ensuring that they are available, now and for generations to come.

Recent policy has embraced a wide range of approaches. Planting of broadleaves has been expanded. Access to forests has been extended, with improved arrangements for access agreements. New *national forests* are being established in the Midlands and in central Scotland.[27] A *community forests* programme was established in 1990 by the then Countryside Commission but it is now operated as partnerships between local authorities and local and national partners, which aim at 'revitalising derelict land, providing new opportunities for leisure, recreation, and cultural activities, enhancing bio-diversity' as well as contributing to healthier living and developing community involvement (see the England's Community Forests website: www.communityforest.org.uk). Additionally there is a wide range of *woodland initiatives* covering diverse activities including reinstating woodland management, creating new woodlands, and supporting projects that develop community cohesion. Finally, there are many *community woodland* groups run by volunteers for the benefit of local people, with the common denominators being the involvement of volunteers and the objective of protecting and caring for a piece of land within the community.

In 1998, a forestry strategy was published, *A New Focus for England's Woodlands*. This was 'informed by our international commitments', such as the Rio Earth Summit, and identified the two main aims of forestry policy in England as being 'the sustainable management of existing woods and forests, and a continued steady expansion of the woodland area to provide more benefits for society and our environment' (p. 2). The strategy had four components: rural development, economic regeneration, recreation, access and tourism, and the environment and conservation. These objectives are not mutually exclusive: for example, the fact that between two-thirds and three-quarters of the population are estimated to visit woodlands gives potential for economic impacts. The revision of this document published in 2007, *A Strategy for England's Trees, Woods and Forests*, offered a different formulation of priorities, with an emphasis on 'environmental, economic and social benefits now and for future generations', along with objectives related to biodiversity, climate change and quality of life. Additionally, there was an objective to 'improve the competitiveness of woodland businesses' (p. 4). The incoming Coalition government announced an intention of disposing of 15 per cent of the forestry estate, but this raised a huge level of objections, many based on concerns about environmental impacts. In response to this concern, the government set up an *Independent Panel on Forestry*, chaired by the Bishop of Liverpool, to advise it on 'the future of England's forests and woodlands'. The strength of public feeling about the proposals for disposal can be gauged by the fact that the panel received 42,000 representations. The final report, issued in 2012, highlighted the contribution of forests to environmental, economic and social sustainability, in such diverse areas as habitat protection, flood reduction, job creation and improving health and well-being. These views were taken on board in the 2013 Government Forestry and Woodlands Policy Statement, where the Minister stated in the foreword:

> England's Public Forest Estate will remain secured in public ownership – for the people who enjoy it, the businesses that depend on it and the wildlife that flourishes in it. We have rescinded the previous policy of disposing of 15% of the Estate and we will be providing sufficient funding in this

Spending Review to ensure that high levels of public benefit can continue to be delivered.

This policy statement has an emphasis on the protection, improvement and expansion of woodlands in a way that complements the environmental objectives noted above. There thus appears to be a close alignment between this policy and the environmental objectives for spatial planning.

'Joining up' policies

The spread of factors considered in this chapter – many of which can be seen as not being natural bedfellows – focuses on the need for policy approaches in rural areas to be 'joined up' if the best outcomes are to be achieved. The idea of integrating social, economic and environmental concerns chimes with the logic of sustainable development and the logic of 'spatial' planning; it also fits well with the idea of rural proofing and of 'mainstreaming' policy objectives throughout the actions of government. Ideas of policy integration are not new, being evident in the work of the Highlands and Islands Development Board, established in 1965, in the work of the Development Board for Rural Wales, and to an extent in the Rural Development Commission's Rural Development Programmes. The nature of the problem was summed up by the Association of District Councils in its 1979 publication *Rural Recovery: Strategy for Survival*, where it stated:

> the closely inter-woven nature of the various problems facing rural areas, and the inherent weakness which stems from the tendency for the various authorities and agencies . . . to approach problems from a narrow service point of view . . . It is therefore essential, both nationally and locally, to ensure a comprehensive and co-ordinated response which takes full account of the implications of any one area of policy on other policies and problems.

As the references above imply, often the spur for integration came from a focus on regeneration. An interesting experiment in this context can be found in the Peak Park, a project in 1981 to assist three small villages suffering from population and job loss and damage to, or loss of, characteristic environmental features. Through a range of complementary projects, new businesses were started, jobs created, community infrastructure improved, population decline arrested and local environmental assets protected or enhanced. Evaluations (Blackburn *et al.* 2000) of the project suggested that the effort involved in fostering integration was rewarded by sustained benefits to the rural communities. In viewing and developing planning policy frameworks, it is therefore necessary to have a view of the value and place of policy integration, and of the range of actors necessary to ensure successful implementation.

Further reading

The rural context and planning

For a number of years, the Countryside Agency/ Commission for Rural Communities produced an invaluable *State of the Countryside* report which brought together and interpreted evidence on a range of rural issues. The last edition was published in 2011 and remains available. Access to other government-funded research on rural matters and a range of other useful documents produced by CRC is available at www.defra.gov.uk/crc/documents/. Access to a range of other government-funded reports is available via http://randd.defra.gov.uk, although the range of recent material on socio-economic matters included here is limited. For Wales, the Wales Rural Observatory (www.walesruralobservatory.org.uk) has published a range of useful reports, whilst in Scotland the Scottish government has funded and published reports on a range of relevant research (www.scotland.gov.uk/ Topics/Research/About/EBAR).

There is now a text *Introduction to Rural Planning* (Gallent *et al.* 2008) which does offer a good overview of the range of topics relevant to rural planning. Similarly, *A New Rural Agenda* edited by Midgely (2006) offers coverage of a selection of key issues. The OECD

Rural Policy Reviews: England, United Kingdom (2011a) does just what its title implies, whilst Agriculture in the United Kingdom (2013, but an annual publication) gives a detailed review of key agricultural trends.

There is a long history of writing on the countryside and its planning. Important books include Champion and Watkins (1991) People in the Countryside: Studies of Social Change in Rural Britain; Cherry (1994) Rural Change and Planning: England and Wales in the Twentieth Century; Cloke et al. (1994) Lifestyles in Rural England; and Newby (1985) Green and Pleasant Land: Social Change in Rural England. Gilg (1996) Countryside Planning: The First Half Century provides a reflective view on a number of (mainly environmental) matters. Two relatively old books still provide valuable insights: Bradley and Lowe (1984) Locality and Rurality: Economy and Society in Rural Regions gives a social and institutional approach to thinking about issues for rural society, whilst Cloke (1979) Key Settlements in Rural Areas provides a fascinating introduction to rural settlement planning. Amongst more recent material, do not discount material such as the two Rural White Papers (1995 and 2000), the Taylor Review (2008) or, on the important but specific issue of rural housing, the final report of the Affordable Rural Housing Commission (2006). An interesting overview of countryside issues in Northern Ireland is given by Greer and Murray (2003) Rural Planning and Development in Northern Ireland. There are many popular books bemoaning the fate of the countryside: a highly readable and informative one by an agricultural journalist is Harvey (1997) The Killing of the Countryside. The damage done to the countryside by modern methods of farming is analysed in a famous text by Shoard (1980) The Theft of the Countryside.

National parks, access to the countryside and rights of way

For accounts of the background to, and the implementation of, the 1949 Act, see Cherry (1975) National Parks and Recreation in the Countryside (Volume 2 of Environmental Planning 1939–1969) and Blunden and Curry (1989) A People's Charter? Forty Years of the National Parks and Access to the Countryside Act 1949. Two major reviews of national parks policies are the Sandford Report (1974) and the Edwards Report (1991). An up-to-date review of the policy and context for national parks is provided by the DEFRA 2010 circular English National Parks and the Broads. A review of the economies of England's national parks is provided in Valuing England's National Parks (Cumulus Consultants Ltd and ICF GHK 2013). A passionate critique of the restrictions on access (and much else) is given by Shoard in This Land is our Land (1987) and A Right to Roam (1999).

Natural England has published a useful Guide to Definitive Maps and Changes to Public Rights of Way (2008), which can appear to be a complicated topic. Official publications on access to the countryside proliferated in 1998 and 1999. Access to the Open Countryside: Consultation Paper (1998), Options on Access (1999), Improving Rights of Way in England and Wales (1999) and Access to Open Countryside of England and Wales: The Government's Framework for Action. All of these provide different elements of insight into the issue of access. Access codes published in England (English Nature) and Scotland (SNH) provide useful insights into practical aspects of access.

Coastal issues and waterways

On coastal issues, for an overview of policy approaches, see PPG 20 Coastal Planning (1992) and PPS 25 Supplement Planning and Coastal Change (2010). Two EU reports were issued in 1999: Lessons from the European Commission's Demonstration Programme on Integrated Coastal Zone Management and Towards a European Integrated Coastal Management Strategy: General Principles and Policy Options. These were followed by the Recommendation (2002/413/ EC) (6) on the implementation of ICZM. The four UK administrations have produced ICZM strategies and these are also useful. The relationship between planning and waterways is dealt with in BWB (2003) Waterways and Development Plans. A useful overview of the history of British waterways is provided in a note produced by the House of Commons Library in 2011 (SN 3184), whilst the focus of activity of its successor, the Canals and Rivers Trust, is given in its Shaping our Future: Strategic Priorities (2012).

Recreation

Recreation in the countryside is described in some detail in a regular report *Monitor of Engagement with the Natural Environment: The National Survey on People and the Natural Environment* (Natural England 2013). Natural England has also produced a report, *The State of the Natural Environment* (2008b), Chapter 4 of which provides a good overview of countryside recreation activities. For a review of the impact of different recreation activities on the environment, see Sidaway (1994) *Recreation and the Natural Heritage*. A 1995 report by the Countryside Commission and others is devoted to exploring the concept of *Sustainable Rural Tourism* and the ways in which it can be translated into practice. See also PPG 21 *Tourism* (1992), Segal Quince Wicksteed (1996) *The Impact of Tourism on Rural Settlements*, Curry (1997) 'Enhancing countryside recreation benefits through the rights of way system in England and Wales' and Bell (1997) *Design for Outdoor Recreation*.

Nature conservation and biodiversity

The report by Natural England referred to above (*The State of the Natural Environment*) contains some valuable material on landscape, biodiversity, pressures and risks, and policy responses, and forms a good introduction to the topics.

Current official guidance on nature conservation in England is given in the NPPF, but the superseded PPS 9 *Good Practice Guide* gives valuable illustrations and case studies of approaches that can be taken to delivering biodiversity objectives. In Scotland, individual items of planning guidance were superseded by an all-embracing *Scottish Planning Policy* in 2010. A revised version is due for adoption in mid-2014. In Northern Ireland, the Department of the Environment's PPS 2 *Natural Heritage* provides planning guidance on this topic. In Wales, *Planning Policy Wales* (2012) and TAN 5 *Nature Conservation and Planning* (2009) give planning policy guidance.

The *UK Biodiversity Action Plan*, published in 1994, was the UK Government's response to signing the Convention on Biological Diversity (CBD) at the 1992 Rio Earth Summit. In July 2012, DEFRA and the devolved administrations jointly published its replacement, *The UK Post-2010 Biodiversity Framework*. *Biodiversity 2020* (2011) is England's strategy. Some guidance on practical aspects of its implementation are available on the Natural England website. For some examples of how biodiversity action plays at the local level, see *Natural Partners: The Achievements of Local Biodiversity Partnerships in England* (2003), published by the Wildlife Trusts.

Despite the special controls, many SSSIs have been damaged: see, for example, reports of the National Audit Office: *Protecting and Managing Sites of Special Scientific Interest in England* (1994); *Protecting Environmentally Sensitive Areas* (1997); and *SSSIs: Better Protection and Management: The Government's Framework for Action* (1999). For some practical guidance on the management of SSSIs, see *Sites of Special Scientific Interest: Encouraging Positive Partnerships*, published by DEFRA in 2003.

Forestry and woodlands

There was a major policy statement issued after the Rio Summit: *Sustainable Forestry: The UK Programme* (1994), following the review reported in *Our Forests: The Way Ahead* (1994). This was followed by forestry strategies produced in 1998 and 2007. Current policy is set out in the Coalition government's *Government Woodlands and Forestry Policy Statement* of 2013.

Notes

1   According to the OECD (2011a), using its internationally applicable standards, England has 'no predominantly rural regions' (p. 14), which partly reflects the fact that most rural settlements are within a relatively short distance of a larger centre. *OECD Rural Policy Reviews: England, United Kingdom 2011* identifies some interesting and distinctive characteristics for England's rural areas and approaches to policy which provide a useful context for considering English rural planning.

2   In addition to the intervention system, there was also: (i) an external protection system, as world prices for

many staple products were below EU minimum prices; (ii) levies and customs duties; (iii) what is described as 'aids to complement the price system', essentially subsidies for processors of products where the EU is unable to impose tariffs because of international agreements such as GATT; and (iv) flat rate aids for some specialised products important to certain localities, with an example being durum wheat.

3 This speech, made in Belfast on 26 February 1990, was very much based on a key Commission document *The Future of Rural Society* (COM(88) 501/2 final) which began the process of placing agricultural policy in a broader rural socio-economic context.

4 The UK did not take part in this element, as the rationalisation it was seeking to promote was not felt to be necessary: the UK already had large, more efficient farms – for example, at the time the average UK farm was 65 hectares, whilst the average Greek farm was 4 hectares.

5 Information from the Farm Business Survey, a sample survey of farms carried out for DEFRA. Data available at: www.farmbusinesssurvey.co.uk/index.html (accessed 15 September 2013).

6 Planning for decline is perhaps most famously represented by the 'D' village policy of the County Durham plan of 1951, where 'D' stood for destruction, principally of former mining villages in the west of the county. It has been described by Pattison (2004: 311–12) in the following terms: 'By the 1950s the pace of decline in mining fostered a perceived planning need to refocus housing provision on areas with greater long-term economic potential, eliminating settlements with poorer prospects. This produced a county planning policy, unique in Britain in that it actively planned for decline.' These policies were, in essence, continued in the revised County Plan of 1964.

7 Gilder found that 'there is no clear-cut relationship between the costs of services and the size of settlement' (1984: 245), whilst Cloke (1979: 204) also points out that the original idea underpinning key settlements – that facilities would be shared by a surrounding rural population – would not take place automatically. However, other studies (for example

Ladd 1992; Powe and Whitby 1994) produced support for the proposition that there is a definable relationship between settlement size and the cost of providing public services, and Powe and Whitby assert (p. 432) that 'the cost of providing services should be an important element in rural settlement planning'. However, the same authors also point out (p. 433) that other factors – economic and non-economic – have an important part to play in developing settlement policy. It is the inclusion of these 'other factors' which contributes to the contestability of decisions in rural settlement planning.

8 In Northern Ireland, all government departments have been required to rural-proof since 2002. A revised guide to rural proofing was issued in 2011, *Thinking Rural: The Essential Guide to Rural Proofing*. In Wales, following some dissatisfaction with approaches to rural matters (see, for example, WRO 2004), an independent review of the rural proofing process was announced in February 2013. The Scottish government claims to have 'mainstreamed the needs of rural Scotland into all of its policies'.

9 'Market towns' is a rather inexact way of characterising such towns, as very few – possibly around one in twenty – has a market. However, it became a useful shorthand descriptor for towns performing a number of functions, including service centres, visitor attractions, employment centres, commuter towns, and retirement centres. Reflecting on the roles such towns perform also offers an opportunity to reflect on the role assigned to larger towns in rural planning policy. For a fuller discussion of how the roles of 'market towns' might be considered, see Powe and Hart 2008.

10 The Rural Development Commission was originally set up by Lloyd George in 1909, with the role of advising on and administering a development fund voted annually by Parliament to benefit the rural economy of England. Whilst its functions changed over the ninety years of its existence – it merged with the Countryside Agency in 1999 – it maintained a major responsibility for the development of rural industries from its inception. In order to target its resources on the rural areas of greatest economic and social need, the commission established a system of

initiatives, sometimes in succession and sometimes occurring at the same time, is partly a reflection of the centralised nature of government in the UK and of the pressure on ministers to maintain their profile and make their mark with new announcements and press releases (Punter 2010). Indeed, the vast range of different initiatives produced by this arrangement is such that it is only possible here to focus on a limited selection of those which have arguably left the biggest mark on the popular imagination.

To this end, this chapter reviews a selection of those urban policy initiatives which have been particularly significant, perhaps because of the extent of their impact on urban areas or the degree to which they signal a change in thinking. The initiatives tell a story of some of the key focuses and changes in urban policy. The first section discusses some of the policy initiatives used to achieve housing improvements during the course of the twentieth century, attending to the expanding role of direct state involvement in the wake of the Second World War and to its subsequent contraction. The significance of poor housing conditions in framing twentieth-century intervention is discussed, as well as the extent to which it has since yielded to affordability concerns. The second section turns from housing to neighbourhood conditions considered in the round and encompassing such issues as employment, crime, environmental quality and connectivity. It charts the rise of 'regeneration' as a loose term often used today to equate the attraction of investment to a place with the interests of its residents. The significance of key initiatives, including the community development projects and the urban programme, are discussed, as well as the shift in New Labour's language from being one of poverty and inequality to social inclusion and equal opportunities. A number of more recent regeneration models are also discussed, ranging from the Single Regeneration Budget to the 'task force' approaches of the urban development corporations, urban regeneration companies and housing market renewal pathfinders. The chapter concludes by considering some of the challenges of evaluating the success of regeneration programmes, especially given the complex and contested question of who regeneration is for.

POOR HOUSING AND SLUM CLEARANCE

Urban replanning in nineteenth-century Britain

It has become fashionable in some quarters to portray planning as a highly or perhaps overly statist activity, but there is nothing inherent in planning that means it must be state-orchestrated. The legislation which began to formalise town planning in the early nineteenth century has already been reviewed (in Chapter 2). During this time, local and central government regulation was extended to cover the setting out of new development and, to a limited degree, the reconstruction of insanitary areas. However, much more ambitious attempts to replan urban areas had taken place prior to this with the impetus coming from the private sector rather than the state. The building of Newcastle's Grainger Town between 1835 and 1842, for example (see Plate 11.1), drew on an entrepreneurial vision of what could be achieved by expanding and improving the trading district above the city's quayside and linking this new space to the affluent residences to the north (Urbed 1998). But Richard Grainger also knew that local political support from the Newcastle Corporation would be essential if key buildings in the city could be acquired and demolished to make way for the scheme. Similarly, the boom in railway construction which occurred during the mid-nineteenth century, and the associated need to drive new stations into the heart of London and the big cities, gradually led to an awareness on the part of railway engineers of the need to balance the 'internal' priorities of the railway network with the implications for wider society, felt in terms of displaced populations, increased pollution and the fragmentation of the built environment (Purcar 2007).

The transfer of responsibilities to the state: urban policy is born

The early twentieth century saw a succession of Acts of legislation which began to transfer more and more

Plate 11.1 Grey Street in Grainger Town, Newcastle upon Tyne. Development of this well-used street, including the Theatre Royal shown on the right, was made possible with the support of Newcastle Corporation (photograph taken by David Webb)

responsibility for urban redevelopment to the state, and efforts to tackle poor housing conditions were a significant contributor to this. There are many reasons why the UK had built up such a large stock of poor-quality, working-class housing by the beginning of the twentieth century, but conditions were worst in areas close to major sites of industrial employment, where long hours and low pay restricted the ability of workers to commute beyond short distances. Demand massively outstripped supply, while the scale of local housing provision was restricted by traditional building methods, high land prices and low wages. The result was often felt in overcrowded, cheaply constructed, poorly serviced and unsanitary buildings. These squalid conditions were a source of general concern, but there was less agreement on the question of who

was to blame for them or who should be responsible for addressing them. The issue was forced by the social fallout and civil unrest which followed the First World War, with the Housing, Town Planning, Etc. Act 1919 (commonly known as the Addison Act) accepting a measure of state-provided housing, not least for fear of a replication of the Russian Revolution two years earlier (Cole and Furbey 1994). The Greenwood Housing Act 1930 significantly increased the scale of earlier, more localised and piecemeal slum-clearance projects, demolishing over a third of a million houses before the Second World War brought the programme to a halt.

By 1938, demolitions had reached the rate of 90,000 a year. Had it not been for the war, over 1 million older houses would (at this rate) have been

BOX 11.1 HOUSING STANDARDS: GENERAL DEFINITIONS

New Labour sought to bring all social housing up to the decent homes standard by 2010, and in 2002 extended this with the aim of getting 70 per cent of vulnerable households (those in receipt of specified means-tested benefits or tax credits) in private-sector housing, whether rented or owner-occupied, into decent homes, by 2010. The Coalition government carried forward a similar definition as a basis for restructuring the financial arrangements governing council housing.

To classify as a 'decent home', social housing should:

- be free of health and safety hazards;
- be in a reasonable state of repair;
- have reasonably modern kitchens, bathrooms and boilers;
- be reasonably insulated.

The current statutory minimum fitness standard is based on that introduced by the Housing Act 2004 – the *Housing, Health and Safety Rating System*. This is not a standard but an evaluation framework of twenty-nine categories of 'housing hazard' which are to be assessed for their impact on *actual* or *potential* residents. The hazards include excess heat and cold, crowding and space, water supply and ergonomics. The extent of a hazard will depend on the type of occupant – for example, stairs will present more of a hazard for elderly people.

The previous fitness standard defined fitness according to the condition of the housing only – in terms of serious disrepair, structural stability, dampness prejudicial to health, and the availability of basic services and facilities. These requirements were a minimum standard, and if one was not met, the house was not considered to be fit for human habitation. In Scotland, the Housing (Scotland) Act 1969 introduced the concept of a tolerable standard, which differs in detail from the fitness standard in England and Wales.

---

contributing factors included regional economic weakness, the development of too much new suburban development, which drew demand from urban areas, a struggling and unregulated private sector and poor or uncoordinated urban management and service delivery (Webb 2010). The housing market renewal initiative is considered in more detail later in this chapter.

Decent homes for all

Many of the initiatives described up until this point concentrated on areas or estates in which the most acute housing problems were found. A characteristic of housing and urban policy, however, is the extent to which discrete projects such as this are buoyed up or undermined by more general political and economic changes. The projects outlined above took place in the context of a broader shift in the means by which social housing in particular was funded and managed. The economic crisis of the late 1970s and its political reaction led not only to a reining back of state programmes for slum clearance and redevelopment but also to the institution of social housing cuts which often stripped the budget for maintenance.

The financial fortunes of social housing eased in the 1990s and then, from 2000, New Labour instituted its more direct approach to improving housing standards. The statutory fitness standard, which had been an instrumental part of earlier clearance and improvement programmes, was overhauled in favour of the housing, health and safety ratings system (HHSRS), which is not a standard, but an assessment of the hazards that homes present for the health and safety of residents. More significantly, the government introduced a target of bringing all social housing up to the decent homes standard by 2010.[4] In 2002, it added a further aim of getting 70 per cent of vulnerable households (those in receipt of specified means-tested benefits or tax credits) into decent, privately rented or owner-occupied homes by 2010. The Coalition government has carried forward a similar definition of decent homes (see Box 11.1) as a basis for implementing their own social housing funding reforms, which seek to achieve the self-financing of social housing stocks.

Implementation of the decent homes standard prompted fears that the standard was being used to push more social housing out of local government control, since in order to meet targets, local authorities had to consider arm's-length management, transferring to other providers and PFI schemes. The drive for decent homes resulted in significant improvements to the condition of the social housing stock, but reports showed that 10 per cent of social rented homes still did not meet the decency standard in March 2010, although it was expected that decency would continue to improve, from 290,070 to 100,973 non-decent local authority homes in 2015 (Bury 2010).

In these times of austerity, we are now all too aware that the improved funding situation for social housing investment in the 2000s was matched by continued liberalisation of private-sector financial regulation. This enabled global investment to surge into the UK housing stock, incentivised improvements and raised prices in the owner-occupied sector. An unfortunate consequence has been growing inequality in housing wealth and, more recently, an increasing proportion of privately rented housing: a new trend which runs counter to the decline of the sector throughout most of the twentieth century. With social housing in very short supply, those on low incomes in the private-rented sector are most likely to experience the worst housing conditions (see Box 11.2).

BOX 11.2 STATE OF HOUSING IN THE UK

England

The housing stock

There were 22.8 million homes in England in 2011 (and 22.0 million households); 79 per cent of homes were houses and 21 per cent flats; 83 per cent in urban and 17 per cent in rural locations.

In total, 65 per cent of homes are owner-occupied (up from 57 per cent in 1980 but down from a peak of 71 per cent in 2003), 17 per cent are privately rented (up rapidly from 10 per cent in 2000: a figure that had stayed reasonably constant for the twenty years before that) and 17 per cent are social-rented (8 per cent rented from local authorities and 9 per cent from registered social landlords; social renting showed a slow but consistent decline of around 0.3 per cent per year following a steeper decline of 8 per cent across the 1980s).

The average size for a home has increased over recent years to 91 square metres. In the owner-occupied sector 48 per cent of properties are over 90 square metres, as compared with 7 per cent in the social-rented sector and 21 per cent in the private-rented sector. The smallest properties are also unevenly distributed, with dwellings under 50 square metres making up just 4 per cent of the owner-occupied sector, as compared with 29 per cent of the social-rented sector.

uncertainties that are less prevalent in many private-sector organisations. Broader social and political changes can make some parts of the city unpopular, harder to manage and more difficult to let. Since housing benefit represents a substantial income stream, there is also great financial sensitivity to any political changes to payments such as these. One area of current uncertainty is the 'removal of the housing benefit spare room subsidy in the social rented sector' which is contained in the Welfare Reform Act 2012 and is more commonly known as the 'bedroom tax'. This has introduced additional charges for social tenants with spare rooms but has not taken into consideration the mismatch between the housing stock on offer and the size of the households it serves. Pledges by the opposition to end the penalty if they win the next election make it difficult for housing providers to decide whether they should pursue alterations to their stock to reflect the demands of the penalty or not.

Empty homes programme

A further housing policy initiative with regeneration potential is the empty homes programme, which aims to increase housing supply (and, in the government's eyes, therefore, reduce cost) by making more use of the existing housing stock. In November 2011, *Laying the Foundations: A Housing Strategy for England* set out £100 million to tackle the problem of empty homes across the country and matched this with a further £50 million to address clusters of empty homes, many of which were located in former housing market renewal areas (more detail on housing market renewal is provided later in this chapter). While empty homes had been recognised as a particular problem in the north of England, the strategy emphasised the potential contribution that bringing them back into use might make towards increasing overall housing supply. In an innovative move prompted by calls from some in the community housing sector, the government agreed to reserve £30 million of the £100 million core fund for community and voluntary groups. Special funding arrangements were put in place to try and

support these groups and encourage small-scale action from organisations which would normally have been deterred from bidding by the costs and complexities involved in applying for registered provider status. The initiative forms part of a wider move towards encouraging different forms of charitable or collectively run organisations which depart from the standard models of profit-driven private-sector or state-controlled services. In other areas, particularly the NHS, there have been fears that such moves are merely a smokescreen to enable private-sector organisations to gain control of public services by the back door. Initial results from the community grants scheme, however, show a wide variety of organisations have benefitted, from large charities to smaller, locally based organisations and Box 11.3 gives details of one of the smaller organisations that benefitted. Nevertheless, given the limited funding available, it is still unclear whether these endeavours can yet be seen as the beginning of an alternative form of organisation capable of making a serious contribution to boosting housing supply and improving conditions.

Housing affordability

The debate on affordability in the UK is currently very polarised. The rise in house prices during the 2000s has meant that those who do not own a house are finding it increasingly difficult to afford one. This has compounded parallel increases in income and wealth inequality, as pointed out by a recent Shelter report:

> It is not an exaggeration to claim that we are moving towards a situation in which this country's children will be divided more by wealth than has been the case since at least Victorian times. For the children of the poor there will be large parts of the country to which they cannot consider moving in the future even if they should wish to. When they have problems in their lives there will not be recourse to family wealth to bail them out, to help with a time when they cannot work or find work, to help pay their way through university studies (for the minority from poor areas who go) without

BOX 11.3 GIROSCOPE HOUSING ASSOCIATION IN HULL

Giroscope housing project in Hull, a self-help housing project, began in the 1980s as a practical response to problems of homelessness and as a reaction to some of the social and economic changes brought in under Thatcherism. Initially constituted as a workers' cooperative, the organisation has grown and now owns close to seventy houses in the city. The small scale of the organisation, its social focus and its hands-on approach to management and lettings have all helped it to grow in a difficult economic environment which has sometimes been described as one of housing market failure. Giroscope's focus is on being financially self-sufficient, but funding from the empty homes programme has allowed it to increase the scale of its activities by working with local tradesmen to bring hard-to-let properties back into productive use. At the same time, Giroscope's independence means it can continue to promise all its tenants security of tenure and protection from Right to Buy, while putting sustainability at the heart of its renovation work.

Plate 11.2 Giroscope housing project in Hull. Combines a mixture of political passion and practical action

(a) Housing in areas such as Victoria Avenue off Wellsted Street in Hull has become run-down and abandoned as traditional forms of employment have declined and private landlords have moved in (photograph courtesy of Giroscope)

Plate 11.2 (b) **This housing in Raywood Villas, Wellsted Street has been successfully brought back into use by Giroscope (photograph courtesy of Giroscope)**

Plate 11.2 (c) **Giroscope's roots lie in a practical expression of 'self-help' politics (photograph courtesy of Steve Morgan Photography)**

Plate 11.2 (d) **Giroscope's workers today (photograph courtesy of iD8 photography)**

BOX 11.4 THE USE OF HOUSING IN THE UK

England

- In 2011, average gross household disposable income per head (after taxes and national insurance contributions) was £16,251 in England and £20,509 in London.
- The average house price in England was £256,558 in 2011, up from £71,741 in 1996. The average house price in London was £403,605 and £296,567 in the South East.
- The average monthly social rent was £327 a month in 2011, with the highest rent in London at £418. The average private rent was £714 a month in England and £1,369 in London.
- In total, 4.4 per cent of homes were vacant in 2011 (967,000 homes). Of these, 288,000 were privately owned, long-term empty and 59,000 were empty local authority and housing association homes.
- Underoccupation currently runs at 49 per cent of households in the owner-occupied sector, 10 per cent in the social-rented sector and 16 per cent in the private-rented sector. There are currently 711,000 second homes in England. The most common reasons for owning a second home are as a holiday home (48 per cent) or as a long-term investment (43 per cent).
- Overcrowding rates in the owner-occupied sector decreased from 1.8 per cent to 1.3 per cent between 1995–6 and 2011–12 compared with a small increase in the social-rented sector from 5.1 per cent to 6.6 per cent and a big increase in the private-rented sector, from 3.2 per cent to 5.7 per cent.

Northern Ireland

- In 2011, average gross household disposable income per head (after taxes and national insurance contributions) was £13,966.
- The average house price in Northern Ireland in 2011 was £143,284, up from £47,894 in 1996.
- In total, 6.4 per cent of homes are vacant.

Scotland

- In 2011, average gross household disposable income per head (after taxes and national insurance contributions) was £15,654.
- The average house price in Scotland in 2011 was £180,213, up from £56,358 in 1996.
- In total, 4.5 per cent of homes are vacant.

Wales

- In 2011, average gross household disposable income per head (after taxes and national insurance contributions) was £14,129.
- The average house price in Wales in 2011 was £164,146, up from £53,007 in 1996.
- In total, 6.0 per cent of homes are vacant.

Sources: DCLG (2013) *Regulated Mortgage Survey Live Tables*; Office for National Statistics (2013) *Statistical Bulletin: Regional Gross Disposable Household Income (GDHI) 2011*; VOA Private rental market data; NIHE (2013) *2011 House Condition Survey*; The Scottish Government (2012) *Housing Statistics for Scotland: Key Information and Summary Tables*; Office of National Statistics (2013) *The 2011 Census*.

working as well, to help when they have children of their own and so on.

(Thomas and Dorling 2004: 3)

In historical terms, house prices are still geared highly when compared with income levels, especially in London. The continuation of high and rising house prices encourages speculation and increased consumption of housing per person; it helps a poorly capitalised banking sector, because its customers feel more affluent and the assets on which its debt is secured appear secure. However, high house prices are making it increasingly difficult for younger generations to buy housing, especially in London and the South East, while large mortgages and high rents reduce the amount of expendable income available to the rest of the economy. The converse is also true: a long-term decline in house prices may cause problems for the banking system and, as speculators leave the market, it could open up housing to owner-occupiers and reduce the need to provide affordable housing. The Coalition's policy on the subject appears to have shifted. In 2011, housing minister Grant Shapps stated that 'Over time we want to move to a position where house prices continue to grow but people's ability and purchasing power increases quicker' (*Daily Mail*, 13 October 2010). In 2013, however, the government announced its Help to Buy scheme which will provide low-cost, guaranteed deposits to assist first time buyers to enter the housing market. Despite scepticism, the government claims that Help to Buy will not lead to further increases in house prices and is presumably intended to assist the private sector to sell new housing by increasing the number of transactions (Mason *et al.* 2013). The relationship between private-sector housing delivery and the ability to secure affordable housing through planning gain is dealt with in more detail in Chapter 6.

Scottish housing

Scottish housing is different from that south of the border in significant ways. There is a high proportion of tenement properties, dwellings tend to be smaller, rents are lower and a slightly higher proportion of the housing stock is owned by public authorities. These and other differences reflect history, economic growth and decline, local building materials and climate. Scotland has historically faced a major problem of poor-quality tenement housing.

The 2011 Scottish House Condition Survey shows that 22 per cent of properties in Scotland are tenements, but the National Home Energy Rating assessment shows that these are, on average, more energy efficient than detached or semi-detached properties. However, older properties are also more likely to be in poor condition, with 8 per cent of pre-1919 properties of a condemnatory standard in which people should not reasonably be expected to continue to live (p. 38)

Historically low council rents in Scotland contributed to relatively low demand for private housing, which, coupled with massive public house building programmes, gave Scotland the highest proportion of public-sector housing in Western Europe. For a time, Glasgow City Council was the largest public-sector landlord in Europe (McCrone 1991). However, it has now transferred over 80,000 houses to smaller providers (Gibb 2003). (It should be noted that the Scots use the term 'house' in the English sense of 'dwelling', i.e. it embraces a flat or tenement.) Right to Buy has shifted the balance between the owner-occupied and public housing sectors: owner-occupation rose from 35 per cent in 1979 to 57 per cent in 1995, plateaued at 62 per cent between 2004 and 2009 and then dropped to 60 per cent, while the social-rented sector fell from 54 per cent to 30 per cent and is now at 24 per cent. Despite social-rented stock running at more than half of all housing at the beginning of the 1980s, the figure is now just seven percentage points ahead of England. The Scottish government is committed to passing legislation in 2013 which will end the Right to Buy from 2016 onwards.

In 2001, the Housing (Scotland) Act introduced a requirement for all local authorities to prepare a local housing strategy (LHS), including an action plan showing how it will be implemented. The strategies provided a broader assessment of housing markets as well as the condition of housing. At the time, Scotland's only benchmark was the 'tolerable standard' below which houses should be condemned. Relatively few houses failed this standard, so the Scottish Executive in 2004 adopted the broader Scottish Housing Quality Standard, with a target that all social housing should meet the standard by 2015. Figures for 2012 show that 26 per cent of social-rented homes did not meet the standard. However, this was a significant improvement from 53 per cent in 2009. The worst conditions were in local authority stocks, 36 per cent of which did not meet the standard, followed by housing associations predominantly tasked with receiving stock transfers: 25 per cent of their properties did not meet the standard. Poor energy efficiency is the single largest reason why properties did not meet the standard. This reflects the much greater implications for Scotland of raising fuel bills and the higher risks of households finding themselves in fuel poverty.

Scotland's approach to planning and housing since the 2010 election has been markedly different from that of England. Despite support for community-led delivery, there has not been the strong rhetoric of localism that has been used to justify reforms in the rest of the UK. Scotland has retained a more plan-led style of governance, both through its national plan, which is effectively equivalent to England's abolished regional spatial strategies, and through continued support for the pursuit of regeneration and sustainable communities. Despite the centre-left feel to many of these policies, concerns have also been raised that efforts to drive up standards in the social-rented sector have come at the cost of demolition and site preparation work to lever in private sector interest. This has led to the loss of nearly 5,000 social homes a year from the social-rented sector between 1992 and 2009 (Glynn 2012).

AREA-BASED REGENERATION AND STATE-SUBSIDISED DEVELOPMENT

The problem is that for too long urban policy has acted as 'a filler in of gaps', mopping up the worst cases of fallout produced by wider economic and policy changes. It has functioned as both a form of symbolism and crisis management.

(Atkinson 1999a: 84)

So far, this chapter has focused on the initiatives which have been used to promote minimum standards for housing, first through large-scale demolition, redevelopment and improvement programmes and latterly through area-level action to address disinvestment and decline in areas dominated by social housing. As the twentieth century progressed, the pursuit of decent housing became an ever more complex task. Deindustrialisation and long-term economic shifts led to serious reductions in investment in particular neighbourhoods and regions. Housing policy initiatives targeting these areas came up against social issues that became hardened by the persistent decline of manufacturing employment and its slow replacement by, often low-paid, service-sector work. And yet, these issues were not principally ones of poor housing: they reflected the changing global role of the UK economy, which could no longer be relied upon to provide full employment as it had in the post-war years.

It was in the context of these changes that the nature of area-based policy began to shift. The programmes of housing redevelopment and improvement in the post-war years were concerned with addressing both absolute and relative poverty by redistributing the proceeds of economic growth in the pursuit of a minimum standard of low-income housing. From the 1980s onwards, and as dependable economic growth and full employment began to falter, the focus shifted onto the provision of financial incentives to private-sector companies in an attempt to persuade them to locate in economically weaker areas. The election of New Labour in 1997 then brought with it a focus on 'social inclusion' as a term which came to replace poverty as the signifier of what urban policy was intended to achieve. As with all far-reaching, political buzzwords, 'social inclusion' contains ambiguities which helpfully appeal to different audiences (Levitas 1998); the key difference with poverty, however, lies in the level of conditionality implied. In other words, inclusion was used to imply a need to reintegrate low-income people and places into the mainstream economy both by addressing the systemic disadvantages that left them with poorer life chances and by making individuals' claims on social support conditional on their efforts to develop skills and find employment. This was deemed to be good not only for society, but also for business, since it would boost the number of people in employment and make full use of talents that would otherwise be wasted (Giddens 1998). However, in a broader context of rising inequality it did not extend to restricting or redistributing the returns from, or the additional privileges being exploited by, those at the opposite end of the social spectrum.

A second and related shift which can be seen in the form of the urban policies reviewed in this part of the chapter is epitomised in the quotation from Atkinson above. It has been suggested that the area-wide basis of urban policy interventions evolved to limit the scale (and cost) of interventions which might otherwise have had a broader social and geographical impact. One consequence of this is that individuals who may be very poor, but who live in an area of relative affluence, are not identified or targeted by policy (Mason 1979). It also makes it possible to announce area-based initiatives in the context of wider changes to social and economic policy which have regressive consequences on the distribution of income and employment, consequences which often hit the poorest hardest. For David Harvey, the use of area-based initiatives in this way allows politicians to present themselves as 'righters of wrongs', even in the context of rising intergenerational inequality, by vocally addressing highly visible and spatially bounded instances of socio-economic fallout (Harvey 1974). During the 1980s and 1990s in particular, when local government control was often politically at odds with that of central government, the fact that the terms of area-based initiatives were often established by central government meant that they could be used, in combination with additional control over local government financing, to prevent local government raising taxes to support more generalised welfare services and to focus their activities instead on those issues defined by central government.

During the 2000s, there was an attempt to align funding for urban intervention with the planning system to achieve a democratically coordinated public and private investment that would mirror the capabilities of the more state-led approach taken in the 1960s and 1970s. This, however, occurred within a context of significant central control of local authority finances and could not avoid the reality that such centralised funding for urban policy intervention has always been driven by particular logics – often politically opportune and short-termist – which differ from the longer-term and highly managed provisions set out in development plans. New Labour's attempt to integrate planning and regeneration through regional co-ordination and a friendlier approach to partnership between central and local government ultimately came to raise questions about whether accountability for controversial schemes lay with local communities or central government. The Coalition government, drawing on its associated agenda of localism, has since formulated a politically effective critique of these processes as centralised and target-driven.

The lead-up to area-based intervention

During the 1960s, a number of reports raised the prospect of centrally funded and area-based policy action, but their proposals were not accepted by government. The Milner Holland Committee (1965) looked favourably on the idea of designating the worst areas as *areas of special control* in which there would be wide powers to control sales and lettings, to acquire, demolish and rebuild property, and to make grants. The National Committee for Commonwealth Immigrants (NCCI 1967) argued for the designation of *areas of special housing need* to control overcrowding, insanitary conditions and the risk of fire. The Plowden Committee (1967) reported on primary education in very broad terms, but underlined the complex web of factors which produced seriously disadvantaged areas. It recommended positive discrimination for designated schools in the most deprived areas and coincided with the establishment of the *educational priority areas programme* in 1966 (Halsey 1972). The Seebohm Committee (1968) reported on personal and family social services and recommended designation of *areas of special need* to be given priority in the allocation of resources.

The Community Development Projects

Many places in the north of England, Liverpool or Newcastle for example, have a heritage of heavy industry that went through periods of decline throughout much of the twentieth century. By the late 1960s, however, some areas began to suffer more acutely from closure or out-migration of traditional industries, leading to increased competition for employment. In 1969, and in the wake of related tensions over immigration and race relations, the Wilson government announced funding, via the Home Office, for a programme which was described as 'a neighbourhood-based experiment aimed at finding new ways of meeting the needs of people living in areas of high social deprivation' (Community Development Project Working Group 1974: 162). A key assumption of the initiative was that the solutions to deprivation in these areas lay in either improving the relationship between public services or tackling pathological (i.e. self-defeating) cultures of poverty. In response, twelve local projects were initiated between 1969 and 1972 in locations ranging from small towns to industrialised, urban areas. Small-grant funding was made available for community projects in an attempt to promote self-help responses to social unrest, race relations and urban poverty.

Based in universities and polytechnics, the research teams working on the Community Development Projects employed a range of methods in an attempt to understand, measure and react to the problems associated with deprivation. This resulted in a range of innovative measures, from connecting housing groups with planners in the hope of working towards a more coordinated approach to renewal, to helping residents bargain for more investment in their area, to working with schools to develop 'constructive discontent' by encouraging 'a critical stance among children towards their environment' (p. 178). A consistent finding, however, was that many of the problems afflicting the study areas – industrial closures and relocations and

depopulation in particular – were symptoms of wider structural changes which could not simply be overcome through better-organised community action at the local level. It was on this point that some of the projects, drawing on neo-Marxist theories, saw more radical solutions to the problems encountered, which were much less amenable to the 'pathological' position taken by the Home Office. This tone can be felt in the title of Newcastle's CDP report for the Benwell area: *The Making of a Ruling Class: Two Centuries of Capital Development on Tyneside* (Benwell Community Project 1978), while in Liverpool the Vauxhall Report concluded with *Government against Poverty? Liverpool Community Development Project 1970–75* (Topping and Smith 1977).[5]

The CDPs were successful in promoting an illuminating political debate on how to deal with post-industrial environments, a debate which is all too often shrouded by today's parties' concerns to channel resources into particular places, and by the consensus in favour of appealing to private enterprise. In the 1970s, it was possible for those such as Professor Ray Pahl to suggest that a cheaper alternative for government than spending money on technical and managerial fixes would be to provide free heating, lighting and house maintenance to deprived areas as a 'pro-poor' means of luring in population and low-key enterprise. Through this, he hoped to encourage an informal, self-help household economy based on cooperative organisation of activities such as bulk buying of food, home brewing, and passing on children's clothes (Pahl 1978). Others, however, drawing on the tradition of slum clearance and housing-focused intervention, promoted more technocratic approaches based on comprehensive, long-term planning, while there were also those who believed that freeing up the market with tax and welfare cuts would be the most effective way of promoting a more 'sink or swim' approach to individual self-sufficiency in these areas (Gibson and Langstaff 1982).

Urban programme

The *urban aid programme* (later recast as the *urban programme*) was established in 1972 as the principal source of funding in a growing policy focus on deprived areas. It initially funded mainly social schemes, but it was progressively widened in scope to embrace voluntary organisations and to cover industrial, environmental and recreational provision. The 1977 White Paper *Policy for the Inner Cities* and the Inner Urban Areas Act 1978 brought about significant changes. The legislation sought to 'give additional powers to local authorities with severe inner area problems so that they may participate more effectively in the economic development of their areas'. There was also a growing awareness that the loss of employment and population from particular areas – generally areas of older housing immediately surrounding city centres – was being compounded not only by the upheaval and blight associated with ongoing slum clearance, but also by the loss of employment and more affluent households to the suburbs and the new towns. This outward expansion of the urban form had made sense in a context of grow- ing prosperity and full employment, and it was even a specific intention of forerunners such as Ebenezer Howard that outward expansion would lead to the decline of inner-city environments, and ultimately to redevelopment on their citizens' own terms (Howard 1902). But the decline in employment in these relatively central, urban areas was not foreseen and by 1977 there was pressure for a more active approach to managing cities and urban areas.

One of the forums through which the argument for this new approach came to be articulated was a series of Inner Area Studies (IASs), announced by the Conservative government in 1972. Consultants, rather than academics, were responsible for the studies and, while many of the reports had a critical edge, the conclusions of earlier analyses – which placed the blame for deprivation on government and the capitalist economy – were no longer advanced. Nevertheless, the IASs did contribute to a broader shift throughout the 1970s, away from an individualisation of the reasons for deprivation and towards recognition of the economic and systemic causes of poverty and unemployment. In place of damning critiques, the IASs adopted a more measured stance in favour of channelling additional resources to inner-city areas,

stressing that these should complement, not substitute for, more general economic policies aimed at poverty alleviation, such as higher pensions and child benefits (Gibson and Langstaff 1982). In support of this position, a new argument emerged, which served to justify this selective, area-based focus by differentiating 'collective deprivation' from 'personal deprivation' (ibid.). Thus, while personal deprivation (i.e. poverty) was acknowledged as something spread far beyond particular urban areas, collective deprivation referred to supposed additional characteristics of these areas which compounded the personal deprivation of their inhabitants. This was effectively the origin of the controversial thesis of 'area effects', which has since provoked intense academic investigation but little agreement on the scale and significance of such effects (Friedrichs *et al.* 2003).

In many ways, the story of the IASs charts the steps towards early professionalisation of a non-threatening knowledge (for government at least) which could be used as a basis for further government action to address areas which were seen to be 'failing' in terms of their contributions to housing and serving the economic demands of, the population at large. Many of the themes which would later become established mainstays of regeneration thinking were developed at this time: a desire for comprehensive area manage- ment, attempts to strengthen local industries and connect local labour with them through improved training opportunities, and measures to restore market confidence and to 'join up' public services. There are particular parallels between many of these approaches and the housing market renewal initiative (discussed later), which also sought a comprehensive approach to dealing with the role of particular neighbourhoods within wider systems of housing choices.

The end of the Wilson government in 1979 did not spell the end of the urban programme, and by 1990, the number of individual programmes had swollen to thirty-four (NAO 1990), albeit in the very context – of wider cuts to welfare – that the IASs had warned against. At its height, about 10,000 projects were funded each year in the fifty-seven programme areas, at a cost of £236 million in 1992–3. In its later years, almost half the expenditure was devoted to economic objectives and the rest was shared roughly equally between social and environmental objectives. Urban programme funding was largely taken over by the Single Regeneration Budget (SRB) operated from the Government Offices for the Regions from 1995.

Policy for the inner cities and *Action for Cities*

The increasing emphasis on economic regeneration objectives came to dominate urban policy during the 1980s and 1990s. The return of a Conservative government led to a review of inner-city policy, which concluded that a much greater emphasis needed to be placed on the potential contribution of the private sector. Ten years later the rhetoric was much the same. After the 1987 election, Margaret Thatcher, then Prime Minister, announced her intention to 'do something about those inner cities'. The immediate result was the publication of a glossy brochure entitled *Action for Cities* (Cabinet Office 1988). This maintained that, although the UK had benefited during the 1980s by embracing the ethic of enterprise, this change in attitude had not reached the inner city. The aim of urban policy, therefore, was to establish 'a permanent climate of enterprise in the inner cities, led by industry and commerce'. It was claimed that *Action for Cities* programmes involved central government expenditure of £3 billion in 1988–9 and £4 billion in 1990–1, but, in fact, little additional money was involved. Critics argued that the package merely gave the appearance that a determined effort was being made to get to grips with the problems. However debatable this might be (Lawless 1989: 155), it did indicate some reorientation of thinking on urban policy.

Following Michael Heseltine's 1981 visit to Merseyside, in the wake of the Toxteth riots, the Merseyside Task Force Initiative was created. Initially, this was a task force of officials from the DoE and the then Department of Industry and Employment, established to work with local government and the private sector to find ways of strengthening the economy and improving the environment in Merseyside. This proved exceptionally difficult, partly because of the multiplicity of agencies involved.

As a result, an attempt was made in later initiatives to obtain a greater degree of coordination through *city action teams* and *inner city task forces*. First set up in 1985, city action teams (CATs) were to take a broader, even regional, view of the coordination of government programmes. Each team was chaired by the regional director of one of the main departments involved: the DoE, the DTI and the Training Agency.[6] Their funding was limited, which reflected their role as coordinators rather than direct providers. Lawless (1989: 61) argues that they were 'unable to devise anything that might be termed a corporate central-government strategy towards inner-city areas'. They were disbanded at the time of the setting up of the Government Offices for the Regions, which took on the coordinating role.

A total of sixteen inner-city task forces were established in 1986–7. Their role was essentially one of trying to bend existing programmes and private-sector investment and priorities to the inner city. They were initially expected to have a life of two years, but in the event the lifespan has been variable. Their general objective was to increase the effectiveness of central government programmes in meeting the needs of the local communities.

One of Heseltine's early initiatives was the establishment of the *Financial Institutions Group* under his leadership and staffed by twenty-five secondees from the private sector. Their most important proposal was for an *urban development grant* (UDG) on the lines of the American *urban development action grant*. This was introduced in 1982 'to promote the economic and physical regeneration of inner urban areas by levering private sector investment into such areas'. It was flexible in terms of the area covered, local authorities contributed 25 per cent of grant aid and the private sector contribution to a project had to be significant. It was replaced by the *urban regeneration grant* in 1986. In its lifetime, it supported 296 projects at a cost of £136 million, with a corresponding private-sector investment of £555 million. This represented a leverage ratio of about 1:4.

The urban regeneration grant supported ten schemes at a cost of £46.5 million, with private-sector investment of £208 million. Together, they provided an estimated 31,966 jobs, 6,750 new homes and 1,456 acres of land brought back into use (Brunivells and Rodrigues 1989: 66). *City grant* was launched by the *Action for Cities* initiative in 1988, and replaced several existing grants including the urban development grant. Its aims were similar, but it was paid directly to the private-sector developers to support the provision of new or converted property for industrial, commercial or housing development. The final evaluation of these grant regimes noted that they were successful in assisting private-sector investment but the focus on job creation in evaluation did not 'address the basic causes of poor regional or local growth' (Price Waterhouse 1993: 63).

Urban development corporations

The 1977 White Paper on inner cities considered the idea of using new town-style development corporations to tackle inner areas, but the then Labour government concluded that it was inappropriate for the inner cities and that local government should be the prime agency of regeneration. The Conservative government from 1979 thought differently, mainly because it had little faith in the capabilities of local government. The manifest argument, however, was that the regeneration of major areas of our cities was 'in the national interest, effectively defining a broader community who would benefit from the regeneration' (Oc and Tiesdell 1991: 313). The effect of imposing centrally directed agencies into the hearts of major cities created considerable conflict (although not all the local authorities were against the designations), and long-lasting bad feeling about the role of central government in urban regeneration.

The Local Government, Planning and Land Act 1980 made the necessary legislative provisions and defined the role of an urban development corporation (UDC) as being

> to secure the regeneration of its area . . . by bringing land and buildings into effective use, encouraging the development of existing and new industry and commerce, creating an attractive environment, and ensuring that housing and social facilities are available to encourage people to live and work in the area.
>
> (Section 136)

Though their structure and powers were based on the experience of the new town development corporations, the UDCs were different in several important respects. Their task was a limited one and they had relatively short lives, of ten years or so. The first were designated in 1981 and three subsequent 'generations' followed, the last in 1993. All four generations of UDCS had wound up operations by 1998, but a 'fifth generation' was born under New Labour to help drive forward the implementation of the *Sustainable Communities Plan* and its growth areas in the South East (see Table 11.3). (Chapter 6 discusses the *Sustainable Communities Plan*.) The designation procedure was rapid, being made by statutory instrument. Fourteen UDCs were created initially, twelve in England, one in Northern Ireland (Laganside) and one in Wales (Cardiff Bay).[7] Except for the London Docklands, their areas were not large, though they suffered from especially severe derelict land or plant closure problems (Bovaird 1992). The UDCs had extraordinary powers of land acquisition and vesting of public-sector land. They also (unlike the new town development corporations) usurped the local authority's development control functions (except in Wales), for determining planning applications (including their own proposals), enforcement and other matters. In short, they had very wide planning responsibilities and freedom from local authority controls. This was not accidental or incidental: it was an essential feature of their conception. Furthermore, they were run by Boards of Directors drawn primarily from business, and they were accountable only to central government.

Expenditure on UDCs rose to a peak of over £600 million in 1990–1, though in most years it stood around £200 million. At one time, this was by far the largest share of spending on regeneration: in 1992–3 it amounted to half of all inner-city spending, reflecting the significant shift away from local authority directed expenditure. The major share of public funding, and

Table 11.1 Selected regeneration and inner-city expenditure and plans 1987–8 to 2001–2 (£m)

	1987–8	1988–9	1989–90	1990–1	1991–2	1992–3	1993–4	1994–5	1995–6	1996–7	1997–8	1998–9	1999–2000	2000–1	2001–2
UDCs and DLR	160.2	255	476.7	607.2	601.8	515	341.2	287.1	217.4	196.1	168.8	0	0.2		
Estates Action	75	140	190	180	267.5	348	357.4	372.6	315.9	251.6	173.5	95.7	66.9	63.9	39.4
Housing Action Trusts				10.1	26.5	78.1	78.1	92	92.5	89.7	88.3	90.2	86.4	88.4	88.4
City Challenge						72.6	240	209.0	204.9	207.2	142	9.8	1.7		
Challenge Fund									125	265.1	483.1				
New Deal for Communities												0.2	48.5	120.7	450.0
City Grant/Derelict Land Grant[a]	103.5	95.7	73.5	100.6	113.8	145.0	128.7								
English Partnerships (URA)[b]				16.8	–16.4	6.5	24.2	191.7	211.1	224.0	258.8	294.2	225.5	212.2	78.2
Urban Programme[c]	245.7	224.3	222.7	225.8	237.5	236.2	166.5	67.8							
City Action Teams			4	7.7	8.4	4.6	3.4	0.2							
Inner City Task Forces	5.2	22.9	19.9	20.9	20.5	23.6	18	15.4	11.9	8.7	6.0	1.6	1.2		
Single Regeneration Budget[d]									136.4	277.5	458.8	560.9	181.4		
Regional Development Agencies[e]												12.2	593.0	628.5	820.0

(Continued)

Table 11.1 (Continued)

	1987–8	1988–9	1989–90	1990–1	1991–2	1992–3	1993–4	1994–5	1995–6	1996–7	1997–8	1998–9	1999–2000	2000–1	2001–2
London Development Agency														241.8	228.5
Manchester Regeneration Fund/ Olympic bid/city centre					0.8	12.2	26.8	30.2	2	0.3	1.1	5.4	1.7		
Coalfield Areas Fund							2.3	2	0.4	0	0	0	10.0	15.0	10.0

Sources: DoE *Annual Report 1993*, Figure 45; DoE *Annual Report 1995*, Figure 43; DoE *Annual Report 1996*, Figure 31, p. 50; and DETR *Annual Report 2000*, Table 10a

Notes:

a City Grant and Derelict Land Grant became part of English Partnerships funding from November 1993 and April 1994 respectively

b Regional funding for English Partnerships transferred to the regional development agencies from April 2000

c Urban Programme funding became the Single Regeneration Budget from 1995–6

d The Single Regeneration Budget transferred to regional development agencies from April 1999, and to the London Development Agency from April 2000

e Regional Development Agencies funding also includes rural development programmes funding from April 1999

Table 11.2 Selected regeneration and inner-city expenditure and plans 2002–3 to 2013–14 (£m)

	2002–3	2003–4	2004–5	2005–6	2006–7	2007–8	2008–9	2009–10	2010–11	2011–12	2012–13	2013–14
UDCs (Northamptonshire, Thurrock, Thames Gateway)		0.2	2.4	26.2	83.4	120.6	101.7	101.4	76.8	52.7	12.6	18.3
London Legacy Dev Corp									37.4	102.9		
Estates Action	14.2	5.0	2.6									
Housing Action Trusts	95.5	69.4	59.4	16.4	7.8	2.8						
New Deal for Communities	176.4	245.2	281.9	272.8	255.2	254.1	234.8	177.8	73.8	–5.3	–3.5	
Housing Market Renewal	3.9	69.6	197.7	303.6	309.7	404.7	380.8	346.0	261.0	35.5		
Regional Development Agencies	1321.9	1523.7	1455.7	1486.7	1563.3	1588.4	1553.8	1591.1	871.3	439.0		
Local Enterprise Partnerships											5.1	20.6
Regional Growth Fund										465.0	161.2	589.5
London Development Agency		19.9	31.1	14.1	1.2	3.5						
Coalfields Area Fund	15	17.2	27.7	18.5	21.7	17.4	18.8	17.7	19.1	17.5	15.5	13.5
Olympics Delivery Agency							357.5	490.0	438.0	5	14	

Source: Freedom of information request to DCLG

Note: Figures are for spend for all years except 2013–14, where the figure shows the budget for each activity

similarly high levels of private investment, went to London Docklands, and within that area to major projects. Just one project, the Limehouse Link road, required the demolition of over 450 dwellings and cost £150,000 per metre, which made it the most expensive stretch of road in the UK (Brownill 1990: 139).

Final evaluation of the first urban development corporations confirms that property development, land reclamation and physical improvement dominated their objectives and outputs. They made a significant impact on the geography of our cities through the transformation of largely derelict and degraded land largely through massive industrial restructuring. Powers of land acquisition, site assembly, reclamation, and financial incentives were important tools. But they have not met any wider objectives. Furthermore, their attention rarely strayed from the designated areas to considerations about the wider economies in which they were situated, a problem exacerbated by narrow performance indicators. Most UDCs could quote impressive figures (Roger Tym and Partners 1998) but their underpinning assumption, that it is in the public interest to use public money to incentivise high value-added private-sector land uses, has inevitably proved controversial. There is a real question as to how far the 'new' uses have simply been moved from elsewhere in the city or the region. The subsidisation of private investment in this way makes it more difficult for other regions or countries to attract their own investment without resorting to similar choices about how to use public money. This is particularly the case now that the approach has been taken up in most of Europe's largest cities – even its most competitive ones. There is also often a visible disconnect between those able to access the new jobs and those who felt the brunt of earlier rounds of economic restructuring. The early assumption that wealth would trickle down from the new migrant employees to the existing community has been contested, particularly in cases such as the development of Canary Wharf, where the effect has been to create an environment with an 'us and them' appearance, and to push up house prices and demand further beyond the research of those on lower incomes (Minton 2009).

The latest round of UDCs – a fifth generation – exhibits a different relationship between central and local government from that of the 1980s and 1990s. The antagonism between an unyielding Margaret Thatcher and often equally militant local authorities has been replaced by a more managerial form of local government in which the financial rewards and opportunities brought by UDCs seem to matter much more than any ideological stance on the principles that underpinned them.

In 2003, the ODPM began consultation on three new urban development areas and UDCs to support the urban growth areas around London. Thurrock UDC was designated in 2003 and was followed by London Thames Gateway and West Northamptonshire in 2004. A partnership approach – characteristic of New Labour – meant that local authorities were represented on the boards of UDCs and that a corporation acted as the local planning authority only for 'applications directly relevant to its purposes (which are defined as strategic and significant)'. Development planning powers stay with the councils, although the UDCs will prepare their own development strategies that take account of the development plan.

This approach can be seen as part of a wider strategy to deal with the chronic housing shortage in London by casting these issues as being of national importance and therefore warranting a more assertive, top-down approach sweetened with the promise of significant public investment. The measures introduced in Thurrock hope to achieve 8,895 new jobs and 5,969 new homes, while in the Thames Gateway the hope is for 5,000 new jobs and 10,500 new homes, with the drawing in of £1.2 billion of private investment. The West Northamptonshire UDC will have spent nearly £100 million, with a greater focus on industrial investment. However, the Coalition government ended funding for the Thurrock and Thames Gateway corporations two years early as part of a so-called 'bonfire of the quangos', which also saw the end of the regional development agencies. In Thurrock, this meant that responsibility for ongoing projects returned to the local authority. The London Development Agency was able to transfer some of its responsibilities to the Greater London Authority – effectively the only regional body

Table 11.3 Urban development corporations in England: designation, expenditure and outputs

			Annual gross expenditure (£m)		Cumulative outputs			Comments
			1992–3	1995–6	Jobs created	Land reclaimed (ha)	Private investment (£m)	
First generation	London Docklands	1981	293.9	129.9	63,025	709	6,084	
	Merseyside	1981	42.1	34	14,458	342	394	
Second generation	Trafford Park	1987	61.3	29.8	16,197	142	915	
	Black Country	1987	68	43.5	13,357	256	690	
	Teesside	1987	34.5	52.8	7,682	356	837	
	Tyne and Wear	1987	50.2	44.9	19,649	456	758	
Third generation	Central Manchester	1988	20.5	15.5	4,909	33	345	
	Leeds	1988	9.6	0	9,066	68	357	
	Sheffield	1988	15.9	16.4	11,342	235	553	
	Bristol	1989	20.4	8.7	4,250	56	200	
Fourth generation	Birmingham Heartlands	1992	5	12.2	1,773	54	107	
	Plymouth	1993	0	10.5	8	6	0	
Total			621.4	398.2	165,716	2,713	11,240	

The new urban development corporations

			End date	Peak spending	Comments
Fifth generation: sustainable communities	Thurrock	2003	2012	£81.6 m (2011–12)	Powers passed back to local authority
	London Thames Gateway	2004	2013	£55 m (2011–12)	Succeeded by London Legacy DC (A mayoral development corporation)
	West Northamptonshire	2004	2014	£14.5 m (2013–14)	Three separate areas around the towns of Northampton, Daventry and Towcester Designated UDA but managed by a Partnership Board of the local authority, English Partnerships and others

Sources: DoE *Annual Reports* 1993, 1995 and 1996; fifth generation data sourced from Annual Reports and Accounts and London Thames Gateway Development Corporation (2012) 'Passing the Baton'
Note: Leeds UDC was wound up on 31 March 1995. Bristol UDC was wound up on 31 December 1995. The figures for outputs are cumulative over the lifetime of the UDCs and include all activity within the urban development area, not just those of the UDC. Designation orders proposed at time of writing.

to survive localism – and had already passed responsibility for the Olympic Games to the Olympic Delivery Authority. Boris Johnson drew on powers given to him in the Localism Act 2011 to set up a new mayoral development corporation – the London Legacy Development Corporation – to manage part of the Thames Gateway portfolio as well as the physical legacy of the Olympic games. A mayoral development corporation is also being set up in Liverpool. This comes in the wake of efforts by both New Labour and the Coalition to concentrate local government responsibilities in city mayors capable of making bold decisions in how to attract private enterprise.

City Challenge

A major switch in regeneration funding mechanisms was announced in May 1991, in the form of City Challenge. This marked a significant change from the top-down, privatised forms of comprehensive planning and redevelopment which favoured the early UDCs in favour of a more social focus that returned leadership to the local authority but promoted active input from citizens, community groups and the private sector. An explicit aim of City Challenge was to restructure local governance, as well as to promote a pro-market attitude and a more entrepreneurial approach towards exploiting opportunities for enterprise and social improvement in the eleven areas which were initially awarded funding. Thus it took the more opportunistic approach of the UDCs – in terms of their focus on the potential to attract high added-value uses – and sought to apply this in residential areas which had often not seen direct investment as part of the UDC programme. This more entrepreneurial focus was reflected in a competitive bidding process that rewarded those local authorities which were judged as submitting the best tenders to central government.

The return to a more social focus, however, was not matched by additional money for such interventions. Projects received £7.5 million for each five-year period they were awarded funding. Total central government expenditure amounted to £1.15 billion over eight years, which was top-sliced from the Urban Programme budget. In contrast with the Urban Programme, more emphasis was placed on using public money as leverage to attract private-sector investment, so as to direct expenditure towards market opportunities rather than state provision.

As Davoudi and Healey (1994) noted from their work in the West End of Newcastle upon Tyne, the initiative also took place in the context of a challenging economic environment:

> A further and perhaps the most important contextual problem is the impact of the economic recession and the cut backs in local authority resources which are weakening the ability of both the public and private sector to devote time or contribute resources to community development. Economic recession also affects the availability of jobs for local communities. It is within this context that City Challenge efforts have to be justified as useful in themselves (building up community capacity and confidence per se), rather than as an instrumental strategy to get local people into jobs.
>
> (Davoudi and Healey 1994: 8)

This was something that would continue to afflict later regeneration projects, which continued to fall between a central government call for more efficiency and better coordination of services and the need to take practical, local decisions about whether to use regeneration funding to temporarily reinstate services which had been withdrawn because of mainstream cuts. For example, still in the West End of Newcastle upon Tyne, but a decade later, Dargan reported on debates about whether to reinstate a city-wide arson response service to cover the New Deal for Communities regeneration area, which was one of the areas worst affected by the problem (Dargan 2002). The New Deal for Communities initiative has some similarities to the community focus of City Challenge and is discussed later in the chapter.

The government's intention to use City Challenge to shift local governance cultures onto a more entrepreneurial basis was, in the event, constrained by a number of recurrent issues which have beset regeneration programmes since. These included late and unclear central guidance about what was expected, short lead-in times and uncertainty about when funding for the programme would finish. These issues, having emerged from the politics of departmental control, coordination and maintaining the party line in central government, often lead to hasty or insufficient partner engagement in the formulation of bids and the design of governance structures.

A further problem was revealed by tensions within the 'new right' ideology which underpinned City Challenge. As was stated above, local communities were seen to be important partners in a shift away from the 1980s approach. But while there was certainly a desire for some 'downward' accountability in this way, the place of communities in management structures was ultimately variable: sometimes they were involved in management; sometimes only as consultees (Bell 1993). 'Upward' accountability, however, was better established. Although central government cheif not quite achieve a payment-by-results approach, its control of the programme was highly attentive to recording 'outputs', which, in a sense, are politically valuable statistics that are tradable in exchange for the City Challenge funding. But this marketised approach to oversight is blind to any improvements secured by the programme which do not show up in statistical improvements, just as it is poor at understanding the connections between different output figures or the causes for changes in outputs. Davoudi and Healey thus referred to a

> tension between the objectives of the initiative and the Central Government's mode of control. The continuation of the City Challenge funds depends on the authorities' performance which is to be judged against output targets such as jobs and training places provided, stock built and refurbished, and trees planted rather than changes relating to building up the links in routes to jobs, or transforming institutional capacity, or empowering the local community.
>
> (1994: 8)

Ironically, this new right approach to governance would go on to inform programmes under New Labour, and criticism of it would then form the basis on which 'localist' devolution of services to private-sector and community-sector control would be advanced under the 2010 Coalition government.

Fifteen authorities were invited to bid for City Challenge funds in the first round, eleven of which were selected to start on their five-year programmes in 1992.[8] Subsequently, a second round was opened to all urban programme authorities and a further sixteen five-year programmes were approved in 1993. Most City Challenge partnership programmes were in inner-city locations, but a few were on the urban fringe. The Dearne Valley was unique in being 52 square kilometres in area, covering a number of smaller settlements in the South Yorkshire conurbation. City Challenge encouraged an integrated approach, with a focus on property development, but cutting across a range of topic areas, including economic development, housing, training, environmental improvements, and social programmes relating to such matters as crime, and equal opportunities. Evaluations (Russell *et al.* 1996; Oatley and Lambert 1995; KPMG Consulting 1999) have emphasised the value of partnership involving local communities and business, and concluded, that it was relatively good value, with leverage ratios of challenge funding to private investment and to other public funding of 1:3.78 and 1:1.45. Needless to say, many areas had particular problems of sustaining improvement after the end of the programme.

The housing and regeneration agency

The Homes and Communities Agency is tasked with working towards the following aims:

- improve the supply and quality of housing in England
- secure the regeneration or development of land or infrastructure in England
- support in other ways the creation, regeneration or development of communities in England or their continued well-being
- contribute to the achievement of sustainable development and good design in England, with a

view to meeting the needs of people living in England.

(Homes and Communities Agency 2013)

The organisation acts as the housing regeneration agency for England, which has its statutory basis in the Housing and Regeneration Act 2008. The act provides the organisation with powers to assist regeneration through the reclamation or redevelopment of land and buildings, the reuse of vacant, derelict and contaminated land and the provision of floorspace for industry, commercial and leisure activities, and housing. The agency has its own compulsory-purchase powers, allowing it to assemble sites independently of local government. It also now has the power, not just to fund the gap between the costs of undertaking development and the end value but to set up special-purpose companies as well. As Table 11.1 shows, this approach has historically enabled English Partnerships

to draw in significant receipts from land sales to offset, or in the short term sometimes to exceed, its costs. Many of these functions were previously enabled by the city grants and derelict land grants outlined in Table 11.1 and after that by English Partnerships.

In its earlier incarnation as English Partnerships, the national housing and regeneration agency worked on developing and managing its portfolio of strategic sites (many of which originate from the new towns programme), creating development partnerships (often with regional bodies and local authorities) and finding new sources of funding to match public resources. It played a key supporting role in the housing market renewal, urban development corporation and coalfields areas. High-profile projects include the regeneration of the Greenwich Peninsula (£180 million) with the Millennium Dome, the Millennium Villages and the Middlehaven development in Middlesbrough (see Plate 11.3).

Plate 11.3 Middlesborough College on the edge of Middlehaven (photograph taken by David Webb)

English Partnerships led the Millennium Communities Programme,[9] which was launched in 1997 and involved about 6,000 homes in seven locations across England. They were demonstration projects aimed at influencing the practice of the house building industry and local authorities by showing how the design and construction of new communities can incorporate sustainability principles, including good public transport links, energy efficiency, a mix of housing and employment opportunities and community involvement. An 'action research project' on sustainable communities (Llewelyn Davies *et al.* 2000) compared two of the millennium communities (Greenwich and Allerton Bywater) with three other places where sustainability had been a key criterion of regeneration and development. The conclusions pointed to the importance of initial site selection in determining many of the sustainability characteristics of built development, especially in relation to the local provision of services and car dependency. Both compared well on resource consumption with, for example, Poundbury. But the programme overall was criticised for 'seeking trend-breaking results through a fairly conventional large scale top-down commercial development process', such that non-commercial outcomes could be achieved only through the imposition of conditions and the use of subsidies.

Single regeneration programme

In response to criticisms of the fragmented nature of funding programmes for regeneration, and building on the competitive and partnership approach of City Challenge, came the Single Regeneration Budget (SRB). This was introduced in 1994 with the intention of promoting integrated economic, social and physical regeneration through a more flexible funding mechanism. It was administered for the first of six rounds by the government offices and for the last two (one in London) by the regional development agencies (RDAs), although central government maintained considerable influence in both writing the rules and making final decisions on funding allocations. The programme enabled local regeneration partnerships to bid for funding against a list of priorities with considerable flexibility in the specific objectives and measures.[10]

From 2002, responsibility for allocating this regeneration funding had been devolved to RDAs within their single programme funding. So SRB had developed first as an integrated approach to funding regeneration administered by the centre, through regional administration and central decision, to complete devolution to the regions. The mid-term evaluation noted that regional delivery meant that SRB funds could be coordinated with other funds and concentrated on particular areas and interventions. This approach contained, in effect, a practical critique of trickle-down logic which sought to marry job creation with targeted communities:

> Early indications are that the RDAs are deploying funds geographically according to a broad needs/ opportunity formula where relative need is assessed according to the Index of Local Deprivation and opportunity is equated, at least in part, to the ability of local areas to encourage and accommodate new economic development on available sites.
>
> (DTLR 2002: 12)

Four further principles underpinned SRB's design and execution: the need for a strategic approach, partnership among the public, private, community and voluntary sectors, competitive bidding for available funds, and payment by results. The SRB addressed criticisms of short-termism and narrow compartmentalised approaches, and brought together twenty previously separate funding programmes. Schemes could include projects needing funding for up to seven years (or much less), but they had to be 'strategic' and establish links to other investment plans, such as those required for European structural funds and economic development strategies. Ambitious projects were promoted, requiring in bids a long list of anticipated outputs in terms of jobs created or safeguarded, number of people trained, new business start-ups, and hectares of land to be improved (Foley *et al.* 1998).

From Round 5 (the first under the Labour administration) there was an attempt to refocus and target the funding more directly to places in most need. A two-tier funding approach was adopted which required 80 per cent of funding to be channelled into large comprehensive schemes in the fifty most deprived local authority areas defined by four measures on the *Index of Local Deprivation*. The remaining 20 per cent was targeted at pockets of deprivation in the coalfields, rural areas and coastal towns. A further innovation was the provision for financial support for capacity building. Indeed, from Round 6 it was envisaged that this will be a component of most bids and that much of the first year of operation should be devoted to capacity building so that local communities can play an active and effective role in the creation and management of schemes. Even so, the SRB investments were still relatively minor compared with the main programmes of participating departments (Hill and Barlow 1995). Only 22 per cent of the resources required for each project came from the SRB, which put considerable pressure on other public and private sources which they may not be able to meet (Hall and Nevin 1999). On the positive side, SRB promoted more strategic thinking and an increase in operational partnerships (Fordham *et al.* 1998). The value of the funds incorporated into the SRB varied considerably from year to year: £6 billion in 1993–4 but falling to £1.36 billion in 1995–6. Funding increased under the Labour administration and the expenditure in 2001–2 was £1.7 billion. Total funding over six rounds was £17 billion for 1,027 approved bids

The ODPM (now the DCLG) commissioned monitoring of the SRB Challenge Fund through the examination of twenty case study partnerships over an eight-year period.[11] The interim findings (Rhodes *et al.* 2002: 13) indicated that most partnerships covered very small areas 'consisting of a number of wards', although a further 20 per cent covered the whole local authority area and in just over half of cases the local authority was lead partner. Earlier evaluations of the SRB noted the domination of employment and economic development-related output measures (though this is largely in the form of human resource development); the centralised and opaque decision-making on bids; the limited resources in comparison with the scale of the problem, which leads to a thin distribution of funding; and the need for more effort to involve local communities (Hall 1995; Brennan *et al.* 1998; Hall and Nevin 1999). Nevertheless, they agreed that the Challenge Fund and the SRB have had a positive impact on promoting more strategic thinking and partnership working in regeneration. Rhodes *et al.* (2002) make a direct comparison of City Challenge and SRB value for money (noting the difficulties in making such comparisons) and conclude that

> each net additional job created in the City Challenge basket was costing approximately £28,000 alongside a cost per qualification provided to a trainee of £3,450. The broad SRB equivalents are £25,000 per net additional job and £4,200 respectively per qualification provided and, on this broad basis, VFM looks very similar between the two schemes.
>
> (p. 22)

New Deal for Communities and neighbourhood renewal

The incoming Labour administration of 1997 sought to break from the mixture of Conservative and neo-liberal thinking which former administrations had drawn on. Rhetorically, there was a much stronger focus on place-bound social exclusion, with efforts across the policy spectrum to reconnect low-income households with paid (albeit often poorly paid) work. At the same time, New Labour maintained a free-market approach to the economy at large, sticking to the Conservative's spending plans for the first term and also continuing much of the approach to urban policy that it inherited, including the competitive elements in the SRB and the emphasis on partnership with the private and community sectors. But a raft of new initiatives were also brought forward, with the intention of making a concerted attack on social exclusion in the worst parts of cities and redirecting funding through local authorities. In 1998, the Secretary of State announced extra funding for housing renewal and urban regeneration amounting to £5 billion over three years, most of which was to be at the disposal of local authorities.[12] An extra £3.6 billion over three years was allocated to local authorities through their housing investment programmes to start tackling the esti-mated £10 billion backlog in housing renovation, with the priority of improving the quality of the public housing stock. The enhanced urban regenera-tion programme had two main elements – the refo-cused SRB (explained previously) and the *New Deal for Communities* (NDC).

The initiative for the latter came from the Cabinet Office's Social Exclusion Unit report *Bringing Britain Together: A National Strategy for Neighbourhood Renewal* (1998).[13] The report set out the intention to concen-trate regeneration efforts on the most deprived neigh-bourhoods. At the same time it introduced 'floor targets', which meant that the performance of govern-ment departments and agencies would be evaluated on the basis of the worst cases in their areas as well as the averages. As well as the NDC, the National Strategy brought forward the Community Empower-ment Fund (to help people take part in local govern-ance), the Community Chest Scheme to provide small grants for local projects and numerous other schemes to support tenant participation, involvement of faith communities and more (all of which were generally targeted at the eighty-eight most deprived areas). The overall goal was to deepen involvement of communi-ties in urban policy (Chanan 2003).

The NDC was New Labour's flagship programme, championing community-led regeneration as a means of highlighting and deriding the top-down and property-led elements of earlier Conservative policies. It provided £1.9 billion for seventeen first-round (pathfinder) partnerships and twenty-two second-round partnerships to be spent over ten years, with oversight from community boards and beefed-up community representation.[14] The NDC was therefore an instrumental part of a shift to providing considerable sums over an extended period and with an emphasis on prior capacity building, moves which explicitly addressed the problems of earlier programmes such as City Challenge.

Eligible local authority areas were invited to establish a partnership and prepare a delivery plan for neighbour-hoods of between 1,000 and 1,400 households. The principal objectives of the scheme were to tackle 'work-lessness' (poor job prospects), high levels of crime, educational underachievement and poor health, but the pathfinder partnerships have in fact addressed a much wider range of issues. The revised guidance includes supplementary objectives, including a better physical environment, improved sports and leisure opportuni-ties, and better facilities for access to the arts. The pro-gramme gives flexibility to the local partnership to define its objectives (within the priorities listed), its ways of working, and its actions, though its plan also requires approval by central government, and political sensitivities as the programmes progressed led to increasing pressure to respond to floor target measures (Imrie and Raco 2003), which thus mirrored the gov-ernance dynamics of the earlier City Challenge.

The first evaluations reveal delays in community involvement because no structures were in place, a lack of trust among stakeholders and agencies, and difficulties in even understanding what has already been spent in the neighbourhood under other mains-tream programmes.[15] Nevertheless, the New Deal for Communities generated considerable interest and enthusiasm in communities, not least for the freedom it gave to the neighbourhoods to define their own approach. But it needs to be remembered that the opportunities and resources it provided were limited to areas with a population of less than 60,000. One way of looking at this is to say that it was a targeting of effort on the most deserving cases. Another is to say that it was no more than an experiment.

Delivering an urban renaissance

A concerted push for 'urban renaissance' emerged as one of the defining features of New Labour's first term in office. 'Our Towns and Cities: the Future', otherwise known as the Urban White Paper (see Box 11.5), set out the formal policy basis for this. The report was strongly influenced by the Urban Task Force, which drew on leading planners and architects to come up with workable suggestions for modernising the ailing economies of urban and northern England and bring-ing people back to the cities. Its report, *Delivering an Urban Renaissance: The Report of the Urban Task Force Chaired by Lord Rogers of Riverside* (1999),[16] stressed the

BOX 11.5 TACKLING 'SOCIAL EXCLUSION': POLICY OBJECTIVES DURING NEW LABOUR'S FIRST TERM IN OFFICE

The Single Regeneration Challenge Funding Round 6 (1999)

The overall priority is to improve the quality of life of local people in areas of need by reducing the gap between deprived and other areas, and between different groups. The objectives are:

- improving the employment prospects, education and skills of local people
- addressing social exclusion and improving opportunities for the disadvantaged
- promoting sustainable regeneration, improving and protecting the environment and infrastructure, including housing
- supporting and promoting growth in local economies and businesses
- reducing crime and drug abuse and improving community safety.

Our Towns and Cities: The Future – Delivering the Urban Renaissance (2000)

Our vision is of towns, cities and suburbs which offer a high quality of life and opportunity for all, not just the few.
    We want to see:

- people shaping the future of their community, supported by strong and truly representative local leaders
- people living in attractive, well-kept towns and cities which use space and buildings well
- good design and planning which makes it practical to live in a more environmentally sustainable way, with less noise, pollution and traffic congestion
- towns and cities able to create and share prosperity, investing to help all their citizens reach their full potential
- good-quality services – health, education, housing, transport, finance, shopping, leisure and protection from crime – that meet the needs of people and businesses wherever they are.

importance of strategy and cross-cutting or 'joined-up' action throughout the Urban White Paper as a means of addressing the varied economic experience across England. A supportive report on *Competitive European Cities* provided the argument why this was needed. It compared the competitiveness of the eight English core cities with their European neighbours.[17] The findings showed that the English cities were recovering from a period of decline, but that they 'lag behind their European counterparts . . . [in] innovation, workforce qualifications, connectivity, employment rates, social composition and attractiveness to the private sector' (Parkinson *et al.* 2004: 50).

These issues were similarly raised by *The State of English Cities* report which accompanied the Urban White Paper.[18] To address them, the report argued that the principles of policy integration, partnership and local authority leadership that have been developed in urban policy since the early 1980s were also right for the future. However, it also stressed the need for 'rethinking scales of intervention' so as to 'tie city policies to the broader frameworks of regional and subregional strategy' and for consideration of urban–rural interdependencies (p. 46). Urban policy, it argued, should be based on relatively large areas – perhaps with specific planning and fiscal regimes as also promoted by the Urban Task Force. Such recommendations gathered political momentum over subsequent years, leading to a referendum for the introduction of an elected regional assembly for the North East. Ultimately, however, this was ineffective and confederate assemblies were developed in their place. This history no doubt contributed to the subsequent fate of regional government. Rather than being seen as critical to making cities the 'dynamos of the UK national economy', as suggested by Parkinson and Boddy (2004), they would instead come to be labelled as bureaucratic, unaccountable and ineffective by the 2010 Coalition government.

The Urban White Paper itself considers most areas of government policy and their impacts on urban areas, though a significant part is related to the planning system and explaining the need for a 'complete physical transformation' of our towns and cities. The main challenges are:

- to accommodate the new homes we will need by 2021, making best use of brownfield land;
- to encourage people to remain in and move back into urban areas;
- to tackle the poor quality of life and lack of opportunity in certain urban areas;
- to strengthen the factors in all urban areas which will enhance their economic success;
- to make sustainable urban living practical, affordable and attractive.

A White Paper *Implementation Plan* was published in 2001 with a long list of short-term actions, many of which illustrate the difficulty of turning vague but ambitious objectives into practical action. On the physical transformation, much of the actions proposed in the White Paper reflect the findings of the Urban Task Force. This report presented 105 recommendations to government, many of which, as the White Paper explains, have found a positive response. Recommendations of particular significance for the planning system are shown in Box 11.6.

BOX 11.6 SELECTED RECOMMENDATIONS OF THE URBAN TASK FORCE 1999

- Require local authorities to prepare a single strategy for the public realm and open space, dealing with provision, design, management, funding and maintenance.
- Revise planning and funding guidance to discourage local authorities from using 'density' and 'overdevelopment' as reasons for refusing planning permission, and to create a planning presumption against excessively low-density urban development.
- Make public funding and planning permissions for area regeneration schemes conditional upon the production of an integrated spatial master plan.
- Develop and implement a national urban design framework, disseminating key design principles through land use planning.
- Place local transport plans on a statutory footing and include explicit targets for reducing car journeys.
- Introduce home zones using tested street designs, reduced speed limits and traffic calming.
- Set a maximum standard of one car parking space per dwelling for all new urban residential development.
- Develop a network of regional resource centres for urban development, coordinating training and encouraging community involvement.

- Produce detailed planning policy guidance to support the drive for an urban renaissance.
- Strengthen regional planning – provide an integrated spatial framework for planning, economic development, housing and transport policies; steer development to locations accessible by public transport; and encourage the use of subregional plans.
- Simplify local development plans and avoid detailed site-level policies.
- Devolve detailed planning policies for neighbourhood regeneration into more flexible and targeted area plans, based upon the production of a spatial master plan.
- Review employment land designations and avoid over-provision.
- Reduce the negotiation of planning gain for smaller developments with a standardised system of impact fees.
- Review planning gain to ensure developers have less scope to buy their way out of providing mixed-tenured neighbourhoods.
- Oblige local authorities to carry out regular urban capacity studies as part of the development plan-making process.
- Adopt a sequential approach to the release of land and buildings for housing.
- Require local authorities to remove allocations of greenfield land for housing from development plans, where they are no longer consistent with planning objectives.
- Retain the presumption against development in the green belt and review the need for designated urban green space in a similar way.
- Prepare a scheme for taxing vacant land.
- Modify the General Development Order so that advertising, car parking and other low-grade uses no longer have deemed planning consent.
- Streamline the compulsory purchase order legislation and allow an additional 10 per cent payment above market value to encourage early settlement.
- Launch a national campaign to 'clean up our land' with targets for the reduction of derelict land over five, ten and fifteen years.
- Introduce new measures to encourage the use of historic buildings left vacant.
- Facilitate the conversion of empty space over shops into flats.

Source: Towards an Urban Renaissance: The Report of the Urban Task Force Chaired by Lord Rogers of Riverside (1999). The Urban Task Force report makes 105 recommendations in total, and many not listed here are of interest to planning. The full list of recommendations together with explanations of the government's response is given in an annexe to DETR (2000) *Our Towns and Cities: The Future – Delivering an Urban Renaissance.*

The most significant planning-related action on the task force report was the revision in 2000 of PPG 3 *Housing* (discussed in Chapter 6), which marked a sea change in the way housing density was perceived, funnelled new development back to urban areas and promoted the reuse of brownfield sites. The report also led to the return of master plans in regeneration; the creation of home zones; initiatives on the training of planners, designers and developers; the creation of regional centres of excellence to address skills improvement in each region; and the creation of urban regeneration companies (as described below). It even went as far as to argue for more local capture of business rate and council tax increases in a move that foreshadowed Coalition changes by over a decade

The Urban Task Force published a review of progress under the title *Towards a Strong Urban Renaissance* in 2005, noting that government initiatives and 'sustained economic growth and stability' had led to some success and that 'For the first time in 50 years there has been a measurable change in culture in favour of towns and cities, reflecting a nationwide commitment to the Urban Renaissance' (p. 2). The review made more than fifty recommendations, including the provision of more guidance on highways design, enforcing reviews of parking standards and measures to promote the development of small infill sites especially for affordable housing. Others, such as the imposition of an energy efficiency obligation on developers, were more ambitious but gained support over time.

In hindsight, there is no denying the radical impact which the Urban Task Force had on the way planning was carried out and on the level of ambition for what it was possible to achieve in post-industrial urban areas. In contrast with the permissive, private sector-led approach to design that had become the norm by the 1990s, the Urban Task Force inspired a new generation of planners and designers to do better and to reimagine unashamedly urban areas, rather than pseudo-rural idylls, as desirable places to live. Some of the reforms of that period, such as the direction to prioritise urban brownfield sites, have since been diluted for fear that gardens were being lost and that costs were becoming prohibitive, but these fears are in many ways the logical outcome of a policy direction that successfully brought development back to the cities. A raft of private-sector companies now specialise in urban redevelopment and many continue to promote the values of the urban task force even under a changing policy framework. On the downside, there are those who have called attention to the gentrifying impact of the return to the city movement (Lees 2003), something that has beset London to a far greater extent than elsewhere. Others have viewed the initiative as something of a missed opportunity, exciting in its original conception but falling well short of its aim of providing a contemporary version of the high-quality historic environments found on the Continent. For Hatherley (2010), what was built ultimately reflected a wider ideological compromise between permissiveness towards private-sector speculative development and a desire to build environments fit for people and communities to live in for the long term. Others have voiced similar fears that the undercurrent of urban renaissance was its use to build large quantities of flats, often with poor space standards and of dubious quality and sometimes to the point of undermining demand for the existing housing stock (Haughton 2009).

Urban regeneration companies

This proposal from the Task Force Report has been taken forward energetically and the initial White Paper target of fifteen urban regeneration companies (URCs) was easily met. A first phase of three pilot companies in England – Liverpool Vision, New East Manchester and Sheffield One – was followed by thirteen more accepted applications by mid-2004. Twenty-one URCs were approved in England and Wales by early 2005 (see Table 11.4). However, only four organisations – Gloucester Heritage, New East Manchester, North Northants Development Company and Barrow Regeneration – now operate as URCs. Ten further organisations are either city development companies (CDCs) or economic development companies (EDCs), their changing names and roles reflecting their geographical and organisational focus. These are:

- 1NG Newcastle Gateshead
- Creative Sheffield (formerly Sheffield One URC)
- Liverpool Vision
- Prospect Leicestershire (formerly Leicester Regeneration Company)
- Coast (formerly ReBlackpool)
- Tees Valley Unlimited (formerly Tees Valley Regeneration)
- Forward Swindon (formerly The New Swindon Company)
- Opportunity Peterborough
- CPR EDC (formerly CPR Regeneration)
- Regenerate Pennine Lancashire.

The type of approach to regeneration which the above organisations promote was well established prior to the Urban Task Force report, though the arrangements may have been more informal. A regeneration company was established in Nottingham in 1998, and some years before a number of councils established strong interagency working arrangements, as in the case of the Heartlands initiative in Birmingham (Wood 1994). In Scotland, the approach is particularly well established through the eight local development companies (LDCs) in Glasgow designated from 1986 and involving a partnership of the City Council, Scottish Enterprise, the social inclusion partnerships and other stakeholders. Following the report on the three pilot companies in England, in 2003 the Scottish Executive consulted on a Scottish variant of the urban regeneration company. Three pathfinder URCs were designated in Scotland in 2004 with a share of £20 million additional funding from the Scottish Executive. Six Scottish URCs are now in existence: Raploch, PARC Craigmillar, Clydebank Rebuilt, Riverside Inverclyde, Irvine Bay and Clyde Gateway. Similar arrangements are also in place in Londonderry, Northern Ireland.

The companies are established by a partnership of local authorities, the regional development agency and other business and community stakeholders. Local strategic partnerships and local inclusion partnerships (Scotland) are also considered important partners. In England, English Partnerships was also a partner in many URCs and provides guidance and support to them all. URCs are independent companies whose operation costs (revenue funding) come from the partners. Private-sector funding of running costs has been encouraged through a tax relief incentive. They do not have capital funding separate from their sponsoring bodies and do not undertake development activity themselves. Essentially, the URCs provide a focus for existing funding rather than generating new money

The original rationale for URCs flows from the analysis of market failure and barriers to implementing physical urban regeneration noted in the Urban Task Force report and numerous other sources. Their main task was to create a favourable climate for private-sector investment in places where there are 'latent development opportunities'. Since then, Coalition ministers have been reluctant to offer a single definition of regeneration, arguing that it should be down to local people to decide, and have resisted the suggestion that the role of regeneration is to tackle market failure. The Homes and Communities Agency's role is to support and assist URCs by providing best-practice support, guidance and training, by providing advice on how to set up new URCs, EDCs and CDCs, possibly with some financial assistance, and by assisting with site assembly and joint-venture agreements.

Evaluations of urban policy and planning implementation over many years point to the importance of leadership, skills and capacity, coordination of activity, and building market confidence, as well as the need for funding. URCs address barriers by engaging with the private sector in regeneration within a wider strategic framework (independent from the local authority), champion the potential of areas, and coordinate plans and actions. Buy-in and support from the wider statutory planning system are often an important factor if URCs are to work effectively (Amion Consulting 2001).

URCs, CDCs and EDCs present an interesting example of how national government has slowly been able to influence the agendas and behaviour of local governance. As with most of the other partnership-based initiatives during the 1990s and 2000s, the proliferation of these companies says much about national government-led attempts to restructure the role and purpose of local authorities away from the delivery of professionalised public and welfare services in favour of projecting them as entrepreneurial agents capable of drawing on community and market opportunities. The 2000s were marked by copious guidance on URCs and the requirement for formal approval from the ODPM and DTI. The fact that the Coalition now feels able to take a hands-off approach says much about the success of moves towards more entrepreneurial local governance. This is not to deny that such an approach can be effective as a way of encouraging markets to meet more of the employment and consumption needs of local citizens. Raco *et al.* (2003) explain how it is argued that a more local focus for urban regeneration policy can provide more flexibility and responsiveness. It may help to deliver a more effective management role in local service

Table 11.4 Urban regeneration companies in the UK

Company name	Established
Bradford Centre Regeneration	2002
Catalyst Corby	2001
Central Salford	2004
Clydebank	2004
CPR Regeneration (Camborne, Poole and Redruth)	2002
Craigmillar, Edinburgh	2004
Derby Cityscape	2003
Gloucester Heritage	2004
Hull Citybuild	2002
Leicester Regeneration Company	2001
Liverpool Vision*	1999
New East Manchester*	1999
Newport Unlimited	2002
Peterborough City	2005
Raploch, Stirling	2004
ReBlackpool	2005
Regenco Sandwell	2003
Renaissance Southend	2005
Sheffield One*	2000
Sunderland Arc	2004
Tees Valley Regeneration	2002
The New Swindon Company	2004
Walsall Regeneration Company	2003
West Lakes Renaissance (Furness and West Cumbria)	2003
Other local initiatives have the same characteristics as URCs	
ILEX (Londonderry and the Derry City Council area)	2003
Nottingham Regeneration Limited	1998
Eight local development companies in Glasgow City Council	1986 on

Notes: List updated February 2005
    * indicates the first three pilot companies in England
    Those organisations in bold have now become either city development companies or economic development companies
    Those organisations which are underlined indicate those which are still operating as UDCs

Plate 11.4 Housing stands empty and boarded up in the Gresham area of Newcastle. It cannot be compulsory-purchased, as the council can no longer demonstrate the viability of the development planned to replace it. However, the council also refuses to make it available for community-led refurbishment (photograph taken by David Webb).

premature ending of financial support for the programme means that for many areas this aim has not been achieved or that the scale of investment may not be sufficient to achieve confident and lasting change. The potential leverage rates on offer to further public investment from 2010 onwards were therefore attractive and the decision to cut funding was heavily criticised on these grounds by the cross-party Regeneration Select Committee in 2011. Nevertheless, there are still hopes from some (e.g. Rydin 2013; Steele 2012) that a reduction in regeneration funding and the rise of a smaller number of opportunities targeted at the third sector might lead to a shift towards community-based area development.

Urban policy under the Coalition

Two watchwords underpin the Coalition's approach to spatially targeted regeneration funding: growth and localism. Of these, the approach to growth is easiest to define since it is essentially a return to the doctrine of market liberalisation popularised under Margaret Thatcher. New Labour's attempts to manage markets, by addressing market failures and steering urban development towards more socially and environmentally sensitive forms, often entailed the use of strategic, spatial planning together with regional coordination of supporting investment. While the Coalition has not tended to disagree vocally with the social and economic aims advanced by this machinery, it has strongly objected to the means of achieving them. The removal of this machinery has been couched in the now familiar language of removing red tape in order to free up markets and to promote short-term economic gains over long-term place management. This has been accompanied by strong resistance to the use of the word 'region' within government, with the favoured alternative being 'localism'. However, localism under the Coalition means something very specific. In practice it has not meant making the private sector more local, for example by breaking up large banks into smaller units and countering oligopoly as proposed by Vince Cable, the Secretary of State for Business, Innovation and Skills. Rather, as the current Conservative chairman puts it, it is about working with the grain of markets rather than trying to manage them (DCLG 2012c). This in itself implies a different relationship between central and local government. One interpretation of localism is, therefore, that it is an attempt to empower localities that agree to buy into this project for a new relationship between the state and the market.

Incentives: the New Homes Bonus and City Deal proposals

The abolition of regional planning, housing and economic development functions removed the principal mechanism by which government had sought strategic control over spatial patterns of development. And yet it has been important to retain a form of control over priority issues, particularly housing and economic development, which have implications both for the national economy and the cost of living. In addition to planning policy changes, and in the context of significant cuts to the revenue support grant provided to local authorities, government has looked towards financial penalty and incentive structures, organised at the national level, in the hope of achieving compliance with national agendas.

The New Homes Bonus is the principal scheme for incentivising the delivery of new housing and runs at £2.2 billion for the four years between 2011 and 2015, of which £950 million is grant funding which replaces the previous government's planning delivery grant fund, while just over £300 million a year is top-sliced from the revenue support grant. The scheme is intended to increase rates of house building by compensating local authorities for the additional demands placed on local services by new areas of housing and is also intended as a sweetener to local authorities faced with opposition from local residents to new housing (much in the same way as the 'Boles Bung' is intended as an incentive to neighbourhood plan making; see Chapter 4). The use of incentives to facilitate new housing delivery assumes that local authorities will change their behaviour in order to maximise their financial return and that this behaviour change will be enough to affect the number of new homes built in the locality. However, the delivery of new homes is about more than just planning and is dependent on consumer demand, available finance and the behaviour of landowners and developers.

If local authorities feel that it is not within their power to affect housing delivery significantly, then the implementation of the New Homes Bonus may simply result in a further complication for public-sector budgets, a boon for some and loss for others that is simply managed rather than planned for. Indeed, it is recognised that the New Homes Bonus leads to a transfer of funding from the north to the south (Nevin and Leather 2011). In 2013, a report from the Public Accounts Committee argued that there was no credible evidence that the New Homes Bonus was working:

> The Department keeps basic records of how much Bonus has been paid to individual authorities, for how many homes and of what type. However, even with the Bonus now in its third year, there is little reliable evidence about its impact on the creation of new homes.
>
> (House of Commons Committee of Public Accounts 2013: para. 8)

The report also notes that the government had yet to provide disaggregated figures from which the cumulative spatial impact of the New Homes Bonus, the Regional Growth Fund and changes to the revenue support grant could be determined.

Not all elements of the government's economic development strategy, however, have relied on incentives. In December 2011, the policy paper *Unlocking Growth in Cities* set out an economic strategy for the core cities, building on arguments going back to the Urban Task Force report that English cities underperform when compared with their European counterparts. Pages eight and nine of the paper set out a menu of indicative things that the government would be willing to negotiate 'as part of the deal making process' of empowering cities (Cabinet Office 2011b: 2). Local authorities were then invited to select and innovate from this menu and to compete with one another for central government agreement. Unlike the New Homes Bonus, the agreement of City Deals has similarities to the 'single conversation' approach adopted by New Labour. As before, the City Deals encourage local authorities to get on board with the central government agenda or risk losing out. While City Deals do not have a dedicated funding stream, they incorporate opportunities to access other funding streams, such as the Regional Growth Fund and money for superfast broadband. They also offer the opportunity to retain a higher proportion of any future increase in business rates, and to borrow against projected increases to fund investment through tax increment financing.

Regional Growth Fund, local enterprise partnerships and enterprise zones

A further plank of the government's support for economic development is its Regional Growth Fund (RGF), chaired by Lord Heseltine and designed to help restructure the economy in those parts of the country which rely on public-sector employment and which were expected to be worst hit by cuts in public expenditure. To some extent, the RGF replaces RDA funding, which was abolished along with regional planning by the Coalition. As might be expected, the RGF is focused on adding value to private businesses where this will allow increased employment or where it will unlock development, rather than on supporting more strategic and longer-term projects such as infrastructure improvements. Another difference lies in the scale of the funding made available, with an effective cut from £2 billion a year at the peak of RDA funding to a total of £3.2 billion of RGF funding to run from 2011–17. Unlike RDA funding, however, government figures do show that this money has been disproportionately spent on more deprived and economically vulnerable parts of the country, with the north-west and north-east regions accounting for a representative 31 per cent of funding in round five (BIS 2014).

The first four years of the Coalition government have seen the emergence of a diverse array of central government funding pots in lieu of the more stable programmes seen during the 2000s. These initiatives stand in contrast to the conscious efforts made by New Labour to pursue long-term regeneration funding and are marked instead by almost constant new announcements and shifting of budgets. A new, £2 billion subregional development pot, for example, will be formed by shifting money from elsewhere, including the New Homes Bonus, which will lose £400 million. We are rapidly returning to the days of 'initiativitus', which beset both John Major's government and the first term of Tony Blair's premiership. Against this backdrop, local authorities have been encouraged to set up local enterprise partnerships (LEPs) with local businesses in readiness to form bids for City Deal powers and other opportunities as they emerge. The approach to neighbourhood regeneration has been similar, with residents encouraged to set up neighbourhood forums and use a 'regeneration toolkit' (DCLG 2012e) to help them draw on powers and funding pots which may be relevant to their area. While LEPs in particular do offer some means of co-ordination, the dangers associated with initiativitus include a loss of capacity to secure added value by joining up investment and a much reduced ability to evaluate whether spending is actually being effective.

Whether such dangers will emerge to challenge the programme of economic development funding laid down by the Coalition is uncertain, since the scale of funding available is much reduced. In the words of one commentator, we have moved from a funding feast to funding famine (Healey 2013). Furthermore, the

BOX 11.8 CITY DEALS: A COMPARISON OF THREE CITIES

Bristol

- Business rates will be retained locally across a network of enterprise zones.
- A city growth hub in the Temple Quarter Enterprise Zone with an indication of funding for a metro system to cover Greater Bristol.
- Establishment of a job search hub in the Temple Quarter with support for emerging small enterprises.
- Private sector leadership of colleges' funding and £5 million of extra investment.
- Pooling and sweating of public-sector land assets

Nottingham

- Establishment of a creative quarter to target technology companies with an accompanied community-interest company which will employ 250 apprentices.
- Moves to align training more closely with employment opportunities.
- Tax increment funding for public transport improvements to connect the Creative Quarter with the region.
- Superfast broadband improvements.

Newcastle

- Establishment of four accelerated development zones to assist development of key sites, including the Gateshead Quayside, retail development in the city centre, development around the railway station and the Science Central area on land that was formerly used to brew Newcastle Brown Ale.
- More apprenticeships and a strategy to tackle young people not in employment, education or training (NEETs).
- Strategic highway improvements to the A1.
- Work on a North East combined authority to assist further devolution to the region.
- Superfast broadband improvements.

So far, the eight biggest cities outside of London have agreed City Deals. These are:

- Greater Birmingham
- Bristol Region
- Greater Manchester
- Leeds City Region
- Liverpool City Region
- Nottingham City Region
- Newcastle Region
- Sheffield City Region.

Twenty further cities have submitted City Deals under wave two.

local community organisations, the Scottish Office, Scottish Enterprise, and other agencies operating in the areas, but were criticised for being inward-looking with little strategic context (Hall 1997). The final evaluation of the urban partnerships (Tarling *et al.* 1999) argues that the £485 million invested over ten years was cost-effective: 3,726 new homes were built and 9,253 improved, employment has improved in two of the partnerships (but fallen in one), and crucially, the partnerships have improved the image of the estates. But the fundamental problems of poverty and disadvantage remain, and a key reason is the 'churning' of the population as newly employed residents move on (to more attractive locations), to be replaced by unemployed households. In 1993, small urban regeneration initiatives (SURIs) were set up to take forward the partnership approach in eleven council estates across Scotland. The depth of the 'partnership' thinking is reflected in key publications at that time, including *Progress in Partnership* (1993), *Programme for Partnership* (1995) and the report *Partnership in the Regeneration of Urban Scotland* (1996). As in England and Wales, a competitive system of allocating funds to partnership projects was established. The main share of urban regeneration resources was ring-fenced for twelve priority partnership areas.[19] In 1999 the urban programme in Scotland was replaced by the Social Inclusion Partnership Fund with £1.3 million extra funding and extra money for the priority partnership areas to develop much needed dedicated support units. As in England, there were significant attempts to join up services and budgets across the public sector during the 2000s, with greater emphasis on social rather than physical forms of policy action.

Northern Ireland

Urban policy in Northern Ireland has followed a similar pattern to that in the rest of the UK, with early emphasis on social problems giving way to a concentration on property-led regeneration and, in recent years, a shift of attention to place marketing, public–private partnerships and community development (Berry and McGreal 1995a). However, the special circumstances, notably the violence and subsequent central government control of policy and implementation, have been important in shaping the problem and responses. Urban conditions and unfairness in employment and housing allocations were important factors in the start of 'the troubles' from 1969, and terrorism has subsequently accentuated the difficulties of tackling them.

Regeneration in Northern Ireland is the responsibility of the Department for Social Development, which was set up in 1999, and assistance is provided by the Northern Ireland Housing Executive. In June 2003, the department published *People and Place: A Strategy for Neighbourhood Renewal*, which identified neighbourhoods for renewal, concentrating on those in the top 10 per cent most deprived areas. The approach to neighbourhood regeneration differs from that of the Coalition in that there is more focus on building capacity within communities and on linking and integrating communities. This more managerial stance no doubt responds to the distinctive and difficult nature of community in Northern Ireland, with its much greater emphasis on sectoral territory and division. However, it also recognises the interdependency between 'community' action and public-sector support which is often glossed over by localism arguments in England, albeit potentially at the expense of the statutory community powers or 'rights' given to English communities by the Localism Act 2011. Another difference is the extensive monitoring of progress towards the aims of the People and Place strategy, with neighbourhood-based statistical indicators being used in an attempt to identify what works across different neighbourhood renewal areas. To date, some of the biggest challenges in Northern Ireland have revolved around the pursuit of mixed communities, dealing with speculative private landlords and the need for a general improvement in the standard of the housing stock. Town centre reinvigoration has also been a priority.

Urban policy in the most deprived areas of Belfast is managed by the Belfast Regeneration Office (BRO) and implemented through four regeneration teams which coordinate activities in thirteen urban renewal areas. The BRO manages the allocation of the

Neighbourhood Renewal Fund, urban development grants and EU funding. Urban renewal in Belfast has a long history: the BRO previously managed the programme *Making Belfast Work*, which covered the thirty-two most deprived wards in Belfast and committed £275 million on 350 projects between 1988 and 2004. A separate Londonderry Initiative was established in 1988 and ran until 2004, spending £42 million in the most deprived parts of the city. Regeneration here and in neighbouring towns is now managed by the North West Development Office (NWDO). A separate People and Place renewal strategy is being developed for Londonderry. Both Belfast and Londonderry also have comprehensive renewal schemes which allow for the compulsory acquisition of land and property after consultation. Elsewhere in Northern Ireland the Regional Development Office (RDO) coordinates regeneration activity. There is one urban development corporation, Laganside, which was established in 1989 following the closure of shipbuilding yards. After development investment of £800 million and considerable change to the areas alongside the River Lagan, the corporation closed in March 2007.

Local government in Northern Ireland is currently undergoing reform, with twenty-six local authorities being reduced to just eleven. This process is expected to be complete by April 2015 and will see the transfer of regeneration responsibilities to these councils. Housing functions will be maintained at the Northern Ireland Housing Executive. The urban policy budget in Northern Ireland amounted to more than £48 million between 1990–2000, while the *Urban Initiative for Peace and Reconciliation* provides over £10 million for urban regeneration to improve the quality of life, enhance the environs of sectarian interface areas and support a wide range of regeneration projects. To this, however, must be added the considerable funding (£63 million p.a.) that came by virtue of the designation of the whole of the Province as a European Regional Development Fund (ERDF) Objective 1 region until 1999 (with transitional funding to 2006) and the significant inward cash flows for employment and housing investment through mainstream funding programmes.

Evaluation of urban and area-based policy

Evaluation of the effectiveness of government initiatives has been a feature of urban policy since the early 1970s, but has gained particular importance since the mid-1990s as part of the search for 'evidence-based policy'. Modern management techniques have also come to the fore with the formulation of more specific objectives, targets and indicators which can be monitored and evaluated. This has made the easy option of vagueness unacceptable in principle, but practice is a different matter, as specific targets are mostly poor substitutes for the rhetorical and vague goals of 'regeneration', 'renaissance', 'liveability' or 'sustainable development', the meaning of which is often fluid and may not necessarily stay the same during the course of a particular regeneration programme (Gunder and Hillier 2009). Measures or outputs such as training places, jobs, visitors, roads and reclaimed land can be counted but are often disputed. Worse still, they do not give a real assessment of outcomes. To what extent have the fundamental objectives of regeneration been realised? And do such measures give a real account of the improvement of places, and social and economic life?

In answering these questions, there is an extraordinary difficulty at the outset: how to define clearly what the objectives of policy are. The problem is very familiar to policy analysts (Rittel and Webber 1973), but it has to be constantly tackled anew by policymakers. A good example (and this discussion has to be illustrative rather than comprehensive) is the apparently simple matter of increasing employment. The problem is one of finding a satisfactory definition (or even concept) of the term 'new employment'. Other complications arise when account is taken of the 'life' of new jobs created: many of the jobs created in the course of regional development programmes later disappeared (Hughes 1991). Moreover, there may be a lag in the growth of jobs which could be difficult to take into account. Again, policy may be directed not to the objective of short-term job creation, but towards increasing the long-term competitiveness of the area in a changing national and world economy. This poses obvious difficulties of evaluation. To take this further, if the aim of policy is wealth creation (a term that was popular for a short time in the mid-1980s) or sharing prosperity (as used in *Delivering an Urban Renaissance*), any thought of evaluation becomes mind-bogglingly difficult. How is wealth or prosperity to be defined (particularly in these environmentally conscious days)? Is the object to raise the average level of wealth, or the level of those who are the poorest? Such questions quickly undermine attempts at evaluation. Similar questions can be asked of current concerns with equality of opportunity and social cohesion.

Despite all the conceptual and practical difficulties, researchers have been able to draw some important conclusions from their evaluations of urban policy. For example, many of the jobs that have been 'created' would have arisen without any intervention. Indeed, if the projects that have been evaluated are typical, 'such programmes are unlikely to make more than a modest contribution to the economic regeneration of the inner cities' (Martin 1989: 638). On reflection, this is perhaps unsurprising. In a complex interdependent society (and, increasingly, an interdependent world), 'local' issues are elusive. In Kirby's words (1985: 216), 'we cannot attempt to understand the complexities of local economic affairs *in situ*'. Much research corroborates this view. The experience of the SDA in Glasgow showed that, though the provision of premises attracted some firms to the area, this was 'at the expense of other parts of the city, and most jobs were filled by inward commuters anyway' (Turok 1992: 372).

Another issue in which difficulties of assessment abound (and in which myths live on) is that of the impact of property-led development. Much urban policy has been based on the assumption that property development will somehow or other stimulate economic growth and social improvements. How this is to happen has not been articulated, and there has been little detailed research on the subject. Such research as has been undertaken offers no clear conclusions, though studies 'suggest that access to markets, management abilities, and the availability of finance are more important than buildings' and 'levels of investment in product development and production

technology, together with differences in the way human resources are managed, are most significant' (Turok 1992: 367).

Many factors other than property will play a part in successful regeneration, such as the availability of a skilled labour force, ready finance and an attractive environment. Moreover, property development can present its own problems. This became clear in 1990, when rental values fell as the economy dipped. The slowing down of property investment in 1990 turned into a spectacular collapse, with catastrophic effects on the construction industry and local economies.

Urban development corporations above all have failed to consider the relationship between the local economy and property development in adopting a policy of privatism, 'the attracting into the inner city of private developers whose activities can in turn demonstrate that regeneration is taking place' (Edwards and Deakin 1992: 362). We have learned that this type of urban policy can have a detrimental effect on local economic activity, as, for example, when the precarious position of small local firms is challenged with competition from outside the locality. Urban policy has been characterised by short-term thinking centred on getting the best return from particular sites (CLES 1992). There is little to support the view that property-led urban regeneration produces a trickle-down of benefits for the local disadvantaged community as the local economy improves. Conclusions from a comprehensive evaluation of urban policy suggest other policy priorities would have greater impact (Robson *et al.* 1994).

These observations on property-led urban regeneration are not to suggest that physical improvements are not needed, and to some extent there has been a return to a concern with the physical quality of places in government policy. It will need to remain a central part of urban regeneration. The point is to understand the ways in which physical regeneration opportunities link to social and economic development. This entails ensuring that 'non-physical policy interventions . . . keep the momentum of regeneration rolling forward once physical rebuilding is complete' (Carley and Kirk 1998).

Much policy takes the form of targeting resources in the most deprived areas. In practice this has had only a marginal impact, because the cuts in mainstream public spending tended to fall more heavily on some of the most deprived areas, leaving them even worse off per capita (Robson *et al.* 1994). Not surprisingly therefore, there was increasing concentration of the most disadvantaged in the worst-off areas. Also, as some people in these areas benefit from intervention and gain employment, they are more likely to move out and be replaced by others in greater need. Housing allocation policies tend to reinforce these effects, which leads to increasing concentration of unemployment in the worst areas. Thus there is growing polarisation within the conurbations, with the benefits of targeting being felt most by the surrounding areas that are better placed to take advantage of it.

Providing an overview of urban policy impact is daunting. Researchers have underlined the interlocking nature of urban policy, continual change in programmes and the difficulty of identifying a single unambiguous set of objectives against which to measure progress as the main difficulties. On the positive side, Robson *et al.* (1994) found that regeneration funding has had a positive impact on residents' perceptions of their area. There has also been some general 'limited success for government policy', particularly in smaller cities and the outer districts of conurbations. Where well-coordinated multi-agency approaches have been taken, some policy instruments have worked well. But 'the amount of money going into urban policy is minuscule compared to the size of the problems which are being tackled . . . many of the poorer areas are not improving or at least not nearly as much as the better-off areas within the districts'.

Does government take heed of the messages from evaluation? The speed at which new initiatives are introduced would suggest not. Ho (2003: 2) explains how 'new regeneration initiatives have often been introduced before the evaluation studies of an ongoing research initiative were completed'. City Challenge replaced the Urban Programme before evaluation was completed; the New Deal was announced while SRB evaluation was still underway. Moreover, there is little evidence to support some aspects of policy and action. A review of the evidence base for regeneration policy found that there was much evidence for physical and economic outcomes and little for health and education outcomes. Moreover, they found that much of the evidence base was not robust and relied on a small number of case studies of 'best practice' (Tyler *et al.* 2001, quoted in Ho 2003: 3). Moreover, the influence of learning 'what works where' is probably greater now than ever, and the large number of initiatives and speed of change arise in part from a desire to experiment. The major initiatives are routinely accompanied by complementary (action) research projects, and other steps have been taken to improve feedback on performance, such as the *Urban Sounding Board*, development of town and city indicators (Wong 2002), research seeking transferable lessons from 1990s initiatives, such as the enterprise zones, and the *Towns and Cities: Partners in Urban Renaissance Project* (URBED 2002). This last project was intended 'to take the pulse of urban renaissance delivery' and involved twenty-four cities and towns in action research to identify and tackle barriers to progress. All very well, but after five main reports and yet more case studies were published for the 2002 Summit, no more was heard. As Ho reminds us: 'the purpose of evaluation is not necessarily to learn lessons' (2003: 4).

Some lessons were taken on board by the post-1997 government. In a review of the prospects for change, Lawless (1996) points to the need for a context which encourages local strategy building rather than centrally directed ad hoc responses. Extraordinary as it may seem, and admittedly with the important exception of jobs and employment, problems which most affect the urban disadvantaged have received minimal attention. But this in turn calls for a new form of urban governance which would be stronger at both the regional and the local community levels, and a more politically mature programme. Bailey *et al.* (1995: 229) conclude more radically that

> cities need to be seen as an important element of the national economy and that the growth, redevelopment and improvement of these assets can and should be linked with redistributive welfare policies as part of a strategic and comprehensive national economic policy driven by the public sector.

This consensus of opinion on the need for change prompted more attention for the quality of cities and their regions seeking a more integrated view of their physical, economic, social and governance qualities). The Single Regeneration Budget and integrated government offices for the regions started a trend to more coordination and consistency in action among departments. City Challenge, too, took a longer-term view and gave a leading role to local authorities, and sought to incorporate (although not without some difficulty) local communities. The experience of promoting partnerships has been taken forward such that the partnership working, first introduced in urban policy in 1995, and now managed competition have become requirements for almost all urban initiatives in England (Oatley 1995; Atkinson 1999b). In Scotland, the competitive element is not so fierce, but the process of negotiation between partnerships and central government is still competitive and has suffered from a lack of transparency. The Labour administration not only continued with competition but also made a much more concerted effort with the areas of worst deprivation, put an equal emphasis on investment in people as well as the physical environment, introduced moves towards more strategic thinking and joined-up policy on regeneration, and to some extent increased resources.

By contrast, the Coalition government has taken a much blunter stance on evaluation, reducing funding and support for the collection of statistical information under the banner of its drive towards localism and away from targets. The results of evaluations such as that undertaken at the end of the New Deal for Communities programme have been largely sidelined as policymakers' relationship with evidence has moved from a 'what works', evidence-based policy stance to the pursuit of principles-based evidence in which experts are on tap, not on top. This again is a reaction to the managerialism of New Labour linked with the idea that local communities should be responsible for defining what regeneration means and that its pursuit or evaluation are topics which should concern local rather than central routes to accountability. Such views clearly jar with the fact that, historically, most of the funding streams for

regeneration have been devised and delivered by central government.

In one sense, Coalition policy on regeneration chimes with the conclusions of a body of work which has sought to interrogate the relationship between area-based initiatives and poverty. This work has often highlighted a discrepancy between, on the one hand, a long-term view of changes to public funding, which sees levels of public-service provision declining, with the reductions concentrated in poorer areas, and on the other, the use of area-based initiatives to prop up services temporarily or mop up the consequences of service reductions. A commonly cited deficiency of urban policies is, therefore, that they assume the issue to be tackled lies in particular parts of particular cities, but 'inner-city' problems is a misleading abstraction. To adapt a passage from Fried (1969) (who was commenting upon the concept of poverty), 'the inner city [is] an empirical category, not a conceptual entity, and it represents congeries of unrelated problems'. Concerns about the limited effectiveness of area-based initiatives at reducing poverty have, it seems, found their way into government policy in the form of a large reduction in regeneration funding. However, this has not been matched by the restructuring of service provision along more redistributive lines, and the effects of post-2010 cuts will again be felt most severely in the more deprived parts of the country.

Most of the problems identified in inner cities or in poor suburban estates are matters of national policy relating to all areas. Thus, though poverty is undoubtedly a problem that is spatially concentrated in particular neighbourhoods, most of the residents in them are not in poverty; and most poverty is not in those neighbourhoods. The arguments *against* any area-based policy are strong (Townsend 1976). Oatley (2000: 89) makes the point most strongly in his critique of the *Neighbourhood Renewal Strategy*:

> Area-based policies are notoriously unsuccessful in addressing 'people poverty'. Concentrating resources on a small number of neighbourhoods is both administratively and politically convenient, masking the widespread nature of deprivation within society and allowing us to feel that the

problem is being dealt with. These responses may at best concentrate resources in areas with high need for the wrong reasons, and at worst, seriously mislead us into thinking that we are tackling the problems when in fact we are only producing palliatives to alleviate the worst symptoms.

To the extent that the problems relate to the deprived, it makes more sense to channel assistance to them directly, irrespective of where they live. Only to the extent that the problems are locationally concentrated should remedies focus on specific locations – as in the case of renewal areas. Oatley goes on to call for radical alternative solutions that might 'break out of the dismal cycle of unfulfilled promises' (2000: 94). None of this is to deny the importance of directly tackling those problems of decay and disadvantage which are all too apparent in many deprived neighbourhoods. Nor is there any argument against the desirability of attempting better organisation of services at local levels, or improved coordination both within and between agencies. But such approaches are not going to solve the problem. Kintrea and Morgan (2005: 6) remind us that

> the character of problem neighbourhoods has not changed much over 25 years, in spite of many different types of intervention. At best, neighbourhood renewal policy has perhaps stopped the worst areas becoming even worse than they would otherwise have been, and helped to sustain the quality of life for residents at a basic level.

While some might argue that we should therefore give up on urban action (Leunig and Swaffield 2008), to do so in the context of rising inequality merely compounds problems of environmental degradation and adds to the trials of those who must live in those areas with the highest crime rates, poorest schools and fewest jobs. Perhaps what is needed is less an attempt to manage and moderate these problems 'from above' and more of an attempt to mobilise such communities against the social and political forces which are often intrinsically tied up with perpetuating the problems they face.

Further reading

Housing renewal

The classic commentaries on post-war housing renewal include Young and Wilmott's *Family and Kinship in East London* (1957), Dennis's *People and Planning* (1970) and Davies's *The Evangelistic Bureaucrat* (1972), all of which continue to have contemporary relevance as enlightening accounts of state attempts to intervene in urban environments. Some broader-ranging context is provided by Gibson and Langstaff (1982) *An Introduction to Urban Renewal*. The more recent policy-related literature on housing runs alongside a counter-current of critical gentrification research. The ODPM commissioned a series of reports, *The Evaluation of English Housing Policy 1975–2000* from a consortium of universities between 2003 and 2004. Of particular value here are Kintrea and Morgan (2005) *Housing Quality and Neighbourhood Quality* and the general overview of findings by Stevens and Whitehead (2005) *Lessons from the Past, Challenges for the Future for Housing Policy*. Bramley *et al.* (2004) *Key Issues in Housing: Policies and Markets in 21st Century Britain* consider the relationships between housing, regeneration and planning; Balchin (1995) *Housing Policy* places renewal in its wider context. See also Balchin and Rhodes (1997) *Housing: the Essential Foundations* and Wood (1991) 'Urban renewal: the British experience', in Alterman and Cars (eds) *Neighbourhood Regeneration: An International Evaluation. The Gentrification Reader* by Lees *et al.* (2010) offers an excellent introduction to the critical commentary on this subject, and related themes are frequently dealt with by a range of journals including *City: Analysis of Urban Trends, Culture, Theory, Policy, Action, The International Journal of Urban and Regional Research* and *Area*.

General reading on urban policy and regeneration

There is a wide literature on urban policies. A number of introductory texts offer overviews of New Labour-era policy initiatives, and these include Jones and Evans (2013) *Urban Regeneration in the UK* and Andrew Tallon's book of the same name (2009). Slightly older volumes

guidance and sat alongside Circular 17/96 *Private Sector Renewal: A Strategic Approach*, Circular 5/03 *Housing Renewal Guidance* and *Running and Sustaining Renewal Areas: Good Practice Guide* (1999). See also ODPM (2002) *What Works? Reviewing the Evidence Base for Neighbourhood Renewal*.

2   The figures are from Kintrea and Morgan (2005), who draw on Wilcox (2002). The DETR published *Sustainable Estate Regeneration: A Good Practice Guide* in 2000, which compares success through Estate Action, the SRB and mainstream funding.

3   ODPM Update Newsletter 5(2004): 8.

4   In 2002, the target was amended to include the goal that 70 per cent of vulnerable households in private housing would be in decent homes by 2010. A series of publications on the decent homes standard is available on the gov.uk website. The 2000 policy statement was *Quality and Choice: A Decent Home for All – The Way Forward for Housing*.

5   There were twelve community development projects in all: in Birmingham, Coventry, Cumbria, Glamorgan, Liverpool, Newcastle, Newham, Oldham, Paisley, Southwark, Tynemouth and West Yorkshire. Additionally, 1974 saw the introduction of a small number of *comprehensive community programmes* in areas of 'intense urban deprivation'. In the wake of these, large numbers of studies were undertaken.

6   There were eight CATs: Birmingham, Cleveland, Leeds/Bradford, Liverpool, London, Manchester/Salford, Nottingham/Leicester/Derby, and Tyne and Wear.

7   The twelve English UDCs were (with their date of designation) London Docklands (1981); Merseyside (1981); Trafford Park (1987); Black Country (1987); Teesside (1987); Tyne and Wear (1987); Central Manchester (1988); Leeds (1988); Sheffield (1988); Bristol (1989); Birmingham Heartlands (1992); and Plymouth (1993). Details of their expenditure and outputs are given in the twelfth edition of this book.

8   The first-round authorities completed their five-year programmes in 1997 and were Bradford, the Dearne Valley Partnership (Barnsley, Doncaster and Rotherham), Lewisham, Liverpool, Manchester, Middlesbrough, Newcastle, Nottingham, Tower Hamlets, Wirral and Wolverhampton. The second-round five-year programme (which was open to all fifty-seven urban programme areas) ended in 1998. The authorities were Barnsley (the only authority to win in both rounds), Birmingham, Blackburn, Bolton, Brent, Derby, Hackney, Hartlepool, Kensington and Chelsea, Kirklees, Lambeth, Leicester, Newham, North Tyneside (ended in 1999), Sandwell, Sefton, Stockton-on-Tees, Sunderland, Walsall and Wigan.

9   The Millennium Communities are at Allerton Bywater (near Leeds), Greenwich Millennium Village, Hastings, New Islington (Manchester), Oakgrove (Milton Keynes), South Lynn (King's Lynn) and Telford.

10   This wording is taken from the SRB Bidding Guidance Round 6 (para. 1.3.1). The guidance for the earlier Rounds 1 to 6 was extensive and sometimes complex to work with.

11   The first bidding round was evaluated by the HC Select Committee on the Environment in its 1995 report *Single Regeneration Budget*, in which the concept was generally supported, subject to a number of recommendations to reduce bureaucracy, improve consistency, and increase the involvement of voluntary and community groups.

12   Most government claims about new funding were contested, and it became increasingly difficult to untangle government funding streams, as reports on spending tended to be linked to cross-cutting objectives rather than programmes.

13   The Social Exclusion Unit produced a *National Strategy for Neighbourhood Renewal* based on the findings of eighteen Policy Action Teams which undertook an intensive programme of policy development.

14   The seventeen first-round pathfinder partnerships were in the local authority areas of Birmingham, Bradford, Brighton and Hove, Bristol, Hackney, Hull, Leicester, Liverpool, Manchester, Middlesbrough, Newcastle-upon-Tyne, Newham, Norwich, Nottingham, Sandwell, Southward and Tower Hamlets. The twenty-two second-round partnerships are in Birmingham, Brent, Coventry, Derby, Doncaster, Hammersmith and Fulham, Haringey, Hartlepool,

Islington, Knowsley, Lambeth, Lewisham, Luton, Oldham, Plymouth, Rochdale, Salford, Sheffield, Southampton, Sunderland, Walsall and Wolverhampton.

15 Findings from the pathfinder authorities are given in the DETR report *New Deal for Communities: Learning Lessons: Pathfinders' Experiences of NDC Phase 1* (1999).

16 Another important influence on the White Paper was the New Commitment to Regeneration approach developed by the Local Government Association and the Cabinet Office. See Russell (2001) for an overview.

17 The core cities are Birmingham, Bristol, Leeds, Liverpool, Manchester, Newcastle, Nottingham and Sheffield.

18 Robson *et al.* (2000). See also Parkinson *et al.* (2005) and the ODPM's glossy *A Tale of Eight Cities* (2004).

19 The Partnership Priority Areas were designated by the Scottish Office in response to proposals from local authority-led partnerships, covering the whole of their area. The twelve areas are Great Northern (Aberdeen), Ardler (Dundee), Craigmillar and North (Edinburgh), East End, North and Easterhouse (Glasgow), Inverclyde, Motherwell North (North Lanarkshire), Paisley (Renfrewshire) and North Ayr (South Ayrshire).

12 Infrastructure planning

Introduction

This chapter is concerned with infrastructure. It focuses particularly on transport and mobility issues, as these are more closely linked with planning practice. Infrastructure can have broad and narrow definitions. In a planning context, some definitions encompass what is often termed social infrastructure, including things like hospitals and schools. Infrastructure can be defined as the physical assets that underpin and constitute the networks for transport, energy generation and distribution, electronic communications, solid waste management, water distribution and waste water treatment. The focus in this chapter is on this latter definition, on forms of networked infrastructure, examples of which are set out in Table 12.1.

As Table 12.1 suggests, infrastructure encompasses not only the physical networks, the roads and tracks, but also the buildings associated with them – the airports, the cycle sheds, etc. We should note just how pervasive such infrastructures are for daily life, even though they often go unnoticed, at least until they fail (Graham 2009).

Approaches to infrastructure planning

For much of the nineteenth and twentieth centuries, infrastructure planning in western cities was practised according to a model of 'facilitating supply'. As cities and economies grew, so the challenge was to 'roll out' networks to make cities healthier (in relation to water and sewerage particularly) and more efficient (through electricity grids, roads and rail, for example). Such tasks were led often by engineers, but planners contributed to a process of 'constructing the modern networked city' (Graham and Marvin 2001). Such a process was about 'facilitating the supply' of an infrastructure for a predicted demand, often by extrapolating the past increase in demand for a particular service.

Table 12.1 Defining infrastructure

Transport infrastructure	Utilities	Electronic
Footpaths and walkways	Gas and electricity networks: pylons, pipes, cables, etc.	Telecoms masts
Cycle paths, lanes and parking	Power generation facilities	Cable networks
Roads and parking facilities	Water and sewerage networks	Broadband/wireless communication
Canals and river infrastructure		Surveillance technologies, e.g. CCTV cameras
Railways and tramways		
Airports and seaports		

Such a process of 'predict and provide' became more and more challenged in the last quarter of the twentieth century. Once a basic level of demand for an infrastructure had been attained, the relationship between demand and supply was more problematic. In response, an alternative position began to emerge, one of managing the demand for infrastructure. Rather than seeing services such as energy, water and transport as being simply 'derived' demand, researchers began to see them as responses to a range of opportunities and social practices. For example, the levels of use of energy and water in the home are related to the ways the house is designed and the social practices that occupants develop in that context (Shove *et al.* 2012).

The shift from policies that emphasise supplying infrastructure to meet a predicted demand to one of focusing also on managing the potential demands for that infrastructure is most visible in the area of transport, and the following sections deal with this shift by reviewing UK policies from 1960 onwards.

Hypermobility and automobility

The distance travelled in the UK by the average citizen increased enormously through the twentieth century.[1] Much of this was underpinned by a growth in car ownership and, particularly since the 1990s, increasing opportunities for air travel. Between 1950 and 1960, the number of vehicles on the roads of Britain more than doubled, from 4.0 million to 8.5 million. The number more than doubled again by the end of 1980, to 19 million. By 2012, the number had risen to 34.5 million. The most dramatic increase was in cars, from around 2 million in 1950 to 28.7 million in 2012. The proportion of households owning a car rose from 14 per cent in 1951 to 74 per cent in 2011. In terms of total road traffic (measured in billion vehicle kilometres), the increase has been from 53 in 1950 to a peak in 2007 of 502 (see Figure 12.1). Figure 12.1 also shows why inner urban areas have congestion problems. Many were largely built before the arrival of mass motorisation and accommodating the demand for travel proved difficult.

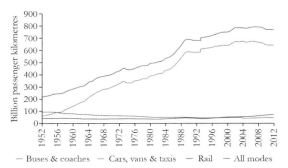

Figure 12.1 Transport trends in the UK
Source: HMSO

However, from 2007 to 2012 motor vehicle traffic fell. This trend was also visible in other western countries and prompted the idea that demand for motor travel may have peaked, so called 'peak car' (Goodman 2013; Metz 2013). Indeed, road traffic in London peaked in the mid-1990s and has been falling ever since, and car ownership has been falling there too. Figure 12.2 provides an explanation for this, with a saturation of demand levels, given the high levels of accessibility now widely available and the lack of speed increases with which people can travel farther within relatively fixed time budgets. There is also evidence that young adults in particular are less enamoured of

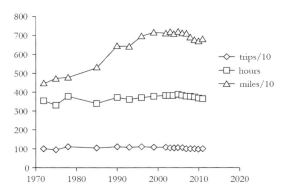

Figure 12.2 The slow down in hypermobility and 'Peak Car'. Average travel time (hours per person per year), distance (miles per person per year) and trips (miles per person per year).
Source: Metz 2013

car ownership because of rising costs and the capabilities of electronic means such as smartphones to enable them to remain connected socially without the need for so much physical travel.

Since the 1990s, this 'automobility' (Sheller and Urry 2000) has been conjoined with large increases in air travel, facilitated through the rise of low-cost airlines,[2] and the increasing use of rail travel. Automobility is only part of the story, therefore, and some authors conclude that we are in a 'hypermobile' age (Adams 2001), with increased amounts of travel for most, but not all, in society a feature of who we became as citizens in the second half of the twentieth century.

Indeed the focus on mobility is something of a red herring. Mobility is not important in itself: its importance is in providing access. Thus large amounts of mobility may be indicators of a spatial system that is not working very well. That is, if people are driving several miles, rather than say, walking a few hundred metres, to access a service, then this is not an efficient system. Transport should thus not be considered in isolation from the social and economic activities that drive the demand for it and the spatial distribution of activities which determines the pattern of journeys. Yet the debate on transport often forgets this elementary point, and focuses on mobility: faster roads, faster trains, and more frequent buses.

The advantage of focusing on accessibility rather than mobility is that it opens up the possibility of alternative means of facilitating transport supply, which usually entails building new infrastructure which is often expensive: changing land use relationships, for example. The greater the accessibility is, the less need for 'transport'. Thus, transport planning is much more than the building of infrastructure. The presentation of statistics is important in this. Figure 12.1 appears to show that only private vehicles are very important in daily travel. But this is because it focuses on distance travelled, which is often by road, air and rail. In planning terms, what is more significant is the number of journeys undertaken, as this relates to accessibility and the daily needs of citizens. For example, 22 per cent of all trips are walking trips, and yet walking has rarely been afforded much attention historically in transport policy.

The (hidden) importance of walking as a mode of transport relates also to the levels of mobility that have resulted from government policy and planning. Attempts to meet actual and projected demand levels have often led to further increases in demand levels. The policy response to increasing demands for mobility in the second half of the twentieth century was to build more road infrastructure. A large road-building programme was initiated from the 1950s, such that there was 392,000 kilometres of road in the UK in 2012, an increase of about a third over the 1951 total of 297,00 kilometres. The primary aim of UK transport policy was to facilitate the rapid, free-flowing movement of people. Thus, although the level of road building proposed by national governments fluctuated throughout the post-war period, the principal transport concern of governments, at least until the mid-1990s, was to roll out a roads programme, largely disconnected from considerations associated with other transport modes and from other forms of spatial development. This approach became known as 'predict and provide', where the demand for travel by various modes was extrapolated and then attempts were made to match the supply of infrastructure to that potential demand. The main mode to which this approach was applied was roads, but similar 'predict and provide' models were used in planning for public transport, where declining passenger numbers in the 1950s were projected into the future and used to justify cuts in rail and bus networks. Successive governments' approach to air travel has also remained curiously wedded to a 'predict and provide' model, with attention focused on where to provide for projected demand, especially in South East England.

However, the 'predict and provide' approach came under increasing challenge in the 1980s and 1990s as its theoretical underpinnings were undermined and the consequences of such a policy became more acute, widely known and understood (Vigar 2002). In its own theoretical terms, such an approach was judged to be deficient. First, it typically ignored the impacts of policy interventions themselves. By investing in road infrastructure and not in other modes, the attractiveness of that infrastructure versus others became clear. Second, building any infrastructure that is free at the point of use, which roads largely are in the UK, fuels an increase in the use of that infrastructure, i.e. increases in supply release so-called 'latent' demand (SACTRA 1994). Third, the influence of land use planning was found to be greater than previously thought. In the longer term, people and businesses change where they live and locate in response to changing travel opportunities. Thus, if new infrastructures are built, these will lead to different behaviours, i.e. if a road from an attractive market town to a city is 'upgraded' from single to dual carriageway then some city dwellers may move from the city to the town and commute along it. Infrastructure, in this case roadspace, thus fills up with people making these longer-term decisions (Dargay and Goodwin 2000). To put it simply, technology, society and mobility are bound together in complex ways, and people will change aspects of their lifestyles in the medium and longer term in ways that are hard to predict.

In addition, a growing awareness of a variety of physical, social, environmental and health-related effects arising from unfettered growth in the use of private cars also contributed to policy change. Recognition of the need to pursue a different trajectory from that suggested by 'predict and provide' thus led to the emergence of what has been termed the new realism, i.e. a realistic idea of what demand levels can be met (Goodwin *et al.* 1991; Vigar 2002), latterly part of a 'sustainable mobility paradigm' (Banister 2008). Such approaches focus on accessibility, not mobility. There is evidence that the most economically successful European cities are characterised by their emphasis on such an approach over a number of decades (CfIT 2001).

Thus the historic huge increases in car traffic and the relative decline in bus and rail travel did not arise from a mass transfer from public transport (see Figure 12.1). On the contrary, most of the increase in car use was newly generated traffic. Much of this trend is down to transport policy – i.e. investing in roads to the exclusion of other modes, at least until recently. But it is also due to the approach of land use planning. In the 1950s, people lived, worked and played in close proximity. The increased separation of land uses and the planning of out-of-town retail and commercial

activities, along with new low-density suburbs, all meant that using a car became more convenient, or indeed a necessity. In other nations, such as the Netherlands, a rather different approach to transport policy, with less emphasis on road building, and with a stronger regulatory approach to land use planning, with tighter control on out-of-town development and parking standards, etc., meant that the transport outcomes were more sustainable (Haq 1997).

The next section deals chronologically with the development of transport policy through a series of reasonably distinct phases, to help us understand how we came to an era of hypermobility.

Transport policy and planning in the 1960s and 1970s

The biggest problem facing governments, both local and national, in the 1960s was the rapid increase in car ownership and use. Towns and cities started to feel the strain of accommodating private cars and in response the government commissioned the Buchanan Report (1963) on *Traffic in Towns* to examine the issue. One of the failings of the government of the time was to commission this report separately from the Beeching Report into the future of the rail network and, less significantly, the Jack Report into rural bus services. All reports exhibited the tendency to use 'predict and provide' techniques, extrapolating existing trends into the future. Any chance to look at the three modes together was lost. In the 1960s, the assumption was often made that public transport would shrink to become a 'residual' mode, of use only to a small proportion who could not or would not use a private vehicle. Cars would be for the vast majority. Buchanan himself was unaware of the emerging Beeching Report:

I still find it difficult to understand why it never seems to have occurred to Ernest Marples [the Transport Minister who commissioned both reports] to bring Beeching and myself together. Had we, for example, but had a discussion on the problems of freight transport I daresay Beeching might have seen the need for retaining freight on rail as far as humanly possible instead of consigning it to already overcrowded roads.

(Buchanan 1993: 70–71)

The problem of traffic in towns

The 1963 Buchanan Report, *Traffic in Towns*, was an eloquent survey of the problem of trying to accommodate traffic in urban areas, and attempted to show the necessity of bringing land use planning and transport planning together, a problem that the UK continues grapple with (Hull 2011). The report showed that accommodating all the traffic likely to want to enter the urban centre could not be achieved without a radical transformation of the physical environment. It showed how this could be achieved, and at what cost to the existing fabric, exposing the costs to the physical environment of what many engineers' departments in local government were proposing. The principal design components of such an approach were the separation of pedestrians and vehicles vertically in space through raised pedestrian walkways and subways and through vast expenditure in urban motorway construction, in tunnels, flyovers and cuttings in particular (see Plate 12.1[3]). Buchanan was acutely aware of the consequences of many of these actions. He proposed the canalisation of traffic movements onto 'primary networks' that would service areas within which places suitable for a 'civilised urban life' could be developed. These 'environmental areas' or 'urban rooms' would be developed through pedestrianisation of high streets and traffic calming in neighbourhoods, to a large extent under the influence of the emerging experience in Germany at the time (Buchanan 1993). In this way, many streets would be closed to allow for what we now think of as 'filtered permeability', whereby cars can only access neighbourhoods at one or two suitable points (Melia 2012).

The great danger, in Buchanan's view, lay in 'muddling through', 'trying to cope with a steadily increasing volume of traffic by means of minor alterations, resulting in the end in the worst of both worlds: poor traffic access and a grievously eroded environment' (Buchanan 1963: 79). Many cities were already embarking on large programmes of urban motorway

ON THE completion of Newcastle's Central Motorway East, vehicle traffic moving on a multi-level road system will be able to avoid the city centre. The ILLUSTRATION ABOVE shows a view of the Great North Road, looking south, with the levels of the proposed motorway intersection rising above it. The ILLUSTRATION BELOW is of Sandyford Road, passing through a cutting below the motorway. Crossing at ground level is the southbound carriageway of the motorway, and above is the northbound carriageway.

Plate 12.1 Newcastle's central motorway plans

Source: Diversion: Newcastle upon Tyne Roadways Report No 1, 1972, Newcastle City Council

building, but were emboldened by Buchanan's report as they picked the elements and choices that fitted their existing world view. The engineering profession in particular was strongly biased at the time in favour of such an approach and local politicians also often seemed convinced of the necessity of road building as an intuitive response to solve a very visible problem.

The emphasis thus became on shaping the 'modern city' to fit the car in response to the predicted demand for motorised travel (Starkie 1982). However, 'it is now apparent that the transportation studies of the 1960s incorporated wildly incorrect, highly optimistic assumptions on such fundamental matters as population change, the growth of disposable incomes, and car ownership' (p. 68). But citizens in particular began to see that the resultant towns and cities were not ones that they liked and protests that began in the 1960s escalated. What Hajer (1995) terms the 'sensory experience' of a problem thus proved vital. Such protest was fuelled by a sympathetic media that generally found favour with residents rather than transport planners. Opposition to the ringway system of orbital motorways in London was the most celebrated example. In Nottingham, plans for urban motorways were regarded by a government inspector as too large, and it was argued that 'what was urgently needed . . . was a much greater emphasis on the need to control, rather than attempt simply to accommodate, town traffic' (Starkie 1982: 85). So, by the late 1960s, a countermovement of opposition groups was suggesting that the maintenance of homes and communities should be given precedence over the building of roads and provision for cars. In addition, some professionals in the field were highlighting the need to manage traffic growth rather than simply catering for it, at least in towns and cities. There was, however, 'little evidence at the beginning of the seventies that the full significance of these changes of public attitude had caught up with the Ministry [of Transport's] road planners' (p. 82). The response of central government was to alleviate the consequences of roads through compensation and mitigation measures. The need for the roads was not questioned, underpinned as it was by the results of land use transportation studies (LUTS) and the visible evidence of increasing congestion in urban areas. Most cities abandoned the radical changes often midway through implementation. Abandonment was a result of both public protest and financial reasons, and Buchanan's 'great danger of muddling through' did indeed come to pass.

The present desirability of UK cities as places to live, work, shop and play would be lessened if some of the grand plans had come to pass. The physical environment left following construction of the Central Motorway illustrated in Plate 12.1, as in many cities, continues to disadvantage pedestrians and cyclists, making it particularly unsafe in subways and on overpasses after dark. Indeed, some have argued that the decline of cities such as Liverpool was hastened by such attempts to improve access to central areas through demolition of inner suburbs, as this also increased the ability to leave the inner city – of both people and jobs. Inner urban housing clearances that moved the workforce to the outer suburbs contributed to the decline.

Buchanan acknowledged also that public transport 'may prove to be the key to the problem in the long-term' (Buchanan 1963: 241). This prescient observation underpins a model of using public transport, often fixed light or heavy rail, at the centre of solutions to urban planning that has developed globally since the 1990s as part of a package of new ideas centred on demand management and 'sustainable mobility' (Banister 2008).

Roads policies in the 1970s

While intra-urban road building (i.e. building roads within urban areas not between them) fell out of favour in the 1970s, inter-urban construction moved on apace. In the 1950s and 1960s, citizens almost exclusively welcomed the arrival of a new inter-urban motorway network.[4] However, in the 1970s a series of high-profile protests against specific projects emerged; that had the effect of challenging the direction of national policy and the assumptions of 'predict and provide' (Tyme 1978). The inquiry process into individual schemes could not deal with these, and protests became more vociferous as a result. In 1977, an independent assessment by the Advisory Committee on Trunk Road Assessment of the methods used for assessing the need for roads was highly critical. The conventional methodologies (i.e. 'predict and provide') were judged to be essentially 'extrapolatory', 'insensitive to policy changes' and partly self-fulfilling. Public concern about road planning was shown to be well founded. Although it was the opposition to specific roads that received most publicity (the Westway in London, and Airedale, West Yorkshire were two high profile examples), there was a lack of confidence in the system by which the need for roads was addressed. The way forward lay in a more 'balanced' appraisal and more openness with regard to the uncertainties inherent in the process. This is a recurring theme in the transport field, the Standing Advisory Committee on Trunk Road Assessment (SACTRA) report on Urban Road Appraisal in 1986 concluded similarly, and the introduction of the New Approach to Appraisal (NATA) was introduced in 1998 to address the continuing critique of appraisal methods (see below).

The Labour government's response was positive, arguing that national policies should be for parliamentary debate in White Papers so that these would provide the background against which local issues could be examined at public inquiries into particular schemes. It was hoped that this would avoid the confusion at local inquiries between national policies and their application in specific areas. It was pointed out, however, that this would work only if the methods of assessing national needs (what the Committee termed 'a highly esoteric evaluation process') were acceptable. Since these methods could not be properly examined at local inquiries (or, indeed, by Parliament), they were to be subject to 'rigorous examination' by SACTRA. The Committee's 1979 report, *Trunk Road Proposals: A Comprehensive Framework for Appraisal*, criticised the narrowness of the Cost Benefit Analysis approach, and certain changes followed. For example, instead of using only one traffic forecast, high and low levels were introduced.

Public transport planning 1960–79

The Labour governments of the period 1964–70 had acted to a degree on the recommendations of the Buchanan Report and encouraged local authorities to invest in public transport in towns and cities, mainly to prevent large-scale destruction of the urban environment through urban motorway building. The movement of central government resources away from intra-urban road construction supported this view, in part driven by economic problems culminating in the devaluation of the pound in 1967. However, local authorities had more locally derived funds to deploy then than latterly and so many continued to pursue the plans derived in the 1960s. The rhetorical emphasis in the Transport White Paper of 1966 and attempts at coordinating public transport in the Transport Act 1968, particularly in advocating the creation of passenger transport authorities, were also manifestations of this approach and reflected a recognition that the maintenance of local rail services, for example, was essential for social reasons (Docherty 1999). Much of this new emphasis was down to a recognition by

government, and particularly the Transport Minister Barbara Castle, that sections of society without access to cars (at this time the majority of households) were being disadvantaged as public transport services were cut back and urban environments became more hostile to walking and cycling. Nevertheless, many suburban rail lines were closed throughout the 1960s following the recommendations of the Beeching Report and this undermined the urban public transport emphasis.

The House of Commons Expenditure Committee of 1972–3 began an investigation because of 'complaints about inadequate train and bus services . . . public demonstrations with traffic congestion, and the swamping of city streets by private motor cars and intrusive heavy lorries' (HCEC 1973, quoted in Starkie 1982: 86). The final report of the Committee recommended improvements to public transport and exhorted local authorities once again to discourage the use of private cars in urban areas. To this end the Committee identified the commuting trip as one that caused particular difficulties. The principal restraint measure advocated to tackle car commuting was to restrict the numbers of parking spaces in central areas. The report made complete a shift from providing for the inevitability of traffic growth to one where policy intervention itself was acknowledged as making a real difference in restraining traffic growth in urban areas. The House of Commons Expenditure Committee Report was well received in government and it led to direct change in methods of funding local authority transport schemes which aimed to reduce a perceived bias toward roads spending. The 1977 White Paper *Transport Policy* further promoted an emphasis on public transport in larger urban areas and the subsequent Transport Act 1978 required Metropolitan County Councils to devise and publish policies for public transport in the form of passenger transport plans.

Infrastructure policy 1979–97

Two features of the Conservative administrations of 1979–97 with regard to infrastructure planning are notable. First, in the late 1980s the government became interested in making a step change in the provision of road infrastructure and to some degree reversed the emphasis on public transport of the previous decade. Second, many utilities and transport services were privatised.

Roads planning 1979–97

The first two Tory governments of this period (1979–87) steadily expanded the road network. Between 1980 and 1990, the road network increased by 18,400 kilometres and by 1989, investment in trunk roads was nearly 60 per cent higher in real terms than ten years earlier. Projects such as the M25 were hailed as a great success by Margaret Thatcher and her colleagues, and policies emphasised the importance of roads to economic growth (despite little evidence to this day that roads contribute much to growth in developed economies).

However, policy change was most evident in 1989, when a White Paper, *Roads for Prosperity*, announced a massive increase in road building, 'the biggest since the Romans'! A set of traffic forecasts underpinned the Paper, based on the assumption that the demand for roads would be met with an appropriate supply, that demand could not be managed, and that attitudes and costs of motoring would not change (DoT 1989). This was in spite of the fact that such arguments had apparently been won in previous years.

As citizens became aware that the White Paper often proposed a scheme in their back yard, increasing public concern about the impact of specific road construction schemes emerged. This became coupled with a more general concern about traffic congestion and the idea of 'building our way out of it'. Research evidence was increasing about the self-fulfilling nature of infrastructure policy that simply followed current trends, and a 'new realism' about the ability of infrastructure supply to keep pace with unmanaged demand was emerging (Goodwin *et al.* 1991; see Vigar 2002 for a detailed assessment of the emergence of 'new realist' thinking and its political traction).

In addition, environmental issues were in the ascendance at the time, with the scale of global environmental challenges becoming clear and being addressed in government, e.g. the White Paper *This*

Common Inheritance in 1990. This White Paper was in direct contradiction of *Roads for Prosperity* and there was open hostility in government between the Departments of Environment and Transport. Into this debate came citizen voices and many unlikely alliances emerged, between younger radical environmentalists and Tory voters, both eager to defend the countryside (see Bryant 1996[5] and Wall 1999 for two insider perspectives from these different constituencies). The effect of this protest, and a squeeze on public expenditure following the recession of 1992–4, was the abandonment of the road-building programme almost before it had begun and road building returned to previous levels.

A contributor to the change in policy was increasing evidence of the extent to which new roads generate traffic. Where does this traffic come from? Downs (2004) suggests that 'triple convergence' occurs, whereby, first, motorists switch from other routes to the new one ('spatial convergence'); second, some motorists who avoided the peak hours will travel at the more convenient peak hour ('time convergence'); third, travellers who had used public transit will switch to driving, as the new road now makes the journey faster ('modal convergence'). So, intuitively, the more congested and difficult a road journey is, the more likely a potential traveller will seek an alternative, which includes not travelling at all or combining journeys in time. Conversely, ease of road journeys must generate increased trips. A SACTRA report of 1994 confirmed that this phenomenon of induced traffic was real and a *Guidance Note on Induced Traffic* was issued (DoT 1/95). Completed road schemes of the time such as the Newbury Bypass confirmed the phenomenon of induced traffic, as they filled up a lot faster than expected and traffic relief elsewhere on the local networks was lower than expected.

There is a wider point, too, made by SACTRA and others: new roads not only induce traffic, but also encourage car ownership and use. As car use increases, other methods of transport are used less and, as a result, standards of service fall (which further increases the attraction of car use). Moreover, road building and ease of car use increase the pressure for suburban development of housing, employment, shopping and leisure. Many new locations are car-dependent and therefore may increase the demand for road travel, and hence the need for more road construction. In this cumulative way, roads certainly generate more traffic. We now know that these longer-term effects are poorly accounted for in most modelling processes and they may be greater than first thought (Goodwin 1999). It was thus clear by the mid 1990s that some fundamental changes in transport policy were needed.

Thus, in 1994, the Departments of the Environment (then responsible for planning) and Transport jointly published PPG 13 *Transport*, which recognised once again the importance of land use planning in shaping transport outcomes. It proposed curtailing certain out-of-town developments, for example, in favour of those that could be better served by public transport. The key aim of PPG 13 was to ensure that local authorities carried out their land use policies and transport programmes in ways which helped to 'reduce growth in the length and number of motorised journeys; encourage alternative means of travel which have less environmental impact; and hence reduce reliance on the private car'. Policy had turned full circle once again.

PPG 13 was important, as transport planning explicitly became linked both to land use planning and to environmental policy and the relatively new at this time *UK Sustainable Development Strategy*. It was stressed throughout PPG 13 that the relationships between transport and land use planning had to be carefully examined at all levels, and that integration and coordination had to be promoted by regional planning guidance and in development plans. Strategies were required that would reduce the need to travel and maximise the opportunities for travel by public transport. Car parking was thought highly significant, with maximum standards introduced for new developments, set through development plans. Other matters dealt with included plans for safe and attractive areas for pedestrians; provision for cyclists; traffic management; provision of park and ride schemes; and 'accessibility profiles for public transport in order to determine locational policies designed to reduce the need for travel by car' (paras 4.23 and 4.24).

The Trunk Roads Review 1994 also signalled a significant shift in road-building policy. It detailed a reduced road programme and focused on the improvement of sections of existing key routes which were likely to experience congestion in the near future (primarily by adding lanes to existing motorways) and on providing 'urgently needed' bypasses. As in the late 1960s, the squeeze on public spending lay as much behind this policy shift (Norris 1996). This was in part due to the need to secure sites from protestors. For example costs for the M11 link road in London saw already high costs double as a result.[6]

Infrastructure privatisation

The second feature of the Tory governments of 1979–97 was the privatisation of state-owned and state-run infrastructure. Many infrastructure assets were nationalised immediately after the Second World War for a number of reasons (Baird and Valentine 2007). First, money for investment was in short supply and export markets were weak and so industries were starved of funds. Second, the Labour Party was committed to taking control of the means of production and indeed the owners of many industries sought assistance as they ran into difficulties in this economic climate. Third, in the case of transport, it was a way of securing more coordinated services free of the negative aspects of competition.

By the 1980s, many perceived these state-run industries as an anachronism. They argued, with little actual evidence it should be said, that the private sector was better placed to run them. In many cases this was relatively uncontroversial: state intervention in road freight distribution, in long-distance coach travel and British Airways, for example. But many argued that privatisation of other infrastructures would remove the ability to coordinate service provision across time and space, especially given the high costs of providing new infrastructure, including borrowing costs, and the long payback on such investment.

We can see these arguments played out in an early privatisation, that of buses. Here privatisation was seen as an attack on urban local government, most likely Labour-controlled, rather than a belief that a wave of private-sector innovation would transform the bus market and arrest the historic decline in bus use (Whitehead 1995). Local authorities keenly contested this battle, not least because many wanted to retain their interests in transport operation. This antagonism between local and central government culminated in court battles over fares policies[7] and the privatisation of the bus network. The Conservatives maintained that they believed in collectivised transport as part of an intra-urban transport policy, although Margaret Thatcher herself was notoriously ambivalent about public transport, seeing it as a safety net for all those unable to participate in her 'car-owning democracy' (Riddell 1991). The next few years would show, however, that privatisation policies did nothing to halt a decline in public transport ridership, and in cities where cheap subsidised fares policies had operated previously, public transport use fell dramatically (Sutton 1988). Against this backdrop, some large-scale investment in urban public transport did occur. The funding of light rail schemes in Croydon, Manchester, Nottingham, South Yorkshire and West Midlands looks rather odd in this context but it should be noted.

Privatisation has to varying degrees occurred in all infrastructure sectors (see Table 12.2). It has been controversial, with varying conclusions drawn as to its effects. For example, passenger numbers on the railways have increased greatly, but this was a trend started under British Rail and has happened in many other countries in response to a range of societal trends (Wolmar 2005). Rail safety under Railtrack was also much criticised, and this led to the network operator Railtrack being renationalised. Indeed the principle criticism has been the separation of track from train operations in a hurried privatisation process (Wolmar 2005). This fragmentation also means that the cost of building new rail infrastructure is almost twice as high in the UK as in mainland Europe (Preston 2012).

The flipside of privatisation is the need for regulation of the industry by government. The effects of different regulatory regimes can be seen by reference to bus use. Buses in London, where privatisation has been limited, have outperformed bus networks elsewhere in England (see Figure 12.3). This has led to other metropolitan

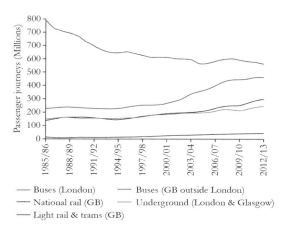

Figure 12.3 Trends in passenger travel
Source: HMSO

areas reintroducing tighter regulation, enabled under the Transport Act 2008 (see below). Whether this will work is debatable. London's success derives in part from the limitations of car use in the capital and the high levels of subsidy; more than 60 per cent of the entire subsidy for buses in England went to London in the early and mid-2000s (Knowles and Abrantes 2008).

In more general terms, there has been great criticism of private rolling stock operating companies (ROSCOs), train and bus operators, airport and seaport owners, and some utility companies for charging higher than EU average prices and for taking large profits. Shaw and Docherty (2014) note that a Glasgow-to-Edinburgh season ticket costs as much as an equivalent card for the entire German network, despite rail subsidy from government twenty years after privatisation being twice what it was at the time of privatisation. All sectors are seen as good places for investment vehicles such as private equity funds, the activities of which are rarely in the long-term interests of the consumer (e.g. Baird 2013). Where regulation is tight, the consumer can benefit, as with the privatisation of British Telecoms, but where regulation is lighter, any efficiency savings tend to flow out in the form of profits to shareholders (Newbery 1997). This

Table 12.2 Major privatisations of UK infrastructure

Infrastructure	Privatisation	Now
Seaports	Transport Act 1981 provided for the sale of the British Transport Docks Board and the creation of Associated British Ports; 8 further ports privatised in the 1990s.	15 of the top 20 ports are privately owned, 3 are trust ports and 2 are municipally owned (both in Scotland) (Baird and Valentine 2007).
Rail	Railways Act 1993 paved the way for sale of British Rail outside Northern Ireland, which remains state-owned. The infrastructure sold as Railtrack in 1996; 25 franchises were sold to train-operating companies; three leasing companies (ROSCOs) hired out the rolling stock; 13 maintenance and renewal units maintained the railway.	Infrastructure now owned and maintained by Network Rail in what is effectively a renationalisation under the Railways Act 2005; as of end 2013 there were 24 TOCs, 4 ROSCOs and 7 freight operators. Eurostar part publicly owned, but government stake likely to be sold.
Buses and light rail	Transport Act 1985 privatised bus networks outside London and Northern Ireland (NI); and part-privatised London's network. All bus operations beyond NI and London were privatised by 1994.	Industry has consolidated into three main players which frequently control up to 90 per cent of local markets (Knowles and Abrantes 2008). Some metropolitan areas reinstating stronger regulation under Transport Act 2008.
Airports	Airports Act 1986 led to creation of BAA plc, owning 7 airports.	Of the 7, 2 brought back into public ownership, Stansted is now majority-owned by Manchester local authorities; Prestwick effectively renationalised when sold to Scottish government.
Roads	One significant toll road in the West Midlands opened 2003; since 1994 some trunk roads run under private finance, 'design, build, finance and operate' (DBFO); some bridges and tunnels privately operated.	Conservative government published a White Paper in 2013 to turn the Highways Agency into a 'government-owned company'.
Water	Privatisation of 10 largest water and sewerage companies in England and Wales in 1989.	Companies remain, although ownerships have often changed.
Energy	Gas Act 1986 privatised British Gas; open markets introduced in 1996. Electricity Act 1989 broke the Central Electricity Generating Board England and Wales into 3 companies; Scottish industry privatised 1991; Northern Ireland in 1992.	Energy supply heavily transformed. Energy prices and issues of long-term supply have been politically contentious.

'accumulation by dispossession' ultimately promotes deeper societal inequalities by concentrating wealth in fewer hands (Harvey 2004).

We should also note that privatisation itself is a complex picture, as other national governments now own large parts of the UK infrastructure. For example, Dutch, French and German national rail companies operate significant elements of Britain's public transport, and many state-owned firms own parts of other utility networks. This all suggests that the ideology that drove privatisation, that public organisations should not run such companies, has proved rather contradictory. Many thus conclude that early privatisations were mostly about the raising of short-term finance by Conservative governments looking to reduce taxes (Baird and Valentine 2007). Perhaps the best indicator of the success, of transport privatisations at least, is that while many countries came to the UK to look at its privatisation experiences, not one copied it (Shaw and Docherty 2014).

Transport policy under Labour 1997–2010

Labour's 'radical' first term

Labour began its three terms of office with some fairly radical statements on transport that deepened the line taken by the previous administration. The 1998 White Paper *A New Deal for Transport: Better for Everyone* promised to improve the urban environment by creating the conditions for people to move around more easily. The White Paper was short on implementable detail, but it did put demand management in place as the key rhetorical aim of transport policy. The document is insistent that 'predict and provide didn't work' (DETR 1998b: 10). More roadspace and priority were to be given to pedestrians, cyclists and public transport, and the emphasis was on 'integration' (see Box 12.1), with the merger of the Departments of Environment and Transport to create the Department of Environment, Transport and the Regions a way of achieving this at national level. Separate, yet similar, papers were published for Northern Ireland, Wales and Scotland. Also at this time, a private members bill became law through the Road Traffic Reduction Act 1997. In practice, this has had little effect but was interesting in that it introduced the idea that local and national governments should aim to *reduce* traffic levels in their jurisdictions.

The government found integration rather hard to achieve. Despite positive attempts to unite land use planning and transport, through PPG 13 in particular, policies in other areas of government, such as education and health, increasingly went against attempts to make travel more sustainable. Principal among these were the contractual obligations of privately financed schools and hospitals, which often led to the implementation of a design that did not factor in good walking and cycling access; the trend towards building larger education and health facilities on the edge of town which were hard to reach by public transport; and the 'choice' agenda, which led to more long-distance travel, especially for schools. Even in transport itself, integration between modes was undermined by the lack of control over deregulated public transport services.

The White Paper established a new Independent Commission for Integrated Transport (abolished 2010), introduced local transport plans and revised regional policy (abolished 2010), promoted extensive partnerships between transport providers and strengthened local authority powers to secure integration. It also revised the appraisal process for transport projects by introducing the New Approach to Appraisal (NATA), which aimed for a more 'balanced' approach

BOX 12.1 INTEGRATION OF TRANSPORT POLICY

An integrated transport policy means:

- integration within and between different types of transport – so that each contributes its full potential and people can move easily between them
- integration with the environment – so that our transport choices support a better environment
- integration with land use planning – at national, regional and local level, so that transport and planning work together to support more sustainable travel choices and reduce the need to travel
- integration with our policies for education, health and wealth creation – so that transport helps to make a fairer, more inclusive society.

Source: DETR 1998b

to transport appraisals, in terms of giving more weight to public transport schemes and increasing the significance of a wider range of environmental impacts. Importantly this methodology did not try to quantify all the issues into numeric data, but presented essentially qualitative matters, such as the destruction of a valued environmental asset, in a range of ways.

A revised roads programme (DETR 1998b) emerged at the same time as the White Paper. It confirmed the basic principles laid down in the Paper and contained some significant substantive proposals. First, it devolved a sizeable portion of the trunk road network into the hands of local authorities and the planning of the network into regional planning guidance. Second, it redefined the role of the Highways Agency as a 'network operator' rather than as a road builder, a potentially important distinction that depended a great deal on changing the culture of this organisation as much as the rebadging itself. Third, it introduced a seven-year investment plan to overcome the short-termism of previous years. In the summer of 2000 this was revised into a ten-year plan, which is discussed below. The programme proposed that new roadspace could only be justified on the grounds of road safety, and the removal of traffic from towns and villages for regeneration purposes and for providing 'breathing space' until other policies took effect.

A revision of PPG 13 followed in 1999. The new PPG 13 had a stronger emphasis on guaranteeing access by public transport to new developments, and on ensuring forms of development that encouraged non-motorised transport, with implementation through green transport plans, transport assessments and national car parking standards. The guidance also included preferred locations for particular types of development (housing, shopping, leisure and services), but the guidance was by necessity very general, mostly in the form of situating new development where it was accessible, and relied on local authorities taking the issues seriously in forward planning and development management.

Labour's transport policies drew a hostile response from the tabloid press, which lampooned its promoter, John Prescott, in particular, and the government found itself labelled 'anti-car'. Continuing difficulties with

the railway industry and a lack of improvement in many bus services made the government's attempts to curb car use look unrealistic to most people. A rural lobby drew political capital from continuing tax increases on fuel duty and their impact on rural car users. The White Paper attempted to reconfigure the issue by saying that past policies of deregulation and privatisation had diminished choice. There was a logic to this argument and it represented a useful counter-argument to the usual view that demand management policies limit personal choice by trying to shape demand patterns. Nevertheless, people had configured their lives around unrestricted car access and their experiences of local public transport (where there was any) were such that they remained unconvinced of the new rhetoric.

As a consequence, subsequent government announcements became less insistent on a demand management rhetoric, a fuel tax escalator (petrol prices should rise 6p per gallon above inflation) introduced by the Conservatives in 1993 at a rate of 3p was scrapped in 2000 following protest by hauliers, and some roads schemes that had been put on hold were reinstated. There was also disquiet over the lack of expenditure being made available to implement better public transport in the early years of the Labour government, which stuck to Tory spending plans. This was not, then, in practice a radical change from the last years of the previous Tory administration. Indeed, the White Paper itself, often perceived as being anti-car, cut just twenty-one road schemes out of a total of 156 inherited from the Conservatives.

The Transport Act 2000 and the ten-year plan

The Transport Act 2000 covered a large amount of ground, including paving the way for a partial privatisation of air traffic control and the creation of a Strategic Rail Authority. It also enabled local authorities to introduce road user charging and to levy workplace parking charges. Subsequent schemes were implemented in London and Nottingham (workplace parking charge), with a tiny one in Durham. In London, Mayor Ken Livingstone was boldly radical and autocratically implemented charging, asking voters to vote him out if it did not work. The public largely thought it a bad idea, but after its implementation most changed their minds. Nottingham has also been bold in charging businesses for the parking spaces available to employees and using this to finance public transport in the city. Edinburgh and Manchester both proposed charging policies which were rejected by the public.

A £180 billion transport investment programme to implement the policies was set out in a subsequent document, *Transport 2010: The 10-Year Plan* (DETR 2000a). While much of the headline figure was already committed spending and a sizeable amount was 'expected' investment from the private sector, this was a step change in transport funding to address what was widely seen as a lack of long-term investment in UK infrastructure. The ten-year plan promised £60 billion each for rail, roads and local transport. The objectives in the plan were very ambitious: a 50 per cent increase in passenger use on the railways, twenty-five new light rail projects, a trebling of cycle journeys, and a 10 per cent increase in bus use, among others (see Headicar 2009: 131). Very few targets were met, in part as significant reshaping of the mobility habits of a nation was always likely to take a lot more than ten years. The plan also fell prey to a very British approach to transport policy, which is guided not by ideas of demand management so much as by pragmatic investment, where all modes are invested in equally. The plan included 360 miles of motorway widening and about 350 other road schemes, alongside its public transport 'emphasis'. This was, therefore, likely to continue a societal trend of 'hypermobility', facilitating movement by any means possible and inducing demand on road and rail networks.

Labour's pragmatic second and third terms, 2001–10

The ten-year transport plan was updated through to 2014–15 in the 2004 White Paper *The Future for Transport*, which responded to continued criticism on lack of progress on national transport problems and, particularly, the demise of Railtrack. The main

message is that instant solutions are not possible. It draws attention to the commitment to sustained investment over a long period to make headway in consequence of a lack of investment in the 1980s and 1990s. The White Paper repeatedly explains the difficulty of tackling transport problems in the short term and pushes further on the need to 'manage the demand for transport'. Nevertheless, it is still timid on actual measures, suggesting for example, that national road charging may be 'technically feasible' by 2015 or 2020. The tentative introduction of tolling on trunk roads, realised for the first time in December 2003 with the opening of the Birmingham Northern Relief Road or the 'M6 Toll' in the West Midlands, did not look to go much further. Government did provide funds for local authorities to investigate road user charging and many cities undertook feasibility studies. But few had the political conviction to go ahead and electorates, as Manchester in particular experienced, were hostile (Vigar *et al.* 2011).

Public transport

Labour did put a lot more investment into bus travel, recognising that it is highly significant to many citizens, being responsible for two thirds of public transport journeys. The Rural Bus Challenge was introduced in 1998 to stimulate the provision and promotion of rural public transport. In 2001, an Urban Bus Challenge was aimed at improving public transport for deprived areas. A rural bus subsidy encouraged the development of new public transport services, particularly for areas with little or no public transport provision. Changes were also made to concessionary fares for the elderly, first in Wales and Scotland, with England forced to follow suit. Many criticised these schemes for leading to a lot of nearly empty buses running, particularly in rural areas, and for indiscriminately subsidising pensioners regardless of income. Much of the investment was reduced under the Coalition government that followed in 2010.

Labour's approach to light rail schemes shifted dramatically in the course of its second term. The ten-year plan provided for up to twenty-five new rapid transit lines in major cities and conurbations. But the

2004 *Future of Transport* White Paper confirmed that the government had no intention of funding this many schemes. Although it gives no details, it makes plain its dissatisfaction with the relatively poor performance of some light rail lines, pointing the finger particularly at South Yorkshire Supertram, where passenger numbers did not meet targets. Light rail schemes were considered expensive for the benefits derived, even though they did have greater success in geting drivers out of cars than buses.

Active travel and 'smarter choices'

Interest in 'active travel' comes from the desire to shift people from motorised modes for a number of reasons. First, congestion on infrastructure, at peak times especially, can be alleviated by facilitating a shift from private vehicles to greener modes. However, public transport networks are also often crowded at such times and providing for such peaks can be very expensive. Walking and cycling take up less roadspace than private vehicles and capacity often readily exists. Second, private vehicles are a major contributor to a variety of environmental problems. These are numerous (see Banister 2005; Potter and Bailey 2008), but just five examples will suffice. First, health impacts include around 25,000 people killed or seriously injured each year on UK roads. Air pollution also accounts for many more premature deaths than crashes, with large costs to the NHS. Indeed, the UK Supreme Court ruled against the government in 2013 for failing to tackle air pollution, and the European Commission is prosecuting the UK on its lack of action. Second, an increasingly unhealthy population, with concerns over obesity in particular, led to transport becoming a wider health concern, interestingly reinstating the link between planning and health from the nineteenth century. Third, the noise pollution associated with transport infrastructure should also not be overlooked. Fourth, infrastructure planning has significant consequences for climate change, with road transport contributing 27 per cent of UK total greenhouse gas emissions. Finally, and with direct relevance to land use planning, infrastructures can require a lot of land. Between 2002 and 2011, an average of 1,500 hectares

per annum was used by transport and utility development, most of it for transport infrastructure, and nearly half on previously undeveloped agricultural land (DCLG 2013b).

In this regard, the first issue discussed in the 1998 *New Deal* White Paper is 'making it easier to walk'. Measures included more pedestrian crossings, more direct and convenient routes for walking, and increased pedestrianisation. Speed limits and 20 miles per hour zones were promoted to make walking and cycling more attractive by reducing the perceived danger from moving traffic. The Transport Act 2000 gave local authorities the opportunity to make orders to create 'home zones' or quiet lanes to govern traffic and reduce speeds. It spells out that a strategic view of how walking and cycling can be encouraged should be set out in regional spatial strategies and that implementation of practical measures should be included in local transport plans and development plans. Progress in this regard was patchy with some local authorities very active and others paying lip service to the agenda. Central government too was strong on rhetoric but investment levels in active travel were low.

Indeed, in 2004, the House of Commons Committee on Obesity concluded, 'if the Government were to achieve its target of trebling cycling in the period 2000–2010 (and there are very few signs that it will) that might achieve more in the fight against obesity than any individual measure we recommend within this report' (para. 316). The government did not hit this target, but why not? An analysis undertaken in 2008 by a consortium of over 100 organisations with interests in health, planning and transport policy, showed that of £19.6 billion spent on transport, in England 0.3 per cent and in Wales 0.4 per cent went on cycling (about £1 per capita per annum) and spending on walking could not be identified. In London 0.75 per cent went to walking and cycling together (Sustrans 2008). Thus the most inclusive, environmentally sustainable and equitable forms of transport, and those most health-promoting, were almost totally neglected. In Amsterdam, the budget just for improving cycle parking at railway stations, at €350 million, was more than the entire UK budget for walking and cycling for a year (ibid.). As an all-party parliamentary group reported in 2013,

> Dutch cities reap massive economic benefits because of a consistently high level of investment [in cycling] for several decades (now £24 per person per year). Although London now plans to spend £14, Scotland is up to £4 . . . England outside the capital still spends less than £2 per head.
>
> (All Party Parliamentary Cycling Group 2013: 63)

The Group called for spending of £10 per person per year on cycling in the short term, to rise to £20. A few local authorities were tackling this through diverting internal funds, but these funds were increasingly scarce as the Coalition government of 2010–15 cut local authority budgets throughout their time in office.

The potential of 'green modes' and 'smarter choices' had been noted in the 1990s. What is more, investment in such things was cheap. The 1996 *National Cycling Strategy* had put cycling back into transport policy, where, beyond a little attention in the mid-to-late 1970s, it had not been since before the Second World War. There has been uplift subsequently, notably as a leisure pursuit, and much of the credit for this goes not to government policy but to the charity Sustrans, and its creation of the National Cycle Network. Sustrans' work has shifted more into everyday travel and school, workplace and neighbourhood-level intervention as the national network has been rolled out. Partnerships with local authorities are essential to much of this work.

The 'smarter choices' agenda also includes 'travel planning'. Travel plans are produced for workplaces, schools, hospitals, universities, etc. and sometimes collections of these. Personal travel planning was also extended to residential neighbourhoods, often newer ones, where people may be unaware of bus services and cycling infrastructure, for example. Plans typically include measures that encourage travel to work by public transport, cycling or walking, a flexible benefits package to provide attractive alternatives to a company car, a review of standard working hours, car-sharing

schemes, using IT to reduce business travel, enhancing the fuel efficiency of the vehicle fleet and sometimes restrictions on levels of parking or employers charging employees to park. The targeting of workplaces typically achieved a 10–20 per cent reduction in peak-time car trips for not very much investment (Sloman *et al.* 2010). Such interventions were chiefly about addressing 'information deficits' in the minds of commuters and others; hence such measures became known as 'smarter choices'. The development management system was a key tool for developing and implementing travel plans, with planning permission granted for development on condition that a plan be implemented. The 2004 *Future of Transport* White Paper thus set a target to roll out school travel plans to every school in England by 2010. School travel programmes continue in some form in all the devolved territories, often dependent on partnerships between health authorities, providers such as Sustrans and local authorities.

A further boost to active travel occurred in 2005, as Labour introduced cycling 'demonstration towns' and 'sustainable travel towns', where a modal shift from private vehicles was attempted through concerted efforts with 'smarter choices' tools. Evaluations showed an increase in cycling levels of 27 per cent in cycling towns (Sloman *et al.* 2009) and high levels of value for money from such investment in sustainable travel towns (Sloman *et al.* 2010). The scheme was extended in 2008 by the Department for Transport, the Department of Health and Cycling England investing over £140 million to promote cycling over three years. Over £40 million was used to promote Bristol as a 'Cycling City' and the rest invested in eleven 'Cycling Towns'. The investment targeted journeys to workplaces and schools in particular, to achieve a modal shift from private cars to cycling.

An *Active Travel Strategy* was jointly produced by the Departments of Health and Transport in February 2010 in an attempt to capitalise on the demonstration town success. It prompted local authorities to put walking and cycling at the heart of local transport and public health strategies over the next decade, and specifically released £12.5 million for cycle training for every child. The other aims of the strategy focused on cycle parking at every public building and rail station, and 20 miles per hour zones in residential streets. There was little new investment to make this happen, however.

The Local Transport Act 2008

The 2008 Act was an attempt to deal with traffic congestion through strengthening local powers in relation to public transport. First, the Act amended the Transport Act 1985, which deregulated the bus market outside London, and the Transport Act 2000, which had introduced 'quality partnership schemes' and 'quality contracts schemes' for bus services. These schemes tried to tackle some of the problems of bus deregulation by providing for partnerships between local authorities and bus operators. Typically, a local authority would provide specified facilities such as bus lanes or enhancements to bus stops and stations, while operators agreed to upgrade vehicles or service frequencies. A quality contracts scheme goes further and effectively franchises routes, and thus has the effect of closing down the deregulated market in an area. Voluntary agreements proved more popular than formal contracts.

Second, the Act allowed *passenger transport authorities* (PTAs) to become *integrated transport authorities* (ITAs). ITAs, covering the six English former metropolitan counties, strengthen their powers to regulate bus services and allow them to look at other transport modes. Some PTAs have pursued this, which has led to unsurprising hostility from elements of the bus industry.

The Planning Act 2008

A feature of infrastructure planning in the UK has been disenchantment with the length of time it takes to get major projects built (Hall 1980; Marshall 2013a, 2013b). Impetus to address the problem was given by the Eddington Report (2006) into the UK transport system and the most recent cause célèbre of approving the development of Heathrow Terminal 5. While this project took more than twenty years to complete, for a variety of reasons, attention focused on the inquiry, which lasted nearly four years (and it took

the Secretary of State a further, not untypical as Eddington's report shows, two more years to make a decision on its findings). No doubt the inquiry could have been speeded up, but it was very much a once-in-a-generation exception (Headicar 2009). It is also true that inquiries are where society debates its future, and foreclosing such possibilities will probably lead to debate and protest emerging elsewhere, as experiences with road proposals tell us (see above and Owens and Cowell 2011). Regardless of the counterarguments, the culmination of government perception that planning was holding back large infrastructure projects led to the 2008 Act.

The 2008 Act applied principally to England and Wales and dealt with projects of a certain scale that are deemed *nationally significant infrastructure projects* (NSIPs). Alongside a streamlined consents regime for such projects were two innovations: first, national policy statements on infrastructure sectors (see below), and second, the creation of an Infrastructure Planning Commission (IPC) to take decisions on such project applications. The IPC was an attempt to depoliticise decisions about large infrastructure projects by taking them away from elected politicians and putting them in the hands of a new agency, rather as interest rate setting had been given to a new Monetary Policy Committee. The IPC was abolished by the Coalition government in 2011 with the passing of the Localism Act 2011 and its work transferred back to the Secretary of State, advised by the Planning Inspectorate.

NSIPs are largely an English issue. In Northern Ireland, nationally significant infrastructure projects are considered as major developments under Article 31 of the Planning (Northern Ireland) Order 1991 (SI 1991/1220). Such projects are dealt with centrally by the Strategic Planning Division within Planning DOE. In Scotland, large development is either 'national development', defined in the National Planning Framework for Scotland, which is deemed to have established the need for that development, or 'major development', with nine classes of large-scale development defined in the Town and Country Planning (Hierarchy of Developments) (Scotland) Regulations 2009 (SI 2009/51).

BOX 12.2 AIMS OF (THE NOW DEFUNCT) REGIONAL TRANSPORT STRATEGIES

The RTS should provide:

- regional objectives and priorities for transport investment and management across all modes to support the spatial strategy and delivery of sustainable national transport policies;
- a strategic steer on the future development of airports and ports in the region consistent with national policy and the development of inland waterways;
- guidance on priorities for managing and improving the trunk road network, and local roads of regional or sub-regional importance;
- advice on the promotion of sustainable freight distribution where there is an appropriate regional or sub-regional dimension;
- a strategic framework for public transport that identifies measures to improve accessibility to jobs and key services at the regional and sub-regional level, expands travel choice, improves access for those without a car, and guides the location of new development;
- advice on parking policies appropriate to different parts of the region; and
- guidance on the strategic context for local demand management measures within the region.

Source: PPS 11 *Regional Spatial Strategies*, p. 58

The regional scale

Labour also continued to promote the integration of land use and transport planning begun under PPG 13, through proposing and then integrating *regional transport strategies* (RTS) with *regional spatial strategies* (RSS) in England. The creation of the RTS within an RSS made regional transport policy more visible than previously. RTS did shape investment to a limited degree in their territories but there was an absence of institutional capacity in the organisations charged with making them. They thus fell victim to being a mixed bag of policies and schemes derived from bargaining processes among local authorities and business leaders (Vigar 2006). The Coalition government abolished RSS, and thus RTPs, in 2010.

One potentially important innovation at the regional scale in England was the increased joint working between regions, especially the North West, Yorkshire and Humber and the North East to form 'the Northern Way'. Increasingly vocal on issues of

transport policy, the Northern Way published a series of transport priority documents (e.g. The Northern Way 2007, 2010). These developed plans for improving east-west and north-south connectivity across northern England, and proposed Manchester as a 'super-hub' for northern England where investment in rail in particular might be concentrated as a counterpoint to London. The proposals had some success, with the subsequent inclusion of elements of the strategy in the National Infrastructure Plan.

The Labour Years: a postscript

It is tempting to conclude that Labour's action did not ever match its policy rhetoric with regard to the promotion of green, integrated travel (see Docherty and Shaw 2008 for a thorough evaluation). It did make a step change in transport investment, particularly in public transport, both in rail, notably with the completion of the UK's first High Speed Rail line, and in buses as part of an increase in transport spending of

83 per cent from 2001 to 2006 in England (a large part of which went to London). An increase of 145 per cent in Scottish transport spending is also notable, as the Scottish government used its powers to prioritise transport spending (MacKinnon and Vigar 2008). Many criticised that the money was not well spent: the abandonment of tolls for river crossing in Scotland, rural buses in England, and the huge amount of spending that went to London came in for particular criticism. That said, it was always likely to take a generation to tackle historic underinvestment, not least as people take many years to respond to changes in the opportunities afforded by new infrastructure and services.

UK infrastructure planning from 2010

Arguably the twentieth century was, for the most part, a time for 'heroic engineering' in relation to infrastructure planning. National grids were developed, underpinned by ideas of universal service provision, i.e. that every citizen should have the right to the same service at the same price as others. Privatisation and marketisation in infrastructure provision have transformed this idea. People are no longer citizens, but consumers, often free to choose, assuming they can pay, from an array of possibilities, often provided by the private sector. Often such possibilities include 'premium networks' (Graham and Marvin 2001) that allow a faster service, such as road tolls in the West Midlands to bypass the M6. That is not to say that government is unimportant in setting the policy context and regulatory frameworks for this activity – just that it was less centrally involved in the direct provision of it by the end of the twentieth century than at just about any other point during it.

In the early twenty-first century, arguably the greatest trend was one of devolution of responsibility for setting the policy context away from the central bodies which led the heroic engineering era. Westminster and Whitehall passed many powers to equivalent bodies in Northern Ireland, Scotland and Wales; and many quasi-autonomous government agencies conduct infrastructure planning at arm's length from governments. This has led to the emergence of a much more differentiated picture of infrastructure planning and policymaking than previously and details of such policy divergence are given later in this chapter.

One notable feature is the development of a National Infrastructure Plan in 2010, 2011 and 2013, which has proved a useful exercise in presenting the national picture on infrastructure investment, although its tone is notably 'promotional'. The priorities have changed little between the three iterations of the plan (2010, 2011 and 2013), but the latest ones are worth examining to show the scale of investment and where it is going. Table 12.3 rather indicates that investment is strongly biased towards London, with £36 billion of spending for London's transport system, for example, further strengthening that economy as against the rest of the UK. (It is worth noting that the figures for the South West and Wales are dominated by private-sector investment in new nuclear power plants).

In general terms, the Coalition government (2010–15) increasingly focused government investment on projects that would 'get the economy moving'. They utilised a very narrow definition of this already narrow objective for transport policy (and one with limited evidence beyond the construction phase) by funding projects that were 'shovel-ready'. Underpinning this approach were roads forecasts of the sort discredited several times since the days of Buchanan (see HM Treasury 2013b and Goodwin 2012). Larger projects tended to be favoured, including, the biggest of all, the High Speed 2 rail link, despite increasing evidence that it was poor value for money. Thus 'pragmatic multi-modalism' (Shaw and Walton 2001) was the order of the day, as under Labour, but with roads projects much favoured, as they were easier to get going, as they did not involve the multiple players that privatisation of public transport services led to. In 2009, the Cabinet Office costed the externalities of the UK approach to transport, including congestion, physical inactivity, carbon emissions and local air pollution, noise and crashes, at £38–48 billion per annum in England. Such costs will not be reduced by such

Table 12.3 Infrastructure spending in the pipeline and region of impact, as at 2013

Sector	£ billions	Number of projects
UK	115	61
England	83	14
Offshore	47	73
London	36	48
South West	19	44
Wales	18	27
England and Wales	13	1
Scotland	11	71
South East	10	74
North West	8	65
East of England	6	43
Yorshire and Humber	3	43
West Midlands	3	28
North East	2	27
East Midlands	2	26
Northern Ireland	0	1
Grand Total	377	646

Source: HM Treasury 2013b
Notes:
£ billions data is in 2012–13 prices
Includes public and private investment
UK-wide projects may impact on several regions

bias in transport policy, and funds and organisational attention for public transport, walking and cycling have diminished as many bodies responsible for such areas have been axed, such as the Commission for Integrated Transport and Cycling England.

In contrast to an eschewed aim of 'localism' which might have entailed using local transport plans to deliver local infrastructure, government launched specific pots to deal with smaller-scale issues. Thus the Local Sustainable Transport Fund (LSTF) offered £600 million over four years (2011–15) to English transport authorities for specific projects in an area such as walking and cycling to school. But such proposals often ran counter to the emphasis on 'shovel-ready' projects emerging from the work of local transport boards (LTBs), for example. A consistent approach was needed, with policies and funding established for the long term within a strategic policy framework. The UK was not alone in following an 'automobile' trajectory in the 1950s and 1960s, but most European countries invested consistently in other modes and in their urban public spaces from the 1970s onwards. England was late to this particular party and has been inconsistent in attending ever since, with greater dependence on private cars as a result.

Table 12.4 National policy statements on infrastructure

Sector	NPS	Date adopted
Energy	General Principles; Renewable Energy; Fossil Fuels; Oil and Gas Supply and Storage; Electricity Networks; and Nuclear Power	All 2011
Transport	Ports; Road and Rail Networks; and Aviation	Ports 2012; Transport Networks published in draft for consultation in 2013; Aviation not get published as of April 2014
Water and waste	Water Supply; Hazardous Waste; and Waste Water Treatment	Water Supply not yet published as of April 2014; Hazardous Waste 2013; Waste Water 2012

National policy statements

National policy statements (NPS) set out government policy on different types of national infrastructure development in England and Wales (see Table 12.4). NPS are accorded a large weighting in the activities of the Planning Inspectorate as it makes decisions on NISPs under the terms of the Planning Act 2008 (see above). Their production has been highly politicised and many are vague (Marshall 2013b). Few have any degree of spatiality, only the nuclear power NPS specifies sites, and this contributes to their lack of utility, with debates over, say, the expansion of port facilities likely to keep emerging in the inquiry process as a result. This was in sharp contrast to the experience in Scotland, with key projects identified on a map (see Figure 12.4).

Many of the NPS, such as that for road and rail, seem to drift back towards predictions of growth that seem unlikely to materialise, as in previous eras. Subsequent policies based on capacity increases are then proposed, with little recourse to the now well-rehearsed debates on induced demand and the environmental and social costs incurred.

Subnational transport planning in England

Transport planning has traditionally been a highly centralised feature of English governance, although

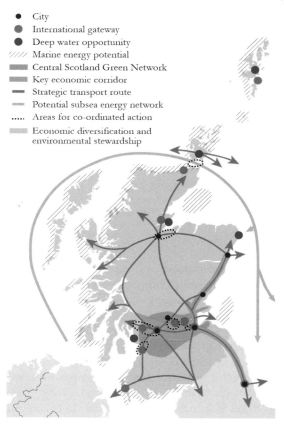

- ● City
- ● International gateway
- ● Deep water opportunity
- ///// Marine energy potential
- ▬ Central Scotland Green Network
- ▬ Key economic corridor
- ▬ Strategic transport route
- ▬ Potential subsea energy network
- ..... Areas for co-ordinated action
- ▬ Economic diversification and environmental stewardship

Figure 12.4 Planning Strategy for Scotland, 2009
Source: Scottish Government 2009

there were signs of change with the freedoms likely to emerge for LTBs. In contrast to European cities, UK local government is highly dependent on national governments for resource allocations in general as it has limited tax-raising powers. Overall, infrastructure spending is thus largely determined at national level. Locally, 60 per cent of the capital funding local authorities spend on transport comes from, and is approved by, the Department for Transport (NAO 2012). In addition, a large element of revenue spending is funded through the formula grant administered by the Department for Communities and Local Government, which is less directed now in terms of what it can be spent on but shrank greatly between 2010 and 2015 such that many authorities diverted money previously spent on transport through the formula to other policy areas.

Subnational transport planning in England has historically taken place at the level of the local authority. That said, there was increasing activity emerging at the subregional scale under the Coalition government. In the six former metropolitan counties in England, and similarly in Strathclyde, residual bodies retained the transport powers after the abolition of the counties in 1985. Chief among them are passenger transport executives and, since the Local Transport Act 2008, integrated transport authorities (formerly passenger transport authorities). ITAs are led by councillors representing the constituent local authorities. They are responsible for making strategic transport policy through local transport plans and they subsidise essential loss-making local transport services, coordinate services, manage bus stations and stops, and publish information including marketing material aimed at growing public transport services. In Merseyside, Strathclyde and Tyne and Wear, some ferry services are also operated by them. The ITAs are funded by negotiating a levy every year that is applied to council tax collected by the local authorities in the areas that they serve. They are likely also to be given devolved funds from central government in the future for their operations (see below).

The previous section noted the abolition of regional transport strategies. However, the pressing need to improve transport infrastructure and to develop awareness of those needs from the bottom up has led to a number of emergent places in which strategic priorities are determined and coordination can be achieved, at least in theory. First, local authorities are increasingly coming together in groupings, encouraged by central government, to create *combined authorities*. In many areas they mirror local enterprise partnerships (LEPs). A combined authority (CA) is enabled by the Local Democracy, Economic Development and Construction Act 2009, with the Greater Manchester Combined Authority established in 2011 leading the way and four others having received approval in 2014. The latter are effectively former metropolitan county council areas, Merseyside, South Yorkshire, Tyne and Wear, and West Yorkshire, some incorporating hinterlands: the North East CA includes Durham and Northumberland County Councils alongside Tyne and Wear councils, for example. They allow a group of local authorities to pool resources and receive certain delegated functions from central government to deliver transport and economic policy. Combined authorities can also be used as an alternative means of receiving additional powers and funding as part of 'City Deals' or 'Growth Deals'. Of particular interest to this chapter is the ability to assume the role of an integrated transport authority.

Second, LEPs have some responsibilities for infrastructure, financed initially through the Growing Places Fund announced in 2011 and extended subsequently. The fund is designed to tackle 'immediate infrastructure investment constraints', with a focus on housing and transport, in a reflection of government concern to get house building moving on the ground at this time. The DCLG press release announcing the fund gave some possible examples, which are quite revealing in terms of their modal bias toward roads:

- Early development of strategic link roads and access works to unlock major mixed-use developments . . .
- Provision of flood storage capacity to enable development of homes, employment space and retail space; and
- Works to improve local connectivity and reduce congestion through interventions such as

extending dual carriageways, enabling developments to be taken forward sustainably [*sic*].

(DfT Press Release, 7 November 2011, '£500M fund to unlock the potential for economic growth')

Funding available was relatively modest, £500 million initially, and more significant were subsequent announcements, including 'City' or 'Growth' deals. The Single Local Growth Fund (SLGF) requires LEPs to develop multi-year local strategic economic plans, which are used for negotiations on 'Growth Deals' with the government. These plans are used as the basis for allocations from a £2 billion a year Single Local Growth Fund which allocates funding to skills, housing and transport from 2015–16. Transport funding makes up more than half of this total. A further £10 billion of funding for the SLGF is planned for the five-year period from 2016–17. LEPs are also able to borrow cheaply through a newly created Public Works Loan Board which allocates funding for local priority infrastructure projects to a maximum value of £1.5 billion and also has responsibility for how €6.2 billion of EU Structural and Investment Funds is spent according to nominations by each LEP (excluding London).

Local transport planning

Local authorities have around 300 statutory responsibilities for transport, such as developing policy and administering the 'national concessionary travel scheme' (NAO 2012). They plan and commission services (including bus and light rail), and provide and maintain infrastructure, increasingly through joint arrangements with others (see above).

A key element was the development of a local transport plan (LTP) or, in London boroughs, local implementation plans, which are subordinate to the London Transport Strategy. These mechanisms replaced transport policies and programmes (TPPs) when the Transport Act 2000 made five-year local transport plans a mandatory requirement for local authorities in England. LTPs were abolished in 2014, with responsibilities transferred to LEPs. They were

useful in priority setting but often fell prey to bargaining among politicians over 'pet projects'. Levels of funding were never high, but dwindled, especially in comparison to the costs of developing them (with requirements for assessments and consultation, for example). The development of a number of one-off financing processes, such as the Local Sustainable Transport Fund and City Deal funding, made them less significant. That said, they have broad professional support (Headicar 2009) and were developed in transparent and rigorous ways that other mechanisms were not (Campaign for Better Transport 2013).

The Department for Transport has announced an intention to devolve funding for local major transport schemes to local transport bodies. It published draft priorities for capital spending from 2015 to 2019 in 2013. Local transport bodies are voluntary partnerships between local authorities, (ITAs and combined authorities, where they exist), LEPs and possibly other organisations. Their primary role is to decide which investments should be prioritised, to review and approve individual business cases for those investments and to ensure effective delivery of the programme. It is intended that the Department for Transport's role will be reduced to a safeguarding one concerned with the appropriate use of public funds, not whether the proposals conform to national priorities. Analysis of the priorities of local transport bodies showed some signs of a lack of learning from the past and a drift back to automobile and hypermobile policy directions, with road-building projects making up 59 per cent of projects and a further 14 per cent of projects consisting of schemes with an element of new road capacity (Campaign for Better Transport 2103).

Infrastructure delivery plans and Community Infrastructure Levy

Infrastructure delivery plans (IDPs) were introduced as part of the 2008 reforms to the planning system. They are designed to support local authority core strategies and perform a wider corporate role, as the definition of infrastructure includes social infrastructure, such as schools. Thus, this wider corporate role involves supporting and informing other strategies relating to capital investment. The IDP provides a framework for determining investment priorities to deliver strategic objectives. IDPs are linked to the Community Infrastructure Levy (CIL), also introduced as part of the 2008 reforms to planning. The CIL allows local authorities to charge developers an amount based on the likely extra demands their development places on local infrastructure networks. It partly replaces section 106 agreements from 2015. Its introduction has been controversial, with areas where viability is often marginal unable or unwilling to implement a tariff.

Scotland

A Transport White Paper and Transport Act emerged in approximate parallel with the equivalents in England in 1998 and 2001. After this, Scotland began to take a somewhat different direction in terms of transport governance. A new agency, Transport Scotland, was created in 2005 and the Scottish rail franchise is controlled by this agency, along with oversight of all other modes (see Table 12.5). The Scottish government has also been more positive in promoting active travel, in part to address public health issues which are greater in Scotland than in England. For example, a Cycling Action Plan for Scotland sets itself the target of achieving 10 per cent of all trips by bicycle by the year 2020.

Of particular note in Scotland is the existence of a National Planning Framework published by the Scottish Government in 2009 which is akin to a genuine national plan, given its explicit identification of nationally important economic locations and the transport and other infrastructure required to serve them (see Figure 12.4). This explication reflects that investment in transport and energy networks in Scotland has been a priority for the devolved government.

Formal, statutory, regional transport planning also exists in Scotland. Scotland has two subnational territorial structures in which transport planning of some kind is conducted. The first of these are the seven regional transport partnerships (RTPs) created by the Transport (Scotland) Act 2005, which cover the entire national territory. Consisting of local authority representatives alongside other stakeholders such as business, these bodies are charged with developing priorities for investment through the development of an RTS. RTPs were designed to overcome problems with executing transport planning in what are often in Scotland very small local authorities, which can imply a limited transport skills base, and problems of cross-boundary competition (Docherty and Begg, 2003). RTPs have disappointed supporters of the concept because of the limited nature of their powers and the low levels of funding available to them (Headicar 2009).

Second, at local authority level, development plans are also expected to make a statement of transport policies, providing a vehicle for integrating infrastructure questions with land use planning concerns and the demands of other policy sectors. Traditionally, the UK planning system has not performed well in this regard compared with those of its European neighbours, and

Table 12.5 Devolved and reserved functions in transport

	Scotland	Wales	Northern Ireland	London
Road	Totally	Substantial	Totally	Totally
Rail	Substantial	Substantial	Totally	Limited
Bus	Totally	Limited	Totally	Totally
Air	Limited	None	None	N/A
Sea	Substantial	Nine	None	N/A

Source: updated from MacKinnon and Vigar 2008

little appears to have changed even under a new regime of 'spatial planning' (Vigar 2009).

Wales

The Welsh Assembly government (WAG) has slowly accrued infrastructure planning powers, with a Wales Bill in 2014 paving the way for tax and borrowing powers to fund infrastructure. Prior to this, Wales had been increasingly active in making transport policy that was distinct from the rest of the UK, starting with a non-statutory transport plan in 2001 delivered by a new parallel institution, Transport Wales. The Transport (Wales) Act 2006 requires the WAG to produce a statutory transport strategy, first published in 2008. The plan is interesting in that, rhetorically at least, it prioritises carbon reduction and acknowledges the role of transport investment in tackling poverty and increasing well-being as well as its role in assisting economic growth – the priority that remains most explicit in England and Scotland. Transport Wales is also responsible for the Welsh railway franchise. As in Scotland, a national spatial plan has implications for infrastructure provision, although such issues are afforded less attention and spatial expression than in Scotland's NPFs 1 and 2.

The National Transport Plan sits alongside regional transport plans in delivering the Wales Transport Strategy. The Transport (Wales) Act 2006 made provision for regional transport plans to be produced by four regional transport consortia (RTCs), to replace the LTPs previously prepared by each local authority. RTCs are made up of members from constituent local authorities and funding is derived from constituent member budgets and from the WAG.

One further innovation is the Active Travel (Wales) Act 2013, which requires local authorities to continuously improve facilities and routes for pedestrians and cyclists. How successful this Act is in practice remains to be seen (Wales, outside Cardiff, starts from a lower base for cycling than the rest of the UK), but it is an important recognition in legislation of the value of good walking and cycling networks for health, social inclusion and economic growth. It was supported by a £5 million Safe Routes in Communities fund in 2014

to kick-start local authority action, especially in relation to active travel to schools. Wales's policy trajectory is thus an interesting one, somewhat paralleling Scotland's, but with a greater emphasis on quality of life issues and social inclusion.

Northern Ireland

Governance generally and transport planning in particular have historically been highly centralised in Northern Ireland. The policy context for the Province is set by the Regional Development Strategy (RDS) developed by the Department of Regional Development (DRD). At the local scale, however, infrastructure development depends on close links to the planning system, and area development plans are prepared by the Planning Service of the Department of the Environment. In this way, local plans have been centrally controlled. These governance arrangements suggest that optimising the effectiveness of transport governance in Northern Ireland rests heavily on cooperation between central government departments.

Thus local government has played a limited role in transport policymaking. It has been in effect a consultee with a useful role in integrating transport and land use proposals on the ground but very limited powers. This will change with a reform of local government in the province which will reduce the number of local authorities from twenty-six to eleven in 2015, although most transport planning matters will remain with the Department for Regional Development Northern Ireland, which prepares the regional transport plan.

London

London has always been somewhat apart in the political geography of the United Kingdom. Its separateness accelerated in the 1990s and the governance of aspects of infrastructure planning also became distinct from the rest of England. Through the London Plan, the city is effectively the one place in England where a regional strategy still exists. The presence, too, of Transport for London, a large well-resourced organisation, also sets it apart. It produces the London Transport

Strategy which is more akin to an investment programme than the equivalent RTS in England were prior to their abolition. In producing such a programme, the presence of a Mayor benefits London financially, providing as it does a focus for negotiations, as does its privileged position in the economic and political geography of the UK (Massey 2010).

The two London mayors, Ken Livingstone and his successor Boris Johnson's greatest successes have arguably been in persuading the Treasury to fund large capital projects. Most of the UK's biggest infrastructure projects are in London, from sewerage tunnels such as Thames Tideway to high-profile transport projects. Chief among the latter are Crossrail, with Crossrail 2 proposed, Thameslink, and Tube modernisation. These projects, alongside continued high levels of bus subsidy, all put London in a unique position with regard to investment. Also important was the ability of London through devolution to pursue its own policy trajectory, and astutely Livingstone and Johnson both appointed good technocrats to guide the planning process and to deliver projects on time and on budget (Shaw and Docherty 2014).

Both Mayors sought to improve conditions for cyclists and pedestrians in the capital. Livingstone backed pedestrianisation schemes such as Trafalgar Square and kick-started the development of a genuine cycle *network* in London. A combination of congestion charging, the July 2007 London bombings on public transport and a change in the perception of cycling, in part a result of Livingstone's policies, led to a large increase in cycling in many parts of London. Johnson extended activity in this area through the 'Boris Bike' cycle hire scheme and 'cycle superhighways'. As a consequence of such policy action, in combination with cultural factors (Aldred 2013), the number of people cycling to work more than doubled (+144 per cent) between 2001 and 2011 (Goodman 2013).

The London Assembly also devised a walking strategy (central governments had successively baulked at this) – 'Feet First, Improving Pedestrian Safety in London' (2013) noted that pedestrian casualties were rising again and recommended changes to 'Green Man' crossing times and a timescale for rolling out 20 miles per hour speed limits in London. In this regard, many

London boroughs were following initiatives in other cities (see www.20splentyforus.org.uk/index.htm for an up-to-date list of cities with default neighbourhood 20 mile per hour limits).

Future prospects

Over half a century, from the 1960s to the 2010s, infrastructure planning followed some distinct trajectories. In the early and mid-1960s, an optimistic era of heroic engineering and economic prosperity sought to provide universal access to all manner of infrastructures. But changes in economic fortune, growing awareness of the effects of such an approach on the physical environment in particular and a realisation that experts alone were ill-equipped to do urban planning led to a shift in approach.

New issues have emerged in this time. Climate change has been a significant factor in shaping infrastructure policy and investment, to varying degrees, depending on the government in office, since the late 1980s. Public health concerns have also emerged to shape transport policy. 'Predict and provide' is rhetorically shunned in many policy areas, but the devil is often in the policy detail, with network expansion in roads, airports and energy all exhibiting signs of this discredited approach.

That said, the demand for many infrastructure services continues to rise. Price is one effective way to manage the demand but there are limits as to the acceptability of such increases. In areas of price, as in others, the issue for governments, both local and central, now as opposed to fifty years ago, is that they have fewer levers of power now than then. Privatisation and marketisation mean that governing infrastructure is a more complex task involving many more actors, and government may have little direct control over prices, especially in the short term.

In many ways, then, ideas of facilitating supply, 'predict and provide', are too expensive, typically in economic, social and environmental terms. But too much emphasis on managing demand is also unpopular with a relatively wealthy public (in the main) which has grown used to abundant energy, and a freedom to move both locally and internationally. It is not surprising, then, that a pragmatic muddling through of the sort warned of by Buchanan in relation to traffic has emerged both in transport and in energy planning. The voices of the majority are very present in such debates, whether locally or nationally, and are very much in the minds of politicians not wanting to be seen to be 'anti-car, or too 'hair-shirt' with regard to energy policy. But 22 per cent of UK households have no car. Where two-person households have one car, it is often not available to the other member at a particular time of day. For the young and the old, the most excluded in society, good public transport is essential; for the old, the young and the partially sighted, clear, well-laid pavements are a serious transport and public health issue.

Greater attention to these issues is notable in the territories of London, Scotland and Wales, which highlights the significance of devolution in infrastructure planning in the twenty-first century. Such attention demonstrates that transport, and indeed energy, policy can be more socially and environmentally progressive, centred on giving more choices – of better public transport and conditions for walking and cycling, of warmer, better-insulated homes powered by renewable energy (see Shaw and Docherty 2014 for one such manifesto in transport). Policies and investment that recognise such issues would mimic our more successful European neighbours to deliver infrastructures that are indeed 'better for everyone'.

Further reading

For up-to-date guidance on travel trends and policies, see www.gov.uk and the Department for Transport's own website, www.dft.gov.uk. Trends can be derived from a very useful DfT annual publication, *Transport Statistics Great Britain*.

Textbooks on transport planning include Banister (2005) *Unsustainable Transport*; Headicar (2009) *Transport Policy and Planning in Great Britain*; Knowles *et al.* (2008) *Transport Geographies*; and a very accessible text by Shaw and Docherty (2014) *The Transport Debate*. All of these

consider the environmental, economic, cultural and social issues associated with transport planning to varying degrees. In addition, Hickman and Banister (2014) *Transport, Climate Change and the City* looks at the relationships between transport trends, carbon dioxide and city form. Marshall (2013b) *Planning Major Infrastructure* comparaes UK policies across a range of infrastructure sectors with those of four other nations.

The Buchanan Report remains a worthwhile read for its ideas, many of which have contemporary resonance. The Rank Films organisation made several films to explain the issues, two of which convey the ideas, issues and thinking prevalent in the 1960s: www.youtube.com/watch?v=VhSXNr4_hUA and www.youtube.com/watch?v=tMamOIdcS9A (accessed 3 March 2014).

There is growing interest in the field of 'mobilities' as a way of understanding travel behaviour in a wider context. Such an approach emphasises a more sociological and psychological understanding of, for example, the importance of habits and routines in understanding travel behaviour, issues which are long neglected in transport studies. An article by Shaw and Hesse (2010) 'Transport, geography and the "new" mobilities' provides background to the debate and the work of Tim Cresswell and John Urry is very useful. Such work crosses over into writings on the 'affective' dimensions of travel, for example Jain and Lyons (2008) 'The gift of travel time'. Such writing emphasises that transport is largely a derived demand: we travel to perform a function; we also travel for enjoyment or at least with the idea that we may prefer one mode more than another, perhaps as it enables us to achieve other things – meet people, get work done in a wifi-equipped train carriage, have 'thinking time', de-stress, etc.

There is a substantial literature on transport and social exclusion. Karen Lucas's work exposes the main issues, for example (2004) *Running on Empty: Transport, Social Exclusion and Environmental Justice*. There are specific literatures on how UK transport policy typically disadvantages the old, including, frequently older drivers, and the young. Much research confirms that children are especially disadvantaged by the transport policies of the past and their continuing legacies, the landmark publication, still relevant today, being Hillman *et al.*

(1990) *One False Move* (1990). Such work has shown how children want to perform their own independent mobility, by walking, cycling and scootering, but are prevented from doing so by parental insistence on escorting, usually using a car. The 'choice' agenda in education has not helped with many children travelling farther to go to school and cars being considered the only safe option.

On public transport, Peter White's work is very relevant for planners, especially his work on deregulation and privatisation, and his textbook '*Public Transport*' (Routledge). For comment on the latest proposals, Christian Wolmar writes for a range of sources and many articles are published on his website: www.christianwolmar.co.uk. *Planning* magazine and *Local Transport Today* are also good sources of information and comment on developments in infrastructure planning.

On cycling, a major research initiative is detailed in Pooley *et al.* (2012) *Promoting Walking and Cycling*. The work of Rachel Aldred, Tim Jones and Justin Spinney complements this well.

The pressure groups Campaign for Better Transport (www.bettertransport.org.uk) and Sustrans (www.sustrans.org.uk) provide much useful information, comment and primary research material on their websites.

Finally, infrastructures are often closely interrelated. The key text in this area is Graham and Marvin (2001) *Splintered Urbanism*, which argues that 'infrastructure unbundling' is part of a wider shift to the neo-liberal, splintered city. A great deal of work also explores the relationships between telecommunications technologies, virtual mobility and physical mobility. Despite a hoped-for substitution of physical travel by virtual travel, the relationships appear more complementary, with activity in one fuelling activity in the other. More recently there are signs of the decoupling of this relationship with the Bankers' Recession of 2007–8 onwards forcing companies to address costs and to use improved telecoms opportunities, such as teleconferencing and Skype. In parallel, young adults appear to be less enamoured of car ownership and use than previous generations and this is attributed to telecommunications possibilities, such as smartphones.

On infrastructure policy and economics generally, and energy policy in particular, Dieter Helm's work is informative and accessible: see www.dieterhelm.co.uk.

Notes

1   It should be noted that the average varies a great deal, with the most mobile group being middle-aged men in employment, who travel much farther than other groups, such as retired women without a driving licence. The responsibility for the externalities arising from such movement, such as climate change, also varies (see for example Brand and Boardman 2008).

2   The relationship between physical transport and electronic communications infrastructure is important here, as the low-cost airline model is underpinned by the availability of e-ticketing and Internet booking.

3   These plans were partly realised with three levels constructed at this point and pedestrians squeezed into narrow spaces and a series of subways. Notice that cycling and public transport are not catered for, or even considered, here.

4   The first motorway-standard road in the UK was an 8 mile section of the M6, opened in 1958.

5   Barbara Bryant, a Conservative Party councillor, appears to have had her belief in British government shaken when her peaceful protest against the proposed M3 extension at Twyford Down led to her being placed under surveillance by the security services.

6   Traffic data company INRIX reported in 2013 that this road was the ninth most congested road in the UK, further adding support to the futility of urban road building.

7   Some local authorities such as Sheffield and London had shown how very cheap fares could lead to large increases in bus travel, a reduction in car use and thus reduced journey times for all on less busy roads. Such policies also led to improved air quality, a wider geographic search for jobs and education among citizens and subsequent labour-market and economic benefits. They also led to long-term behavioural changes, such as greater numbers of children, used to using buses for free, maintaining such behaviours into adulthood (Goodwin and Dargay 2001).

13 Planning, the profession and the public

Planning proposals are generally presented to the public as a *fait accompli*, and only rarely are they given a thorough *public* discussion

(Cullingworth 1964: 273).

Introduction

The right of the public to have a direct say in planning matters and the inherently political nature of planning are now taken for granted. The formal machinery for objections and appeals, initially devised only for specified uses by a restricted range of interests, is now employed much more widely. Many informal mechanisms have been created by planning authorities and others to improve the capacity of, and opportunity for, local communities and interest groups to play a part or the lead in formulating and implementing planning policy. Even so, many questions remain about the effectiveness of public participation, whose interests are served by planning, and the relationship between professional and political decisions. The first part of this chapter explains the history of public participation in planning, the nature of 'interests' and the mechanisms that enable them to influence the planning process. The second part then considers the professionalisation of planning and the relationship between the profession, politicians and the public.

Participation in planning: from consultation to co-production?

The lack of concern for public participation indicated in the quotation above (which is taken from the first edition of this book) was a result in part of the political consensus of the post-war period and in part of the trust that was accorded to 'experts' – which, by definition, included professionals. While exhibitions of planning proposals were common in the first half of the twentieth century, especially with regard to post-war reconstruction, it is doubtful that the pubic was ready to criticise and comment on expert-driven material such as this (Larkham and Lilley 2012). The time was perceived to be one of rapidly expanding scientific achievement, and the methods that had made such progress in the physical sciences were thought to be transferable to the problems of social and political organisation (Hague 1984). This, together with the advent of new social security, health and other public services, led to a rapid growth in professions and the 'welfare-bureaucracies' in which they worked. In the same spirit as that of the established professions, these newer professions sought to establish a strong, scientific and objective knowledge base. Armed with the right techniques in manipulating the environment, they were to address the physical spatial development problems of the nation and, at least by implication, the underlying social and economic forces which drive physical development. So the possibilities of public comment on plans suggested in the 1940s diminished subsequently.

In retrospect, the approach implied a depoliticising of issues which were later appreciated to be of intense public concern. This was further obscured by professional techniques and language which the public could not be expected to understand (Glass 1959). Of course, planners were not alone in this: on the contrary, they simply took the same stance as other 'disabling professions' – to use Illich's (1977) term. At the time, however, the lack of political debate and participation was not widely recognised as a problem. Professionals were perceived as acting in everyone's interest – the general public interest.

It was in the 1960s that these ideas were effectively challenged in the UK. This closely followed the experience in the USA, (see, for example, Broady (1968) on the UK, and Gans (1968, 1991) and Jacobs (1961) on the USA). By this time, there was widespread dissatisfaction both with the lack of access to decision-making within government and with the way in which benefits were being distributed. Although it claimed to serve the public interest, the planning system began to be seen as an important agent in the distribution of resources – frequently with regressive effects (Hall *et al.* 1973; Pickvance 1982). The physical bias of the planning system had failed to address social and economic problems. Perhaps it sometimes even made them worse. There was growing concern for a new type of 'social planning' which would seek to redress the imbalance in access to goods, services, opportunities and power. To achieve this, some saw the need for 'advocacy planning' (Goodman 1972), which would provide experts to work directly with disadvantaged groups. This critique has had consequences for planning practice of greater permanence than that achieved by the intellectual arguments themselves. Changes were made in the statutory planning procedures, and consultation and participation gradually became an important feature of the planning process.

The Skeffington Report (1969) is sometimes celebrated as the turning point in attitudes to public participation in UK planning, though its recommendations are mundane and rather obvious, for example, on keeping people informed throughout the preparation of plans and asking them to make comments. This is testimony to the distance which British local government had to go in making citizen participation a reality. The Committee was aware of, but did not report

on, more fundamental issues, noting, for example, the need for participation not to become solely identified and associated with formal planning procedures.

Skeffington's proposals for the appointment of 'community development officers . . . to secure the involvement of those people who do not join organisations' and for 'community forums' had little impact at the time (1969: App 1, para. 3). What was conspicuously lacking in the debate on public participation was an awareness of its implications for the transfer of some power from elected members to groups of electors, from representative to more direct forms of democracy. But it was not Skeffington, but the Seebohm Committee (1968) which highlighted the tension between participation and traditional representative democracy.

Participation cannot be effective unless it is organised, but this, of course, is one of the fundamental difficulties. Though a large number of people may feel vaguely disturbed in general about the operation of the planning machine, and particularly upset when they are individually affected, it is only a minority who are prepared to do anything. The idea of a 'silent majority' who may be broadly in favour of a development has been mobilised extensively to counteract what are seen as vociferous anti-development groups. The propensity of citizens to engage with planning is typically in direct proportion to the distance from their home, and often manifests only when a development becomes visible, i.e. when a bulldozer or planning application notice appears on site. Interest in the development of a plan is less forthcoming, even though this may be important in setting the policy context for a future development that might affect them.

Despite their failure to address more fundamental questions, the reforms introduced as a result of Skeffington were generally acclaimed by the profession. It was the academic commentators who first questioned the underlying assumptions. Neo-Marxists drew attention to the more fundamental divisions of power in the political and economic structure of capitalist society, and how these continued to be evident in the outcomes from the planning system. Practitioners, too, began to see that extensive participation exercises produced only limited gains, and some advocate planners were among the first to reject the approach for its weaknesses. The critics argued that, like all 'agents of the state', planners operate within a 'structural straitjacket' and, irrespective of their own values, will inevitably serve the very interests which they are supposed to control (Ambrose 1986). This was supported by research findings demonstrating that planning had operated systematically in the interests of property owners (Hall *et al.* 1973). There was also substantial theoretical work concerned with the role played by the planning system in the interests of capital (Paris 1982).

Sherry Arnstein also published her influential idea of a 'ladder of participation' at this time (see Figure 13.1). This also contains within it the idea that where participation had been attempted, it was often token-istic. While it can be criticised for suggesting that planners always aim for the top rungs of 'empower-ment', when this may not be desired by citizens or appropriate to the issue at hand, it was a significant critique.

The critiques were powerful but, by their very nature, they could offer little guidance to planners working in a professional, politically controlled system. Indeed, how could they respond to allegations that a fundamental purpose of planning in society is the legitimation of the existing order? If participation

merely supports a charade of power-sharing, leaving entrenched interests secure, what alternatives do planners have? Planning and planners became the primary explanation for the failures of urban and rural development: the post-war housing estates that were built as quickly as possible (and with limited resources left over for 'amenities'); the urban motorways that were built to cope with the great increase in traffic, but which destroyed the social and physical fabric of towns; the demise of village amenities in areas of development restraint; and the participation processes which raised hopes which were dashed by the outcomes. Certainly, planning played a part in all these, but was it the determining factor or was it, in Ambrose's (1986) words, the scapegoat?

In response to these failures, some planners and community activists tried more radical approaches. A movement for 'popular planning' aimed to

> democratise decision-making away from the state bureaucrats . . . [to] people who live in a particular area . . . empowering groups and individuals to take control over decisions which affect their lives, and therefore to become active agents of change.
> (Montgomery and Thornley 1990: 5)

The reality of participation in most planning practice has been somewhat different. The Planning Act 1968 (and its Scottish equivalent of 1969) made public participation a statutory requirement in the preparation of development plans. The main stimulus for this came, not from the grass roots, but from central government, which was keen to divest itself of the responsibility for considering all development plans and appeals and what the then permanent secretary called 'the crushing burden of casework' for the government department. From 1968, many powers were devolved and with this came requirements for opportunities for other interests (predominantly development interests) to participate and object during the plan-making and adoption process.

The new consultation provisions had limited effect on much of the country, because, during the 1970s and 1980s, many local authorities avoided preparing statutory development plans, in part because they

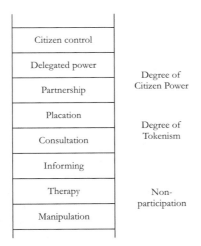

Figure 13.1 Arnstein's Ladder of Participation

believed the costs of taking a plan through the formal procedures of consultation and objection outweighed any benefits (Bruton and Nicholson 1983). As a result, the legitimacy which plans provided to decision-making was limited. The failure of local authorities to keep plans up to date exacerbated this. When planning authorities did seek public 'involvement', they tended to adopt a 'decide, announce, defend' strategy (Rydin 1999: 188).

Other planning authorities demonstrated a commitment to enabling community participation in planning in the 1970s, notably through *community development plans*, which tried to address some of the structural critique identified above. But during the 1980s, the dominant influence was 'business'. Housing, industrial, commercial and minerals interests all effectively enjoyed special treatment through the planning system. The then Secretary of State Nicholas Ridley argued that in the interests of the country as a whole, local concerns needed to be set aside to allow for a presumption in favour of new development. Many decisions were taken out of the hands of local planning authorities (never mind communities) so as to allow for major development of all forms. There was 'a consistent diminution of the significance accorded to general public participation in policy formulation, as part of an effort to "streamline" the system and reduce delays' (Thomas 1996: 177).

Disregard of local opinion gave rise to fierce criticism of the Conservative government from its own party members in the shires, and it was forced into an about-turn on community involvement under the slogan 'local choice'. This was not a signal to local authorities that they could respond as they wished to local demands. Local autonomy was to be exercised only where it was within parameters laid down by the centre in policy statements.

During the 1980s and 1990s, numerous amendments were made to plan-making and development control procedures. Their general effect was to reduce the emphasis on early public participation and consultation in the statutory procedure, while increasing opportunities for formal objection. In the early stages of the process, a statutory requirement for consultation before the planning authority has adopted its preferred view was replaced with discretion to decide on the appropriate publicity for individual plans. In contrast, the rights of formal objection after deposit of the plan were extended.

This approach was somewhat reversed in the 2000s. In development management in the 1990s, the idea of speeding up decision-making led to the introduction of time limits for local authorities to make decisions. There was much criticism of targets for leading to poor, but quick, decision-making. While there was truth in this, it did force poorly performing local authorities to address procedures which could leave planning decisions unsettled for long periods of time. These deadlines were tightened and linked to financial incentives in the 2000s. The effect of these, however, was to continue the trend towards sacrificing speed for democracy. However, in the area of forward planning, central government dictated a different approach. It suggested that 'front-loading' community consultation could save delays later on (see below).

In Scotland, experience in both the principle and practice of participation followed a similar trajectory, although it is often recorded anecdotally that the small size of the country and the proximity of central government, local authorities and communities have led to closer informal linkages. Scotland has for some time given greater emphasis to the benefits to be gained from early consultation of a wide range of interests.

The Labour government of 1997–2001 set about 'modernising government'. Part of this was to introduce the 'community strategy', a new instrument for providing more local community involvement in policy-making and implementation. The Local Government Act 2000 placed a duty on all local authorities in England and Wales to prepare one, although this authority has now transferred to local enterprise partnerships (LEPs). The community strategy should 'allow local communities to articulate their aspirations, needs and priorities; coordinate [and focus] the actions of the council, and of the public, private, voluntary and community; locally [. . .] and private sector organisations that operate contribute to the achievement of sustainable development both locally and more widely' (DETR 2000b: para. 10). Later statements also referred to engaging with 'hard-to-reach groups'. A community strategy should include a long-term vision across all sectors of activity, together with projected outcomes and an action plan with priorities. The process should deliver a shared commitment to implementation, and implementation of the strategy must be monitored. The government recognised that it might be difficult to engage particular interests and build up mutual trust in the process, so the community strategy is seen in a developmental way with early publication of an initial document, then reviews of performance and continued efforts to engage a wider range of interests and to embed the strategy into partners' plans and programmes. These appear to have 'withered on the vine' since 2010 (Lowndes and Pratchett 2012), and indeed their value was questioned in many quarters because of the bland nature of much of their contents. They were in many instances a useful vehicle for policy integration, with planning, policing, health and education, for example. But the scale at which they operated and the lack of any obvious benefits to participation meant that engagement with citizens was always limited. Indeed the name 'community strategy' was something of a misnomer.

Community involvement and the 2004 reform of the planning system

A primary rhetorical component of the planning reform agenda for the Labour governments 1997–2010 and their programme for 'sustainable communities' was improving community involvement. A key document that heralded reform was *Community Involvement in Planning: The Government's Objectives* (ODPM 2004b). A notable feature was that plans should be 'front-loaded' – involving participation at the earliest stages of a proposal, when it might make a difference. A further innovation was to require a statement on community involvement (SCI) from local authorities and this had to be approved as part of the inquiry process into the development plan.

The Coalition government withdrew the requirement to produce an SCI but the idea of 'front-loading'

remained. Thus the National Planning Policy Framework (NPPF) required in relation to local plans,

> early and meaningful engagement and collaboration with neighbourhoods, local organisations and businesses . . . A wide section of the community should be proactively engaged, so that Local Plans, as far as possible, reflect a collective vision and a set of agreed priorities for the sustainable development of the area, including those contained in any neighbourhood plans that have been made.
>
> (para. 155)

Subsequent NPPG did not take this much further, and organisations such as the Town and Country Planning Association argued for greater guidance along the lines of the 2004 *Community Involvement* paper cited above. The NPPF also rather reshaped the idea of front-loading to place a greater emphasis on dialogue between local authorities and developers.

Community, parish and neighbourhood planning

England and Wales have a great deal of experience of making plans for parishes. Parishes are the smallest element of local government and since the 1970s have often developed a plan which sets out a future vision for the area. Town councils can make town plans on a similar basis. Interest in such plans gathered pace in the 1990s, and the Labour Party gave impetus to such efforts, first through the 2000 Rural White Paper, which advocated parish plans and other mechanisms to give local communities a more effective voice, and drew attention to successful experience with village appraisals, action plans and village design statements. Such plans did not, however, overlap much with traditional land use plans, as they rarely specified land for development (Bishop 2010). The practice of preparing village design statements was, by the end of the 1990s, well established and understood.

Parish planning provided one impetus for major change in 2010, with the publication of the Localism Bill, followed by the Localism Act 2011. This granted statutory authority to 'neighbourhood plans' prepared either by parish councils or by 'neighbourhood fora'.[1] Neighbourhood planning proved popular after a slow start and by spring 2014 over 1,000 communities had begun (and some had abandoned) the process of preparing a plan. Money was available from central government to help with plan preparation and intermediaries were also funded to help with capacity-building. At the end of the preparation process, the plan has to be approved by more than 50 per cent of voters at a referendum. By the end of March 2014, all ten plans taken to referenda had been passed with very large margins, around 90 per cent on average, by local voters.

It should not be assumed that more locally oriented plans necessarily mean wider or deeper involvement. Neighbourhood plans, parish plans and other very local instruments are just as susceptible to manipulation by powerful interests as any other plan. The mechanisms (and political will) to engage people in the process are critical, whether those in control are planners or community members. The optional nature of neighbourhood plans and the large amount of resources needed to make them explained why early neighbourhood plans were being prepared in predominantly wealthy rural parishes (Vigar *et al.* 2012). That and the withdrawal of much urban regeneration funding left many concerned that the Coalition government's approach to localism, with its emphasis on self-help, was socially regressive.

Public participation in development management

Local authorities have a degree of flexibility over how to advertise a planning application. Major developments require *either* site notices *or* neighbour notification, *and* a newspaper advertisement (see Chapter 5). Developments involving listed buildings or conservation areas require newspaper advertisements in lieu of neighbour notification. As was noted earlier in this chapter, governments have continually stressed that obligations to publicise applications should not hold up decision-making. In this case,

speed is thus apparently to have higher priority than public participation.

The Coalition government's main innovation from 2010 was to focus on the pre-application stage. While legislation was not amended to require consultation before an application is submitted, local authorities were to 'encourage any applicants who are not already required to do so by law to engage with the local community before submitting their applications' (NPPF 2012, para. 189). There was emerging evidence that some developers were doing this in any case, often taking the local authority with them. Such an approach was endorsed by the RTPI, which asserted that

> early, collaborative discussions between developers, public sector agencies and the communities affected by a new development can help to shape better quality, more accepted schemes and ensure improved outcomes for the community. These discussions also avoid wasted effort and costs.
>
> (RTPI 2014: 4)

Scotland and Northern Ireland have different arrangements for neighbour notification and publicity about planning applications. In the Scottish system, notification is the responsibility of the applicant, who certifies to the local authority that neighbours, as well as owners and lessees, have been notified. This can be problematic for the applicant, and can lead to false certification (whether inadvertent or deliberate) and has been a constant source of complaint to the Ombudsman. The Northern Ireland system of notification is a two-tier approach. A non-statutory system requires the Planning Service to notify neighbours, but the identification of neighbours (through presentation of a list of 'notifiable interests') is undertaken by the applicant.

Over recent years, there has been a considerable increase in opportunities for presentations to be made to planning committees by both the applicant and objectors, with evidence of very positive results in terms of the 'customer's' perception of the service (Shaw 1998; Darke 1999; Manns and Wood 2001). However, planning committees often have remarkably little time during a meeting in which to come to a decision, with objectors and applicants often limited to three minutes to present their case. Also, many applications are now dealt with through delegated powers and will not be discussed by committees (see Chapter 5). Limited open discussion of applications may be a reflection of harmonious relationships between councillors and officers, although there have been a number of well-publicised cases where elected members have consistently acted against the recommendations of officers.

The value of public participation in development control needs to be considered in the light of the very different attitudes that people will have about a planning principle that they generally agree with (for example, preventing unnecessary development in the countryside, when it affects to their own land). A whole host of acronyms, some of them rather pejorative, have arisen to describe the values people may hold and the ways they sometimes are contradictory. That said, there is no reason why people should not act in a self-interested way in relation to a development. It is then up to a planning officer to balance public comments against other material considerations (see Box 13.1).

Mediation

The more extensive introduction of mediation in the planning process has been on the agenda for some time, though with little concrete progress. A series of studies from the Labour governments (1997–2010), the Scottish Government, the Planning Inspectorate and the RICS between 2000 and 2010 all highlighted the use of mediation to resolve or reduce objections to planning proposals. Such approaches are often seen to be most relevant at the local level in dealing with particular proposals and have proved effective.

Rights of appeal

An unsuccessful applicant for planning permission can appeal to the Secretary of State, and a large number do so. While the development plan has primacy, there is a wide area of discretion legally allowed to the planners in the operation of planning controls. For Planning Inspectors faced with a planning appeal, their role is quasi-judicial: decisions are taken not on the basis of legal rules, as in a court of law, or in accordance with case law, but on a judgement as to what course of action is, in the particular circumstances and in the context of ministerial policy and other material considerations, desirable, reasonable and equitable. The presence of an appeals system is a significant check on the operations of local government. Local authority councillors can be prone to making decisions for reasons that might make sense in the realm of local politics but not with regard to exising planning policy.

Third-party interests

While planning appeals are an important part of the system for developers, others with a stake in a planning

BOX 13.1 THE TERM NIMBY AND SOME COMMON DERIVATIVES

NIMBY – Not In My Back Yard: opponents of a development close to their home, even though they may accept the need for the development to happen (just not here!)

NIABY – Not In Anyone's Back Yard: opposition to certain developments as inappropriate anywhere, for example the building of nuclear power plants

BANANA – Build Absolutely Nothing Anywhere Near Anything (or 'Anyone'). The opposition of certain advocacy groups to all development

decision to not have the same opportunities. Third parties are those affected by planning decisions, but who have no legal interest, not being the applicant or the authority. Over the years, there have been calls to extend to them the right to appeal, should planning permission be granted, as is already the case in Ireland. Indeed, in Ireland over half of all third-party appeals result in a refusal of the original permission, which suggests that such a system can be an important check on the quality of decision-making (Ellis 2006).

Use of public inquiries and examinations

Public inquiries into major planning decisions and called-in planning applications have been subject to a great deal of critique from inside and outside the planning profession, particularly those held in connection with infrastructure (see Chapter 12). The planning inquiry/examination is a microcosm of the land-use planning system, and it reflects many of its competing positions and underlying conflicts of interest. It is perhaps here that the clash of planning ideologies, and indeed societal values, is most easily seen (Owens and Cowell 2011).

Inquiries into plans are known as examinations. The inspector's report is binding and the examination addresses both objections and the overall soundness of the plan. Such tests of soundness became a major issue for planning in the 2000s and broadened the work of the Planning Inspectorate. The Inspectorate became less of a distant quasi-judicial body and honest broker, and more central to defining what could and should be done in the name of planning. Taking on the national infrastructure consents regime further strengthened its role (see Chapter 12).

But these changes do not challenge the fundamental rights of objectors to make their views known, to appear at the examination, and to question the planning authority. The procedure is a long-standing feature of British government administration, with its origin in the Parliamentary Private Bill procedure that provided an opportunity for objections to government proposals to be heard by a Parliamentary Committee

(Wraith and Lamb 1971). The procedure has grown as much by accident as design, since it has been successively amended to take into account changes elsewhere in the system. Like appeal inquiries, local plan inquiries involve the same balancing of private and public interests through a procedure which, although essentially administrative, has many of the hallmarks of judicial courtroom practice. However, the essential nature of the planning procedure is administrative. Final decisions are taken by government, at either the central or the local level. The 2004 reforms changed the balance, in that the inspector's report was binding in all cases, whereas previously it was only so for appeals.

The inspector (and Secretary of State) has wide discretion. The legitimacy of decisions rests with the political accountability of the decision-maker (Parliament or the local council) rather than on the weighing and testing of evidence, as in a court of law. Objectors have a statutory right to appear, and although round-table and more informal hearings are commonly used, the evidence may be tested through a process of adversarial questioning before the inspector. There is inherent ambiguity in a system which has as its main objective the gathering of evidence to assist in the making of a governmental decision, while at the same time operating in the manner of a judicial hearing (Wraith and Lamb 1971).

The Franks Committee, advocated the application in the inquiry procedure of the principles of 'openness, fairness and impartiality' (Bruton *et al.* 1980: 377). These three principles have guided inspectors with some success, and the courts have played a relatively small part in the planning process (although this is growing). Nevertheless, each of the three principles requires some qualification. The Franks Report (1957) itself recognised that impartiality needed to be qualified, since, in some circumstances, central government was both a party to the debate, perhaps putting forward a proposal, and at the same time the decision-maker. How, in this situation, can the procedure be impartial? Here, one of the parties to the dispute will make the final decision, giving at least the appearance of being the judge and jury in its own court. Indeed, there was some ambiguity in the 2000s about whether the Inspectorate qualified under European law as suitably independent of government to perform its function (see below).

The openness and fairness of the inquiry also need to be qualified. First, there is widespread misunderstanding of the procedure, especially the respective roles of inspector and local authority. The adversarial nature of the inquiry, with the inspector playing a passive role, while objectors and the local authority exchange evidence and questions, has important implications for the way in which the agenda is structured, and it limits potential outcomes.

The 'plan-led system' introduced in 1991 meant that more development interests, neighbouring authorities and service providers became concerned to influence the content of plans and thus the outcome of inquiries. The number of objections to plans increased dramatically in the 1990s, and the reforms of 2004 and 2008 were partly a response to this pressure. Government's response has been to emphasise the need for involving stakeholders early in plan preparation processes, so-called 'front-loading.

A difficulty with many inquiries or examinations is determining where the boundaries of discussion are to be drawn: there is always the danger that argument will spill over into a broader policy framework. It is common at inquiries into particular matters for the most general questions of policy to arise. This is hardly surprising, since typically the development being debated is, in fact, the application of one or more policies to a particular situation. This readily offers the opportunity for questioning whether the policy is intended to apply to the case at issue – or whether it should. Even wider issues arise, such as the desirability of supporting a particular way of generating energy, or the need for more roads or the role of the planning system in providing affordable housing. Pressure groups which, for example, may be opposed to the building of new roads or out-of-town shopping centres anywhere, irrespective of the merits (or otherwise) of particular projects, will want to use the inquiry as a platform on which to make their wider case. That they are able to do this is sometimes a reflection of the lack of a national policy on certain issues. The Heathrow Terminal 5 inquiry was the latest in a long line of

Plate 13.1 Local–national tensions
Source: Friends of the Earth

inquiries which have spent much time and public money debating what the national policy should be. The introduction of national policy statements on infrastructure and the nationally significant infrastructure projects regime is one response to this, effectively front-loading debates on overall policy direction, and putting parliament in a stronger position. Many have criticised this as undemocratic, with the inquiry process reduced to irrelevant detail (see Plate 13.1 and Owens and Cowell 2011).

This raises the question as to whether the provisions for national policy debate are adequate. It makes sense to argue that Parliament should be the arena for the national policy debate, and the local authority for debate on local policies. It also seems reasonable to maintain that it is quite inappropriate for major issues of principle to be argued when they are simply being applied locally. But issues are not so easily packaged. In reality there is a sharing of competences between different jurisdictional levels. Some site-specific proposals raise acute issues of national policy which cannot be settled or adequately discussed.

Human Rights Act 1998

The Human Rights Act 1998 came into force on 2 October 2000. The Act enables citizens to take action under the European Convention on Human Rights (dating from 1950), through the domestic courts. The Convention was used in court cases involving planning well before the Act came into force. Thus, the Act does not confer new rights but its enactment raises awareness about the rights set out in the Convention and should ensure that they are taken into account by government and the courts. All public bodies are required to act in accordance with the Convention. Local planning authorities and other planning agencies, including the Planning Inspectorate, are thus bound by the Act.

The main issue for planning is the Convention right arising from Article 6 which guarantees that 'In the determination of his civil rights and obligations or of any criminal charge against him, everyone is entitled to a fair and public hearing within a reasonable time by an independent and impartial tribunal established by law.' This raised questions for the UK, since ministers make policy, apply it and decide on appeals. While this duty is mostly undertaken through the Planning Inspectorate or the Recorder's Office in Scotland, their independence from ministers is in question (inspectors represent and 'stand in the shoes of the minister'). In any event, the minister has the right to 'recover' appeals for his or her decision.

The Scottish Minister accepted that neither ministers nor reporters are independent and impartial (in the meaning of the Convention) but argued that the provision to a further right to challenge decisions in the courts does meet the Convention. Although the right of challenge in the courts is severely restricted (being limited, generally, to procedural issues rather than the policy merits of the case), this argument has been accepted in at least one case so far. Nevertheless, the future of the Planning Inspectorate was put in question for a while.

Race and planning

Questions of equal racial opportunities have figured increasingly on the town planning agenda since the early 1970s, though with a questionable impact on practice. The Race Relations Act 1976 places a duty on all local authorities to eliminate unlawful racial discrimination in their activities and to promote 'good relations between persons of different racial groups'. In 1978, the RTPI established a joint working party with the Commission for Racial Equality (CRE) to investigate the multiracial dimension of planning, and to make recommendations for any necessary changes in practice. Their 1983 report, *Planning for a Multi-Racial Britain*, was a frank assessment of the inadequacies of the then current thinking on race and planning (RTPI and CRE 1983). It is perhaps only a little less relevant today, and its recommendations apply to all planners, wherever they work.

During the 1990s, various surveys showed that there continued to be little progress. A survey for the Local Government Association revealed that, in general, equal opportunity issues had a lower priority than in the 1980s, that there was little policy guidance on the issue and the colour-blind approach still dominated practice. The effect has been to institutionalise indirect discrimination in the planning system (Loftman and Beazley 1998a, 1998b) and to stereotype different groups by oversimplifying their internal diversity (Ratcliffe 1998).

It was hoped that new duties on local authorities under the Race Relations (Amendment) Act 2000 would provide added impetus for effective action. Public bodies now have a 'general duty' to promote race equality. Public bodies, including local authorities, must eliminate unlawful discrimination, promote equality of opportunity and promote good race relations. This is a shift from reaction to instances of discrimination to a more active role in promoting change for better relations. As part of the duty, each body must prepare a race equality scheme (RES) which sets out its corporate approach to promoting race equality and assesses its performance.

Thomas (2004) has surveyed Welsh local authority race equality schemes and found that the planning service remains peripheral to race relations issues and in one authority was not even recognised as relevant to the new statutory duty. He is pessimistic about the impact of the 2000 Act:

the assessment presents a sobering picture which is consistent with other early assessments (CRE and Schneider-Ross, 2003). It finds schemes which are, at best, formally competent, but . . . the spirit of which seems to have left them unmoved. Perhaps most worrying of all is the lack of transparency in the schemes, and in particular the almost universal lack of reasons for choices.

(p. 44)

A particular area of discrimination in planning relates to the accommodation of gypsies and travellers. Governments have changed the demands on local authorities in relation to providing suitable sites. The consequences have been dire, not least for the travellers themselves. Guidance now suggests:

Local planning authorities should, in producing their Local Plan: a) identify and update annually, a supply of specific deliverable sites sufficient to provide five years' worth of sites against their locally set targets; b) identify a supply of specific, developable sites or broad locations for growth, for years six to ten and, where possible, for years 11–15.

(DCLG 2012f: para. 9)

The political ramifications of allocating sites are hard for some local authorities to deal with, as land allocations typically provoke strong negative responses from the 'settled community'. The Dale Farm case in the UK became a cause célèbre in this regard, with Basildon Council clearing a site, at great expense, while not providing suitable alternative accommodation (Home 2012; also see Box 5.7). In 2013, the Hartlepool Local Plan passed a soundness test only to be withdrawn by the local authority solely over the allocation of a site for travellers.

Women and planning

Greed (1994) has described the male domination of the profession as 'only a temporary intermission'. Women were primary contributors to the social movement which promoted town planning early in the twentieth century. She argues that the professionalisation of planning, its institutionalisation within the government structure, and the limited access to qualifying courses has 'kept most women out'.

The skewing of the profession towards men has profound effects. Awareness of the specific nature of women's issues in planning grew strongly during the 1980s and was reflected in important and influential reports from the GLC (1986) and the RTPI (1987, 1988). The negative impacts of mobility and planning policies, which often assumed access to a private vehicle at all times of day, and which failed to account for feelings of safety and for the particular demands of escorting young children in the city (still an area where women assume greater responsibilities typically), came in for particular criticism (Hamilton 1999). Effectively the city is planned from the perspective of most planners, who until recently tended to be white, middle-aged men of relative affluence. The complexities of what are for majority populations relatively simple movements in space and time are often difficult for minorities (Miciukiewicz and Vigar 2013). Race and gender can often become overlapping concerns here.

One difficulty (in addition to the power of traditional attitudes and ways of thinking and perceiving) is the general lack of explicit social policies in plans. Thus, the provision of sporting facilities and the open-space standards applied to them are routinely regarded as legitimate land use matters. These predominantly male activities are contrasted by Greed (1993: 237) with crèches, which are commonly regarded as social issues, even though they 'may have major implications for central area office development' – a fact which is explicitly taken into account in some US cities (Cullingworth 1993).

Concerns for the needs of women are generally much less developed than those of race or disability (Davies 1996). Even where they are, childcare is the predominant 'women's issue', which reflects the assumption that women's primary role is as carers, and demonstrates 'the limited nature of the majority of planning initiatives which it may be argued are designed to ameliorate current constraints on women rather than to challenge the status quo in the drive for greater equality for women' (Little 1994b: 266).

Partly in response to awareness of the issues cited above, the RTPI (2003) developed a gender mainstreaming toolkit which remains an important reference. It fleshed out the detail of how UK and EU legislation could be met in practice. However such considerations have been drowned out by a series of competing requirements on local government (Greed 2006).

Planning and people with disabilities

The Planning Act 2004 requires local authorities to draw to the attention of planning applicants the need to consider the requirements of the Chronically Sick and Disabled Persons Act 1970, the Disability Discrimination Act 1995, and associated Regulations. Disability has a broad definition. The 1995 Act defines a disabled person as someone with 'a physical or mental impairment which has a substantial and long-term adverse effect on his ability to carry out normal day-to-day activities' (section 1). As Gilroy with Marvin (1993: 24) points out, most people are or will be physically impaired at some time during their lives. That impairment or disability can be turned into a handicap by the environment. This has led to an interest in creating older person-friendly neighbourhoods (DCLG 2007d). This publication acknowledged that planning in the UK had to adapt to an increasing ageing population, with specific approaches worked out on a neighbourhood-by-neighbourhood basis.

The needs of people with disabilities have been considered largely in terms of the design of the built environment, and planning has been at the centre of this. From 2004 (2002 for education providers), regulations made under the 1995 Act require employers and service providers to alter their premises or services when 'a physical feature makes it impossible or unreasonably difficult for disabled persons to access its service' (Part III of the 1995 Act).

It is primarily the Building Regulations and not the planning system which are the means by which access requirements are enforced. Important though this is,

it leads to an overly simplistic stereotyping of the problems faced by individuals with disabilities. Thomas (1992) has made a strong critique of typical attitudes:

> The 'regs' can become a checklist which defines the needs of disabled people, ignoring, indeed disallowing, the possibility that individual professionals dealing with particular cases need to learn from the experience of disabled people themselves. The British legislation which relates specifically to planning with its references to practicality and reasonableness, reinforces a strand in planners' professional ideologies which emphasises the role of the planner in reaching optimum solutions in situations involving competing needs or interests. Thus might a fundamental right to an independent and dignified life be reduced to an 'interest' to be balanced against the 'requirements' of conservation or aesthetics.
>
> (p. 25)

One example of such tensions lies in the idea of 'shared space', where pedestrians, cyclists and motorists negotiate their way through a space devoid of road markings. Such practice is driven in part by the desire to promote active travel, a potentially socially and environmentally progressive policy. Shared space has, however, come under strong criticism for disabling vision-impaired people, where significant progress had been made in recent decades through the use of tactile paving, etc. (Imrie 2012). This demonstrates not only the difficulties planners may face in designing schemes, but also the need to avoid easy labelling of people or issues.

Maladministration, the Ombudsman and probity

Most legislation is based on the assumption that the organs of government will operate efficiently and fairly. This is not always the case but, even if it were, provision has to be made for investigating complaints by citizens who feel aggrieved by some action (or inaction). As modern post-industrial society becomes more complex, and as the rights of electors and consumers are viewed as important, pressures for additional means of protest, appeal and restitution grow.

The *Local Government Ombudsmen* examine complaints about councils and some other authorities and organisations. In most cases, the body complained about should be addressed first and have a chance to sort out the complaint before the Ombudsman can consider it.

Typically around 15 per cent of Ombudsman cases concern planning matters – around 3,000 of the more than 20,000 complaints received in England in 2010–11 and 2011–12, for example. The vast majority of complaints concern planning applications and allege that local authorities have lacked attention to detail in assessing an application, typically where it impacts on a neighbour's amenity; and a failure of a local authority to comply with its own policy. In some cases, the Ombudsmen concluded that the failures did not significantly affect the outcome, but do have an effect on public trust in councils and their planning procedures. The Ombudsmen have no power to deal with the merits of planning decisions, but they have difficulty in explaining the difference between a planning decision which constitutes maladministration and one which is simply disputed (Hammersley 1987). Where maladministration is found, the authority is usually asked to value the cost of the failure (for example, in loss of amenity) and to pay a sum for the trouble taken in bringing the case.

Despite improvements to procedure in many local authorities, there have been recent well-publicised cases of extreme maladministration leading to fraud and corruption. Although small in number, such cases raise more general concerns about the integrity of officers and the probity of councillors. In recent decades, the most high-profile cases have been those of North Cornwall District Council and Doncaster. In North Cornwall, complaints were first taken up by the Ombudsman, then the district auditor, the police and Channel 4 television. Finally, the DoE set up an official inquiry (Lees 1993), which unequivocally condemned local councillors for granting permissions to local people for development in the open countryside. In

Doncaster, five people were prosecuted in relation to bribes and planning applications, with three going to jail, including the Chair of the Planning Committee and a developer, for a total of eleven years. The whistle-blower in what became known as Donnygate was a senior planning officer, Ken Burley, and his reflections on it make fascinating reading (Burley 2005).

The Local Government Act 1992 and the Code of Conduct for Councillors established the principle that councillors should not take part in proceedings if they have a direct or indirect pecuniary interest in the issue under discussion. The Localism Act 2011 makes this a legal matter. But such a simple distinction does not cover the many ways that councillors and officers can be influenced or themselves influence decisions. At the heart of the issue is lobbying. Applicants and objectors lobby councillors (and sometimes officers). Councillors may lobby colleagues (although not taking part in the decision themselves). Committee chairs can put pressure on officers, and so on. Whether or not such practices constitute improper activity is not always easy to discern.

Because of the number of cases of alleged impropriety or 'sleaze' in government generally during the 1990s, the Nolan Committee was established to consider Standards in Public Life. The Nolan Report (1997) on local government included a chapter on planning. In 2001, a Standards Board for England was established to promote ethical standards and to investigate allegations where elected members' behaviour may have fallen short. In 2003–4, 3,566 allegations of breaches of the code were referred to the Board, which alone seemed to justify its existence, but the Coalition government (2010–15) abolished it in 2012. Planning figured significantly in its problem cases, so much so that planning was seen to require extra measures, including the need for councillors to undertake training in planning because of the difficulties in dealing fairly with planning law and its implementation.

Since 2012, local government standards are prescribed in the Localism Act 2011. The Act requires councils to maintain a code of conduct which must be based on the (Nolan) Committee on Standards in Public Life's seven principles of public life. Example codes are given in DCLG guidance. Failure to disclose a pecuniary interest in a planning application is made a criminal matter under the new legislation.

In Scotland and Wales, different regimes apply. In Scotland, a 1998 consultation paper on the Nolan Report, *A New Ethical Framework for Local Government in Scotland*, broadly accepted its recommendations, but took issue with a number of them. The outcome was one of the first pieces of legislation enacted by the Scottish Parliament, the Ethical Standards in Public Life etc. (Scotland) Act 2000. It established a national Standards Commission, created with objectives similar to those of the Standards Board in England. A single code has been published for all local government (instead of a model code), together with a guidance note for local authorities.

In Wales, the Local Government Act 2000 created a power for the National Assembly for Wales to issue a model code of conduct to apply to members of all relevant authorities in Wales. This power was transferred to Welsh Ministers by the Government of Wales Act 2006. In 2008, Welsh Ministers issued the current Model Code of Conduct which all relevant authorities are required to adopt. Wales now has a unique approach that is a hybrid of the current and previous regimes in England, that is, while local authorities can make additions to the Code, provided these are consistent with the Model, the existence of a single Model gives certainty as to what standards are expected. It helps to ensure consistency for service users such as developers and the public.

The professionalisation of planning

Planning's professional body, the Royal Town Planning Institute (RTPI), has fundamentally changed throughout its existence of over 100 years. Such change has involved redefining the core of its expertise, an enlarging and broadening of its membership, and reforming the way that planners are educated. Other professional bodies have found the same need for major reform in response to major challenges to the professions. Some background is needed on how the planning profession has evolved in the UK before coming back to these recent changes.

The emergence of the planning profession in the early part of the twentieth century has been well documented (for example, Cherry 1974; Healey 1985). The profession emerged in part as a response to the fragmentation of policy, expertise and professions that were contributing to urban and regional development (Sutcliffe 1981). The emphasis of the movement for a planning profession was on the need for coordination of actions among different sectors and disciplines, largely through physical plan-making and design skills. The expansion of planning activity created demands for planners who initially were provided by other professions, although planning education and the profession expanded rapidly. The impetus for establishing planning practice slowly moved from social reform movements to the profession. However, Healey (1995: 496), among others, argues that there was little intellectual underpinning to the conception at that time of town planning in the context of social and economic development (see also Davoudi and Pendlebury 2010 for an updated assessment).

The strength of the professional body in the UK owes much to its place in government. Planning is a state activity, and the state has given legitimacy to the profession through formal recognition in the designation of the Royal Charter. Wilding (1982) describes this as a 'profession–state alliance' in which the state uses professions to assist with fulfilling its responsibilities and to legitimate state intervention, while state sponsorship has enabled the occupation of planning to 'gain control of the substance of its own work' (quoted in Low 1991: 26). Thus the RTPI is incorporated by Royal Charter 'to advance the science and art of town planning for the benefit of the public'. It exhibits many of the recognised 'traits' of professions: a body of knowledge which it seeks to consolidate and reproduce, control over the recruitment, education and training of its members, a measure of autonomy and self-regulation, and maintenance of a common code of ethics. Working as a planner is not restricted to members of the RTPI, but many employers require membership, and through its Code of Conduct it provides the public with a degree of recourse against alleged malpractice.

The professionalisation of planning has been challenged by numerous authors who draw on wider debates about the role of professions in society. Reade (1987) and Evans (1995) have been most critical from outside the profession in the UK. Healey (1985), Vigar (2012) and Campbell (2006) have made constructive proposals from within. There are three main parts to the critique: professions are self-serving and act in their own interest, not a wider public interest; the whole idea of the public interest which planning argues to promote is flawed, especially given the increased diversity in society over the second half of the twentieth century; and the lack of a distinctive knowledge base for the profession. The argument that planners work in the general public interest is more important to planning than most other professions, and it is increasingly difficult to make. This was certainly a stronger argument in the immediate post-war years under conditions of a strong political consensus (Hague 1984). Reade (1987) has made a comprehensive case against the professional mystique that is created around planning by claims that it provides objective technical expertise in the general interest on matters which, he says, should be resolved through the political process. Low (1991: 26) concludes that 'in practice urban planning is a disguised form of political decision-making'. Nevertheless, Campbell (1999: 302) suggests that 'the ethic of neutrality . . . is still deeply rooted in conceptions of the planner's professional role'. Reade (1987) has also challenged claims to an identifiable distinctive intellectual base or competence for planning. Vigar contends that

> planning shares with professions such as accountancy a basis in an 'esoteric *collection of areas of knowledge'*, . . . It is thus inherently multidisciplinary . . . Information has never been so ubiquitous but with that comes a role in turning information into usable knowledge, rebutting some claims and foregrounding others.
>
> (2012: 373; see also Davoudi and Pendlebury 2010)

Despite uncertainties about the political objectives of planning and the intellectual basis of planning expertise, planning as a profession in some ways strengthened in the later years of the twentieth century.

Planning services have continued to be in demand and planning has retained its professional status. But societal trends are tending to question claims for, and the value of, professionalism. The role of the state as a provider of goods and services has been eroded, and so too the professions which provide public services. Planning action is now much more overtly concerned with action in public and private sectors and with cooperation between the two, which is a further challenge to the public interest goals of planning. Traditional departments of local government dominated by single professional groups have been broken down in favour of cross-authority working. The relations and tensions between planning officers and elected members are changing, with the Nolan Report (1997) firmly placing the politicians' interests before professional judgement (Campbell 1999).

The impacts of the rise of neo-liberalism during the 1980s, its antagonism toward the welfare state and its adherence to the primacy of individual freedom and personal choice have also undermined notions of the public interest. Hague (1997: 142) argues that the public-interest ethos 'must be revamped to protect the interest of minorities and to deliver equal opportunities', and similar thoughts can be found in the work of Heather Campbell. Planning practice has a major challenge in dealing with changed public attitudes that reject collectivist solutions and wider forces of marketisation and consumerism.

Into this arena have come an emphasis on localism and the 'big society' under the Coalition government (2010–15). Such emphases support theoretical claims to collaborative planning (Healey 1997) where planning is something done among a group of individuals with no single claim to an overriding expertise or knowledge base. Such an ethos demands an approach that plans *with* communities not *for* them, as was the dominant approach in the twentieth century. Such an approach may also provide a way out for public sector planners who found themselves engaged in little more than 'bureaucratic proceduralism' (Tewdwr-Jones 1996).

After the critique, it is perhaps surprising to note the remarkable extent and success of the professionalisation of planning in the UK. Planning practice and education in the UK are more professionalised than any other European country (Healey 1985: 493). The RTPI has over 23,000 members (March 2012) of whom more than 1,000 are based overseas in more than eighty countries. By 2008, 30 per cent of members worked in the private sector, up from 18 per cent in 1990. Women and ethnic minorities are under-represented, although there has been growth in the proportion of women members from 15 per cent in 1990 to 27 per cent in 2003 and 35 per cent in 2012. The RTPI, with financial assistance from government, also supports the Planning Aid service, which provides free independent advice on planning (through the voluntary effort of members) to people, communities and voluntary groups.

The vision for planning

Partly in response to the above critique, a programme of fundamental reform of the RTPI was fashioned which complemented and contributed to the Labour government's reform of the planning system in the early 2000s. The objective of the reform was described in the Institute's *A New Vision for Planning* as a 'radical evolution which will lead to a body so different that it will be seen as a New Institute' (RTPI 2000: 1). The *Vision* document goes on to say that it is 'built around the core ideas of a planning that is spatial, sustainable, integrative, inclusive, value-driven, and action-oriented'. This conceptualisation influenced the reform of the UK planning system in 2004 but has not resulted in the radical evolution espoused.

The use of terms such as 'spatial action' and 'mediating place' did not make the Institute any more accessible, but they did try to summarise a view of planning that stresses its role in addressing the interrelationships of all government policy with spatial impacts by bringing together a mix of knowledge and skills to do this. The *New Vision* sees the professional planner as a facilitator. The planner will recognise and work with conflicting sets of values, and competing objectives. Therefore most attention is given to the process of planning as an inclusive procedure working towards negotiated outcomes.

The objectives of planning were given much less attention in the *New Vision* and the Institute filled this gap with a bold and campaigning *Manifesto for Planning* (RTPI 2004). The *Manifesto* presented ten campaign priorities, including a national spatial development framework (a cause also championed tirelessly by the Town and Country Planning Association) and 'an end to simplistic targets based on the speed of decision-making'.

While a desire for change existed among RTPI members, radical change has been harder to realise in practice. There are many reasons for this, not least the constant changes to the planning system, especially in England, in the period 2000–12 (see Gunn and Hillier 2012).

Planning education

The planning profession upholds its status through controlling entry to the profession and the qualifications that provide entry. The RTPI held its own examinations for membership until 1992 but, from the earliest days, specific courses were also set up to train planners (the first being at Liverpool University in 1909, followed by University College London in 1914) (Batey 1993; Collins 1989). By 1945, there were nine courses in town and country planning, all of which were postgraduate, with fifty-seven (twenty undergraduate) by 1981.

In recent times, course and student numbers have fluctuated. A government-inspired review of *Manpower Requirements for Physical Planning* forced some courses to cease recruitment during the 1980s (Amos *et al.* 1982), although the impact on the number of planning students was reduced because of increased intakes in the remaining schools. In 1988, a total of 766 students were recruited onto thiry-one courses. Student numbers have increased substantially since then, spurred on by the promotion of participation in higher education generally. In 2013, twenty-six institutions had planning courses which were accredited by the RTPI.

In 2001, the RTPI instituted the most thorough-going review of town planning education, qualifications and training in fifty years (Brown *et al.* 2003). The Commission's report reflected the changes in planning practice and the profession noted above. Its most radical recommendation was to reduce the minimum duration of study for postgraduate courses leading to membership from two academic years full-time to bring it into line with the standard masters requirement in the UK of one calendar year. This was complemented by more emphasis on learning in the workplace and continuing professional development.

More mundanely, the Institute publishes guidelines setting out policy on the education of planners. All accredited courses must demonstrate how a potential graduate of the course would be able to demonstrate a set of specific learning outcomes. These learning outcomes are applicable to all courses, though each course can address the learning outcomes in different ways. Students completing accredited courses can (after two years of practical experience) apply for corporate membership of the Institute. The reaccreditation of courses is now done in a light-touch way with an annual visit from a Partnership Board.

The demand for planning graduates fluctuates a little with the property market. Times were tough for graduates in the recessions of 1992–4 and 2007–10 but generally the market is buoyant. A shortage of planning graduates in the early and mid-2000s led central government to provide bursaries for postgraduate study. More recently, the numbers of students have fallen, in part because of the signals sent in the 2007–10 period that employment in planning was hard to find. A further shortage of graduates seems likely as property markets return to some semblance of normality as the 2010s progress.

In conclusion

Planning has been dominated for many years by a veritable orgy of institutional change. The pace of change accelerated under the reforming Blair administration and continued under the Coalition. Although all this was intended as a means of facilitating better planning, it often had the opposite effect of preventing good local planning from emerging tailored to local circumstances and reflecting citizen priorities. The discretionary spaces open to public-sector planners to make a difference narrowed first under the demands of a spatial planning system delivered in a too top-down manner (Gunn and Hillier 2012), and then as public sector cuts undermined the planning resource and the recession undermined some private funds for development. What does seem clear is that the faith in the efficacy of institutional change was misplaced. As well as changes to the system and its procedures, consecutive attempts at the reorganisation of subnational government have also undermined attempts, particularly at strategic planning. The development of a regional infrastructure, its subsequent abolition in 2011, and the hasty partial replacement of institutional infrastructure at the city-region/subregion scale, principally through LEPs, are one case in point; as is significant reform of local government.

The basic problems lie deeper: they relate to the functions, scope and practicability of 'town and country planning'. The crucial issues with which 'planning' is concerned do not fall within the responsibility or competence of the planning authority, or even within that of local government – jobs and poverty being the two most obvious ones. Hence central government wrestles with the political pressures to which problems in such areas give rise, though typically with disappointing results. More positively, there has been a continued discussion of the limits, role and purpose of planning. There has been a steady succession of reviews and studies, and the RTPI *Vision* and *Manifesto* of the early 2000s are genuinely thoughtful and radical. We should also not forget that lots of good planning work continues in the unsung but extremely important worlds of development management and in the private sector.

Coalition government reforms could open the door to more radical change. Local government has come under great pressure, from cuts to funding and from an emasculation of its authority in many areas, including planning. The narrow emphasis on economic viability has real implications for the choices society might make under the guise of planning. A greater degree of marketisation in tools such as the Community Infrastructure Levy served to entrench the growth-dependent approach to planning further (Rydin 2013). It is well known that the UK planning system favours the well-off, landowners in particular (Hall *et al.* 1973; Hastings and Matthews 2012). The increasing dependence of the public sector on the private in governance arrangements has entrenched this further. That the control of the land and the taxation of value uplifts due to the granting of planning permission is not more of a political issue is a fascinating area of study.

Similarly, the Coalition's narrow definition of sustainable development meant that where citizens did avail themselves of their new 'community rights', through developing neighbourhood plans, for example, they found themselves heavily constrained by central government framing of what they could and could not determine. The positives from this are that localism pushed planners, many of whom have been reluctant to see community engagement as much more than an annoyance and/or an irrelevance, to consider communities on more equal terms. The democratic problem here is that the statutory processes of local government such as plan making are of much less relevance than the political manoeuvrings associated with City and Growth Deals and the activities of LEPs and combined authorities. The complexity of the array of institutions and mechanisms is now seemingly a given, even though the detail constantly evolves. Indeed, such state forms facilitate the penetration of neo-liberalism and market practices through their flexibility and variability (Haughton *et al.* 2013). As such, they present few opportunities for citizen input, such as the planning system has developed for several decades. Indeed, the challenges this presents to planning and to society are one reason why they are the preferred way of doing things.

Further reading

Planning and politics

This is an enormous topic and many of the central texts on planning address it. Early work remains required reading, including: Ambrose (1986) *Whatever Happened to Planning?*; Reade (1987) *British Town and Country Planning*; and Thornley (1993) *Urban Planning under Thatcherism*. In addition, two books by Healey, (1997)

Collaborative Planning and (2010) *Making Better Places*, are very useful, as are two by Rydin ((2010) *Governing for Sustainable Urban Development* and (2013) *The Future of Planning*). A series of articles by Haughton and Allmendinger (e.g. 2013) debate recent changes to planning in the context of 'post-politics'.

Participation in planning

An overview of the historical development of participation in planning is given by Thomas (1996) 'Public participation in planning' and Bishop (2010). Early titles include Dennis (1970) *People and Planning*; Davies (1972) *The Evangelistic Bureaucrat: A Study of a Planning Exercise in Newcastle-upon-Tyne*; and, from the USA, Jacobs (1961) *The Death and Life of American Cities* and Gans (1968) *People and Plans*. One successful example often still quoted is Coin Street on the south bank of the Thames in central London, where the local community prepared their own plan for a highly valued commercial site including affordable housing and other amenities (Brindley *et al.* 1996).

Reviews of practice in participation are given by Coulson (2003) 'Land-use planning and community influence'; Brand and Gaffikin (2007) 'Collaborative Planning in an Uncollaborative World'; and the work of Sue Brownhill and Gavin Parker independently and together, e.g. Brownill and Parker (2010) 'Why bother with good works? The relevance of public participation(s) in a post-collaborative era'. A collection by Davoudi and Madanipour (2014) *What is Local in Localism?* updates many of the issues in the context of localism.

Community, neighbourhood and parish planning

A useful starting point explaining the meaning of 'community involvement' is the Community Development Foundation's (2003) *Searching for Solid Foundations: Community Involvement in Urban Policy*. For a Scottish perspective, see also Stevenson (2002) *Getting 'Under the Skin' of Community Planning*.

On parish planning, Gavin Parker's work is useful (e.g. 2008; 2012 with Murray). See also Gallent (2012)

Neighbourhood Planning: Communities, Networks and Governance and work by practitioner Jeff Bishop (e.g. (2010) 'From Parish Plans to Localism in England: Straight Track or Long and Winding Road?'). There are good examples of neighbourhood planning emerging, in relation to both process and product: plans for Thame and Upper Eden were both shortlisted for RTPI awards in 2013; see www.rtpi.org.uk/events/awards/rtpi-awards-for-planning-excellence-2014/rtpi-awards-for-planning-excellence-winners-and-shortlist/ (accessed 15 April 2014).

Race and planning

Reeves (2005) *Planning for Diversity* covers much of the material for this and other sections. Thomas (2000) provides a thorough examination in *Race and Planning: The UK Experience*. On housing traveller communities, Robert Home's work is highly readable. Yasminah Beebeejaun's work on ethnicity, and also on women and participation is important and useful.

Women and planning

Two textbooks provide a comprehensive analysis of theory and practice. Greed (1994) *Women and Planning: Creating Gendered Realities* is a mine of interesting examples, and provides a guide to reading. Little (1994a) *Gender, Planning and the Policy Process* links the issue to wider debates. A series of publications by the Women's Design Service are accessible and excellent. The organisation is no more, but publications are online at www.wds.org.uk.

People with disabilities

DCLG (2006g) *Planning and Access for Disabled People: A Good Practice Guide* sets out many of the issues facing planners. It lacks the nuance much of the research literature demands. Rob Imrie's work is a good starting point for this.

Maladministration, the Ombudsman and probity

The annual reports of the separate Commissioners for England, Scotland and Wales provide all the facts, together with a flavour, through thumbnail sketches, of the cases being heard.

The planning profession

The intellectual development of the planning profession is explained by Healey (1985) 'The professionalisation of planning in Britain'; see also Taylor, N. (1992) 'Professional ethics in town planning' and Evans (1993) 'Why we no longer need a town planning profession'. Clifford and Tewdwr-Jones (2013) *The Collaborating Planner* is a good review of the literature and contains a survey of public-sector planners. Heather Campbell's work is useful as an entry into deeper theoretical debates and for pointers to the future.

Note

1   Neighbourhood fora are self-defining where no parish council exists, i.e. the majority of urban areas. They should consist of twenty-one persons with an established interest in a place. They propose a boundary within which to undertake a plan. Both the forum and the plan area have to be approved by the local authority. This has proved contentious in some areas, with competing fora emerging.

Bibliography

A note on official publications

Government policy and other documents are now routinely published online, and planning policy was one of the first areas to migrate to the recently developed www.gov.uk website. Here can be found all the important current planning policy documents, such as the *National Planning Policy Framework* (NPPF), current consultations and the like. Following the Taylor Review and the quest for simplification of government guidance, the new *Planning Practice Guidance* was developed, originally in pilot form, but formally launched in March 2014. It can be accessed at: http://planningguidance.planningportal.gov.uk. These two developments have the virtue of making it easier to ensure that you have the up-to-date position on any particular topic.

However, the historical context is often of interest. Indeed, it is felt to be one of the important qualities of this book that recent developments are placed in their historical context, and the very brevity which has been an objective for recently developed guidance can sometimes leave questions about matters of detail. For these reasons, it is often useful to gain access to superseded policy and guidance. In the authors' experience, the ready availability of such documents via the Internet can be a little patchy: whilst some documents can be easily located on departmental sites, on the gov.uk site or in the National Archives, this is not always the case and persistence can be required.

The bibliography does not include NPPF or its predecessor PPSs and PPGs, nor circulars, etc. which have been superseded by the Planning Practice Guidance.

Abercrombie, P. (1945) *Greater London Plan*, London: HMSO

Abercrombie, P. and Abercrombie, L. (1923) *Stratford upon Avon: Report on Future Developments,* Liverpool: Liverpool University Press

Adams, D., Croudace, R. and Tiesdell, S. (2011) 'Design codes, opportunity space and the marketability of new housing', *Environment and Planning B* 38(2): 289–306

Adams, J. (2001) 'Hypermobility and its consequences: a lecture to the RSA', available at: http://john-adams.co.uk/wp-content/uploads/2006/hypermobilityforRSA.pdf

Adams, W. M. (1996) *Future Nature: A Vision for Conservation*, London: Earthscan

Adamson, G. and Pavitt, J. (2011) *Postmodernism: Style and Subversion 1970–1990,* London: V&A Publishing

Addison and Associates (2004) *Evaluation of the Planning Delivery Grant 2003–04*, London: ODPM

Affordable Rural Housing Commission (2006) *Final Report,* London: ARHC

Ahlfeldt, G., Holman, N. and Wendland, N. (2012) *An Assessment of the Effects of Conservation Areas on Value*, London: LSE/English Heritage

Ahmed, K. and Sánchez-Triana, E. (2008) *Strategic Environmental Assessment for Policies: An Instrument for Good Governance*, Washington, DC: World Bank

Akkerman, A. (2001) 'Urban planning in the founding of Cartesian thought', *Philosophy and Geography* 4(2): 141–67

Albrechts, L. (2003) 'Reconstructing decision making: planning versus politics', *Planning Theory* 3(3): 249–68

Alden, J. and Romaya, S. (1994) 'The challenge of urban regeneration in Wales: principles, policies and practice', *Town Planning Review* 65: 435–61

Aldous, T. (1975) *Goodbye Britain?* London: Sidgwick & Jackson

Aldous, T. (1992) *Urban Villages: A Concept for Creating Mixed-use Developments on a Sustainable Scale*, London: Urban Villages Group

Aldred, R. (2013) 'Who are Londoners on Bikes and what do they want? Negotiating identity and issue definition in a "pop-up" cycle campaign', *Journal of Transport Geography* 30: 194–201

Aldridge, H. R. (1915) *The Case for Town Planning*, London: National Housing and Town Planning Council

Alexander, A. (2009) *Britain's New Towns: Garden Cities to Sustainable Communities*, Abingdon: Routledge

Allen, C. (2008) *Housing Market Renewal and Social Class*, London: Routledge

Allen, H. J. B. (1990) *Cultivating the Grass Roots: Why Local Government Matters*, The Hague: International Union of Local Authorities

Allinson, J. (1999) 'The 4.4 million households: do we really need them anyway?' *Planning Practice and Research* 14(1): 107–13

Allmendinger, P. (2009) *Planning Theory,* Basingstoke: Palgrave Macmillan

Allmendinger, P. (2011) *New Labour and Planning: From New Right to New Left*, London: Routledge

Allmendinger, P. and Gunder, M. (2005) 'Applying Lacanian insight and a dash of Derridian deconstruction to planning's "dark side"', *Planning Theory* 4(1): 87–112

Allmendinger, P. and Haughton, G. (2009a) 'Soft spaces, fuzzy boundaries, and metagovernance: the new spatial planning in the Thames Gateway', *Environment and Planning A* 41(3): 617–33

Allmendinger, P. and Haughton, G. (2009b) 'Critical reflections on spatial planning', *Environment and Planning A* 41(11): 2544–49

Allmendinger, P. and Haughton, G. (2010) 'Post-political spatial planning in England: a crisis of consensus?' *Transactions of the Institute of British Geographers* 37(1): 89–103

Allmendinger, P. and Haughton, G. (2013) 'The evolution and trajectories of English spatial governance: "Neoliberal" episodes in planning', *Planning Practice and Research* 28(1): 6–26

Allmendinger, P. and Tewdwr-Jones, M. (2000) 'New Labour, new planning? The trajectory of planning in Blair's Britain', *Urban Studies* 37(8): 1379–402

Allmendinger, P. and Tewdwr-Jones, M (2002) *Planning Futures: New Directions for Planning Theory,* London: Routledge

Allmendinger, P. and Thomas, H. (eds) (1998) *Urban Planning and the British New Right*, London: Routledge

All Parliamentary Cycling Group (2013) *Get Britain Cycling,* available at: http://allpartycycling.org/inquiry/ (accessed 14 April 2014)

Alterman, R. and Cars, G. (eds) (1991) *Neighbourhood Regeneration: An International Evaluation*, London: Mansell

Ambrose, P. (1986) *Whatever Happened to Planning?*, London: Methuen

Amery, C. and Cruikshank, D. (1975) *The Rape of Britain*, London: P. Elek

Amion Consulting (2001) *Urban Regeneration Companies: Learning the Lessons,* London: DTLR

Amos, F. J. C., Davies, D., Groves, R. and Niner, P. (1982) *Manpower Requirements for Physical Planning*, Birmingham: Institute of Local Government Studies

Apps, P. (2013) 'Sharp rise in households with two or more families, figures show', *Inside Housing*, available at www.insidehousing.co.uk/care/sharp-rise-in-households-with-two-or-more-families-figures-show/6529312.article (accessed 11 March 2014)

Armstrong, H. and Taylor, J. (2000) *Regional Economics and Policy,* Oxford: Blackwell

Arnstein, S. R. (1969) 'A ladder of citizen participation', *Journal of the American Institute of Planners* 35(4): 216–24

Arup (2004) *Review of the Publicity Requirements for Planning Applications*, London: ODPM

Arup Economics and Planning with the Bailey Consultancy (2002) *Resourcing of Local Planning Authorities*, London: DTLR

Arup with Nick Davies Associates (2004) *Standard Application Forms: Final Report*, London: ODPM

Ashby, E. and Anderson, M. (1981) *The Politics of Clean Air*, Oxford: Clarendon Press

Ashworth, W. (1954) *The Genesis of Modern British Town Planning*, London: Routledge and Kegan Paul

Association of British Insurers (ABI) (2004) *A Changing Climate for Insurance*, London: ABI

Atkinson, R. (1999a) Discourses of partnership and empowerment in contemporary British urban regeneration', *Urban Studies* 36(1): 59–72

Atkinson, R. (1999b) 'Urban crisis: new policies for the next century', in Allmendinger, P. and Chapman, M. (1999) *Planning Beyond 2000*, Chichester: Wiley

Atkinson, R. and Mills, L. (2005) 'The thematic strategy on the urban environment: will it make a difference?', *Town and Country Planning* 74(3): 106–8

Atkinson, R. and Moon, G. (1994) *Urban Policy in Britain: The City, the State and the Market*, London: Macmillan

Audit Commission (1992) *Building in Quality: A Study of Development Control,* London: HMSO

Audit Commission (1998) *Building in Quality: A Study of Development Control*, London: Audit Commission

Audit Commission (2002) *Development Control and Planning*, London: Audit Commission

Audit Commission (2009) *Housing Market Renewal: Programme Review*, London: Audit Commission

Australia ICOMOS (2013) *The Burra Charter: The Australia ICOMOS Charter for Places of Cultural Significance*, Burwood: Australia ICOMOS

Bailey, N. and Barker, A. (eds) (1992) *City Challenge and Local Regeneration Partnerships: Conference Proceedings*, London: Polytechnic of Central London

Bailey, N. and Turok, I. (2001) 'Central Scotland as a polycentric urban region: useful planning concept or chimera?' *Urban Studies* 38(4): 697–715

Bailey, N., Barker, A. and MacDonald, K. (1995) *Partnership Agencies in British Urban Policy*, London: UCL Press

Bains Report (1972) *The New Local Authorities: Management and Structures,* London: HMSO

Baird, A. (2013) 'Acquisition of UK ports by private equity funds', *Research in Transport Business and Management* 8: 158–65

Baird, A. and Valentine, V. (2007) 'Port privatisation in the United Kingdom', in Brooks, M. and Cullinane, K. P. B. (eds) *Devolution, Port Governance and Port Performance, Research in Transportation Economics*, Amsterdam: Elsevier, Vol. 17, pp. 55–84

Baker Associates (1999) *Proposals for a Good Practice Guide on Sustainability Appraisal of Regional Planning Guidance,* London: DETR

Baker Associates (2001) *Review of the Use Classes Order and Part 4 of the GDPO (Temporary Uses)*, London: DTLR

Baker, L. (2006) 'Listings create local shield', *Planning*, 17.

Baker, M. (1998) 'Planning for the English regions: a review of the Secretary of State's regional planning guidance', *Planning Practice and Research* 13(2): 153–69

Baker, M. and Roberts, P. (1999) *Examination of the Operation and Effectiveness of the Structure Planning Process: Summary Report*, London: DETR

Baker, M. and Wong, C. (2013) 'The delusion of strategic spatial planning: what's left after the Labour government's English regional experiment?' *Planning Practice and Research* 28(1): 83–103

Baker, M., Roberts, P. and Shaw, D. (2003), *Stakeholder Involvement in Regional Planning*, London: TCPS

Baker, M., Hincks, S. and Sherrif, G. (2010) 'Getting involved in plan making: participation and stakeholder involvement in local and regional spatial strategies in England', *Environment and Planning C* 28(4): 574–94

Baker, S., Kousis, M., Richardson, D. and Young, S. (1997) *The Politics of Sustainable Development*, London: Routledge

Balchin, P. (1995) *Housing Policy: An Introduction*, London: Routledge

Balchin, P. and Rhodes, M. (eds) (1997) *Housing: The Essential Foundations*, London: Routledge

Baldock, J. (2012a) *Retail Planning under the NPPF,* Farnborough: Association of Convenience Stores

Baldock, J. (2012b) *Education for Retail Planning: An Initial Scoping Paper,* Aylesbury: National Retail Planning Forum

Balfour, J. (1983) *A New Look at the Northern Ireland Countryside*, Belfast: Department of the Environment, Northern Ireland

Balloch, S. and Taylor, M. (2001) *Partnership Working: Policy and Practice*, Bristol: Policy Press

Bandarin, F. and van Oers, R. (2012) *The Historic Urban Landscape: Managing Heritage in an Urban Century*, Chichester: Wiley-Blackwell

Banister, D. (1999) 'Review essay: the car is the solution, not the problem?', *Urban Studies* 36(13): 2415–19

Banister, D. (2005) *Unsustainable Transport*, London: Spon

Banister, D. (2008) 'The sustainable mobility paradigm', *Transport Policy* 15(2): 73–80

Banister, D. and Anable, J. (2009) 'Transport policies and climate change', in Davoudi, S., Crawford, J. and Mehmood, A. (eds), *Planning for Climate Change*, London: Earthscan, pp. 55–70

Barker, A. (1982) *Quangos in Britain: Government and Networks of Public Policy-making*, London: Macmillan

Barker, K. (2004) *Delivering Stability: Securing our Future Housing Needs. Barker Review of Housing Supply – Final Report – Recommendations*, Norwich: HMSO

Barker, K. (2006) *Barker Review of Land Use Planning – Final Report – Recommendations*, Norwich: HMSO

Barlow, J., Cocks, R. and Parker, M. (1994) *Planning for Affordable Housing,* London: Department of the Environment

Barlow, M. (chair) (1940) *Report of the Royal Commission on the Distribution of the Industrial Population*, Cmd 6153, Barlow Report, London: HMSO

Barrington, T. (1991) 'Local government reform: problems to resolve' in Walsh, J. (ed.) *Local Economic Development and Administrative Reform,* Dublin: The Regional Studies Association

Barton, H. (2009) 'Land use planning and health and well-being', *Land Use Policy* 26(S): 115–23

Batey, P. (1993) 'Planning education as it was', *The Planner* 79(4): 25–6

Batty, E., Beatty, C., Foden, M., Lawless, P., Peason, S. and Wilson, I. (2010) 'The new deal for communities experience: a final assessment' London: CLG, available at http://extra.shu.ac.uk/ndc/ (accessed 24 March 2014)

Batty, M. and Marshall, S. (2009) 'The evolution of cities: Geddes, Abercrombie and the new physicalism', *Town Planning Review* 80(6): 551–74

BDP Planning and Leighton Berwin (1998) *The Use of Permitted Development Rights by Statutory Undertakers*, London: TSO

Beck, U. (1998) *Democracy without Enemies*, Cambridge: Polity Press

Beebeejaun, Y. (2012) 'Including the excluded? Changing understanding of ethnicity in contemporary English planning', *Planning Theory and Practice* 13(4): 529–48

Bell, J. L. (1993) *Key Trends in Communities and Community Development*, London: Community Development Foundation

Bell, S. (1997) *Design for Outdoor Recreation*, London: Spon

Bentley, I. (1990) 'Ecological urban design', *Architect's Journal* 192: 69–71

Bentley, I., Alcock, A., Murrain, P., McGlynn, S. and Smith, G. (1985) *Responsive Environments: A Manual for Designers,* London: Architectural Press

Benwell Community Project (1978) *The Making of a Ruling Class: Two Centuries of Capital Development on Tyneside*, Newcastle upon Tyne: Project, B. C.

Berry, J. and McGreal, S. (eds) (1995a) *European Cities, Planning Systems and Property Markets*, London: Spon

Berry, J. and McGreal, S. (1995b) 'Community and inter-agency structures in the regeneration of inner-city Belfast', *Town Planning Review* 66: 129–42

Beveridge, W. (chair) (1942) *Social Insurance and Allied Services*, Cmd 6404, Beveridge Report, London: HMSO

Bingham, M. (2001) 'Policy utilisation in planning control: planning appeals in England's "plan-led" system', *Town Planning Review* 72(3): 321–40

Binney, M. (2005) *SAVE Britain's Heritage 1975–2005: Thirty Years of Campaigning*, London: Scala

Birmingham City Council (2012) *Shopping and Local Centres: Supplementary Planning Document,* Birmingham: BCC

Bishop, J. (2010) 'From parish plans to localism in England: straight track or long and winding road?', *Planning Practice and Research* 25(5): 611–24.

Bishop, K., Tewdwr-Jones, M. and Wilkinson, D. (2000) 'From spatial to local: the impact of the European Union on local authority planning in the UK', *Journal of Planning and Environmental Management* 43(3): 309–34

Bishop, K. D. and Phillips, A. C. (1993) 'Seven steps to market: the development of the market-led approach to countryside conservation and recreation', *Journal of Rural Studies* 9: 315–38

Blackburn, S., Errington, A., Lobley, M., Winter, M. and Selman, P. (2000) *Two Villages, Two Valleys, Too Early? A Review of the Peak District Integrated Rural Development Project, 11 Years On,* Plymouth: Department of Land Use and Rural Management, University of Plymouth

Blackman, T. (1985) 'Disasters that link Ulster and the North East', *Town and Country Planning* 54: 18–20

Blackman, T. (1991) 'People-sensitive planning: communication, property and social action', *Planning Practice and Research* 6(3): 11–15

Blackman, T. (1995) *Urban Policy in Practice*, London: Routledge

Blackmore, R., Wood, C. and Jones, C. E. (1997) 'The effect of environmental assessment on UK infrastructure project planning decisions', *Planning Practice and Research* 12(3): 223–38

Bloodworth, A. J. (2011) *Evaluation of Minerals Policy Statements,* London: DCLG

Blowers, A. (ed.) (1993) *Planning for a Sustainable Environment,* London: Earthscan

Blowers, A. and Leroy, P. (1994) 'Power, politics and environmental inequality: a theoretical and empirical analysis of the process of "peripheralisation"', *Environmental Politics* 3(2): 197–228

Blowers, A., Boersema, J. and Martin, A. (2009) 'Whatever happened to environmental politics?' *Journal of Integrative Environmental Sciences* 6(2): 97–101

Blühdorn, I. and Welsh, I. (2007) 'Eco-politics beyond the paradigm of sustainability: a conceptual framework and research agenda', *Environmental Politics* 16(2): 185–205

Blundell, V. H. (1993) *Labour's Flawed Land Acts 1947–1976*, London: Economic and Social Science Research Association

Blunden, J. and Curry, N. (1989) *A People's Charter? Forty Years of the National Parks and Access to the Countryside Act 1949*, London: HMSO

Boardman, B. (2007) 'Examining the carbon agenda via the 40% house scenario', *Building Research and Information*, 35(4): 363–78

Boddy, M. and Hickman, H. (2013) 'The demise of strategic planning? The impact of the abolition of regional spatial strategy in a growth region', *Town Planning Review* 84(6): 743–68

Boden, T. A., Marland, G. and Andres, R. J. (2010) *Global, Regional, and National Fossil-Fuel CO$_2$ Emissions*, Oak Ridge, TN: US Department of Energy

Bogdanor, V. (1999) *Devolution in the United Kingdom*, Oxford: Oxford University Press (Opus Books)

Bolan, P. (1999) 'The role of local lists', *Context* 61: 26–8

Bolton, N. and Chalkley, B. (1990) 'The rural population turnround: a case study of North Devon', *Journal of Rural Studies* 6(1): 29–43

Booth, P. (1996) *Controlling Development: Certainty and Discretion in Europe, the USA, and Hong Kong*, London: UCL Press

Booth, P. (2002) 'A desperately slow system? The origins and nature of the current discourse on development control' *Planning Perspectives* 17(4): 309–23

Booth, P. and Huxley, M. (2012) '1909 and all that: reflections on the Housing, Town Planning, Etc. Act 1909', *Planning Perspectives* 27(2): 267–283

Borchardt, K. (1995) *European Integration: The Origins and Growth of the European Community* (fourth edition), Luxembourg: Office for the Official Publications of the European Communities

Bosworth, J. and Shellens, T. (1999) 'How the Welsh Assembly will affect planning', *Journal of Planning and Environment Law* March: 219–24

Botkin, D. B. (1990) *Discordant Harmonies: A New Ecology for the Twenty-first Century*, New York: Oxford University Press

Bourne, F. (1992) *Enforcement of Planning Control* (second edition), London: Sweet and Maxwell

Bovaird, T. (1992) 'Local economic development and the city', *Urban Studies* 29(3–4): 343–68

Bowley, M. (1945) *Housing and the State 1919–1944*, London: Allen and Unwin

Boyle, R. (1993) 'Changing partners: the experience of urban economic policy in West Central Scotland, 1980–90', *Urban Studies* 30(2): 309–24

Bradbury, J. (ed.) (2008) *Devolution, Regionalism and Regional Development: The UK Experience*, London: Routledge

Bradley, T. and Lowe, P. (eds) (1984) *Locality and Rurality: Economy and Society in Rural Regions*, Norwich: Geo Books

Bramley, G., Munro, M. and Lancaster, S. (1997) *The Economic Determinants of Household Formation: A Literature Review*, London: DETR

Bramley, G., Munro, M. and Pawson, H. (2004) *Key Issues in Housing: Policies and Markets in 21st Century Britain*, London: Palgrave Macmillan

Bramley, G., Pawson, H., White, M., Watkins, D. and Pleace, N. (2010) *Estimating Housing Need*, London: DCLG

Brand, C. and Boardman, B. (2008) 'Taming the few – the unequal distribution of greenhouse gas emissions from personal travel in the UK', *Energy Policy* 36(2): 224–38

Brand, R. and Gaffikin, F. (2007) 'Collaborative planning in an uncollaborative world', *Planning Theory* 6(3): 303–34

Breheny, M. J. (ed.) (1992) *Sustainable Development and Urban Form*, London: Pion

Breheny, M. J. (1997) 'Urban compaction: feasible and acceptable?', *Cities* 14(4): 209–17

Breheny, M. J. (ed.) (1999) *The People: Where Will they Work?*, London: TCPA

Breheny, M. J. and Hall, P. (1996) *The People – Where Will they Go? National Report of TCPA Regional Enquiry into Housing Need and Provision in England*, London: TCPA

Brennan, A., Rhodes, J. and Tyler, P. (1998) *Evaluation of the Single Regeneration Challenge Fund Budget: A Partnership for Regeneration – An Interim Evaluation*, London: DETR

Briggs, A. (1952) *History of Birmingham* (2 vols), Oxford: Oxford University Press

Brindley, T., Rydin, Y. and Stoker, G. (1996) *Remaking Planning* (second edition), London: Routledge

British Waterways Board (BWB) (2003) *Waterways and Development Plans*, Rugby: BWB

Broady, M. (1968) *Planning for People*, London: Bedford Square Press

Brooks, J. (1999) '(Can) modern local government (be) in touch with the people?', *Public Policy and Administration* 14(1): 42–59

Brotherton, I. (1993) 'The interpretation of planning appeals', *Journal of Environmental Planning and Management* 36(2): 179–86

Brown, C., Claydon, J. and Nadin, V. (2003) 'The RTPI's Education Commission: context and challenges', *Town Planning Review* 74(3): 333–45

Brownill, S. (1990) *Developing London's Docklands: Another Great Planning Disaster?*, London: Paul Chapman

Brownill, S. and Downing, L. (2013) 'Neighbourhood planning – is an infrastructure of localism emerging?', *Town and Country Planning* 82(9): 372–6

Brownill, S. and Parker, G. (2010) 'Why bother with good works? The relevance of public participation(s) in a post-collaborative era', *Planning Practice and Research* 25(3): 275–82

Brundtland, G. H. (chair) (1987) *Our Common Future* (World Commission on Environment and Development), Brundtland Report, Oxford: Oxford University Press

Brunet, R. (1989) *Les Villes européennes*, Rapport pour la Délégation à l'Aménagement du Territoire et à l'Action Régionale (DATAR), Paris: la Documentation française

Brunivells, P. and Rodrigues, D. (1989) *Investing in Enterprise: A Comprehensive Guide to Inner City Regeneration and Urban Renewal*, Oxford: Blackwell

Bruton, M. J. (1980) 'PAG revisited', *Town Planning Review* 51(2): 134–44

Bruton, M. J. and Nicholson, D. J. (1983) 'Non-statutory plans and supplementary planning guidance', *Journal of Planning and Environment Law* 1983: 432–43

Bruton, M. J. and Nicholson, D. J. (1985) 'Supplementary planning guidance and local plans', *Journal of Planning and Environment Law* 1985: 837–44

Bruton, M. J., Crispin, G. and Fidler, P. (1980) 'Local plans: public local inquiries', *Journal of Planning and Environment Law* 1980: 374–85

Bryant, B. (1996) *Twyford Down: Roads, Campaigning and Environmental Law* (first edition) London: Spon

Buchanan, C. (1963) *Traffic in Towns: A Study of the Long Term Problems of Traffic in Urban Areas, Reports of the Steering and Working Groups to the Minister of Transport,* London: HMSO

Buchanan, C. (1993) *I Told You So: An Autobiography*, unpublished.

Buckingham-Hatfield, S. and Evans, B. (eds) (1996) *Environmental Planning and Sustainability*, London: Wiley

Building for Life Partnership (2012) *The Sign of a Good Place to Live: Building for Life 12*, available at www.made.org.uk/images/uploads/BfL12booklet2.pdf (accessed 14 November 2013)

Bulkeley, H. (2006) 'A changing climate for spatial planning?', *Planning Theory and Practice*, 7(2): 203–14

Bulkeley, H. and Newell, P. (2010) *Governing Climate Change,* London: Routledge

Burley, K. (2005) 'The darker side of communities; is this the real world of planning?', *Planning Theory and Practice* 6(4): 517–41

Burwood, S. and Roberts, P. (2002) *Learning from Experience: The BURA Guide to Achieving Effective and Lasting Regeneration*, London: BURA

Bury, R. (2010) 'Decent homes goal missed by 10 per cent' *Inside Housing*, 26 August, available at www.insidehousing.co.uk/decent-homes-goal-missed-by-10-per-cent/6511343.article (accessed 24 March 2014)

Business Innovation and Skills (BIS) (2010a) *Local Growth: Realising Every Place's Potential,* Cm 7961, London: TSO

Business Innovation and Skills (BIS) (2010b) *Penfold Review of Non-planning Consents,* London: BIS

Business Innovation and Skills (BIS) (2010c) *Government Response to the Penfold Review of Non-planning Consents,* London: BIS

Business Innovation and Skills (BIS) (2013) *Implementation of the Penfold Review of Non-planning Consents: Progress Update March 2013,* London: BIS

Business Innovation and Skills (BIS) (2014) *Regional Growth Fund For Large Businesses*, available at www.gov.uk/understanding-the-regional-growth-fund (accessed 22 January 2014)

Byrne, S. (1989) *Planning Gain: An Overview – A Discussion Paper*, London: Royal Town Planning Institute

Byrne, T. (1990) *Local Government in Britain* (fifth edition), London: Penguin

Byrne, T. (2000) *Local Government in Britain: Everyone's Guide to How it All Works,* London: Penguin

Cabinet Office (1988) *Action for Cities,* London: HMSO

Cabinet Office (1999) *Professional Policy Making for the 21st Century,* London: Cabinet Office

Cabinet Office (2011a) *Open Public Services White Paper*, Cm 8145, London: Cabinet Office

Cabinet Office (2011b) *Unlocking Growth in Cities*, London: Cabinet Office

Cambridge Policy Consultants (1998) *Regenerating London Docklands*, London: DETR

Campaign for Better Transport (2013) 'Where the money's going: are the new Local Transport Bodies heading in the right direction?', available at www.bettertransport.org.uk/files/admin/LTB_report_250913_web_FINAL.pdf (accessed 3 April 2014)

Campbell, H. (2006) 'Just planning: the art of situated ethical judgement', *Journal of Planning Education and Research* 26(1): 92–106

Campbell, H., Ellis, H., Henneberry, J. and Gladwell, C. (2000) 'Planning obligations, planning practice and land-use outcomes', *Environment and Planning B* 27(5) : 759–75

Campbell, H., Ellis, H., Hennebury, J., Poxon, J and Rowley, S. (2001) *Planning Obligations and the Mediation of Development,* London: RICS Foundation

Campbell, S. (1999) 'Planning, green cities, growing cities, just cities? Urban Planning and the contradictions of sustainable development', in Satterthwaite, D. (ed.) *Sustainable Cities,* London: Earthscan

Canal and River Trust (2012) *Shaping our Future: Strategic Priorities*, Milton Keynes: Canal and River Trust

Capita Management Consultancy (1996) *An Evaluation of Six Early Estate Action Schemes* (DoE), London: HMSO

Capita Management Consultancy (1997–8) *Housing Action Trusts: Evaluation of Baseline Conditions* (11 vols), London: DETR

Carley, M. and Kirk, K. (1998) *City-wide Urban Regeneration* (Central Research Unit, Scottish Executive), Edinburgh: TSO

Carley, M., Kirk, K. and McIntosh, S. (2001) *Retailing, Sustainability and Neighbourhood Regeneration,* York: Joseph Rowntree Foundation

Carley, M., Campbell, M., Kearns, A., Wood, M. and Young, R. (2000) *Regeneration in the 21ˢᵗ Century: Policies into Practice,* Bristol: Policy Press

Carmichael, P. (1992) 'Is Scotland different? Local government policy under Mrs Thatcher', *Local Government Policy Making* 18(5): 25–32

Carmona, M. (1998) 'Residential design policy and guidance: prevalence, hierarchy and currency', *Planning Practice and Research* 13(4): 407–19

Carmona, M. (1999) 'Residential design policy and guidance: content, analytical basis, prescription and regional emphasis', *Planning Practice and Research* 14(1): 17–38

Carmona, M. (2013a) 'When is guidance not guidance?', *Town and Country Planning* 82(11): 492–4

Carmona, M. (2013b) 'Does urban design add value?', *Urban Design* 126: 47–9

Carmona, M. and Tiesdell, S. (eds) (2007) *Urban Design Reader,* Oxford: Architectural Press

Carmona, M., Heath, T., Oc, T. and Tiesdell, S. (2003) *Public Places – Urban Spaces,* Oxford: Architectural Press.

Carole Millar Research (1999) *Perceptions of Local Government: A Report of Focus Group Research,* Edinburgh: Central Research Unit, Scottish Office

Carson, R. (1962) *Silent Spring,* Harmondsworth: Penguin

Carter, N., Brown, T. and Abbott, T. (1991) *The Relationship between Expenditure-Based Plans and Development Plans,* Leicester: School of the Built Environment, Leicester Polytechnic

Cartwright, L. (1997) 'The implementation of sustainable development by local authorities in the south east of England', *Planning Practice and Research* 12(4): 337–47

Cartwright, L. (2000) 'Selecting local sustainable development indicators: does consensus exist in their choice and purpose?', *Planning Practice and Research* 15(1–2): 65–78

Castree, N. (2008) 'Neoliberalising nature: the logics of deregulation and reregulation', *Environment and Planning A* 40(1): 131–52

Cave, S., Rehfisch, A., Smith, L. and Winter, G. (2013) *Comparison of the Planning Systems in the 4 UK Countries,* Research Paper 13/39, London: House of Commons Library

CB Hillier Parker and Saxell Bird Axon (1998) *The Impact of Large Foodstores on Market Towns and District Centres* (DETR), London: TSO

CEMAT (2000) *Guiding Principles for Sustainable Spatial Development of the European Continent,* Brussels: Council of Europe

Centre for City and Regional Studies (CCRS) (2005) *Policy Interactions and Outcomes in Deprived Areas,* London: BIS

Centre for Local Economic Strategies (CLES) (1990) *Inner City Regeneration: A Local Authority Perspective,* Manchester: CLES

Centre for Local Economic Strategies (CLES) (1992) *Social Regeneration: Directions for Urban Policy in the 1990s,* Manchester: CLES

Centre for Public Scrutiny (2012) *Musical Chairs: Practical Issues for Local Authorities in Moving to a Committee System,* London: CfPS

Centre for Urban Policy Studies (1998) *The Impact of Urban Development Corporations in Leeds, Bristol and Central Manchester,* London: DETR

Champion, T. (2003) 'Testing the differential urbanisation model in Great Britain 1901–1991', *Tijdschrift voor Economische en Sociale Geografie* 94(1): 11–22

Champion, T. and Shepherd, J. (2006) *Demographic change in rural England* London: Rural Evidence Research Centre

Champion, T. and Watkins, C. (1991) *People in the Countryside: Studies of Social Change in Rural Britain,* London: Paul Chapman

Chanan, G. (2003) *Searching for Solid Foundations: Community Involvement in Urban Policy,* London: ODPM

Chandler, J. A. (2007) *Explaining Local Government: Local Government in Britain since 1800*, Manchester: Manchester University Press

Chelmsford Borough Council (CBC) (2008) *Core Strategy and Development Control Policies, Chelmsford Borough Council Local Development Framework 2001– 2021,* Chelmsford: CBC

Cherry, G. E. (1974) *The Evolution of British Town Planning*, London: Leonard Hill

Cherry, G. E. (1975) *National Parks and Access to the Countryside: Environmental Planning 1939–1969*, Vol. 2, London: HMSO

Cherry, G. E. (1988) *Cities and Plans: The Shaping of Urban Britain in the Nineteenth and Twentieth Centuries,* London: Edward Arnold

Cherry, G. E. (1994) *Rural Change and Planning: England and Wales in the Twentieth Century*, London: Spon

Cherry, G. E. (1996) *Town Planning in Britain since 1900: The Rise and Fall of the Planning Ideal*, Oxford: Blackwell

Cherry, G. E. and Penny, L. (1986) *Holford: A Study in Architecture, Planning and Civic Design,* London: Mansell Publishing Limited

Cheshire, P., Hilber, C. and Kaplanis, I. (2011) *Evaluating the Effects of Planning Policies on the Retail Sector: Or Do Town Centre First Policies Deliver the Goods?,* London: Spatial Economics Research Centre, LSE

Cheshire, P., Leunig, T., Nathan, M. and Overman, H. (2012) *Links between Planning and Economic Performance: Evidence Note for LSE Growth Commission,* London: LSE

Chesman, G. R. (1991) 'Local authorities and the review of the definitive map under the Countryside and Wildlife Act 1981', *Journal of Planning and Environment Law* 1991: 611–14

Chisholm, M. (1995) *Britain on the Edge of Europe*, London: Routledge

Christie, I. and Warburton, D. (eds for Real World Coalition) (2001) *From Here to Sustainability*, London: Earthscan

Church, C. and McHarry, C. (eds) (1999) *One Small Step: A Guide to Action on Sustainable Development in the UK*, London: Community Development Foundation

City of London Corporation (CLC) (2007) *Rising to the Challenge: The City of London Corporation's Climate Adaptation Strategy*, London: City of London Corporation

Claydon, J. (1998) 'Discretion in development control: a study of how discretion is exercised in the conduct of development control in England and Wales', *Planning Practice and Research* 13(1): 63–80

Claydon, J. and Smith, B. (1997) 'Negotiating planning gains through the British development control system', *Urban Studies* 34(12): 2003–22

Cleator, B. and Irvine, M. (1995) *Review of Legislation Relating to the Coastal and Marine Environment of Scotland*, Perth: Scottish Natural Heritage

Clifford, B. and Tewdwr-Jones, M. (2013) *The Collaborating Planner: Practitioners in the Neoliberal Age*, Bristol: Policy Press

Cloke, P. (1979) *Key Settlements in Rural Areas,* London: Methuen

Cloke, P., Milbourne, P. and Thomas, C. (1994) *Lifestyles in Rural England*, Salisbury: RDC

Cocks, R. (1998) 'The mysterious origin of the law for conservation.' *Journal of Planning and Environment Law* March: 203–9

Cole, I. and Furbey, R. (1994) *The Eclipse of Council Housing*, London: Routledge

Cole, I., Green, S., Hickman, P. and McCoulough, E. (2003) *A Review of NDC Strategies for Tackling Low Demand and Unpopular Housing,* Sheffield: CRESR

Collar, N. (2010) *Planning Law*, Edinburgh: W. Green

Collins, M. P. (1989) 'A review of 75 years of planning education at UCL', *Planner* 75(6): 18–22

Commission for Architecture and the Built Environment (CABE) (2003) *Protecting Design Quality in Planning*, London: CABE

Commission for Architecture and the Built Environment (CABE) (2005a) *Housing Audit: Assessing the Design Quality of New Homes in the North East, North West and Yorkshire & Humber*, London: CABE

Commission for Architecture and the Built Environment (CABE) (2005b) *Making Design Policy Work: How to Deliver Design through your Local Development Framework*, London: CABE

Commission for Architecture and the Built Environment (CABE) (2006) *Design and Access Statements: How to Write, Read and Use Them*, available at www.cabe.org.uk/files/design-and-access-statements.pdf (accessed 14 November 2013)

Commission for Architecture and the Built Environment (CABE) (2007) *Housing Audit: Assessing the Design Quality of New Homes Housing in the East Midlands, West Midlands and South West*, London: CABE

Commission for Architecture and the Built Environment (CABE) and DETR (2001) *The Value of Urban Design*, Tonbridge: Thomas Telford

Commission for Architecture and the Built Environment (CABE) and English Heritage (2007) *Guidance on Tall Buildings*, London: CABE

Commission for Integrated Transport (CfIT) (2001) *Study of European Best Practice in the Delivery of Integrated Transport*, available at http://webarchive.nationalarchives.gov.uk/20110304132839/http:/cfit.independent.gov.uk/pubs/2001/ebp/index.htm (accessed 17 April 2014)

Commission for Rural Communities (CRC) (2010) *State of the Countryside 2010*, Cheltenham: CRC

Commission of the European Communities (CEC) (1991) *Europe 2000: Outlook for the Development of the Community's Territory*, Luxembourg: Office for Official Publications of the European Communities

Commission of the European Communities (CEC) (1992) *Towards Sustainability: A European Community Programme of Policy and Action in Relation to the Environment and Sustainable Development*, Brussels: CEC

Commission of the European Communities (CEC) (1996) *EIA – A Study on Costs and Benefits*, Brussels: CEC

Commission of the European Communities (CEC) (1999a) *ESDP: European Spatial Development Perspective: Towards Balanced and Sustainable Development of the Territory of the European Union*, Luxembourg: Office for Official Publications of the European Communities

Commission of the European Communities (CEC) (1999b) *Global Assessment: Europe's Environment: What Directions for the Future?* Brussels: CEC

Commission of the European Communities (CEC) (2004) *A New Partnership for Cohesion: Convergence, Competitiveness, Cooperation – Third Report on Economic and Social Cohesion*, Luxembourg: Office for Official Publications of the European Communities

Commission of the European Communities (CEC) (2007) *Green Paper: Adapting to Climate Change in Europe: Options for EU Action*, COM (2007) 354 Final, Luxembourg: Office for Official Publications of the European Communities

Commission of the European Communities (CEC) (2008) *The Roadmap for Maritime Spatial Planning*, Brussels: CEC

Commission of the European Communities (CEC) (2010) *Europe 2020: A Strategy for Smart, Sustainable and Inclusive Growth*, Brussels: CEC

Commission of the European Communities (CEC) (2013a) *Environment: A Healthy and Sustainable Environment for Future Generations* (part of series: the European Union Explained), available at http://europa.eu/pol/env/flipbook/en/files/environment.pdf (accessed 11 April 2014)

Commission of the European Communities (CEC) (2013b) *Beyond GDP*, available at www.beyond-gdp.eu (accessed 20 April 2014)

Commission of the European Communities (CEC) (2014) *Living Well, within the Limits of our Planet*, The Seventh Environment Action Programme to 2020, Brussels: CEC

Committee on the Civil Service (1968) *Report of the Committee on the Civil Service*, Fulton Report, London: HMSO

Committee on Standards in Public Life (1997) *Third Report of the Committee on Standards in Public Life: Standards of Conduct in Local Government in England, Scotland and Wales*, Vol. 1: Report, Cm 3702-1, Nolan Report, London: TSO

Committee on Standards in Public Life (2013) *Standards Matter: A Review of Best Practice in Promoting Behaviour in Public Life*, 14th Report, Cm 8519, London: TSO

Communities and Local Government Committee (2013) *Councillors on the Frontline* (HC432), London: TSO

Community Development Foundation (2003) *Searching for Solid Foundations: Community Involvement and Urban Policy*, London: ODPM

Community Development Project Working Group (1974) 'The British national community development project 1969–1974', *Community Development Journal* 9(3): 162–86

Connal, R. C. and Scott, J. N. (1999) 'The new Scottish Parliament: what will its impact be?', *Journal of Planning and Environment Law* June: 491–7

Conservative Party (2010) *Open Source Planning*, London: Conservative Central Office

Cooke, P. (1983) *Theories of Planning and Spatial Development*, London: Hutchinson

Coon, A. (1988) 'Local plan provision: the record to date and prospects for the future', *The Planner* 74(5): 17–20

Coulson, A. (2003) 'Land-use planning and community influence: a study of Selly Oak, Birmingham', *Planning Practice and Research* 18(2): 179–95

Council for the Protection of Rural England (CPRE) (2010) *England's Hedgerows: Don't Cut Them Out! Making the Case for Better Hedgerow Protection*, London: CPRE

Counsell, D. (1998) 'Sustainable development and structure plans in England and Wales: a review of current practice', *Journal of Environmental Planning and Management* 41(2): 177–94

Counsell, D., Hart, T., Jonas, A. and Kettle, J. (2007) 'Fragmented regionalism? Delivering integrated strategies in Yorkshire & the Humber', *Regional Studies* 41(3): 391–402

Country Land and Business Association (CLA) (2011) *Averting Crisis in Heritage: CLA Report on Reforming a Crumbling System*, London: CLA

Countryside Agency (1989) *Incentives for a New Direction in Farming*, Cheltenham: Countryside Agency

Countryside Agency (1999) *Planning for Quality of Life in Rural England*, Cheltenham: Countryside Agency

Countryside Agency (2003) *A Guide to Definitive Maps and Changes to Public Rights of Way*, Cheltenham: Countryside Agency

Courtney, P., Mills, J., Gaskell, P. and Chaplin, S. (2013) 'Investigating the incidental benefits of Environmental Stewardship schemes in England', *Land Use Policy* 31: 26–37

Cowell, R. and Murdoch, J. (1999) 'Land use and limits to (regional) governance: some lessons from planning for housing and minerals in England', *International Journal of Urban and Regional Research* 23(4): 654–69

Cowell, R. and Owens, S. (1998) 'Suitable locations: equity and sustainability in the minerals planning process', *Regional Studies* 32(9): 797–811

Crook, A. and Whitehead, C. (2002) 'Social housing and planning gain: is this an appropriate way of providing affordable housing?' *Environment and Planning A* 34(7): 1259–79

Crook, A., Monk, S., Rowley, R. and Whitehead, C. (2006) 'Planning gain and the supply of new affordable housing in England', *Town Planning Review* 77(3): 353–73

Cross, D. and Bristow, R. (eds) (1983) *English Structure Planning*, London: Pion

Cullen, G. (1961) *Townscape*, London: The Architectural Press.

Cullingworth, J. B. (1960) 'Household formation in England and Wales', *Town Planning Review* 31(1): 5–26

Cullingworth, J. B. (1964) *Town and Country Planning in England and Wales*, London: George Allen and Unwin

Cullingworth, J. B. (1993) *The Political Culture of Planning: American Land Use Planning in Comparative Perspective*, New York and London: Routledge

Cullingworth, J. B. (1994) 'Alternate planning systems: is there anything to learn from abroad?', *Journal of the American Planning Association* Spring: 162–72

Cullingworth, J. B. (ed.) (1999) *British Planning: 50 Years of Urban and Regional Policy*, London: Athlone Press

Cullingworth, J. B. and Caves, R. W. (2009) *Planning in the USA: Policies, Issues and Processes,* Abingdon: Routledge

Cumulus Consultants Ltd and ICF GHK (2013) *Valuing England's National Parks*, London: National Parks England

Cunningham, A. (ed.) (1998) *Modern Movement Heritage*, London: Spon

Curry, N. (1997) 'Enhancing countryside recreation benefits through the rights of way system in England and Wales', *Town Planning Review* 68(4): 449–63

Danson, M. W., Lloyd, M. G. and Newlands, D. (1989) 'Rural Scotland and the rise of Scottish Enterprise', *Planning Practice and Research* 4(3): 13–17

Dargan, L. (2002) *A New Approach to Regeneration in Tyneside? The Rhetoric and Reality of the New Deal for Communities*, PhD thesis, Newcastle University

Dargay, J. and Goodwin, P. (2000) *Changing Prices: A Dynamic Analysis of the Role of Pricing in Travel Behaviour and Transport Policy*, London: Landor Publishing

Darke, R. (1999) 'Public speaking rights in local authority planning committees', *Planning Practice and Research* 14(2) : 171–83

Darley, G. and McKie, D. (2013) *Ian Nairn: Words in Place,* Nottingham: Five Leaves Publications

Davies, G. and Brooks, J. (1989) *Positioning Strategy in Retailing,* London: Chapman

Davies, H. (2014) 'Liverpool is the hardest hit major UK city in government's latest round of funding cuts' *Liverpool Echo*, 19 January

Davies, H., Nutley, S. and Smith, P. (2000) *What Works? Evidence-based Policy and Practice in Public Services,* Bristol: Policy Press

Davies, H. W. E. (ed.) (1989) *Planning Control in Western Europe,* London: HMSO

Davies, H. W. E. (1998) 'Continuity and change: the evolution of the British Planning System 1947–97', *Town Planning Review* 69(2): 135–52

Davies, H. W. E., Edwards, D., Roberts, C., Rosborough, L. and Sales, R. (1986a) *The Relationship between Development Plans and Appeals*, Reading: Department of Land Management and Development, University of Reading

Davies, H. W. E., Edwards, D. and Rowley, A. R. (1986b) *The Relationship between Development Plans, Development Control, and Appeals*, Reading: Department of Land Management and Development, University of Reading

Davies, H. W. E., Rowley, A. R., Edwards, D., Blom-Cooper, A., Roberts, C., Rosborough, L. and Tilley, R. (1986c) *The Relationship between Development Plans and Development Control*, Reading: Department of Land Management and Development, University of Reading

Davies, H. W. E., Gosling, J. A. and Hsia, M. T. (1994) *The Impact of the European Community on Land Use Planning in the United Kingdom*, London: RTPI

Davies, J. (1972) *The Evangelistic Bureaucrat: A Study of a Planning Exercise in Newcastle upon Tyne,* London: Tavistock

Davies, L. (1996) 'Equality and planning: race' and 'Equality and planning: gender and disability', in Greed, C. (ed.) *Implementing Town Planning: The Role of Town Planning in the Development Process*, Harlow: Longman

Davoudi, S. (1999a) 'Making sense of the European Spatial Development Perspective', *Town and Country Planning Journal* 68(12): 367–9

Davoudi, S. (1999b) 'A quantum leap for planners: the role of the planning system within changing approaches to waste management', *Town and Country Planning* 68(1): 20–3

Davoudi, S. (2000a) 'Planning for waste management: changing discourses and institutional relationships', *Progress in Planning* 53(3): 165–216

Davoudi, S. (2000b) 'Sustainability: A New Vision for the British Planning System', *Planning Perspectives* 15(2): 123–37

Davoudi, S. (2001) 'Planning and the twin discourses of sustainability', in Layard, A., Batty, S. and Davoudi, S., *Planning for a Sustainable Future*, London: Spon

Davoudi, S. (2003) 'Polycentricity in European spatial planning: from an analytical tool to a normative agenda', *European Planning Studies* 11(8): 979–99

Davoudi, S. (2004) 'Territorial cohesion: an agenda that is gaining momentum', *Town and Country Planning* 73(7/8): 224–7

Davoudi, S. (2005a) 'ESPON: past, present and the future', *Town and Country Planning* 74(3): 100–2

Davoudi, S. (2005b) 'Understanding Territorial Cohesion', *Planning Practice and Research,* 20(4): 433–41

Davoudi, S. (2005c) 'The Northern Way: a polycentric megalopolis', *Regional Review* 15(1): 2–4

Davoudi, S. (2006a) 'Evidence-based planning: rhetoric and reality' *DisP165* 2/2006: 14–24

Davoudi, S. (2006b) 'Strategic waste planning: the interface between the "technical" and the "social"', *Environment and Planning C* 24(5): 681–700

Davoudi, S. (2007) 'Territorial cohesion, the European social model and spatial policy research', in Faludi, A. (ed.) *Territorial Cohesion and European Model of Society*, Cambridge, MA: The Lincoln Institute for Land Policy, pp. 81–105

Davoudi, S. (2008) 'Conceptions of the city region: a critical review', *Journal of Urban Design and Planning* 161(DP2): 51–60

Davoudi, S. (2009a) 'Territorial cohesion, European social model and transnational cooperation', in Knieling, J. and Othengrafen, F. (eds) *Planning Cultures in Europe: Decoding Cultural Phenomena in Urban and Regional Planning,* Aldershot: Ashgate, pp. 269–79

Davoudi, S. (2009b) 'Governing waste: introduction to the special issue', *Journal of Environmental Planning and Management* 52(2): 131–6

Davoudi, S. (2009c) 'Scalar tensions in the governance of waste: the resilience of state spatial Keynesianism', *Journal of Environmental Planning and Management* 52(2): 137–57

Davoudi, S. (2011a) 'Localism and the reform of the planning system in England', *disP* 47(187): 92–6

Davoudi, S. (2011b) 'From the Infrastructure Planning Commission to the Major Infrastructure Planning Unit, good news for democracy but is it good news for speed and efficiency?', *Planning in London* 77: 34–5

Davoudi, S. (2012a) 'The legacy of positivism and the emergence of interpretive tradition in spatial planning', *Regional Studies* 46(4): 429–41

Davoudi, S. (2012b) 'Climate risk and security: New meanings of "the environment" in the English planning system', *European Planning Studies* 20(1): 49–69

Davoudi, S. (2012c) 'Resilience, a bridging concept or a dead end?' *Planning Theory and Practice* 13(2): 299–307

Davoudi, S. (2013) 'Climate change and the role of spatial planning in England', in Knieling, J. and Filho, W. L. (eds) *Climate Change Governance*, Berlin: Springer

Davoudi, S. (2014) 'Climate change, securitisation of nature, and resilient urbanism', *Environment and Planning C* 32(2): 360–75

Davoudi, S. and Cowie, P. (2013) 'Are English Neighbourhood Forums democratically legitimate?', *Planning Theory and Practice* 14(4): 562–6

Davoudi, S. and Evans, N. (2005) 'The challenge of governance in regional waste planning', *Environment and Planning C* 23(4): 493–519

Davoudi, S. and Healey, P. (1994) *Perceptions of City Challenge Policy Processes: The Newcastle Case*, Newcastle upon Tyne: Department of Town and Country Planning, University of Newcastle upon Tyne

Davoudi, S. and Healey, P. (1995) 'City challenge: sustainable process or temporary gesture?', *Environment and Planning C* 13(1): 79–95

Davoudi, S. and Layard, A. (2001) 'Sustainable development and planning: an introduction to concepts and contradictions', in Layard, A., Davoudi, S. and Batty, S. (eds) *Planning for a Sustainable Future,* London: Spon, pp. 7–19

Davoudi, S. and Madanipour, A. (eds) (2014) *What is Local in Localism?*, Abingdon: Routledge

Davoudi, S. and Pendlebury, J. (2010) 'Evolution of planning as an academic discipline', *Town Planning Review* 81(6): 613–44

Davoudi, S. and Porter, L. (2012) 'The politics of resilience for planning: a cautionary note', *Planning Theory and Practice* 13(2): 329–33

Davoudi, S. and Stead, D. (2002) 'Urban–rural relationships: an introduction and a brief history', *Built Environment* 28(4): 269–77

Davoudi, S. and Strange I. (2009) 'Space and place in twentieth-century planning: an analytical framework and an historical review', in Davoudi, S. and Strange, I. (eds) *Conceptions of Space and Place in Strategic Spatial Planning*, London: Routledge, pp. 7–42

Davoudi, S. and Wishardt, M. (2005) 'Polycentric turn in the Irish spatial strategy', *Built Environment* 31(2): 122–32

Davoudi, S., Hull, A. and Healey, P. (1996) 'Environmental concerns and economic imperatives in strategic plan-making', *Town Planning Review* 64(4): 421–36

Davoudi, S., Healey, P. and Hull, A. (1997) 'Rhetoric and reality in British structure planning in Lancashire: 1993–95', in Healey, P., Khakee, A., Motte, A. and Needham, B. (eds) *Making Strategic Spatial Plans: Innovation in Europe*, London: UCL Press

Davoudi, S., Crawford, J. and Mehmood, A. (eds) (2009) *Planning for Climate Change: Strategies for Mitigation and Adaptation for Spatial Planners*, London: Earthscan

Davoudi, S., Brooks, L. and Mehmood, A. (2013) 'Evolutionary resilience and strategies for climate adaptation', *Planning Practice and Research* 28(3): 307–22

Day, P. and Klein, R. (1987) *Accountabilities: Five Public Services,* London: Tavistock

Delafons, J. (1995) 'Policy forum: planning research and the policy process', *Town Planning Review* 66(1): 83–95

Delafons, J. (1997) *Politics and Preservation*, London: Spon

de Groot, L. (1992) 'City Challenge: competing in the urban regeneration game', *Local Economy* 7(3): 196–209

Deimann, S. (1994) *Your Rights under European Union Environment Legislation*, Brussels: European Environmental Bureau

Dennis, N. (1970) *People and Planning: The Sociology of Housing In Sunderland*, London: Faber

Department for Communities and Local Government (DCLG) (2006a) *Strong and Prosperous Communities: The Local Government White Paper*, Cm 6939, London: DCLG

Department for Communities and Local Government (DCLG) (2006b) *Key Lessons for Development Control: An Overview of the Evaluation of Planning Standards Authorities 2005/06,* London: DCLG

Department for Communities and Local Government (DCLG) (2006c) *Code for Sustainable Homes: A Step-change in Sustainable Home Building Practice,* London: DCLG

Department for Communities and Local Government (DCLG) (2006d) *Good Practice Guide on Planning for Tourism*, London: DCLG

Department for Communities and Local Government (DCLG) (2006e) *Design Coding in Practice: An Evaluation*, London: DCLG

Department for Communities and Local Government (DCLG) (2006f) *Planning Obligations: Practice Guide,* London: DCLG

Department for Communities and Local Government (DCLG) (2006g) *Planning and Access for Disabled People: A Good Practice Guide*, London: DCLG

Department for Communities and Local Government (DCLG) (2007a) *Eco Towns Prospectus*, London DCLG

Department for Communities and Local Government (DCLG) (2007b) *Using Evidence in Spatial Planning*, London: DCLG

Department for Communities and Local Government (DCLG) (2007c) *Building a Greener Future: Policy Statement,* London: DCLG

Department for Communities and Local Government (DCLG) (2007d) *Towards Lifetime Neighbourhoods*, London: DCLG

Department for Communities and Local Government (DCLG) (2007e) *The Road Ahead: Final Report of the Independent Task Group on Site Provision and Enforcement*, London: DCLG

Department for Communities and Local Government (DCLG) (2008a) *Final Report. Spatial Plans in Practice: Supporting the Reform of Local Planning*, London: DCLG

Department for Communities and Local Government (DCLG) (2008b) *Valuing Planning Obligations in England: Final Report,* London: DCLG

Department for Communities and Local Government (DCLG) (2008c) *The Community Infrastructure Levy,* London: DCLG

Department for Communities and Local Government (DCLG) (2010a) *The Incidence, Value and Delivery of Planning Obligations in England in 2007–08: Final Report*, London: DCLG

Department for Communities and Local Government (DCLG) (2010b) *The Community Infrastructure Levy: An Overview,* London: DCLG

Department for Communities and Local Government (DCLG) (2010c) *Proposals for Changes to Planning Application Fees in England: Consultation*, London: DCLG

Department for Communities and Local Government (DCLG) (2010d) *Affordability and Housing Market Areas*, London: DCLG

Department for Communities and Local Government (DCLG) (2011a) *Presumption in Favour of Sustainable Development*, available at www.communities.gov.uk/planningsystems/planningpolicy (accessed 24 June 2011)

Department for Communities and Local Government (DCLG) (2011b) *A Plain English Guide to the Localism Act*, London: DCLG

Department for Communities and Local Government (DCLG) (2011c) *Planning for Traveller Sites: Consultation*, London: DCLG

Department for Communities and Local Government (DCLG) (2012a) *Extending Permitted Development Rights for Homeowners and Businesses: Technical Consultation,* London: DCLG

Department for Communities and Local Government, (DCLG) (2012b) *Renegotiation of Section 106 Planning Obligations: Consultation,* London: DCLG

Department for Communities and Local Government (DCLG) (2012c) *Technical Review of Planning Appeal Procedures: Consultation,* London: DCLG

Department for Communities and Local Government (DCLG) (2012d) *Parades to be Proud of: Strategies to Support Local Shops,* London: DCLG

Department for Communities and Local Government, (DCLG) (2012e) *Regeneration to Enable Growth: A Toolkit Supporting Community-led Regeneration*, London: DCLG

Department for Communities and Local Government, (DCLG) (2012f) *Planning Policy for Traveller Sites*, London: HMSO

Department for Communities and Local Government, (DCLG) (2012g) *Guidance on Planning Propriety Issues*, London: DCLG

Department for Communities and Local Government (DCLG) (2013a) *Housing Statistical Release: Affordable Housing Supply, April 2012 to March 2013,* London: DCLG

Department for Communities and Local Government (DCLG) (2013b) *Live Data on Land Use Change*, available at www.gov.uk/government/statistical-data-sets/live-tables-on-land-use-change-statistics (accessed 15 April 2014)

Department for Communities and Local Government and Department for Culture, Media and Sport (2009) *Circular on the Protection of World Heritage Sites*, London: TSO

Department for Communities and Local Government and Department for Transport (2007) *Manual for Streets,* London: Thomas Telford

Department for Culture, Media and Sport and Department of Transport, Local Government and the Regions (2001) *The Historic Environment: A Force for Our Future*, London: DCMS

Department for Environment, Food and Rural Affairs (DEFRA) (2002) *Safeguarding Our Seas,* London: DEFRA

Department for Environment, Food and Rural Affairs (DEFRA) (2004) *Rural Strategy 2004,* London: DEFRA

Department for Environment, Food and Rural Affairs (DEFRA) (2005) *Making Space for Water. Taking Forward a New Government Strategy for Flood and Coastal Erosion Risk Management in England: First Government Response to the Autumn 2004 Making Space for Water Consultation Exercise,* available at www.defra.gov.uk/environ/fcd/policy/strategy/firstresponse.pdf (accessed July 2009)

Department for Environment, Food and Rural Affairs (DEFRA) (2006a) *Climate Change: The UK Programme 2006*, London: TSO

Department for Environment, Food and Rural Affairs (DEFRA) (2006b) *Statutory Nuisance from Insects and Artificial Light: Guidance on Sections 101 to 103 of the Clean Neighbourhoods and Environment Act 2005,* London: DEFRA

Department for Environment, Food and Rural Affairs (DEFRA) (2007) *Well-being: International Policy Interventions*, London: DEFRA

Department for Environment, Food and Rural Affairs (DEFRA) (2010) *The Noise Policy Statement for England*, London: DEFRA

Department for Environment, Food and Rural Affairs (DEFRA) (2011a) *The Natural: Choice: Securing the Value of Nature: White Paper*, London: DEFRA

Department for Environment, Food and Rural Affairs (DEFRA) (2011b), *Mainstreaming Sustainable Development – The Government's Vision and What This Means in Practice*, London: DEFRA

Department for Environment, Food and Rural Affairs (DEFRA) (2011c) *A Description of the Marine Planning System for England*, London: DEFRA

Department for Environment, Food and Rural Affairs (DEFRA) (2012a) *UK Climate Change Risk Assessment: Government Report,* London: TSO

Department for Environment, Food and Rural Affairs (DEFRA) (2012b) *Rural Statement 2012,* London: DEFRA

Department for Environment, Food and Rural Affairs (DEFRA) (2013a) *Waste Management Plan for England,* London: DEFRA

Department for Environment, Food and Rural Affairs (DEFRA) (2013b) *Guide to Rural Proofing: National Guidelines,* London: DEFRA

Department for Trade and Industry (DTI) (2004) *The Foresight Future Flooding Project*, available at www. foresight.gov.uk/OurWork/CompletedProjects/Flood/index.asp (accessed July 2009)

Department of the Environment (DoE) and Ministry of Agriculture, Fisheries and Food (MAFF) (1995) *Rural England: A Nation Committed to a Living Countryside,* London: HMSO

Department of the Environment (Northern Ireland) (DoE) (2010) *Planning Bill: Explanatory and Financial Memorandum*, Belfast: DoE

Department of the Environment, Transport and the Regions (DETR) (1990) *This Common Inheritance: Britain's Environmental Strategy,* Cm 1200, London: HMSO

Department of the Environment, Transport and the Regions (DETR) (1998a) *A Mayor and Assembly for London: The Government's Proposals for Modernising the Governance of London*, Cm 3987, London: TSO

Department of the Environment, Transport and the Regions (DETR) (1998b) *A New Deal for Transport: Better for Everyone*, Cm 3950, London: HMSO

Department of the Environment, Transport and the Regions (DETR) (1998c) *Modern Local Government: in Touch with the People*, Cm 4014, London: TSO

Department of the Environment, Transport and the Regions (DETR) (1999) *Local Leadership, Local Choice,* Cm 4298, London: TSO

Department of the Environment, Transport and the Regions (DETR) (2000a) *Transport 2010: The 10-Year Plan*, London: DETR

Department of the Environment, Transport and the Regions (DETR) (2000b) *Preparing Community Strategies: Government Guidance to Local Authorities*, London: DETR

Department of the Environment, Transport and the Regions (DETR) and CABE (2000) *By Design: Urban Design in the Planning System: Towards Better Practice*, London: DETR

Department of Health (DoH) (2013) *Improving Outcomes and Supporting Transparency* (Part 1A), London: DoH

Department of Health and Department for Transport (2010) *Active Travel Strategy*, London: HMSO

Department of Transport (DoT) (1989) *Roads for Prosperity*, London: HMSO

Department of Transport, Local Government and the Regions (DTLR) (2001) *Planning: Delivering a Fundamental Change*, London: DTLR

Department of Transport, Local Government and the Regions (DTLR) (2002) *Lessons and Evaluation Evidence from Ten Single Regeneration Budget Case Studies: Mid Term Report,* Norwich: HMSO

Design Council CABE (2011) *The Bishop Review: The Future of Design in the Built Environment*, London: Design Council CABE

Design Council CABE, Landscape Institute, RTPI and RIBA (2013) *Design Review: Principles and Practice*, London: Design Council CABE.

Devereux, M. and Guillemoteau, D. (2001) 'A spatial vision for north-west Europe', *Town and Country Planning* 71(2): 48–9

Dinan, D. (ed.) (1998) *Encyclopedia of the European Union*, London: Macmillan

Distressed Town Centre Property Task Force (DTCPTF) (2013) *Beyond Retail: Redefining the Shape and Purpose of Town Centres,* London: DTCPTF

Dobry, G. (chair) (1975) *Review of the Development Control System: Final Report*, Dobry Report, London: HMSO

Docherty, I. (1999) *Making Tracks: The Politics of Local Rail Transport*, Aldershot: Avebury

Docherty, I. and Begg, D. (2003) 'Back to the city region?', *Scottish Affairs* 45: 128–56

Docherty, I. and Shaw, J. (2008) *Traffic Jam*, Bristol: Policy Press

Dodd, A. M. and Pritchard, D. E. (1993) *RSPB Planscan Northern Ireland: A Study of Development Plans in Northern Ireland*, Sandy, Beds: Royal Society for the Protection of Birds

Doig, A. (1984) *Corruption and Misconduct in Contemporary British Politics*, Harmondsworth: Penguin

Dower, J. (chair) (1945) *National Parks in England and Wales*, Cmd 6628, Dower Report, London: HMSO

Dowling, J. A. (1995) *Northern Ireland Planning Law*, Dublin: Gill and Macmillan

Downs, A. (2004) *Still Stuck in Traffic: Coping with Peak-hour Traffic Congestion*, Washington, DC: Brookings Institute

Drury, A., Watson, J. and Broomfield, R. (2006) *Housing Space Standards*. London: Greater London Authority

Dryzek, J. (1997) *The Politics of the Earth*, Oxford: Oxford University Press

Dühr, S., Farthing, S. and Nadin, V. (2005) 'Taking forward the spatial visions', *Town and Country Planning* 74(3): 97–9

Dühr, S., Colomb, C. and Nadin, V. (2010) *European Spatial Planning and Territorial Cooperation*, London: Routledge

Dunleavy, P. (1981) *The Politics of Mass Housing in Britain, 1945–75: Study of Corporate Power and Professional Influence in the Welfare State,* Oxford: Oxford University Press

Durrant, K. (2000) 'Making design decisions', *Planning Inspectorate Journal* 18: 7–10

Duxbury, R. (2012) *Telling & Duxbury's Planning Law and Procedure*, Oxford: Oxford University Press

Economist (2001) 'Heathrow T5 saga', 10 November, p. 38

Economist (2011) 'The Anthropocene: a man-made world', 26 May, available at www.economist.com/node/18741749/print (accessed 4 April 2014)

ECOTEC (1993) *Reducing Transport Emissions through Planning*, London: HMSO

ECOTEC (1999) *Scoping Study of RPG Targets and Indicators*, London: ODPM

ECOTEC (2003) *The Economic Impact of the Restoration of the Kennet and Avon Canal*, Watford: British Waterways

ECOTEC (2004) *The Economic and Social Impacts of Cathedrals in England*, London: English Heritage and Association of English Cathedrals

Eddington, R. (2006) *The Eddington Transport Study. Main Report: Transport's Role in Sustaining the UK's Productivity and Competitiveness*, London: HMSO

Edinburgh College of Art and Brodies WS and Halliday Fraser Munro Planning (1997) *Research on the General Permitted Development Order and Related Mechanisms*, Edinburgh: TSO

Edwards, J. (1997) 'Urban policy: the victory of form over substance', *Urban Studies* 34(5–6): 825–43

Edwards, J. and Deakin, N. (1992) 'Privatism and partnership in urban regeneration', *Public Administration* 70(3): 359–68

Edwards, R. (chair) (1991) *Fit for the Future: Report of the National Parks Review Panel*, Edwards Report, Cheltenham: Countryside Commission

Elkin, T., McLaren, D. and Hillman, M. (1991) *Reviving the City: Towards Sustainable Development*, London: Policy Studies Institute

Ellis, G. (2006) 'Third party appeals – pragmatism and principle', *Planning Theory and Practice* 7(3): 330–9

Elson, M. J. (1986) *Green Belts: Conflict Mediation in the Urban Fringe,* London: Heinemann

Elson, M. J., Walker, S. and MacDonald, R. (1993) *The Effectiveness of Green Belts*, London: HMSO

Elvin, D. and Robinson, J. (2000) 'Environmental impact assessment', *Journal of Planning and Environment Law* September: 876–93

Emerson, M. (1998) *Redrawing the Map of Europe*, London: Macmillan

England, J. (2000) *Retail Impact Assessment: A Guide to Best Practice,* Abingdon: Routledge

English Heritage (1999) *The Heritage Dividend: Measuring the Results of English Heritage Regeneration 1994–1999*, London: English Heritage

English Heritage (2000) *Power of Place: The Future of the Historic Environment*, London: English Heritage

English Heritage (2002) *The Heritage Dividend 2002*, London: English Heritage

English Heritage (2004) *Capital Solutions*, London: English Heritage

English Heritage (2005) *The Heritage Dividend Methodology: Measuring the Impact of Heritage Projects*, London: English Heritage

English Heritage (2006) *Shared Interest: Celebrating Investment in the Historic Environment*, London: English Heritage

English Heritage (2007) *Regeneration in Historic Coastal Towns*, London: English Heritage

English Heritage (2008a) *Conservation Principles: Policies and Guidance for the Sustainable Management of the Historic Environment*, London: English Heritage

English Heritage (2008b) *Enabling Development and the Conservation of Significant Places*, London: English Heritage

English Heritage (2008c) *Constructive Conservation in Practice*, London: English Heritage

English Heritage (2009) *The Protection and Management of World Heritage Sites in England: English Heritage Guidance Note to Circular for England on the Protection of World Heritage Sites*, London: DCLG, EH and DCMS

English Heritage (2011) *Valuing Places: Good Practice in Conservation Areas*, London: English Heritage

English Heritage (2012) *Good Practice Guide for Local Heritage Listing*, London: English Heritage

English Heritage (2013a) *The National Heritage Protection Plan Action Plan 2011–15: English Heritage: Revision 2: April 2013 – March 2015*, London: English Heritage

English Heritage (2013b) *Heritage Counts 2013*, London: English Heritage

English Heritage (2013c) *Constructive Conservation: Sustainable Growth for Historic Places*, London: English Heritage

English Heritage and CABE (2001) *Building in Context: New Development in Historic Areas*, London: EH/CABE

English Heritage and DEFRA (2005) *Building Value: Public Benefits of Historic Farm Building Repair in the Lake District*, London: English Heritage and DEFRA

English Partnerships (2004) *Urban Regeneration Companies: Coming of Age*, Milton Keynes: English Partnerships

English Partnerships and Housing Corporation (2000) *Urban Design Compendium*, London: Llewelyn Davies

Entec (2003) *The Relationships between Community Strategies and Local Development Frameworks*, London: ODPM

Environment Agency (EA) (2012) *Preventing Unacceptable Risks from Pollution,* Bristol: EA

Environmental Audit Committee (2011) *Third Report: Sustainable Development in the Localism Bill*, available at www.publications.parliament.uk/pa/cm201011/cmselect/cmenvaud/799/79902.htm (accessed 7 July 2014)

Equality and Human Rights Commission (EHRC) (2009) *Gypsies and Travellers: Simple Solutions for Living Together*, Manchester: EHRC

Essex, S. (1996) 'Members and officers in the planning policy process', in Tewdwr-Jones, M. (ed.) *British Planning Policy in Transition: Planning in the 1990s*, London: UCL Press

Essex County Council (1997) *A Design Guide for Residential and Mixed Area Uses*, Chelmsford: Essex Planning Officers Association, Essex Council Council

Essex County Council (2005) *The Essex Design Guide*, Chelmsford: Essex Planning Officers Association, Essex Council Council

Etzioni, A. (1973) 'Mixed scanning: a "third" approach to decision making', in Faludi, A. (ed.) *A Reader in Planning Theory*, Oxford: Pergamon

European Environment Agency (EEA) (2007) *Climate Change: The Cost of Inaction and the Cost of Adaptation,* EEA Technical Report No. 13/2007, Luxembourg: Office for Official Publications of the European Communities

European Environment Agency (EEA) (2013) *European Environmental Indicator Report: Natural Resources and Human Wellbeing in a Green Economy*, available at www.eea.europa.eu/publications/environmental-indicator-report-2013 (accessed 3 April 2014)

European Resource Efficiency Platform (EREP) (2013) *Action for a Resource-efficient Europe*, available at http://ec.europa.eu/environment/resource_efficiency/documents/action_for_a_resource_efficient_europe_170613.pdf (accessed 3 April 2014)

Evans, B. (1993) 'Why we no longer need a town planning profession', *Planning Practice and Research* 8(1): 9–15

Evans, B. (1995) *Experts and Environmental Planning*, Aldershot: Avebury

Ewbank, M. (2011) *The Blended Separation of Powers and the Organization of Party Groups: The Case of English Local Government*, PhD thesis, University of Birmingham

Experian (2012) *Town Centre Futures 2020*, London: Experian Ltd

Faludi, A. (ed.) (1973) *A Reader in Planning Theory*, Oxford: Pergamon

Faludi, A. (2004) 'Territorial cohesion: old (French) wine in new bottles?' *Urban Studies* 41(7): 1349–65

Faludi, A. (2005) 'Territorial cohesion: an unidentified political objective. Introduction to the special issue', *Town Planning Review* 76(1):1–15

Faludi, A. (2007) (ed.) *Territorial Cohesion and the European Model of Society*, Cambridge, MS: The Lincoln Institute for Land Policy

Faludi, A. (2009) 'A turning point in the development of European spatial planning? The "Territorial Agenda of the European Union" and the "First Action Programme"', *Progress in Planning* 71(1): 1–42

Faludi, A. (2010) *Cohesion, Coherence, Cooperation: European Spatial Planning Coming of Age?* London: Routledge

Faludi, A. and Waterhout, B. (2002) *The Making of the European Spatial Development Perspective: No Masterplan*, London: Routledge

Faludi, A. and Waterhout, B. (2006a) 'Introducing evidence-based planning' *disP* 42(165): 4–13

Faludi, A. and Waterhout, B. (2006b) 'Debating evidence-based planning' *disP* 42(165): 71–72

Farnsworth, D. (2013) 'Incremental plans and self-build cities', *Town and Country Planning* 82(11): 481–4

Fergusson, A. (1973) *The Sack of Bath*, Salisbury: Compton Russell

Findlay, A. and Sparks, L. (2005) *Publications on Retail Planning in 2004*, Stirling: Institute for Retail Studies, University of Stirling

Firn, J. and McGlashan, D. (2001) *A Coastal Management Trust for Scotland: A Concept Development and Feasibility Study*, Edinburgh: Scottish Executive

Flinders, M. and Skelcher, C. (2012) 'Shrinking the quango state: five challenges for reforming quangos', *Public Money & Management* 32(5): 327–34

Flint, A. (2009) *Wrestling with Moses*, New York: Random House

Flynn, N. (2012) *Public Sector Management*, London: Sage

Foley, P., Hutchinson, J. and Fordham, G. (1998) 'Managing the challenge: winning and implementing the Single Regeneration Budget Fund', *Planning Practice and Research* 13(1): 63–80

Ford, J. and Seavers, J. (2000) *Attitudes to Moving and Debt: Household Behaviour in the 1990s*, London: Council of Mortgage Lenders

Fordham, G., Hutchinson, J. and Foley, P. (1998) 'Strategic approaches to local regeneration: the Single Regeneration Budget Challenge Fund', *Regional Studies* 33(2): 131–41

Forester, J. (1982) 'Planning in the face of power', *Journal of the American Planning Association* 48(1): 67–80

Forester, J. (1989) *Planning in the Face of Power*, Berkeley, CA: University of California Press

Forshaw, J. and Abercrombie, P. (1943) *County of London Plan*, London: Macmillan

Forsyth, M. (2013) *Understanding Historic Building Conservation*, Chichester: Wiley-Blackwell

Fothergill, S., Kitson, M. and Monk, S. (1985) *Urban Industrial Change: The Causes of Urban–Rural Contrast in Manufacturing Employment Trends*, London: HMSO

Fried, M. (1969) 'Social differences in mental health', in Kosa, J., Antonovsky, A. and Zola, I. K. (eds) *Poverty and Health: A Sociological Analysis*, Cambridge, MA: Harvard University Press

Friedrichs, J., Galster, G. and Musterd, S. (2003) 'Neighbourhood effects on social opportunities: the

European and American research and policy context', *Housing Studies* 18(6): 797–806

Friend, J. and Jessop, W. (1969) *Local Government and Strategic Choice: An Operational Research Approach to the Processes of Public Planning*, London: Tavistock

Friend, J. and Hickling, A. (1987) *Planning Under Pressure: The Strategic Choice Approach*, Oxford: Pergamon Press

Friends of the Earth (FoE) (2005) *Tackling Climate Change at the Local Level: The Role of Local Development Frameworks in Reducing the Emissions of New Developments*, available at www.foe.co.uk/resource/briefings/ldf_climate_briefing.pdf (accessed July 2009)

Gallent, N. (2012) *Neighbourhood Planning: Communities, Networks and Governance*, Bristol: Policy Press

Gallent, N. and Bell, P. (2000) 'Planning exceptions in rural England: past, present and future', *Planning Practice and Research* 15(4): 375–84

Gallent, N., Juntti, M., Kidd, S. and Shaw, D. (2008) *Introduction to Rural Planning,* London: Routledge

Game, C. (2011) *Hung Councils and Local Coalitions: Where Are They Going, and How?,* Birmingham: Institute for Local Government Studies

Gans, H. J. (1968) *People and Plans: Essays on Urban Problems and Solutions*, New York: Basic Books

Gans, H. J. (1991) *People, Plans and Policies: Essays on Poverty, Racism and Other National Urban Problems*, New York: Columbia University Press

Garlick, R. (2013) 'Deregulation trial must be watched closely', *Planning*, 17 May

Gatenby, I. and Williams, C. (1992) 'Section 54A: the legal and practical implications', *Journal of Planning and Environment Law* 1992: 110–20

Gatenby, I. and Williams, C. (1996) 'Interpreting planning law', in Tewdwr-Jones, M. (ed.) *British Planning Policy in Transition*, London: UCL Press

Gentleman, H. (1993) *Counting Travellers in Scotland*, Edinburgh: Scottish Office

Gibb, K. (2003) 'Transferring Glasgow's council housing: financial, urban and housing policy implications', *International Journal of Housing Policy* 3(1): 89–114

Gibbons, S., Mourato, S. and Resende, G. (2011) *The Amenity Value of English Nature: A Hedonic Price Approach*, SERC Discussion Paper 74, London: LSE

Gibson, L. and Pendlebury, J. (eds) (2009) *Valuing Historic Environments*, Farnham: Ashgate

Gibson, M. and Langstaff, M. (1982) *An Introduction to Urban Renewal*, London: Hutchinson

Giddens, A. (1998) *The Third Way: The Renewal of Social Democracy*, Cambridge: Polity Press

Gilder, I. (1984) 'State planning and local needs', in Bradley, T. and Lowe, P. (eds) *Locality and Rurality: Economy and Society in Rural Regions,* Norwich: Geo Books

Gilg, A. W. (1996) *Countryside Planning: The First Half Century* (second edition) London: Routledge

Gill, S., Handley, J., Ennos, R., Nolan, P. (2009) 'Planning for green infrastructure: adapting to climate change', in Davoudi, S., Crawford, J. and Mehmood, A. (eds), *Planning for Climate Change*, London: Earthscan, pp. 249–62

Gillett, E. (1983) *Investment in the Environment: Planning and Transport Policies in Scotland*, Aberdeen: Aberdeen University Press

Gilroy, R. with Marvin, S. (1993) *Good Practices in Equal Opportunities*, Aldershot: Avebury

Glass, R. (1959) 'The evaluation of planning: some sociological considerations', in Faludi, A. (ed.) (1973) *A Reader in Planning Theory*, Oxford: Pergamon

Glasson, J. (1999) 'The first 10 years of the UK EIA system: strengths, weaknesses, opportunities and threats', *Planning Practice and Research* 14(3): 363–75

Glasson, J., Therivel, T. and Chadwick, A. (1998) *Introduction to Environmental Impact Assessment* (second edition), London: UCL Press

Glendinning, M. (2013) *The Conservation Movement: A History of Architectural Preservation*, London: Routledge

Glynn, S. (2012) 'You can't demolish your way out of a housing crisis'. *City* 16(6): 656–71

GMA Planning, P-E International and Jacques & Lewis (1993) *Integrated Planning and Granting of Permits in the EC* (DoE Planning Research Programme), London: HMSO

Gold, J. (1997) *The Experience of Modernism: Modern Architects and the Future City 1928–1953*, London: Spon

Goldsmith, M. and Newton, K. (1986) 'Central–local government relations: a bibliographic summary of the ESRC research initiative', *Public Administration* 64(1): 102–8

Goodchild, K., Marwick, A., Grant, M., Jones, A., Lyddon, D. and Robinson, D. (1989) 'Directions for change in the British planning system: comments and response on a paper by Pasty Healey in *Town Planning Review*, 60(2) April 1989, pp. 125–49', *Town Planning Review* 60(3): 319–32

Goodlad, R., Flint, J., Kearns, A., Keoghan, M., Paddison, R. and Raco, M. (1999) *The Role and Effectiveness of Community Councils with Regard to Community Consultation*, Edinburgh: Central Research Unit, Scottish Office

Goodman, A. (2013) 'Walking, cycling and driving to work in the English and Welsh 2011 Census: trends, socio-economic patterning and relevance to travel behaviour in general', *Plos One*, DOI: 10.1371/journal.pone.0071790

Goodman, R. (1972) *After the Planners,* Harmondsworth: Penguin

Goodwin, P. (1999) 'Transformation of transport policy in Great Britain', *Transportation Research Part A: Policy and Practice*, 33(7–8): 655–69

Goodwin, P. (2012) 'Due diligence, traffic forecasts and pensions', *Local Transport Today*, 12 April

Goodwin, P. and Dargay, J. (2001) 'Bus fares and subsidies in Britain', in Grayling, A. (ed) *Any More Fares?* London: IPPR

Goodwin, P., Hallett, S., Kenny, F. and Stokes, G. (1991) 'Transport: the new realism', Report to the Rees Jeffreys Road Fund, Oxford: University of Oxford Transport Studies Unit

Government Office for the East of England (GOEE) (2008) *East of England Plan: The Revision to the Regional Spatial Strategy for the East of England*, London: TSO

Government Office for the North West (GONW) (2008) *North West of England Plan: Regional Spatial Strategy to 2021*, London: TSO

Graham, B., Ashworth, G. J. and Tunbridge, J. E. (2000) *A Geography of Heritage*, London: Arnold

Graham, S. (2009) *Disrupted Cities: When Infrastructure Fails*, Abingdon: Routledge

Graham, S. and Marvin, S. (1999) 'Planning cybercities? Integrating telecommunications into urban planning', *Town Planning Review* 70(1): 89–114

Graham, S. and Marvin, S. (2001) *Splintered Urbanism: Networked Infrastructures, Technological Mobilities and the Urban Condition,* London: Routledge

Grant, M. (1996) *Permitted Development* (second edition), London: Sweet & Maxwell

Grant, M. (ed.) (1997) *Encyclopedia of Planning Law and Practice* (6 vols), London: Sweet & Maxwell (looseleaf)

Graves, G., Max, R. and Kitson, T. (1996) 'Inquiry procedure: another dose of reform?', *Journal of Planning and Environment Law* 1996: 99–106

Greater London Authority (GLA) (2009) *The London Plan: Spatial Development Strategy for Greater London – Consultation Draft Replacement Plan,* London: GLA

Greater London Authority (GLA) (2011) *The London Plan: Spatial Development Strategy for Greater London,* London: GLA

Greater London Authority (GLA) (2012a) *Housing Supplementary Planning Guidance,* London: GLA

Greater London Authority (GLA) (2012b) *London View Management Framework SPG*, London: GLA

Greater London Authority (GLA) (2013) *Town Centres: Draft Supplementary Planning Guidance,* London: GLA

Greater London Council (GLC) (1986) *Race and Planning Guidelines*, London: GLC

Greed, C. (1993) *Introducing Town Planning*, Harlow: Longman

Greed, C. (1994) *Women and Planning: Creating Gendered Realities*, London: Routledge

Greed, C. (2006) 'Institutional and conceptual barriers to the adoption of gender mainstreaming within spatial planning departments in England', *Planning Theory and Practice* 7(2): 179–97

Greed, C. and Roberts, M. (eds) (1998) *Introducing Urban Design: Interventions and Responses,* Harlow: Longman

Greenhalgh, L. and Worpole, K. (1995) *Park Life: Urban Parks and Social Renewal,* Stroud: Comedia

Greer, J. and Murray, M. (eds) (2003) *Rural Planning and Development in Northern Ireland*, Dublin: IPA

Grimley J. R. Eve (1992) *Use of Planning Agreements*, London: HMSO

Grimsey, B. (2013) *The Grimsey Review: An Alternative Future for the High Street*, available at www.vanishing highstreet.com/wp-content/uploads/2013/09/GrimseyReview04.092.pdf (accessed 14 February 2014)

Gunder, M. and Hillier, J. (2009) *Planning in Ten Words or Less,* Farnham: Ashgate

Gunn, S. and Hillier, J. (2012) 'Processes of innovation: reformation of the English strategic spatial planning system', *Planning Theory and Practice* 13(3): 359–81

Gunn, S. and Hillier, J. (2014) 'When uncertainty is interpreted as risk: an analysis of tensions relating to spatial planning reform in England,' *Planning Practice and Research* 29(1): 56–75

Gunn, S. and Vigar, G. (2012) 'Reform processes and discretionary acting space in English planning practice, 1997–2010', *Town Planning Review* 83(5): 533–51

Guy, C. (2007) *Planning for Retail Development: A Critical Review of the British Experience,* Abingdon: Routledge

GVA (2012) *Unlocking Town Centre Retail Developments,* London: GVA

GVA Grimley with Environmental Resources Management and University of Westminster (2004) *Reforming Planning Obligations: the Use of Standard Charges*, London: ODPM

Hadrian's Wall Management Plan Committee (2008) *Frontiers of the Roman Empire World Heritage Site: Hadrian's Wall Management Plan: 2008–2014*, available at www.visithadrianswall.co.uk/dbimgs/3_%20Hadrian's%20Wall%202008-2014%20-%20Greyscale%20text%20%26%20appendices.pdf (accessed 2 May 2014)

Hagman, D. G. and Misczynski, D. J. (1978) *Windfalls for Wipeouts: Land Value Capture and Compensation,* St Paul, MN: West

Hague, C. (1984) *The Development of Planning Thought: A Critical Perspective*, London: Hutchinson

Haigh, N. (1990) *EEC Environmental Policy* (second edition), Harlow: Longman

Hajer, M. A. (1995) *The Politics of Environmental Discourse: Ecological Modernisation and the Policy Process*, Oxford: Clarendon Press

Halcrow Group (2004) *Unification of Consent Regimes*, London: ODPM

Hales, R. (2000) 'Land use development planning and the notion of sustainable development: exploring constraint and facilitation within the English planning system', *Journal of Environmental Planning and Management* 43(1): 99–121

Hall, D., Hebbert, M. and Lusser, H. (1993) 'The planning background', in Blowers, A. (ed.) *Planning for a Sustainable Environment: A Report by the Town and Country Planning Association*, London: Earthscan

Hall, J. (2009) 'Integrated assessment to support regional and local decision making', in Davoudi, S., Crawford, J. and Mehmood, A. (eds) *Planning for Climate Change*, London: Earthscan

Hall, P. (1980) *Great Planning Disasters*, London: Weidenfeld and Nicolson

Hall, P. (1992) *Urban and Regional Planning,* London: Routledge

Hall, P. (1997) 'Regeneration policies for peripheral housing estates: inward and outward looking approaches', *Urban Studies* 34(5–6): 873–90

Hall, P. (1999a) 'The regional dimension', in Cullingworth, J. B. (ed.) *British Planning: 50 Years of Urban and Regional Policy*, London: Athlone

Hall, P. (1999b) *Sustainable Cities or Town Cramming?*, London: Town and Country Planning Association

Hall, P. (2002a) *Urban and Regional Planning* (fourth edition), London: Routledge

Hall, P. (2002b) *Cities of Tomorrow*, Oxford: Blackwell

Hall, P. with Tewdwr-Jones, M. (2010) *Urban and Regional Planning*, London: Routledge

Hall, P. and Ward, C. (1998) *Sociable Cities: The Legacy of Ebenezer Howard*, Chichester: Wiley

Hall, P., Gracey, H., Drewett, R. and Thomas, R. (1973) *The Containment of Urban England*, London: Allen and Unwin

Hall, S. (1995) 'The SRB: Taking Stock' *Planning Week* 27 April: 16–17

Hall, S. and Nevin, B. (1999) 'Continuity and change: a review of English regeneration policy in the 1990s', *Regional Studies* 33(4): 447–91

Hall, W. and Weir, S. (1996) *The Untouchables: Power and Accountability in the Quango State,* London: The Scarman Trust

Halsey, A. J. (ed.) (1972) *Educational Priority, Vol. 1: EPA Problems and Policies,* London: HMSO

Hambleton, R. and Sweeting, D. (2004) 'US-style leadership for English local government?', *Public Administration Review* 64(4): 474–88

Hambleton, R. and Thomas, H. (eds) (1995) *Urban Policy Evaluation: Challenge and Change,* London: Paul Chapman

Hamdouch, A. and Zuindeau, B. (2010) 'Sustainable development 20 years on: methodological issues, practices and open issues', *Journal of Environmental Planning and Management* 52(4): 427–38

Hamilton, K. (1999) 'Women and transport: disadvantage and the gender divide', *Town and Country Planning* 68(10): 318–19

Hammersley, R. (1987) 'Plans, policies and the local ombudsman: the Chellaston case', *Journal of Planning and Environment Law* 1987: 101–5

Hampton, P. (2005) *Reducing Administrative Burdens: Effective Inspection and Enforcement,* London: HM Treasury

Handley, J., Pauleit, S. and Gill, S. (2007) 'Landscape, Sustainability and the City', in Benson, J. F. and Roe, M. (eds) *Landscape and Sustainability* (second edition), London: Routledge, pp. 167–95

Hanley, N., Whitby, M. and Simpson, I. (1999) 'Assessing the success of agri-environmental policy in the UK', *Land Use Policy* 16(2): 67–80

Haq, G. (1997) *Towards Sustainable Transport Planning,* Aldershot: Avebury

Hargreaves, M. and Brindley, M. (2011) *Planning for Gypsies and Travellers: The Impact of Localism,* London: Irish Traveller Movement in Britain

Harris, N. and Thomas, H. (2009) 'Making Wales: spatial strategy making in a devolved context', in Davoudi, S. and Strange, I. (eds) *Conceptions of Space and Place in Strategic Spatial Planning,* New York: Routledge, pp. 43–71

Harrison, J. (2012) 'Life after regions? The evolution of city-regionalism in England', *Regional Studies* 46(9): 1243–59

Harrison, M. (1992) 'A presumption in favour of planning permission?', *Journal of Planning and Environment Law* 1992: 121–9

Hart, T., Haughton, G. and Peck, J. (1996) 'Accountability and the non-elected local state: bringing Training and Enterprise Councils to local account', *Regional Studies* 30(4): 429–41

Harvey, D. (1974) 'On Planning the Ideology of Planning', in Burchall, J. (ed.) *Planning in the 1980s: Challenge and Response,* New York: Rutgers University Press

Harvey, D. (2004) 'The "new" imperialism: accumulation by dispossession', *Socialist Register* 40: 63–87

Harvey, G. (1997) *The Killing of the Countryside,* London: Cape

Harwood, R. (2012) *Historic Environment Law: Planning, Listed Buildings, Monuments, Conservation Areas and Objects,* Builth Wells: Institute of Art and Law

Hastings, A. and Matthews, P. (2012) 'Connectivity and conflict in periods of austerity', available at http://eprints.gla.ac.uk/60140/1/60140.pdf (accessed 7 July 2014)

Hatherley, O. (2010) *A Guide to the New Ruins of Great Britain,* London: Verso

Haughton, G. (2009) 'Celebrating Leeds as it builds the slums of tomorrow?', *Yorkshire and Humber Regional Review* 19(4): 28–32

Haughton, G. and Allmendinger, P. (2013) 'Spatial planning and the new localism', *Planning Practice and Research* 28(1): 1–5

Haughton, G. and Counsell, D. (2004) *Regions, Strategies and Sustainable Development,* London: Routledge

Haughton, G., Allmendinger, P. and Oosterlynck, S. (2013) 'Spaces of neoliberal experimentation: soft spaces, postpolitics, and neoliberal governmentality', *Environment and Planning A* 45(1): 217–34

Haughton, G., Deas, I. and Hincks, S. (2014) 'Commentary. Making an impact: when agglomeration boosterism meets antiplanning rhetoric', *Environment and Planning A* 46(2): 265–70

Haughton, G., Allmendinger, P., Counsell, D. and Vigar, G. (2010) *The New Spatial Planning: Territorial Management with Soft Spaces and Fuzzy Boundaries*, Abingdon: Routledge

Hayton, K. (1992) 'Scottish Enterprise: a challenge to local land use planning?', *Town Planning Review* 63(3): 265–78

Hazell, R. (ed.) (1999) *Constitutional Futures: A History of the Next Ten Years*, the Constitution Unit, Oxford: Oxford University Press

Headicar, P. (2009) *Transport Policy and Planning in Great Britain*, Abingdon: Routledge

Healey, P. (1983) *Local Plans in British Land Use Planning*, Oxford: Pergamon

Healey, P. (1985) 'The professionalisation of planning in Britain', *Town Planning Review* 56(4): 492–507

Healey, P. (1986) 'The role of development plans in the British planning system: an empirical assessment', *Urban Law and Policy* 8: 1–32

Healey, P. (1989) 'Directions for change in the British planning system', *Town Planning Review* 60(2): 125–49

Healey, P. (1990) 'Places, people and politics: plan-making in the 1990s', *Local Government Policy Making* 17(2): 29–39

Healey, P. (1992) 'Planning through debate: the communicative turn in planning theory', *Town Planning Review* 63(2): 143–62

Healey, P. (1993) 'The communicative work of development plans', *Environment and Planning B* 20(1): 83–104

Healey, P. (1997) *Collaborative Planning: Shaping Places in Fragmented Societies*, Basingstoke: Macmillan

Healey, P. (1998) 'Collaborative planning in a stakeholder society', *Town Planning Review* 69(1): 1–21

Healey, P. (2004a) 'The treatment of space and place in the new strategic spatial planning in Europe', *International Journal of Urban and Regional Research* 28(1): 45–67

Healey, P. (2004b) 'Towards a "social democratic" policy agenda for cities', in Johnstone and Whitehead (eds) *New Horizons in British Urban Policy*, Aldershot: Ashgate

Healey, P. (2007) *Urban Complexity and Spatial Strategies*, London: Routledge

Healey, P. (2010) *Making Better Places: The Planning Project in the Twenty-first Century*, Basingstoke: Palgrave Macmillan

Healey, P. (2013) *What Can Community Initiative Achieve?* Unpublished Working Paper

Healey, P. and Shaw, T. (1994) 'Changing meanings of "environment" in the British planning system', *Transactions of the Institute of British Geographers*, 19(4): 425–38

Healey, P., McNamara, P., Elson, M. and Doak, A. (1988) *Land Use Planning and the Mediation of Change*, Cambridge: Cambridge University Press

Healey, P., Davoudi, S., O'Toole, M., Tavsanoglu, S. and Usher, D. (1992) *Rebuilding the City: Property-led Urban Regeneration*, London: Spon

Healey, P., Cameron, S., Davoudi, S., Graham, S. and Madani-Pour, A. (eds) (1995) *Managing Cities: The New Urban Context*, Chichester: Wiley

Hebbert, M. (2009) 'The three P's of place-making for climate change', *Town Planning Review* 80(4–5): 359–70

Hendry, J. (1989) 'The control of development and the origins of planning in Northern Ireland', in Bannon, M. J., Nowlan, K. I., Hendry, J. and Mawhinney, K. (eds) *Planning: The Irish Experience 1920–1988*, Dublin: Wolfhound Press

Hendry, J. F. (1992) 'Plans and planning policy for Belfast: a review article', *Town Planning Review* 63(1): 79–85

Herbert-Young, N. (1995) 'Reflections on section 54A and plan-led decision-making', *Journal of Planning and Environment Law* 1995: 292–305

Herington, J. (1990) *Beyond Green Belts: Managing Urban Growth in the 21st Century*, London: Jessica Kingsley

Heritage Link (2003) *Volunteers and the Historic Environment*, London: English Heritage

Heseltine, M. (1979) 'Secretary of State's address', *Report of Proceedings of Town and Country Planning Summer School: 8–19 September 1979*, London: RTPI

Heseltine, M. (2012) *No Stone Unturned in Pursuit of Growth*, London: BIS

Hewison, R. (1987) *The Heritage Industry: Britain in a Climate of Decline*, London: Methuen

Hewitt, L. (2011) 'Towards a greater urban geography: regional planning and associational networks in London during the early twentieth century', *Planning Perspectives* 26(4): 551–68

Hewitt, L. (2012) 'The civic survey of Greater London: social mapping, planners and urban space in the early twentieth century', *Journal of Historical Geography* 38(3): 247–62

Hickman, R. and Banister, D. (2005) 'Reducing travel by design', in Williams, K. (ed.) *Spatial Planning, Urban Form and Sustainable Transport*, Aldershot: Ashgate, pp. 102–22

Hickman, R. and Banister, D. (2014) *Transport, Climate Change and the City,* London: Routledge

Hill, L. (1991) 'Unitary development plans for the West Midlands: first stages in the statutory responses to a changing conurbation', in Nadin, V. and Doak, J. (eds) *Town Planning Responses to City Change*, Aldershot: Gower

Hill, S. and Barlow, J. (1995) 'Single regeneration budget: hope for "those inner cities"?', *Housing Review* 44(2): 32–5

Hillier, B. and Sahbaz, O. (2012) 'Safety in numbers: high-resolution analysis of crime in street networks', in Ceccato, V. (ed.) *The Urban Fabric of Fear and Crime*, New York: Springer

Hillier, J. (2002) *Shadows of Power: An Allegory of Prudence,* London: Routledge

Hillier, J. and Healey, P. (2008) *Critical Essays in Planning Theory,* Aldershot: Ashgate

Hillier Parker and Cardiff University (2004) *Policy Evaluation of the Effectiveness of PPG 6,* London: ODPM

Hillier Parker, Dundas and Wilson, Edinburgh School of Planning and Housing (1998) *Review of Development Planning in Scotland*, Central Research Unit, Scottish Office, Edinburgh: TSO

Hillman, M., Adams, J. and Whitelegg, J. (1990) *One False Move . . . A Study of Children's Independent Mobility*, London: Policy Studies Institute

Hills, J. (2007) *Ends and Means: The Future Roles of Social Housing in England*, London: LSE

Hirsch, D. (ed.) (1994) *A Positive Role for Local Government: Lessons for Britain from Other Countries*, London: Local Government Chronicle and Joseph Rowntree Foundation

Historic Scotland (2011) *Scottish Historic Environment Policy*, Edinburgh: Historic Scotland

HM Government (1985) *Lifting the Burden,* Cm 9571, London: HMSO

HM Government (2009a) *The UK Low Carbon Transition Plan: National Strategy for Climate and Energy*, London: HM Government

HM Government (2009b) *The UK Renewable Energy Strategy*, London: HM Government

HM Government (2010a) *Local Growth: Realising Every Place's Potential*, Cm 7961, Norwich: TSO.

HM Government (2010b) *The Government's Statement on the Historic Environment for England 2010*, London: DCMS

HM Government (2011a) *The Carbon Plan: Delivering our Low Carbon Future,* London: HM Government

HM Government (2011b) *Securing the Future: Delivering UK Sustainable Development Strategy,* London: HM Government

HM Government (2012) *2012 Energy and Emissions Projections: Projections of Greenhouse Gas Emissions and Energy Demand 2012 to 2030*, London: HM Government

HM Government (2013) *The Strategy for Exercising the Adaptation of Reporting Power,* London: HM Government

HM Treasury (2006a) *Investing in Britain's Potential: Building our Long-term Future. Pre-Budget Report*, London: HM Treasury

HM Treasury (2006b) *The Stern Review: The Economics of Climate Change*, London: HM Treasury and Cabinet Office

HM Treasury (2007a) *Review of Sub-national Economic Development and Regeneration*, London: HM Treasury

HM Treasury (2007b) *Meeting the Aspirations of the British People: 2007 Pre-Budget Report and Comprehensive Spending Review*, London: HM Treasury

HM Treasury (2010a) *Total Place: A Whole Area Approach to Public Services,* London: HM Treasury/DCLG

HM Treasury (2010b) *National Infrastructure Plan,* London: HMSO

HM Treasury (2013a) *Government's Response to the Heseltine Review,* London: HM Treasury/BIS

HM Treasury (2013b) *National Infrastructure Plan 3,* London: HMSO

Ho, S. Y. (2003) *Evaluating British Urban Policy: Ideology, Conflict and Compromise*, Aldershot: Ashgate

Hobhouse, A. (chair) (1947) *Report of the National Parks Committee (England and Wales)*, Cmd 7121, Hobhouse Report, London: HMSO

Hobson, E. (2004) *Conservation and Planning: Changing Values in Policy and Practice*, London: Spon

Hodge, I. (1999) 'Countryside planning: from urban containment to sustainable development', in Culling-worth, J. B. (ed.) *British Planning: 50 Years of Urban and Regional Policy*, London: Athlone

Holling, C. S. (1973) 'Resilience and stability of ecological systems', *Annual Review of Ecological Systems* 4: 1–23

Holling, C. S. (1996) 'Engineering resilience versus ecological resilience', in Schulze, P. C. (ed.) *Engineering within Ecological Constraints*, Washington, DC: National Academy Press, pp. 31–45

Holling, C. S. and Gunderson L. H. (2002) 'Resilience and adaptive cycles', in Gunderson L. H. and Holling C. S. *Panarchy: Understanding Transformations in Human and Natural Systems*, Washington, DC: Island Press, pp. 25–62

Holmans, A. (2013) *New Estimates of Housing Demand and Need in England, 2011 to 2013*, Town and Country Planning Tomorrow Series Paper 16, London: TCPA

Holstein, T. (2002) 'Moving to a SEA Green Paper', *Town and Country Planning* 71(9): 202–20

Home, R. (1992) 'The evolution of the Use Classes Order', *Town Planning Review* 63(2): 187–201

Home, R. (2002) 'Negotiating security of tenure for peri-urban settlement: traveller-gypsies and the planning system in the United Kingdom', *Habitat International* 26(3): 335–46

Home, R. (2012) 'Forced eviction and planning enforcement: the Dale Farm Gypsies', *International Journal of Law in the Built Environment* 4(3): 178–88

Homes and Communities Agency (2013) *Statutory Role and History*, available at www.homesandcommunities. co.uk/statutory-role-and-history (accessed 24 March 2014)

Hopkins, L. (2001) *Urban Development: The Logic of Making Plans*, Washington, DC: Island Press

House Builders Federation (HBF) (1998) *Urban Life: Breaking Down the Barriers to Brownfield Development*, London: HBF

House Builders Federation (HBF) (2000) *PPG 3: The Consumer Response*, London: HBF

House of Commons Business, Innovation and Skills Committee (2013) *Local Enterprise Partnerships*, Ninth Report of Session 2012–13, London: TSO

House of Commons Committee of Public Accounts (2013) *The New Homes Bonus: Twenty-ninth Report of Session*, London: TSO

House of Commons Communities and Local Government Select Committee (2011) *Localism: Third Report of Session 2010–12*, Vol. I, HC 547, London: TSO

House of Commons Environment Committee (1995) *First Report: Single Regeneration Budget* (together with the Proceedings of the Committee), London: HMSO

House of Commons Expenditure Committee (HCEC) (1973) *Urban Transport Planning*, House of Commons Paper 57-1, London: HMSO

House of Commons Public Administration Select Committee (2008) *From Citizen's Charter to Public Service Guarantees: Entitlements to Public Services: Twelfth Report of Session 2007–08*, London: TSO

House of Commons Select Committee on Coastal Zone Protection and Planning (1992) *Second Report from the Environment Committee of Session 1991–92*, HC 17, London: HMSO

House of Commons Select Committee on Environmental Audit (1998) *First Report: Pre-Budget Report*, HC 547, London: TSO

House of Commons Select Committee on Health (2004) *Third Report*, London: TSO

House of Commons Select Committee on Transport, Local Government and the Regions (2002) *Planning Green Paper; Thirteenth Report of Session 2001–2002, Vol. 1: Report and Proceedings of the Committee*, London: TSO

Howard, E. (1898) *To-morrow: A Peaceful Path to Real Reform*, London: Swan Sonnenschein; republished in facsimile (2003) with a commentary by Hall, P., Hardy, D. and Ward, C.

Howard, E. (1902) *Garden Cities of Tomorrow*, London: Swan Sonnenschein; reprinted by Faber in 1946, with an introduction by Osborn, F. J. and an introductory essay by Mumford, L.; see also Howard (1898), the original edition

Howard, J. (2009) 'Climate change mitigation and adaptation in developed nations: A critical perspective on the adaptation turn in urban climate planning', in Davoudi, S., Crawford, J. and Mehmood, A. (eds), *Planning for Climate Change*, London: Earthscan

HRH Prince of Wales (1988) *A Vision of Britain: A Personal View of Architecture,* London: Doubleday

Hughes, J. T. (1991) 'Evaluation of local economic development: a challenge for policy research', *Urban Studies* 28(6): 909–18

Hull, A. D. (2011) *Transport Matters: Integrated Approaches to Planning City-regions*, London: Routledge

Huntingdonshire District Council (HDC) (2009) *Report to Development Control Panel, 23rd February 2009*, available at http://applications.huntingdonshire.gov.uk/moderngov/ieListDocuments.aspx?CId=257&MID=2957 (accessed 21 May 2013)

Huxley, J. S. (chair) (1947) *Conservation of Nature in England and Wales*, Cmd 7122, Huxley Report, London: HMSO

Illich, I. (1971) *Deschooling Society*, New York: Harper and Row

Illich, I. (1977) *Disabling Professions*, London: Boyars

Imrie, R. (2012) '"Auto-disabilities": the case of shared space environments', *Environment and Planning A*, 44(9): 2260–77

Imrie, R. and Raco, M. (2003) *Urban Renaissance? New Labour, Community and Urban Policy,* Bristol: Policy Press

Imrie, R. and Thomas, H. (1999) *British Urban Policy: An Evaluation of the Urban Development Corporations,* London: Paul Chapman

Innes, J. E. (1995) 'Planning theory's emerging paradigm: communicative action and interactive practice', *Journal of Planning Education and Research* 14(3): 183–90

Institute of Directors (IOD) (2011) *Freebie Growth Plan,* London: IOD

Institute of Historic Building Conservation (IHBC) (2013) *Scotland's Local Authority Conservation Services: First 'Scoping' Report: 2013*, Tunbridge Wells: IHBC

International Panel on Climate Change (IPCC) (2001) 'Summary Report', *Climate Change 2001: Synthesis Report*, Cambridge: Cambridge University Press

International Panel on Climate Change (IPCC) (2007a) *Climate Change 2007: Synthesis Report, Summary for Policymakers, Contribution of Working Group I to the Fourth Assessment Report of the Intergovernmental Panel on Climate Change*, Cambridge: Cambridge University Press

International Panel on Climate Change (IPCC) (2007b) *Climate Change 2007: Impacts, Adaptation and Vulnerability, Contribution of Working Group II to the Fourth Assessment Report of the Intergovernmental Panel on Climate Change*, Cambridge: Cambridge University Press

International Panel on Climate Change (IPCC) (2013) *Climate Change 2013: The Physical Science Basis, Contribution of Working Group I to the Fifth Assessment Report of the Intergovernmental Panel on Climate Change*, Cambridge: Cambridge University Press

International Panel on Climate Change (IPCC) (2014) *Climate Change 2014: Impacts, Adaptation, and Vulnerability, Contribution of Working Group II to the Fifth Assessment Report of the Intergovernmental Panel on Climate Change. Approved Summary for Policy Makers,* available at http://ipcc-wg2.gov/AR5/images/uploads/IPCC_WG2AR5_SPM_Approved.pdf (accessed 4 April 2014)

Jacobs, J. (1961) *The Death and Life of Great American Cities*, London: Cape and Penguin

Jacobs, M. (1999) *Environmental Modernisation: The New Labour Agenda*, London: Fabian Society

Jagger, M. (1998) 'The planner's perspective: a view from the front', in Warren, J., Worthington, J. and Taylor, S. (eds) *Context: New Buildings in Historic Settings*, Oxford: Architectural Press, pp. 71–82

Jain, J. and Lyons, G. (2008) 'The gift of travel time', *Journal of Transport Geography* 16(2): 81–9

Janin Rivolin, U. (2005) 'Cohesion and subsidiarity: towards good territorial governance in Europe', *Town Planning Review* 76(1): 93–107

Janin Rivolin, U. (2008) 'Conforming and performing planning systems in Europe: an unbearable cohabitation', *Planning Practice and Research* 23(2): 167–86

Jarvis, R. K. (1980) 'Urban environments as visual art or as social settings? A review', *Town Planning Review* 51(1): 50–66

Jay, S. (2012) 'Marine space: manoeuvring towards relational understanding', *Journal of Environmental Policy and Planning* 14(1): 67–81

Johnstone, C. and Whitehead, M. (eds) (2004) *New Horizons in British Urban Policy: Perspectives on New Labour's Urban Renaissance*, Aldershot: Ashgate

Jokilehto, J. (1999) *A History of Architectural Conservation*, Oxford: Butterworth Heinemann

Jones, B. and Norton, P. (2013) *Politics UK,* Harlow: Pearson

Jones, C., Wood, C. and Dipper, B. (1998) 'Environmental assessment in the UK planning process: a review of practice', *Town Planning Review* 69(3): 315–39

Jones, C., Dunse, N., Watkins, N. and Watkins, D. (2010) *Affordability and Housing Market Areas,* London: DCLG

Jones, M. (2013) 'It's like deja vu, all over again' in Ward, M. and Hardy, S. *Where Next for Local Enterprise Partnerships?* London: Smith Institute, pp. 86–94

Jones, P. (1999) 'Groundwork: changing places and agendas', *Town and Country Planning* 68: 315–17

Jones, P. and Evans, J. (2013) *Urban Regeneration in the UK* (second edition), London: Sage

Jordan A (2008) 'The governance of sustainable development: taking stock and looking forwards', *Environment and Planning C* 26(1): 17–33

Jordan A (2009) 'Revisiting the governance of sustainable development: taking stock and looking forwards', *Environment and Planning C* 27(5): 762–5

Jordan, A. and Adelle, C. (eds) (2012) *Environmental Policy in the EU: Actors, Institutions and Processes*, London: Earthscan

Jordan, A., Huitema, D., van Asselt, H., Rayner, T. and Berkout, F. (eds) (2000) *Climate Change Policy in the European Union: Confronting the Dilemmas of Mitigation and Adaptation?,* Cambridge: Cambridge University Press

Joseph Rowntree Foundation (JRF) (1994) *Inquiry into Planning for Housing*, York: JRF

Jowell, J. (1975) 'Development control [review article on the Dobry Report]', *Political Quarterly* 46(3): 340–4

Keeble, L. (1964) *Principles and Practice of Town and Country Planning*, London: Estates Gazette

Keene, Hon. Justice (1999) 'Recent trends in judicial control', *Journal of Planning and Environment Law* January: 30–7

Keith, M. and Rogers, A. (eds) (1991) *Hollow Promises: Rhetoric and Reality in the Inner City*, London: Mansell

Kenny, M. and Meadowcroft, J. (eds) (1999) *Planning Sustainability (Environmental Politics)* London: Routledge

Khan, A. M. (1995) 'Sustainable development: the key concepts, issues and implications', *Sustainable Development*, 3(2): 63–9

Kidd, S. and Ellis, G. (2012) 'Using terrestrial planning to understand the process of marine spatial planning', *Journal of Environmental Policy and Planning* 14(1): 49–67

Killian, J. and Pretty, D. (2008) *Planning Applications: A Faster and More Responsive System: Final Report*, London: DCLG

King, A. and Crewe, I. (2013) *The Blunders of our Governments,* London: Oneworld

Kintrea, K. and Morgan, J. (2005) *Evaluation of English Housing Policy 1975–2000, Theme 3: Housing Quality and Neighbourhood Quality*, London: ODPM

Kirby, A. (1985) 'Nine fallacies of local economic change', *Urban Affairs Quarterly* 21(2): 207–20

Kirkwood, G. and Edwards, M. (1993) 'Affordable housing policy: desirable but unlawful?', *Journal of Planning and Environment Law* 1993: 317–24

Kitchen, T. (1997) *People, Politics, Policies and Plans: The City Planning Process in Contemporary Britain*, London: Paul Chapman

Kitchen, T. (1999) 'Consultation on government policy initiatives: the case of regional planning guidance', *Planning Practice and Research* 14(1): 5–16

Knieling, J. and Filho, W. L. (eds) (2013) *Climate Change Governance*, Berlin: Springer

Knowles, R. and Abrantes, P. (2008) 'Buses and light rail: stalled en route?', in Docherty, I. and Shaw, J., (eds) *Traffic Jam*, Bristol: Policy Press

Knowles, R., Shaw, J. and Docherty, I. (2008) *Transport Geographies*, Oxford: Blackwell

KPMG Consulting (1999) *City Challenge: Final National Evaluation*, London: DETR

Krippendorf, J. (1987) *The Holiday Makers: Understanding The Impact Of Leisure And Travel*, Oxford: Butterworth-Heinemann

Ladd, H. F. (1992) 'Population growth, density and the costs of providing public services', *Urban Studies* 29(2): 273–95

Lafferty, W. and Meadowcroft, J. (1996) *Democracy and the Environment: Problems and Prospects*, Cheltenham: Edward Elgar

Lalenis, K. (2010) 'A theoretical analysis on planning policies', in Tosics, I. (2010) *National Spatial Planning Policies and Governance Typology*, Report 2.2.1 of the PLUREL project of the EU Sixth Framework Programme, available at www.plurel. net/images/D221.pdf (accessed 23 March 2014)

Lambert, C. (2004) 'Local strategic partnerships, community strategies and development planning: negotiating horizontal and vertical integration in an emergent local governance', Paper Presented to the AESOP Conference, Grenoble, July

Lambert, D. (2006) 'The history of the country park, 1966–2005: towards a renaissance?' *Landscape Research* 31(1): 43–62

Lancaster City Council (LCC) (2008) *Local Development Framework Core Strategy 2003–2021 (Adopted July 2008)*, Lancaster: LCC

Land Use Consultants (1995) *Effectiveness of Planning Policy Guidance Notes*, London: HMSO

Land Use Consultants (1999) *Review of National Planning Policy Guidelines*, Edinburgh: Scottish Office

Land Use Consultants (2006) *Review and Evaluation of Heritage Coasts in England*, Cheltenham: Countryside Commission

Land Use Consultants and Wilbraham and Co. (2003) *Formulation of Guidance on the Use of Local Development Orders*, London: ODPM

Larkham, P. J. (1996) *Conservation and the City*, London: Routledge

Larkham, P. J. (2003) 'The place of urban conservation in the UK reconstruction plans of 1942–1952', *Planning Perspectives* 18(3): 295–324

Larkham, P. J. and Chapman, D. W. (1996) 'Article 4 directions and development control: planning myths, present uses, and future possibilities', *Journal of Environmental Planning and Management* 39(1): 5–19

Larkham, P. J. and Lilley, K. D. (2012) 'Exhibiting the city: planning ideas and public involvement in wartime and early post-war Britain,' *Town Planning Review* 83(6): 647–68

Latin, H. (2012) *Climate Change Policy Failures: Why Conventional Mitigation Approaches Cannot Succeed*, Singapore: World Scientific

Laurian, L., Day, M., Backhurst, M., Berke, P., Ericksen, N., Crawford, J., Dixon, J., and Chapman, S. (2004) 'What drives plan implementation? Plans, planning agencies and developers', *Journal of Environmental Planning and Management* 47(4): 555–77

Law Society Gazette (1988) 'Section 52 agreements and planning gain', available at www.lawgazette.co.uk/news/section-52-agreements-and-planning-gain (accessed 1 June 2013)

Lawless, P. (1989) *Britain's Inner Cities* (second edition), London: Paul Chapman

Lawless, P. (1996) 'The inner cities: towards a new agenda', *Town Planning Review* 67(1): 21–43

Lawless, P. and Robinson, D. (2000) 'Inclusive regeneration? Integrating social and economic regeneration in English local authorities', *Town Planning Review* 71(3): 289–310

Layard, A., Batty, S. and Davoudi, S. (2001) *Planning for a Sustainable Future*, London: Spon

Leather, P., Cole, I. and Ferrari, E. (2009) *National Evaluation of Housing Market Renewal Pathfinders 2005–2007*, London: Department for Communities and Local Government

Leather, P., Nevin, B., Cole, I., Eadson, W. (2012) *The Housing Market Renewal Programme in England: Development, Impact and Legacy*, Sheffield: Centre for Regional, Economic and Social Research

Leccese, M. and McCormick, K. (eds) (2000) *Charter of the New Urbanism*, New York: McGraw-Hill

Leeke, M., Sear, C. and Gay, O. (2003) *An Introduction to Devolution in the UK*, Research Paper 03/84, London: House of Commons Library

Lees, A. (1993) *Enquiry into the Planning System in North Cornwall District*, London: HMSO

Lees, L. (2003) 'Visions of "urban renaissance": the Urban Task Force report and the Urban White Paper', in Imrie, R. and Raco, M. (eds) *Urban Renaissance? New Labour, Community and Urban Policy*, Bristol: Policy Press, pp. 61–82

Lees, L. Slater, T. and Wyly, E. (eds) (2010) *The Gentrification Reader*, London: Routledge

LeGates, R. (ed.) (1998) *Early Urban Planning 1870–1940 (9 vols)*, London: Routledge

Leunig, T. and Swaffield, J. (2008) *Cities Unlimited: Making Urban Regeneration Work*, London: Policy Exchange

Levett, R. (2000) 'What counts for quality of life', *EG* 6(3): 6–8

Levitas, R. (1998) *The Inclusive Society? Social Inclusion and New Labour*, Basingstoke: Palgrave

Lewis, N. (1992) *Inner City Regeneration: The Demise of Regional and Local Government*, Buckingham: Open University Press

Lewis, R. P. (1992) 'The Environmental Protection Act 1990: waste management in the 1990s: waste regulation and disposal', *Journal of Planning and Environment Law* 1992: 303–12

Lichfield, N. (2003) *Review of Permitted Development Rights*, London: ODPM

Little, J. (1994a) *Gender, Planning and the Policy Process*, Oxford: Pergamon

Little, J. (1994b) 'Women's initiatives in town planning in England', *Town Planning Review* 65(3): 261–76

Live/Work Network (2012) *Live/Work Business Briefing 2012*, available at www.liveworknet.com (accessed 12 February 2012)

Llewelyn Davies (2000) *Urban Design Compendium*, London: English Partnerships and the Housing Corporation

Llewelyn Davies, CAG Consultants and GHK Economics (2000) *Millennium Villages and Sustainable Communities*, London: DETR

Lloyd, M. G. and Peel, D. (2004) *Evaluation of Revised Planning Controls over Telecommunications Development*, Scottish Executive Research Findings 183/2004

Lloyd, M.G. and Peel, D. (2010) 'National Planning Framework for Scotland 2010–2025', *Planning Theory and Practice*, 11(3): 461–4

Lloyd, M. G. and Peel, D. (2012) 'Planning reform in Northern Ireland: Planning Act (Northern Ireland) 2011', *Planning Theory and Practice* 13(1): 177–82

Lloyd, M. G. and Purves, G. (2009) 'Identity and territory: the creation of a national planning framework for Scotland', in Davoudi, S. and Strange, I. (eds) *Conceptions of Space and Place in Strategic Spatial Planning*, London: Routledge, pp. 71–94

Local Government Association (LGA) (2007) *A Climate of Change: Final report of the LGA Climate Change Commission*, London: LGA

Local Government Association (LGA) (2010) *Place-based Budgets: The Future Governance of Local Public Services,* London: LGA

Local Government Management Board (LGMB) (1995a) *Indicators for Local Agenda 21*, Luton: LGMB

Local Government Management Board (LGMB) (1995b) *Sustainable Settlements: A Guide for Planners, Designers and Developers* (Barton, H., Davis, G. and Guise, R.), Luton: LGMB

Lock, D. (2012) 'The end of inspector's reports that are binding', *Town and Country Planning* 81(2): 58–60

Loftman, P. and Beazley, M. (1998a) *Race, Equality and Planning*, London: Local Government Association

Loftman, P. and Beazley, M. (1998b) 'Racial equality and the planning agenda', *Town and Country Planning* 67(10): 326–7

London Borough of Merton (LBM) (2010) *Local Development Framework Core Strategy* (pre-submission representation publication), London: LBM

London Borough of Redbridge (2011) *Community Infrastructure Plan 2007–2017*, available at www2.redbridge.gov.uk/cms/planning_and_the_environment/planning_policy__regeneration/local_development_framework/community_infrastructure_levy/examination_in_public_eip.aspx (accessed 1 June 2013)

Lord, A. and Hincks, S. (2010) 'Making plans: the role of evidence in England's reformed spatial planning system', *Planning Practice and Research* 25(4): 477–96

Low, N. (1991) *Planning, Politics and the State: Political Foundations of Planning Thought*, London: Unwin Hyman

Lowe, P. and Flynn, A. (1989) 'Environmental politics and policy in the 1980s', in Mohan, J. (ed.) *The Political Geography of Contemporary Britain*, London: Macmillan

Lowe, P. D. (1977) 'Amenity and equity: a review of local environmental pressure groups in Britain', *Environment and Planning A* 9(1): 35–8

Lowndes, V. and Pratchett, L. (2012) 'Local governance under the Coalition government: austerity, localism and the "big society"', *Local Government Studies* 38(1): 21–40

Lucas, K. (2004) *Running on Empty: Transport, Social Exclusion and Environmental Justice*, Bristol: Policy Press

Lutyens, E. and Abercrombie, P. (1945) *A Plan for the City and County of Kingston upon Hull*, Hull: Brown

Lynch, K. (1960) *The Image of the City*, Cambridge, MA: MIT Press

Lyons, M. (2007) *Lyons Enquiry into Local Government. Place-shaping: A Shared Ambition for the Future of Local Government*, London: TSO

McAllister, A. and McMaster, R. (1994) *Scottish Planning Law: An Introduction*, Edinburgh: Butterworths

McAllister, A., McMaster, R., Prior, A. and Watchman, J. (2013) *Scottish Planning Law*, London: Bloomsbury Professional

McAllister, D. (ed.) (1996) *Partnership in the Regeneration of Urban Scotland*, Scottish Office, Edinburgh: HMSO

McAuslan, J. P. W. B. (1991) 'The role of courts and other judicial type bodies in environmental management', *Journal of Environmental Law* 3(2): 195–208

McCarthy, J. (1999) 'Urban regeneration in Scotland: an agenda for the Scottish Parliament', *Regional Studies* 33(6): 559–66

McCarthy, J. and Newlands, D. (eds) (1999) *Governing Scotland: Problems and Prospects – The Economic Impact of the Scottish Parliament*, Aldershot: Ashgate

McCormick, J. (1991) *British Politics and the Environment*, London: Earthscan

McCrone, G. (1991) 'Urban renewal: the Scottish experience', *Urban Studies* 28(6): 919–38

MacDonald, K. M. (1995) *The Sociology of the Professions*, London: Sage

McDonald, N. (2013) *Choice of Assumptions in Forecasting Housing Requirements: Methodological Notes*, Cambridge: CCHPR

Macdonald, S., Normandin, K. and Kindred, B. (eds) (2007) *Conservation of Modern Architecture*, Shaftsbury: Donhead

McEvoy, D., Lindley, S. and Handley, J. (2006) 'Adaptation and mitigation in urban areas: synergies and conflicts', *Municipal Engineer* 159(4): 185–91

MacEwen, A. and MacEwen, M. (1982) *National Parks: Conservation or Cosmetics?*, London: Allen and Unwin

MacGregor, B. and Ross, A. (1995) 'Master or servant? The changing role of the development plan in the British planning system', *Town Planning Review* 66(1): 41–59

MacGregor, S. and Pimlott, B. (eds) (1990) *Tackling the Inner Cities: The 1980s Revisited, Prospects for the 1990s*, Oxford: Clarendon Press

McIntosh, N. (chair) (1999) *Moving Forward: Local Government and the Scottish Parliament*, McIntosh Report, Edinburgh: Scottish Office

MacKinnon, D. and Vigar, G. (2008) 'Devolution and the changing policy spaces of UK transport planning', in Docherty, I. and Shaw, J. (eds) *Traffic Jam*, Bristol: Policy Press

McLaughlin, B. (1986) *Deprivation in Rural Areas*, unpublished study for the Department of the Environment

McLoughlin, J. B. (1969) *Urban and Regional Planning: A Systems Approach*, London: Faber & Faber

Macnaghten, P., Grove-White, R., Jacobs, M. and Wynne, B. (1995) *Public Perceptions and Sustainability in Lancashire*, Report by the Centre for the Study of Environmental Change, Lancaster University, Preston: Lancashire County Council

Macrae, M. (2007) '1967–2007: Milton Keynes: the next 40 years', *Urban Design*, 104: 12–13

Macrory, I. (2012) *Measuring National Well-being – Households and Families, 2012,* Newport: ONS

Mandelker, D. R. (1962) *Green Belts and Urban Growth,* Madison, WI: University of Wisconsin Press

Mandler, P. (1997) *The Fall and Rise of the Stately Home,* London: Yale University Press

Manns, S. and Wood, C. (2001) 'Giving locals a chance to speak', *Planning,* 5 May: 15

Manns, S. and Wood, C. (2002) 'Public representations at development control planning committees', *Public Policy and Administration* 17(1): 39–51

Marks, G., Hooghe, L. and Schakel, A. (2008) 'Patterns of regional authority', *Regional and Federal Studies* 18(2–3): 167–81

Marshall, T. (2013a) 'The remodeling of decision making on major infrastructure in Britain', *Planning Practice and Research* 28(1): 122–40

Marshall, T. (2013b) *Planning Major Infrastructure,* Abingdon: Routledge

Martin, S. (1989) 'New jobs in the inner city: the employment impacts of projects assisted under the urban development grant programme', *Urban Studies* 26(6): 627–38

Martin, S. and Bovaird, T. (2005) *Meta-evaluation of the Local Government Modernisation Agenda: Progress Report on Service Improvement in Local Government,* London: ODPM

Martin, S., Tricker, M. and Bovaird, A. (1990) 'Rural development programmes in theory and practice', *Regional Studies* 24(3): 268–76

Martin, S., Downe, J., Entwistle, T. and Guarneros-Meza, V. (2013) *Learning to Improve: An Independent Assessment of the Welsh Government's Policies for Local Government, 2007–2011,* Cardiff: Welsh Government

Martin and Voorhees Associates (1980) *Review of Rural Settlement Policies 1945–1980,* London: Martin and Voorhees Associates for the Department of the Environment

Mason, R., Osborne, H. and Inman, P. (2013) 'Help to buy will not drive up house prices, says minister', *The Guardian,* 8 October

Mason, T. (1979) 'Politics and planning of urban renewal in the private housing sector', in Jones, C. (ed.) *Urban Deprivation and the Inner City,* London: Croom Helm

Massey, D. (2010) *World City,* Cambridge: Polity Press

Matless, D. (1998) *Landscape and Englishness,* London: Reaktion Books

Maud Committee (1967) *The Management of Local Government,* London: HMSO

Mawson, J. (ed.) (1995) *The Single Regeneration Budget: The Stocktake,* Birmingham: Centre for Urban and Regional Studies, University of Birmingham

Mawson, J. (1996) 'The re-emergence of the regional agenda in the English regions: new patterns of urban and regional governance', *Local Economy* 10(4): 300–26

Meadows, D., Randers, J. and Behrens, W. (1972) *The Limits to Growth: A Report For the Club of Rome's Project on the Predicament of Mankind,* London: Pan

Melia, S. (2012) 'Filtered and unfiltered permeability: the European and Anglo-Saxon approaches', *Project* 4: 6–9

Metz, D. (2013) 'Peak Car and beyond: the fourth era of travel', *Transport Reviews* 33(3): 255–70

Meyerson, G. and Rydin, Y. (1996) 'Sustainable development: the implications of the global debate for land use planning', in Buckingham-Hatfield, S. and Evans, B. (eds) *Environmental Planning and Sustainability,* London: Wiley, pp. 19–34

Miciukiewicz, K. and Vigar, G. (2013) 'Encounters in motion: considerations of time and social justice in urban mobility research', in Henckel, D., Thomaier, S., Könecke, B., Zedda, R., and Stabilini, S. (eds) *Space-Time Design of the Public City,* Dordrecht: Springer, pp.171–85

Midgeley, J. (ed.) (2006) *A New Rural Agenda,* Newcastle: IPPR North

Miller, C. (2000) *Planning and Pollution Revisited: A Background Paper for the Royal Commission on Environmental Pollution Review of Environmental Planning,* London: Royal Commission on Environmental Pollution

Miller, H. (2002) *Postmodern Public Policy,* New York: SUNY Press

Miller, V. (2013) *Voting Behaviour in the EU Council*, Standard Note: SN/IA/6646, London: House of Commons Library

Millichap, D. (1995a) 'Law, myth and community: a reinterpretation of planning's justification and rationale', *Planning Perspectives* 10: 279–93

Millichap, D. (1995b) *The Effective Enforcement of Planning Controls* (second edition), London: Butterworths

Milton, K. (1991) 'Interpreting environmental policy: a social scientific approach', in Churchill, R., Warren, L. M. and Gibson, J. (eds) *Law, Policy and the Environment*, Oxford: Blackwell

Minay, C. L. W. (ed.) (1992) 'Developing regional planning guidance in England and Wales: a review symposium', *Town Planning Review* 63(4): 415–34

Minhinnick, R. (1993) *A Postcard Home,* Llandysul: Gomer Press

Ministry of Agriculture, Fisheries and Food (MAFF) and Department of Environment, Transport and the Regions (DETR) (2000) *Our Countryside, the Future: A Fair Deal for Rural England,* London: TSO

Ministry of Housing and Local Government (MHLG) (1961) *Homes for Today and Tomorrow*, London: HMSO

Minton, A. (2009) *Ground Control: Fear and Happiness in the 21st Century City*, London: Penguin

Montgomery, J. and Thornley, A. (eds) (1990) *Radical Planning Initiatives: New Directions for Urban Planning in the 1990s*, Aldershot: Gower

Moore, V. (2000, 2002) *A Practical Approach to Planning Law* (seventh and eighth editions), London: Blackstone Press

Moore, V. and Purdue, M. (2012) *A Practical Approach to Planning Law,* Oxford: Oxford University Press

Morgan, K. (1995) 'Reviving the valleys? Urban renewal and governance structures in Wales', in Hambleton, R. and Thomas, H. (eds) *Urban Policy Evaluation: Challenge and Change*, London: Paul Chapman

Morphet, J. (2007) *Modern Local Government*, London: Sage

Morphet, J. (2009) 'Local integrated spatial planning – the changing role in England', *Town Planning Review* 80(4–5): 393–414

Morphet, J. (2011) *Effective Practice in Spatial Planning*, London: Routledge

Morris, R. (1998) 'Gypsies and the planning system', *Journal of Planning and Environment Law* July: 635–43

Morris, W. (1877) *Restoration,* reprinted by the Society for the Protection of Ancient Buildings as their Manifesto, London: SPAB

Morrison, N. and Pearce, B. (2000) 'Developing indicators for evaluating the effectiveness of the UK land use planning system', *Town Planning Review* 71(2): 191–211

Morton, A. and Dericks, G. (2013) *21ˢᵗ Century Retail Policy: Quality, Choice, Experience and Convenience*, London: Policy Exchange

Moseley, M. (2003) *Local Rural Development: Principles and Practice,* London: Sage

MTCP (1947) *Advisory Handbook on the Redevelopment of Central Areas*, London: HMSO

Muñoz-Viñas, S. (2005) *Contemporary Theory of Conservation*, Oxford: Elsevier

Murdoch, J. and Abram, S. (2002) *Rationalities of Planning: Development versus Environment in Planning for Housing,* Farnham: Ashgate

Murray, M. (2009) 'Building consensus in contested spaces and places? The Regional Development Strategy for Northern Ireland', in Davoudi, S. and Strange, I. (eds) *Conceptions of Space and Place in Strategic Spatial Planning*, London: Routledge, pp. 125–47

Mynors, C. (1992) *Planning Control and the Display of Advertisements*, London: Sweet & Maxwell

Mynors, C. (1994) 'Planning and the historic environment: the final version of PPG 15.' *Context* 44: 20–2

Mynors, C. (2006) *Listed Buildings, Conservation Areas and Monuments*, Andover: Sweet & Maxwell

Nadin, V. (1999) 'British planning in its European context', in Cullingworth, B. (1999) *British Planning: 50 Years of Urban and Regional Policy*, London: Athlone

Nadin, V. (2007) 'The emergence of the spatial planning approach in England', *Planning Practice and Research* 22(1): 43–62

Nadin, V. and Dühr, S. (2005) 'Some help with Euro-planning jargon', *Town and Country Planning* 74(3): 82–4

Nadin, V. and Shaw, D. (1999) *Subsidiarity and Proportionality in Spatial Planning Activities in the European Union*, London: DETR

Nadin, V., Cooper, S., Shaw, D., Westlake, T. and Hawkes, P. (1997) *The EU Compendium of Spatial Planning Systems and Policies*, Luxembourg: Office for the Official Publications of the European Communities

Nairn, I. (1955) *Outrage,* London: Architectural Press

Nairn, I. (1957) *Counter Attack against Subtopia,* London: Architectural Press.

Nathan, M. and Overman, H. (2011) *What We Know (and Don't Know) about the Links between Planning and Economic Performance*, Policy Paper 10, London: Spatial Economics Research Centre, LSE

National Adaptation Programme (NAP) (2013) *The National Adaptation Programme: Making the Country Resilient to a Changing Climate 2013*, London: TSO

National Audit Office (NAO) (1988) *Urban Development Corporations*, London: HMSO

National Audit Office (NAO) (1990) *Regenerating the Inner Cities: Report by the Comptroller and Auditor General*, London: HMSO

National Audit Office (NAO) (1993) *Regenerating the Inner Cities*, London: HMSO

National Audit Office (NAO) (2007) *Housing Market Renewal: Report by the Comptroller and Auditor General*, London: TSO

National Audit Office (NAO) (2010) *Regenerating the English Regions: Regional Development Agencies' Support to Physical Regeneration Projects*, London: TSO

National Audit Office (NAO) (2011) *Case Study on Integration: Measuring the Costs and Benefits of Whole-place Community Budgets,* London: DCLG

National Audit Office (NAO) (2012) *Funding for Local Transport: An Overview*, London: NAO

National Committee for Commonwealth Immigrants (NCCI) (1967) *Areas of Special Housing Need*, London: NCCI

National Housing and Planning Advice Unit (NHPAU) (2009) *Review of European Planning Systems*, Research for the National Housing and Planning Advice Unit by De Montfort University, available at www.dora.dmu.ac.uk/bitstream/handle/2086/7536/NHPAU %20Planning.pdf?sequence=1 (accessed 3 April 2014)

National Housing Federation (2009) *A Place in the Country: An Inquiry into the North's Rural Housing Challenges*, Manchester: NHF North

Natural England (2008a) *Guide to Definitive Maps and Changes to Public Rights of Way*, Sheffield: Natural England

Natural England (2008b) *The State of the Natural Environment*, Sheffield: Natural England

Natural England (2013) *Monitor of Engagement with the Natural Environment: The National Survey on People and the Natural Environment, Annual Report from the 2012–13 survey*, Sheffield: Natural England

Natural England and CPRE (2010) *Green Belts: A Greener Future*, London: CPRE

Navigant (2011) *From Self-Financing to Self-Determination*, available at www.londoncouncils. gov.uk/policylobbying/housing/finance/hrareformcontext.htm (accessed 11 February 2014)

Nelsen, B. F. and Stubb, A. C.-G. (2003) *The European Union: Readings on the Theory and Practice of European Integration* (third edition), Basingstoke: Lynne Riener

Nevin, B. (2002) *Securing Housing Market Renewal: A Submission to the Comprehensive Spending Review*, Birmingham: Centre for Urban and Regional Studies

Nevin, B. and Leather, P. (2011) 'The bonus culture', *Town and Country Planning* 80(1): 47–8

Nevin, B., Lee, P., Goodson, L., Murie, A. and Phillimore, J. (2001) *Changing Housing Markets and Urban Regeneration in the M62 Corridor*, Birmingham: Centre for Urban and Regional Studies

Newbery, D. (1997) 'Privatisation and liberalisation of network utilities', *European Economic Review* 41(3–5): 357–83

Newby, H. (1985) *Green and Pleasant Land: Social Change in Rural England* (second edition), London: Wildwood House

Newby, H. (1990) 'Ecology, amenity and society: social science and environmental change', *Town Planning Review* 61(1): 3–13

New Economics Foundation (NEF) (2005) *Clone Town Britain,* London: NEF

Newman, I. (2005) *Parish and Town Councils and Neighbourhood Governance,* York: Joseph Rowntree Foundation

Newman, O. (1973) *Defensible Space: Crime Prevention Through Urban Design (CPTUD),* New York: Macmillan

Newman, P. W. G. and Kenworthy, J. R. (1999) *Sustainability and Cities: Overcoming Automobile Dependence,* Washington, DC: Island Press

Niner, P. (1999) *Insights into Low Demand for Housing,* Foundations Report 739, York: Joseph Rowntree Foundation

Niner, P. (2002) *The Provision and Condition of Local Authority Gypsy and Traveller Sites in the English Countryside,* London: ODPM

Niner, P. (2004) 'Accommodating nomadism: an examination of accommodation options for Gypsies and travellers in England', *Housing Studies* 19(2): 141–59

Nolan, Lord (chair) (1997) *Third Report of the Committee on Standards in Public Life: Standards of Conduct in Local Government,* Cm 3702, Nolan Report 3, London: TSO

Norris, S. (1996) *Changing Trains,* London: Hutchison

North, D., Syrett, S. and Etherington, D. (2007) *Devolution and Regional Governance. Tackling the Economic Needs of Deprived Areas: Main Report,* York: Joseph Rowntree Foundation

Northampton Borough Council (2013) *Planning for the Future: Documents, Studies and Publications,* available at www.northampton.gov.uk/info/200205/planning_for_the_future/1739/ (accessed 18 March 2014)

Northern Way (2007) *Strategic Direction for Transport,* Northern Way: Newcastle

Northern Way (2010) *Meeting the Economic Challenge: Delivering the Northern Way's Transport Priorities,* Northern Way: Leeds

Norwich City Council (2012) *Local Development Order for Replacement Windows and Doors in Flats,* available at www.norwich.gov.uk/Planning/Pages/Planning-LocalDevelopmentOrder.aspx (accessed 23 May 2013)

Nuffield Foundation (1986) *Town and Country Planning,* Nuffield Report, London: Nuffield Foundation

Nugent, N. (1999) *The Government and Politics of the European Community* (fourth edition), London: Macmillan

Oatley, N. (1995) 'Competitive urban policy and the regeneration game', *Town Planning Review* 66(1): 1–14

Oatley, N. (ed.) (1998) *Cities, Economic Competition and Urban Policy,* London: Paul Chapman

Oatley, N. (2000) 'New Labour's approach to age-old problems: renewing and revitalising poor neighbourhoods – the national strategy for neighbourhood renewal', *Local Economy* 15(2): 86–97

Oatley, N. and Lambert, C. (1995) 'Evaluating competitive urban policy: the City Challenge initiative', in Hambleton, R. and Thomas, H. (eds) *Urban Policy Evaluation: Challenge and Change,* London: Paul Chapman, pp. 141–57

Oc, T. and Tiesdell, S. (1991) 'The London Docklands Development Corporation 1981–1991: a perspective on the management of urban regeneration', *Town Planning Review* 62(3): 311–30

Oc, T., Tiesdell, S. and Moynihan, D. (1997) 'The death and life of City Challenge: the potential for lasting impacts in a limited-life urban regeneration initiative', *Planning Practice and Research* 12(4): 367–81

Office for Budget Responsibility (OBR) (2011) *Briefing Paper No. 3: Forecasting the Economy,* London: OBR

Office for National Statistics (ONS) (2005) *Focus on People and Migration,* Basingstoke: Palgrave Macmillan

Office for National Statistics (ONS) (2008) *Family Spending,* Basingstoke: Palgrave McMillan

Office for National Statistics (ONS) (2012a) *Supply Side of Tourism Report 2009,* Newport: ONS

Office for National Statistics (ONS) (2012b) *The Geography of Tourism Employment* Newport: ONS

Office for National Statistics (ONS) (2013a) *Employment Characteristics of Tourism Industries 2011,* Newport: ONS

Office for National Statistics (ONS) (2013b) *A Century of Home Ownership and Renting in England and Wales,* Newport: ONS

Office of the Deputy Prime Minister (ODPM) (2002) *Living Places: Cleaner, Safer, Greener*, London: ODPM

Office of the Deputy Prime Minister (ODPM) (2003a) *Sustainable Communities: Building for the Future,* London: ODPM

Office of the Deputy Prime Minister (ODPM) (2003b) *Better Streets, Better Places,* London: DCLG

Office of the Deputy Prime Minister (ODPM) (2004a) *The Planning Response to Climate Change: Advice on Better Practice,* London: ODPM

Office of the Deputy Prime Minister (ODPM) (2004b) *Community Involvement in Planning: The Government's Objectives,* London: ODPM

Office of the Deputy Prime Minister (ODPM) (2004c) *Employment Land Reviews: Guidance Note,* London: ODPM

Oliver, D. and Waite, A. (1989) 'Controlling neighbourhood noise: a new approach', *Journal of Environment Law* 1(2): 173–91

O'Neill, J. (1999) 'Moving and shaking: inquiries and hearings', *Planning Inspectorate Journal*, 16 (summer): 18–20

Orbasli, A. (2008) *Architectural Conservation*, Oxford: Blackwell

Organisation for Economic Cooperation and Development (OECD) (2008) *Sustainability Impact Assessment: Definition, Approaches and Objectives*, available at www.oecd.org/greengrowth/39924538. pdf (accessed April 2014)

Organisation for Economic Cooperation and Development (OECD) (2010) *Guidance on Sustainability Impact Assessment*, available at www. oecd.org/greengrowth/46530443.pdf (accessed April 2014)

Organisation for Economic Cooperation and Development (OECD) (2011a) *OECD Rural Policy Reviews: England, United Kingdom,* Paris: OECD Publishing

Organisation for Economic Cooperation and Development (OECD) (2011b) *Economic Surveys United Kingdom March 2011 Overview*, available at www.oecd.org/eco/surveys/47319830.pdf (accessed 8 July 2014)

Organisation for Economic Cooperation and Development (OECD) (2013) *OECD Better Life Index*, available at www.oecdbetterlifeindex.org (accessed 20 April 2014)

O'Riordan, T. (1985) 'What does sustainability really mean? Theory and development of concepts of sustainability', in *Sustainable Development in an Industrial Economy: Proceedings of a Conference*, Cambridge: UK Centre for Economic and Environmental Development, Queen's College, Cambridge

O'Riordan, T. and Weale, A. (1989) 'Administrative reorganization and policy change: the case of Her Majesty's Inspectorate of Pollution', *Public Administration* 67(3): 277–94

Osmond, J. (1977) *Creative Conflict: The Politics of Welsh Devolution*, London: Routledge

Ove Arup (2010) *Planning Costs and Fees: Final Report,* London: DCLG

Ove Arup and Partners, Regional Forecasts and Oxford Economic Forecasting (2005) *Regional Futures: England's Regions in 2030*, English Regions Network

Owen, S. (2002) 'From Village Design Statements to Parish Plans: some pointers towards community decision making in the planning system in England', *Planning Practice and Research* 17(1): 81–9

Owens, S. (1994) 'Land, limits and sustainability: a conceptual framework and some dilemmas for the planning system', *Transactions of the Institute of British Geographers*, 19(4): 439–56

Owens, S. (1995) 'Land use planning as an instrument of sustainable development', *The Globe* 3: 6–8

Owens, S. and Cowell, R. (1996) *Rocks and Hard Places: Mineral Resource Planning and Sustainability*, London: Council for the Protection of Rural England

Owens, S. and Cowell, R. (2002) *Land and Limits: Interpreting Sustainability in the Planning Process*, London: Routledge

Owens, S. and Cowell, R. (2011) *Land and Limits* (second edition), Abingdon: Routledge

Pacione, M. (1997) 'The urban challenge: how to bridge the great divide', in Pacione, M. (ed.) *Britain's Cities: Geographies of Division in Urban Britain*, London: Routledge

Pahl, R. (1978) 'Will the inner city problem ever go away?' *New Society* 45(834): 678–81

Pantazis, C. and Gordon, D. (eds) (2000) *Tackling Inequalities: Where Are We Now and What Can Be Done?*, Bristol: Policy Press

Paris, C. (ed.) *Critical Readings in Planning Theory*, Oxford: Pergamon

Parker, G. (2008) 'Parish and community-led planning, local empowerment and local evidence bases: an examination of "good practice"', *Town Planning Review* 79(1): 61–85

Parker, G. and Murray, C. (2012) 'Beyond tokenism? Community-led planning and rational choices: findings from participants in local agenda-setting at the neighbourhood scale in England', *Town Planning Review* 83(1): 1–28

Parkinson, M. and Boddy, M. (2004) 'Introduction', in Boddy, M. and Parkinson, M. (eds) *City Matters: Competitiveness, Cohesion and Urban Governance*, Bristol: Policy Press

Parkinson, M. and Robson, B. (2000) *Urban Regeneration Companies: A Process Evaluation*, London: DETR, available at www.ljmu.ac.uk/EIUA/67773.htm (accessed 24 March 2014)

Parkinson, M., Hutchins, M., Simmie, J., Clark, G. and Verdonk, H. (2004) *Competitive European Cities: Where do the Core Cities Stand?*, London: ODPM

Parkinson, M., Hutchins, M., Champion, T., Coombes, M., Dorling, D., Parks, A., Simmie, J. and Turok, I. (2005) *State of the Cities: A Progress Report to the Delivering Sustainable Communities Summit*, London: ODPM

Pattison, G. (2004) 'Planning for decline: the "D"-village policy of County Durham, UK', *Planning Perspectives* 19(3): 311–32

Pawley, M. (1998) *Terminal Architecture*, London: Reaktion Books

Pearce, D., Markandya, A. and Barbier, E. (1989) *Blueprint for a Green Economy: Report for the Department of the Environment*, London: Earthscan

Pelling, M. 2010, *Adaptation to Climate Change*, London: Routledge

Pendall, R., Foster, K. and Cowell, M. (2010) 'Resilience and regions: building understanding of the metaphor', *Cambridge Journal of Regions, Economy and Society* 3(1): 71–84

Pendlebury, J. (1999) 'The conservation of historic areas in the UK: a case study of Newcastle upon Tyne', *Cities* 16(6): 423–34

Pendlebury, J. (2000) 'Conservation, Conservatives and consensus: the success of conservation under the Thatcher and Major governments, 1979–1997', *Planning Theory and Practice* 1(1): 31–52

Pendlebury, J. (2003) 'Planning the historic city: 1940s reconstruction plans in Britain', *Town Planning Review* 74(4): 371–93

Pendlebury, J. (2009) *Conservation in the Age of Consensus*, London: Routledge

Pendlebury, J., Short, M. and While, A. S. (2009) 'Urban World Heritage Sites and the problem of authenticity', *Cities* 26(6): 349–58

Pennington, M. (2002) *Liberating the Land,* London: IEA

Performance and Innovation Unit (PIU) (1999) *Rural Economies*, London: Cabinet Office

Pickard, R. D. (1996) *Conservation in the Built Environment,* Harlow: Longman

Pickles, E. (2010) 'The bonfire of local government targets', *Conservative Home*, available at www.conservativehome.com/localgovernment/2010/10/the-bonfire-of-local-government-targets.html (accessed 31 October 2013)

Pickvance, C. (1982) 'Physical planning and market forces in urban development', in Paris, C. (ed.) *Critical Readings in Planning Theory*, Oxford: Pergamon

Pitt, M. (2008) *Learning Lessons from the 2007 Floods*, Pitt Review, available at http://webarchive.nationalarchives.gov.uk/20100807034701/http:/archive.cabinetoffice.gov.uk/pittreview/thepittreview.html (accessed 5 July 2014)

Pittock, A. B. (2009) *Climate Change: The Science, Impacts and Solutions*, London: Earthscan

Pizarro, R. (2009) 'Urban form and climate change: towards appropriate development patterns to mitigate and adapt to global warming', in Davoudi, S., Crawford, J. and Mehmood, A. (eds) *Planning for Climate Change*, London: Earthscan, pp. 33–46

Planning (2011) 'Minister charges Trust with duping public over reforms', 26 August, p. 5

Planning (2013a) 'Referenda back neighbourhood plans', 17 May, pp. 8–9

Planning (2013b) 'Poorer areas see few local plan applications', 25 March, pp. 4–5

Planning Advisory Group (PAG) (1965) *The Future of Development Plans*, London: HMSO

Planning Advisory Service (PAS) (2008) *Development Management: Guidance and Discussion Document*, London: PAS

Planning Advisory Service (PAS) (2011) *The Culture of Development Management*, available at www.pas.gov.uk/pas/core/page.do?pageId=323621 (accessed 7 May 2013)

Planning Advisory Service (PAS) (2013a) *Probity in Planning for Councillors and Officers*, London: LGA

Planning Advisory Service (PAS) (2013b) *Ten Key Principles for Owning Your Housing Number: Finding Your Objectively Assessed Need*, London: LGA

Planning Exchange (1989) *Evaluation of the Use and Effectiveness of Planning Publications*, Edinburgh: Scottish Development Department

Planning Inspectorate (PINS) (2012) *Planning Inspectorate Good Practice Advice Note 02: The Householder Appeals Service (HAS)*, Bristol: PINS

Planning Practice and Research (2013) 'Deconstructing resilience: lessons from planning practice', *Special Issue*, 28(3)

Planning Theory and Practice (2005) 'Interface: territorial cohesion', 6(3): 387–409

Pooley, C. G., Jones, T., Tight, M., Horton, D., Scheldeman, G., Mullen, C., Jopson, A. and Strano, E. (2013) *Promoting Walking and Cycling*, Bristol: Policy Press

Portas M (2011) *The Portas Review: An Independent Review into the Future of our High Streets*, London: BIS

Porter, L. and Shaw, K. (2009) *Whose Urban Renaissance?* London: Routledge

Porter, M. E. (1996) 'What is strategy?', *Harvard Business Review* 74(6): 61–78

Potter, S. and Bailey, I. (2008) 'Transport and the environment', in Knowles, R., Shaw, J. and Docherty, I. (eds) *Transport Geographies*, Oxford: Blackwell

Poulton, M. (1991) 'The case for a positive theory of planning' (in two parts), *Environment and Planning B* 18(2): 225–32; 18(3): 263–75

Powe, N. and Hart, T. (2008) 'Market towns: understanding and maintaining functionality', *Town Planning Review* 79(4), 1–14

Powe, N. and Whitby, M. (1994) 'Economics of settlement size in rural settlement planning', *Town Planning Review* 65(4): 415–34

Powe, N., Hart, T. and Shaw, T. (eds) (2007) *Market Towns: Roles, Challenges and Prospects*, Abingdon: Routledge

Poxon, J. (2000) 'Solving the development plan puzzle in Britain: learning lessons from history', *Planning Perspectives* 15(1): 73–89

Pressman, J. L. and Wildavsky, A. (1973) *Implementation: How Great Expectations in Washington are Dashed in Oakland*, Los Angeles: University of California Press

Preston, J. (2012) 'High speed rail in Britain: about time or a waste of time?', *Journal of Transport Geography* 22: 308–11

Price Waterhouse (1993) *Evaluation of Urban Development Grant, Urban Regeneration Grant, and City Grant* (DoE), London: HMSO

Prins, G., Galiana, I., Green, C., Grundmann, R., Korhola, A., Laird, F., Nordhaus, T., Pielke, Jnr, R., Rayner, S., Sarewitz, D., Shellenberger, M., Stehr, N. and Tezuko, H. (2010) *The Hartwell Paper: A New Direction for Climate Policy after the Crash of 2009*, Institute for Science, Innovation and Society, University of Oxford; LSE Mackinder Programme, London School of Economics and Political Science, London, UK; available at http://eprints.lse.ac.uk/27939/ (accessed 3 April 2014)

Property Advisory Group (1981) *Planning Gain*, London: HMSO

Punter, J. (2009) *Urban Design and the Britich Urban Renaissance*, London: Routledge

Punter, J. (2010) 'Planning and good design: indivisible or invisible? A century of design regulation in English town and country planning', *Town Planning Review* 81(4): 343–79

Punter, J. and Carmona, M. (1997) *The Design Dimension of Planning*, London: Spon

Punter, K. (2010) 'Urban design and the English urban renaissance 1999–2009: a review and preliminary evaluation', *Journal of Urban Design* 16: 1–41

Purcar, C. (2007) 'Designing the space of transportation: railway planning theory in nineteenth and early twentieth century treatises', *Planning Perspectives* 22(3): 325–52

Purdue, M. (1991) 'Green belts and the presumption in favour of development', *Journal of Environmental Law* 3(1): 93–121

Purdue, M. (1999) 'The changing role of the courts in planning', in Cullingworth, J. B. (ed.) *British Planning: 50 Years of Urban and Regional Policy*, London: Athlone

Qui, W. and Jones, P. J .S. (2011) *The Emerging Policy Landscape for Marine Spatial Planning in Europe*, UCL Department of Geography Working Paper (Version 2), available at www.homepages.ucl.ac.uk/~ ucfwpej/pdf/EPLMSPEU.pdf (accessed 8 July 2014)

Raco, M. (2007) *Building Sustainable Communities: Spatial Policy and Labour Mobility in Post-war Britain,* Bristol: Policy Press

Raco, M., Turok, I. and Kintrea, K. (2003) 'Local development companies and the regeneration of Britain's cities', *Environment and Planning C* 21(2): 277–304

Raemaekers, J. (1995) 'Scots have way to go on strategic waste planning', *Planning* 1106 (17 February): 24–5

Raemaekers, J., Prior, A. and Boyack, S. (1994) *Planning Guidance for Scotland: A Review of the Emerging New Scottish National Planning Policy Guidelines*, Edinburgh: Royal Town Planning Institute in Scotland

Ramidus Consulting Limited (2012) *London Office Policy Review 2012*, London: Greater London Authority

Rao, N. (1990) *The Changing Face of Housing Authorities*, London: Policy Studies Institute

Ratcliffe, P. (1998) 'Planning for diversity and change: implications of polyethnic society', *Planning Practice and Research* 13(4): 359–69

Ravenscroft, N. (2000) 'The vitality and viability of town centres', *Urban Studies* 37(13): 2533–49

Ravetz, A. (1980) *Remaking Cities: Contradictions of the Recent Urban Environment*, London: Croom Helm

Read, L. and Wood, M. (1994) 'Policy, law and practice', in Wood, M. (ed.) *Planning Icons: Myth and Practice* (Planning Law Conference, *Journal of Planning and Environment Law*), London: Sweet & Maxwell

Reade, E. (1987) *British Town and Country Planning*, Milton Keynes: Open University Press

Rees, P. H., Durham, H. and Kupiszewski, M. (1996) 'Internal migration and regional population dynamics in Europe: United Kingdom case study', Working Paper 96/20, Leeds: School of Geography, University of Leeds

Reeve, A. and Shipley, R. (2012–13) 'Assessing the effectiveness of the Townscape Heritage Initiative in regeneration', *Journal of Urban Regeneration and Renewal* 6(2): 189–210

Reeves, D. (2005) *Planning for Diversity: Policy and Planning in a World of Difference*, London: Routledge

Rhodes, J., Tyler, P., Brennan, A., Stevens, S., Warnock, C. and Otero-Garcia, M. (2002) *Lessons and Evaluation Evidence from Ten Single Regeneration Budget Case Studies: Mid-term Report*, London: ODPM

Rhodes, R. A. W. (2011) *Everyday Life in British Government,* Oxford: Oxford University Press

Rhodes, R. A. W. (2013) 'Political anthropology and civil service reform: prospects and limits', *Policy and Politics* 41(4): 481–96

Rice, D. (1998) 'National Parks for Scotland: a major step forward?', *Town and Country Planning* 67: 159–61

Richards, J. (1994) *Facadism*, London: Routledge

Richardson, T. and Jenson, O. B. (1999) *North*, 6: 5, available at www.nordregio.se (accessed 10 July 2002)

Ricketts, S. and Field, D. (2012) *Localism and Planning*, London: Bloomsbury

Riddell, P. (1991) *The Thatcher Era and its Legacy*, Oxford: Blackwell

Rittel, H. W. J. and Webber, M. M. (1973) 'Dilemmas in a general theory of planning', *Policy Sciences* 4(2): 155–69

Robert, J., Stumm, T., de Vet, J., Reincke, C. J., Hollanders, M. and Figueiredo, M. A. (2001) *Spatial Impacts of Community Policies and Costs of*

Non-coordination, Study carried out at the request of the European Commission DG REGIO, Brussels: CEC

Roberts, J. (1995) 'Historic parks and gardens: listing, awareness and the future', *Journal of Architectural Conservation* 1(1): 38–55

Roberts, M. and Greed, C. (eds) (2001) *Approaching Urban Design,* Harlow: Longman

Roberts, P. (1996) 'Regional planning guidance in England and Wales: back to the future?', *Town Planning Review* 67(1): 97–109

Roberts, P. and Hart, T. (1997) 'The design and implementation of European programmes for regional development in the UK: a comparative review', in Bachtler, J. and Turok, I. (eds) *The Coherence of EU Regional Policy: Contrasting Perspectives on the Structural Funds,* London: Jessica Kingsley/ Regional Studies Association

Roberts, P. and Lloyd, G. (1999) 'Institutional aspects of regional planning, management and development: models and lessons from the English experience', *Environment and Planning B* 26(4): 517–31

Roberts, P. and Sykes, H. (1999) *Urban Regeneration: A Handbook*, London: Sage

Roberts, P., Hart, T. and Thomas, K. (1993) *Europe: A Handbook for Local Authorities,* Manchester: CLES

Roberts, T. (1998) 'The statutory system of town planning in the UK: a call for detailed reform', *Town Planning Review* 69(1): iii–vii

Robins, N. (1991) *A European Environment Charter*, London: Fabian Society

Robinson, M. (1992) *The Greening of British Party Politics*, Manchester: Manchester University Press

Robson, B. (1988) *Those Inner Cities: Reconciling the Social and Economic Aims of Urban Policy*, Oxford: Clarendon Press

Robson, B., Parkinson, M., Boddy, M. and Maclennan, D. (2000) *The State of English Cities*, London: DETR

Robson, B., Bradford, M. G., Deas, I., Hall, E., Harrison, E., Parkinson, M., Evans, R., Garside, P. and Harding, A. (1994) *Assessing the Impact of Urban Policy* (DoE), London: HMSO

Rodwell, D. (2007) *Conservation and Sustainability in Historic Cities*, Oxford: Blackwell

Roger Tym and Partners (1995) *The Use of Article 4 Directions* (DoE), London: HMSO

Roger Tym and Partners (1998) *Urban Development Corporations: Performance and Good Practice*, London: DETR

Rogers, S. (2013) 'The England cuts map: what's happening to each local authority and council', *The Guardian*, 11 January, available at www.theguardian. com/news/datablog/interactive/2012/nov/14/local-authority-cuts-map (accessed 24 March 2014)

Romero-Lankao, P. (2007) 'Are we missing the point? Particularities of urbanization, sustainability and carbon emission in Latin American cities', *Environment and Urbanization* 19(1): 159–75

Rowan-Robinson, J. and Durman, R. (1992) *Section 50 Agreements*, Edinburgh: Central Research Unit, Scottish Office

Rowan-Robinson, J. and Lloyd, M. G. (1991) 'National planning guidelines: a strategic opportunity wasting', *Planning Practice and Research* 6(3): 16–19

Royal Commission on Environmental Pollution (RCEP) (2002) *Environmental Planning*, Norwich: TSO

Royal Commission on Environmental Pollution (RCEP) (2007) *The Urban Environment*, Norwich: TSO

Royal Institute of British Architects (RIBA) (2011) *The Case for Space: The Size of England's New Homes,* London: RIBA

Royal Institution of Chartered Surveyors (RICS) (1991) *Britain's Environmental Strategy: A Response by the RICS to the White Paper 'This Common Inheritance'*, London: RICS

Royal Town Planning Institute (RTPI) (1987) *Report and Recommendations of the Working Party on Women in Planning,* London: RTPI

Royal Town Planning Institute (RTPI) (1988) *Planning for Choice and Opportunity*, London: RTPI

Royal Town Planning Institute (RTPI) (1999) *Radical Review of the Development Plan System in England: Process and Procedures,* London: RTPI

Royal Town Planning Institute (RTPI) (2000) *A New Vision for Planning: Delivering Sustainable Communities, Settlements and Places – Mediating Space, Creating Place*, London: RTPI

Royal Town Planning Institute (RTPI) (2001) *A New Vision for Planning: Delivering Sustainable Communities, Settlements and Places: 'Mediating Space – Creating Place': the Need for Action*, London: RTPI

Royal Town Planning Institute (RTPI) (2003) *Gender Mainstreaming Toolkit*, available at www.rtpi.org.uk/ (accessed 14 April 2014)

Royal Town Planning Institute (RTPI) (2004) *A Manifesto for Planning*, London: RTPI

Royal Town Planning Institute (RTPI) (2007) 'Political language and the debate around the green belt', available at www.rtpi.org.uk/briefing-room/news-releases/2007/september/political-language-and-the-debate-around-the-green-belt/ (accessed 16 December 2013)

Royal Town Planning Institute (RTPI) (2013) *The Assessment of Professional Competence: The Route from Associate to Chartered Membership,* available at www.rtpi.org.uk/media/529679/apc_associate_-_guidance_2013_-_copy_-_copy.pdf (accessed 30 March 2014)

Royal Town Planning Institute (RTPI) (2014) *10 Commitments for Effective Pre-application Engagement*, London: RTPI

Royal Town Planning Institute and Commission for Racial Equality (RTPI and CRE) (1983) *Planning for a Multi-Racial Britain*, London: RTPI

Ruiz, Y. and Walling, A. (2005) 'Home-based working using communication technologies', *Labour Market Trends* 113(10): 417–26

Rural Development Commission (RDC) (1997) *The Economic Impact of Recreation and Tourism in the English Countryside,* Salisbury: RDC

Rural Development Commission (RDC) (1998) *Rural Development and Land Use Policies*, Salisbury: RDC

Russel, D., Turnpenny, J. and Rayner, T. (2013) 'Reining in the executive? Delegation, evidence, and parliamentary influence on environmental public policy', *Environment and Planning C* 31(4): 619–32

Russell, H. (2001) *Local Strategic Partnerships: Lessons from New Commitment to Regeneration*, Bristol: Policy Press

Russell, H., Dawson, J., Garside, P. and Parkinson, M. (1996) *City Challenge: Interim National Evaluation*, London: HMSO

Russell, S. (2000) 'Environmental appraisal of development plans', *Town Planning Review* 71(2): 529–46

Rydin, Y. (1999) 'Public participation in planning', in Cullingworth, J. B. (ed.) *British Planning: 50 Years of Urban and Regional Policy*, London: Athlone

Rydin, Y. (2003) *Urban and Environmental Planning in the UK*, Basingstoke: Palgrave MacMillan

Rydin, Y. (2009) 'Sustainable construction and design in UK planning', in Davoudi, S., Crawford, J. and Mehmood, A. (eds), *Planning for Climate Change*, London: Earthscan, pp. 181–91

Rydin, Y. (2010) *Governing for Sustainable Urban Development*, London: Earthscan

Rydin, Y. (2011) *The Purpose of Planning: Creating Sustainable Towns and Cities*, Bristol: Policy Press

Rydin, Y. (2013) *The Future of Planning: Beyond Growth Dependence*, Bristol: Policy Press

Rydin, Y., Home, R. and Taylor, K. (1990) *Making the Most of the Planning Appeals System: Report to the Association of District Councils*, London: Association of District Councils

Sadun, R. (2013) *Does Planning Regulation Protect Independent Retailers?* Working Paper 12-044, Boston: Harvard Business School

Sandford, Lord (chair) (1974) *Report of the National Park Policies Review Committee*, Sandford Report, London: HMSO

Satterthwaite, D. (1999) *The Earthscan Reader in Sustainable Cities* (Earthscan Reader Series), London: Earthscan

Satterthwaite, D. (2008) 'Cities' contribution to global warming: notes on the allocation of greenhouse gas emissions', *Environment and Urbanization* 20(2): 539–49

Satterthwaite, D., Huq, S., Pelling, M., Reid, A. and Romero-Lankao, P. (2007) *Building Climate Change Resilience in Urban Areas and Among Urban Populations in Low- and Middle-income Nations*, commissioned report for the Rockefeller Foundation, International Institute for Environment and Development (IIED) Research Report 112

Saunders, M. (1996) 'The conservation of buildings in Britain since the Second World War', in Marks, S.

Concerning Buildings, Oxford: Architectural Press, pp. 5–33

Saunders, M. (2002) *The Role of the Amenity Societies*, paper presented at the Planning and the Historic Environment Conference, Oxford, May 2002

SAVE Britain's Heritage (1978) *Preservation Pays: Tourism and the Economic Benefits of Conserving Historic Buildings*, London: SAVE

Scott, A. (1999) 'Whose futures? A comparative study of Local Agenda 21 in mid Wales', *Planning Practice and Research* 14(4): 401–21

Scott, L. F. (chair) (1942) *Report of the Committee on Land Utilisation in Rural Areas*, Cm 6378, Scott Report, London: HMSO

Scottish Government (2011) *Scottish House Condition Survey*, available at www.scotland.gov.uk/Publications/2012/12/4995/downloads (accessed 10 December 2013)

Scottish Government (2014) *Our Place in Time: The Historic Environment Strategy for Scotland*, Edinburgh: Scottish Government

Scottish Government and COSLA (2007) *Scottish Budget Spending Review 2007: Concordat Between the Scottish Government and Local Government,* Edinburgh: Scottish Government and COSLA

Scrase, T. (1999) 'The judicial review of local planning authority decisions: taking stock', *Journal of Planning and Environment Law* August: 679–90

Secretary of State for Communities and Local Government (2012) *Government response to the House of Commons Communities and Local Government Committee Report of Session 2010–2012: Regeneration*, Norwich: TSO

Seebohm, Lord (chair) (1968) *Report of the Committee on Local Authority and Allied Personal Social Services,* Cmnd 3703, Seebohm Report, London: HMSO

Segal Quince Wicksteed (1996) *The Impact of Tourism on Rural Settlements*, Salisbury: Rural Development Commission

Selman, P. and Wragg, A. (1999) 'Local sustainability planning: from interest-driven networks to vision-driven super-networks?', *Planning Practice and Research* 14(3): 329–40

Semple, J. (2007) *Review into Affordable Housing,* Belfast: Department of Social Development

Sharp, T. (1935) *A Derelict Area: A Study of the South West Durham Coalfield*, London: Hogarth Press

Sharp, T. (1940, 1945) *Town Planning*, Harmondsworth: Penguin

Sharp, T. (1947) *Exeter Phoenix*, London: Architectural Press

Sharp, T. (1948) *Oxford Replanned*, London: Architectural Press

Sharp, T. (1968) *Town and Townscape*, London: John Murray

Sharp, T., Gibberd, F. and Holford, W. G. (1953) *Design in Town and Village,* London: HMSO

Shaw, D. and Lord, A. (2009) 'From land-use to "spatial planning": reflections on the reform of the English planning system', *Town Planning Review* 80(4–5): 415–35

Shaw, D. and Sykes, O. (2003) 'Investigating the application of the ESDP to regional planning in the UK,' *Town Planning Review* 74(1): 31–50

Shaw, D., Roberts, P. and Walsh, J. (eds) (2000) *Regional Planning and Development in Europe*, Aldershot: Ashgate

Shaw, J. (1998) 'Who's afraid of the double whammy?', *Town and Country Planning* 67(9): 306–7

Shaw, J. and Docherty, I. (2014) *The Transport Debate*, Bristol: Policy Press

Shaw, J. and Hesse, M. (2010) 'Transport, geography and the "new" mobilities', *Transactions of the Institute of British Geographers* 35(3): 305–12

Shaw, J. and Walton, W. (2001) 'Labour's new trunk-roads policy for England: an emerging pragmatic multimodalism?', *Environment and Planning A* 33(6): 1031–56

Shaw, K. (2012) '"Reframing" resilience: challenges for planning theory and practice', *Planning Theory and Practice* 13(2): 308–12

Shaw, K. and Robinson, F. (1999) 'Learning from experience? Reflections on two decades of British urban policy', *Town Planning Review* 70(1): 49–63

Shaw, R., Colley, M. and Connell, R. (2007) *Climate Change Adaptation by Design: A Guide for Sustainable Communities*, London: TCPA

Sheail, J. (1981) *Rural Conservation in Inter-war Britain*, Oxford: Clarendon Press

Sheail, J. (1992) 'The amenity clause: an insight into half a century of environmental protection in the United Kingdom', *Transactions of the Institute of British Geographers* 17: 152–65

Sheller, M. and Urry, J. (2000) 'The city and the car', *International Journal of Urban and Regional Research* 24(4): 737–57.

Shelton, A. (1991) 'The well informed optimist's view', in Nadin, V. and Doak, J. (eds) *Town Planning Responses to City Change*, Aldershot: Gower

Shepley, C. (1999) 'Decision-making and the role of the Inspectorate', *Journal of Planning and Environment Law* May: 403–7

Sherlock, H. (1991) *Cities are Good for Us*, London: sPaladin

Shipley, R., and Newkirk, R. (1999) 'Vision and visioning in planning: what do these terms really mean?', *Environment and Planning B* 26(4): 573–91

Shiva, V. (1992) 'Recovering the real meaning of sustainability', in Cooper, D. E. and Palmer, J. A. (eds) *The Environment in Question*, London: Routledge

Shoard, M. (1980) *The Theft of the Countryside*, London: Maurice Temple Smith

Shoard, M. (1987) *This Land is Our Land*, London: Paladin (updated by Gaia Books in 1997)

Shoard, M. (1999) *A Right to Roam*, Oxford: Oxford University Press

Short, M. (2012) *Planning for Tall Buildings*, London: Routledge

Short, M., Jones, C., Carter, J., Baker, M. and Wood, C. (2004) 'Current practice in the strategic environmental assessment of development plans in England', *Regional Studies* 38(2): 177–90

Shorten, J. (2004) 'New light on country life', *Town and Country Planning* 73(6): 186–91

Shorten, J., Brown, C. and Daniels, I. (2001) *Are Villages Sustainable? A Review of the Literature*, Cheltenham: Countryside Agency

Shove, E., Pantzar, M. and Watson, M. (2012) *The Dynamics of Social Practice: Everyday life and how it changes*, London: Sage

Sidaway, R. (1994) *Recreation and the Natural Heritage: A Research Review*, Perth: Scottish Natural Heritage

Sinclair, G. (1992) *The Lost Land: Land Use Change on England 1945–1990*, London: Council for the Protection of Rural England

Skeffington, A. (chair) (1969) *Report of the Committee on Public Participation in Planning*, Skeffington Report, London: HMSO

Slater, S., Marvin, S. and Newson, M. (1994) 'Land use planning and the water sector', *Town Planning Review* 65(4): 375–97

Slater, T. (2006) 'The eviction of critical perspectives from gentrification research', *International Journal of Urban and Regional Research* 30(4): 737–50

Sloman, L., Cavill, N., Cope, A., Muller, L., and Kennedy, A. (2009) *Analysis and Synthesis of Evidence on the Effects of Investment in Six Cycling Demonstration Towns*, Report for Department for Transport and Cycling England

Sloman, L., Cairns, S., Newson, C., Anable, J., Pridmore, A. and Goodwin, P. (2010), *The Effects of Smarter Choice Programmes in the Sustainable Travel Towns: Research Report*, London: DfT

Smith, H., Ballinger, R. C. and Stojanovic, T. (2012) 'The spatial development basis of marine spatial planning in the United Kingdom', *Journal of Environmental Policy and Planning* 14(1): 29–49

Smith, J. (2013) *Neighbourhood Planning*, Standard Note SN/SC/5838, London: House of Commons Library

Smith, L. (2006) *The Uses of Heritage*, London: Routledge

Smith, R. and Wannop, U. (1985) *Strategic Planning in Action: The Impact of the Clyde Regional Plan 1946–82*, Aldershot: Gower

Smith & Williamson Management Consultants (2002) *Planning Green Paper Report: Processing and Analysis of Responses*, London: DTL

Social Exclusion Unit (1999) *Improving Shopping Access for People Living in Deprived Neighbourhoods*, Report by Policy Action Team 13, Department of Health, London: SEU

Stamp, G. (2007) *Britain's Lost Cities*, London: Aurum Press

Standing Advisory Committee on Trunk Road Assessment (SACTRA) (1994) *Trunk Roads and the Generation of Traffic*, London: HMSO

Starkie, D. (1982) *The Motorway Age*, Oxford: Pergamon

Stead, D. and Nadin, V. (2000) *Urban and Rural Interdependencies in the West of England*, Faculty of the Built Environment Working Paper, Bristol: University of the West of England

Steel, J., Nadin, V., Daniels, R. and Westlake, T. (1995) *The Efficiency and Effectiveness of Local Plan Inquiries*, London: HMSO

Steele, J. (2012) 'Self-renovating neighbourhoods: unlocking resources for the new regeneration', *Journal of Urban Regeneration and Renewal* 6(1): 53–65

Stevens, M. and Whitehead, C. (2005) *Lessons from the Past, Challenges for the Future for Housing Policy: An Evaluation of English Housing Policy 1975–2000*, London: ODPM

Stevens, R. (chair) (1976) *Planning Control over Mineral Working*, Stevens Report, London: HMSO

Stevenson, R. (2002) *Getting 'Under the Skin' of Community Planning*, Report to the Community Planning Task Force, Edinburgh: Scottish Executive

Stewart, J. (2003) *Modernising Local Government: An Assessment of Labour's Reform Programme*, Basingstoke: Palgrave Macmillan

Stewart, M. (1938) *Third Report of the Commissioner for the Special Areas, England and Wales, etc.*, Report of the Commissioner for the Special Areas in England and Wales for the year ended 30 September 1938, Cm 5896, London: Commissioner for the Special Areas, England and Wales

Stiglitz, J. E., Sen, A. and Fitoussi, J.-P. (2009) *Report by the Commission on the Measurement of Economic Performance and Social Progress*, Commission on the Measurement of Economic Performance and Social Progress

Stoker, G. (1991) *The Politics of Local Government*, London: Macmillan

Stoker, G. (ed.) (1999) *The New Management of British Local Governance*, London: Macmillan

Stoker, G. and Wilson, D. (eds) (2004) *British Local Government into the 21st Century*, Basingstoke: Palgrave Macmillan

Sullivan, H. and Skelcher, C. (2002) *Working Across Boundaries: Collaboration in Public Services*, Basingstoke: Palgrave Macmillan

Sustainable Development Commission (2004) *Shows Promise but Must Try Harder*, London: Sustainable Development Commission

Sustrans (2008) *Take Action on Active Travel*, available at www.fph.org.uk/uploads/Take_action_on_active_travel.pdf (accessed 7 July 2014)

Sutcliffe, A. (1981) *Towards the Planned City: Germany, Britain, the United States and France 1780–1914*, Oxford: Blackwell

Sutton, J. (1988) *Transport Co-ordination and Social Policy*, Aldershot: Avebury

Swain, C., Marshall, T. and Baden, T. (eds) (2012) *English Regional Planning 2000–2010: Lessons for the Future*, London: Routledge

Swart, R. and Raes, F. (2007) 'Making integration of adaptation and mitigation work: mainstreaming into sustainable development policies?', *Climate Policy* 7(4): 288–303

Sykes, O. (2013) 'Entering unknown territory?' *Town and Country Planning* 82(6): 302–5

Tait, M. and Campbell, H. (2000) 'The politics of communication between planning officers and politicians: the exercise of power through discourse', *Environment and Planning A* 32(3): 489–506

Talen, E. (1996), 'Do plans get implemented? A review of evaluation in planning', *Journal of Planning Literature* 10(3): 248–59

Tallon, A. (2009) *Urban Regeneration in the UK*, London: Routledge

Tarling, R., Hirst, A., Rowland, B., Rhodes, J. and Tyler, P. (1999) *An Evaluation of the New Life for Urban Scotland Initiative*, Edinburgh: Development Department, Scottish Executive

Taussik, J. and Smalley, J. (1998) 'Partnerships in the 1990s: Derby's successful City Challenge bid', *Planning Practice and Research* 13(3): 283–97

Taylor, M. (2008) *Living Working Countryside: The Taylor Review of Rural Economy and Affordable Housing*, London: DCLG

Taylor, M. (2012) *External Review of Government Planning Practice Guidance: Report submitted by Lord Matthew Taylor of Goss Moor*, London: DCLG

Taylor, N. (1992) 'Professional ethics in town planning', *Town Planning Review* 63(3): 227–41

Taylor, N. (1998) *Urban Planning Theory Since 1945*, London: Sage

Teisman, G. R., and Klijn, E.-H. (2002) 'Partnership arrangements: governmental rhetoric or governance scheme?', *Public Administration Review* 62(2): 189–98

Tewdwr-Jones, M. (ed.) (1996) *British Planning Policy in Transition*: *Planning in the 1990s*, London: UCL Press

Tewdwr-Jones, M. (2002) *The Planning Polity: Planning, Government and the Policy Process*, London: Routledge

Tewdwr-Jones, M. (2012) *Spatial Planning and Governance: Understanding UK planning*, Basingstoke: Palgrave Macmillan

Tewdwr-Jones, M. (2013) 'LEPS and planning – more than a mechanism of convenience', in Ward, M. and Hardy, S. *Where next for Local Enterprise Partnerships?*, London: Smith Institute, pp. 46–53

Tewdwr-Jones, M. and Allmendinger, P. (1998) 'Deconstructing communicative rationality: a critique of Habermasian collaborative rationality', *Environment and Planning A* 30(11): 1975–89

Tewdwr-Jones, M. and Allmendinger, P. (2006) *Territory, Identity and Spatial Planning: Spatial Governance in a Fragmented Nation*, London: Routledge

Tewdwr-Jones, M., Gallentand N. and Morphet, J. (2010) 'An anatomy of spatial planning: coming to terms with the spatial element in UK planning', *European Planning Studies* 18(2): 239–57

Therivel, R. (2010) *Strategic Environmental Assessment in Action* (second edition), London: Earthscan

Thomas, B. and Dorling, D. (2004) *Know Your Place*, London: Shelter

Thomas, H. (1992) 'Disability, politics and the built environment', *Planning Practice and Research* 7(1): 22–6

Thomas, H. (1996) 'Public participation in planning', in Tewdwr-Jones, M. (ed.) *British Planning Policy in Transition*: *Planning in the 1990s*, London: UCL Press

Thomas, H. (2000) *Race and Planning: The UK Experience*, London: UCL Press

Thomas, H. (2004) 'British planning and the promotion of race equality: the Welsh experience of Race Equality Schemes', *Planning Practice and Research* 19(1): 33–47

Thomas, K. (1997) *Development Control: Principles and Practice*, London: UCL Press

Thomas, K. and Littlewood, S. (2010) 'From green belts to green infrastructure? The evolution of a new concept in the emerging soft governance of spatial strategies', *Planning Practice and Research* 25(2): 203–22

Thornley, A. (1993) *Urban Planning under Thatcherism: The Challenge of the Market* (second edition), London: Routledge

Thornley, A. (1998) 'The ghost of Thatcherism' in Allmendinger, P. and Thomas, H. (eds) (1998) *Urban Planning and the British New Right*, London: Routledge

Thurley, S. (2013) *Men from the Ministry: How Britain Saved Its Heritage,* New Haven, CT: Yale University Press

Tibbalds, F. (1988) 'Urban design: Tibbalds offers the Prince his ten commandments', *The Planner Mid-month Supplement* 74(12): 1

Tomaney, J. (2013) *From Regionalism to Localism in England: Conflict, Contradiction, Confusion, Instability,* London: UCL

Tomaney, J. and Colomb, C. (2013) 'Planning for independence?', *Town and Country Planning* 82(9): 377–88

Topping, P. and Smith, G. (1977) *Government Against Poverty? Liverpool Community Development Project 1970–75*, Oxford: Social Evaluation Unit

Town and Country Planning Association (TCPA) (1986) *Whose Responsibility? Reclaiming the Inner Cities*, London: TCPA

Town and Country Planning Association (TCPA) (2007) *Best Practice in Urban Extensions and New Settlements*, London: TCPA

Town and Country Planning Association (TCPA) (2011a) *England 2050? A Practical Vision for a National Spatial Strategy*, London: TCPA

Town and Country Planning Association (TCPA) (2011b) *Giving Local Authorities Flexibility in Setting Planning Application Fees*, TCPA Briefing Paper 21, London: TCPA

Town and Country Planning Association (TCPA) (2011c) *Re-imagining Garden Cities for the 21st Century: Benefits and Lessons in Bringing Forward Comprehensively Planned New Communities*, London: TCPA

Town and Country Planning Association (TCPA) (2012) *The Lie of the Land! England in the 21st Century,* Executive Summary, London: TCPA

Town Planning Review (2005) Special issue on Territorial Cohesion, 76(1): 1–118

Townsend, P. (1976) 'Area deprivation policies', *New Statesman*, 6 August, pp. 168–71

Townshend, T. G. (2014) 'Walkable neighbourhoods: principles, measures and health impacts', in Burton, E. and Cooper, R. (eds) *Well-being and the Environment*, Oxford: Wiley-Blackwell

Tribal (2010) *Evaluation of Planning Performance Agreements,* London: HCA

Truelove, P. (1999) 'Transport planning', in Cullingworth J. R. (ed.) *British Planning: 50 Years of Urban and Regional Policy*, London: Athlone, pp. 198–212

Turok, I. (1992) 'Property-led urban regeneration: panacea or placebo?', *Environment and Planning A* 24(3): 361–79

Turok, I. (2004) 'Scottish urban policy: continuity, change and uncertainty post-devolution', in Johnstone, C. and Whitehead, M. (eds) (2004) *New Horizons in British Urban Policy: Perspectives on New Labour's Urban Renaissance*, Aldershot: Ashgate, pp. 111–28

Turok, I. and Shutt, J. (eds) (1994) 'Urban policy into the 21st century', Special Issue of *Local Economy* 9(3): 211–304

Tyler, P., Rhodes, J., Lawless, P. and Dabinett, G. (2001) *A Review of the Evidence Base for Regeneration Policy and Practice*, London: DETR

Tyme, J. (1978) *Motorways versus Democracy*, Basingstoke: Macmillan

UK Foresight (2010) *Land Use Futures: Making the Most of Land in the 21st Century*, Final Project Report, London: Government Office for Science

UK National Ecosystem Assessment (UK NEA) (2010) *Progress and Steps Towards Delivery*, Cambridge: UNEP-WCMC

United Nations Human Settlements Programme (UNHGR) (2009) *Planning Sustainable Cities*, London: Earthscan

Unwin, R. (1909) *Town Planning in Practice,* London: T. Fisher

Upton, W. (2005) 'Planning reform: the requirement to replace supplementary planning guidance with supplementary planning documents', *Journal of Planning and Environment Law* January: 34–40

Urban Task Force (1999) *Towards an Urban Renaissance,* London: Spon

URBED (1998) *Grainger Town: Investing in Quality*, Newcastle upon Tyne: The Grainger Town Project

URBED (2002) *Towns and Cities: Partners in Urban Renaissance*, London: ODPM

Uthwatt, A. A. (chair) (1942) *Final Report of the Expert Committee on Compensation and Betterment*, Cmd 6386, Uthwatt Report, London: HMSO

van Leeuwen, E. (2010) *Urban–Rural Interactions: Towns as Focus Points in Rural Development,* Heidelberg: Physica-Verlag

van Ravesteyn, N. and Evers, D. (2004) *Unseen Europe: A Survey of EU Politics and its Impact on Spatial Development in the Netherlands*, Rotterdam: NAi Publishers

Varney, D. (2006) *Service Transformation: A Better Service for Citizens and Businesses, a Better Deal for Taxpayers*, London: HM Treasury

Vigar, G. (2002) *The Politics of Mobility: Transport, the Environment and Public Policy*, London: Spon

Vigar, G. (2006) 'Deliberation, participation, and learning in the development of regional strategies', *Planning Theory and Practice* 7(3): 267–87

Vigar, G. (2009) 'Spatial planning and policy integration: evidence from Scottish practice', *European Planning Studies* 17(11): 1571–90

Vigar, G. (2012) 'Planning and professionalism: knowledge, judgement and expertise in English planning', *Planning Theory* 11(4): 361–78

Vigar, G. and Healey, P. (1999) 'Territorial integration and "plan-led" planning', *Planning Practice and Research* 14(2): 153–69

Vigar, G., Shaw, A. and Swann, R. (2011) 'Selling sustainable mobility: the reporting of the

Manchester Transport Innovation Fund bid in the UK media', *Transport Policy* 18(2): 468–79

Vigar, G., Brooks, E. and Gunn, S. (2012) 'The innovative potential of neighbourhood plan-making', *Town and Country Planning* 81(7/8): 317–19

Vigar, G., Healey, P., Hull, A. and Davoudi, S. (2000) *Planning, Governance and Spatial Strategy in Britain: An Institutionalist Analysis*, London: Macmillan

Vogel, D. (1986) *National Styles of Regulation: Environmental Policy in Great Britain and the United States*, Ithaca, NY: Cornell University Press

Wägenbaur, R. (1991) 'The European Community's policy on implementation of environmental directives', *Fordham International Law Journal* 14: 455–77

Waldegrave, W., Byng, J., Paterson, T. and Pye, G. (1986) *Distant Views of William Waldegrave's Speech*, London: Centre for Policy Studies

Wales Rural Observatory (WRO) (2004) *Overview of Policy and Resources Impacting on Rural Wales*, Cardiff: WRO

Wales Rural Observatory (WRO) (2009) *Deep Rural Localities*, Cardiff: WRO

Walker, D. (2000) *Living with Ambiguity: The Relationship between Central and Local Government*, York: Joseph Rowntree Foundation

Wall, D. (1999) *Earth First! and the Anti-Roads Movement: Radical Environmentalism and Comparative Social Movements*, London: Routledge

Walsh, J. (2009) 'Space and place in the national spatial strategy in the Republic of Ireland', in Davoudi, S. and Strange, I. (eds) *Conceptions of Space and Place in Strategic Spatial Planning*, London: Routledge, pp. 95–125

Wannop, U. (1995) *The Regional Imperative: Regional Planning and Governance in Britain, Europe and the United States*, London: Regional Studies Association and Jessica Kingsley

Warburton, D. (2000) 'Comment [on Blair and environment]', *EG* 6(10): 1–2

Warburton, D. (2002) 'Participation: delivering a fundamental change', *Town and Country Planning* 71(3): 82–4

Ward, C. (1989) *Welcome, Thinner City: Urban Survival in the 1990s*, London: Bedford Square Press

Ward, C. (1993) *New Town, Home Town: The Lessons of Experience*, London: Gulbenkian Foundation

Ward, M. and Hardy, S. (2013) *Where Next for Local Enterprise Partnerships?*, London: Smith Institute

Ward, S. V. (2004) *Planning and Urban Change* (second edition), London: Spon

Warren, H. and Davidge, W. R. (eds) (1930) *Decentralisation of Population and Industry: A New Principle in Town Planning*, London: King

Warren, J., Worthington, J. and Taylor, S. (eds) (1998) *Context: New Buildings in Historic Settings*, Oxford: Architectural Press

Waterton, E. (2010) *Politics, Policy and the Discourses of Heritage in Britain*, Basingstoke: Palgrave Macmillan

Webb, D. (2010) 'Rethinking the role of markets in urban renewal: the Housing Market Renewal Initiative in England', *Housing, Theory and Society* 27(4) 313–31

Weir, S. and Hall, W. (1994) *EGO Trip: Extra-governmental Organisations in the United Kingdom and their Accountability,* London: Charter 88 Trust

Weiss, C. H. (1977) *Using Social Research in Public Policy Making,* Lexington: Lexington-Heath

Welsh Assembly Government (WAG) (2008) *People, Places, Futures: The Wales Spatial Plan Update 2008,* Cardiff: WAG

Welsh Assembly Government (WAG) (2014) *Community and Town Council*, available at http://wales.gov.uk/topics/localgovernment/community towncouncils/?lang=en (accessed 24 June 2014)

Welsh Office (1993) *Local Government in Wales: A Charter for the Future*, Cm 2155, London: HMSO

Weston, J. (2000) 'Reviewing environmental statements: new demands for the UK's EIA procedures', *Planning Practice and Research* 15(1–2): 135–42

Whatmore, S. (2008) 'Editorial: remaking environment: history, practices, policies', *Environment and Planning A*, 40(8): 1777–8

Whatmore, S. and Boucher, S. (1993) 'Bargaining with nature: the discourse and practice of "environmental planning gain"', *Transactions of the Institute of British Geographers* 18(2): 166–78

Wheeler, S. (2013) *Planning for Sustainability: Creating Livable, Equitable and Ecological Communities*, London: Routledge

While, A. (2007) 'The state and the controversial demands of cultural built heritage: modernism, dirty concrete, and postwar listing in England', *Environment and Planning B* 34(4): 645–63

Whitbourn, P. (2002) 'World Heritage Sites: the first thirty years', available at www.sal.org.uk/lectures (accessed 8 July 2003)

White, P. (2009) *Public Transport* (fifth edition), London: Routledge

Whitehead, A. (1995) 'Planning in an unplanned environment: the Transport Act 1985 and municipal bus operators', in McConville, J. and Sheldrake, J. (eds) *Transport in Transition*, Aldershot: Avebury, pp. 77–95

Widdicombe, D. (chair) (1986, 1988) *The Conduct of Local Authority Business: Report*, Cmnd 9797, Widdicombe Report, London: HMSO; *Research Volumes*, Cmnd 9798, 9799, 9800 and 9801, London: HMSO; *Government Response to the Report*, Cm 433, London: HMSO

Wilcox, S. (2002) *UK Housing Review: 2002/2003*, Coventry: Joseph Rowntree Foundation, Chartered Institute of Housing and Council of Mortgage Lenders

Wildavsky, A. (1973) 'If planning is everything, maybe it's nothing', *Policy Sciences*, 4: 127–53

Wildavsky, A. (1978) *Speaking Truth to Power: The Art and Craft of Policy Analysis* (second edition), London: Transaction

Wilding, S. and Raemaekers, J. (2000) 'Environmental compensation for greenfield development: is the devil in the detail?', *Planning Practice and Research* 15(3): 211–32

Wildlife Trusts (2003) *Natural Partners: The Achievements of Local Biodiversity Partnerships in England*, Newark: The Wildlife Trusts

Wilkinson, D., Bishop, K. and Tewdwr-Jones, M. (1998) *The Impact of the EU on the UK Planning System*, London: DETR

Williams, G., Strange, I., Bintley, M. and Bristow, R. (1992) *Metropolitan Planning in the 1990s: The Role of Unitary Development Plans*, Manchester: Department of Planning and Landscape, University of Manchester

Williams, K. (1999) 'Urban intensification policies in England: problems and contradictions', *Land Use Policy* 16(3): 167–78

Williams, K., Burton, E. and Jenks, M. (eds) (2000) *Achieving Sustainable Urban Form*, London: Spon

Williams, R. H. (1996) *European Union Spatial Policy and Planning*, London: Paul Chapman

Williams-Ellis, C. (1928) *England and the Octopus*, London: Geoffrey Bles

Willis, M. and Jeffares, S. (2012) 'Four viewpoints of whole area public partnerships', *Local Government Studies* 38(5): 539–56

Wilson, D. and Game, C. (1998) *Local Government in the United Kingdom* (second edition), London: Macmillan

Wilson, D. and Game, C. (2006) *Local Government in the United Kingdom* (fourth edition), London: Macmillan

Wilson, E. (2006) 'Developing UK spatial planning policy to respond to climate change', *Journal of Environmental Policy and Planning*, 8(1): 9–25

Wilson, E. and Piper, J. (2010) *Spatial Planning and Climate Change*, London: Routledge

Winter, M. (1996) *Rural Politics: Policies for Agriculture, Forestry and the Environment*, London: Routledge

Wolmar, C. (2005) *On the Wrong Line*, London: Aurum Press

Wong, C. (2002) *Development of Town and City Indicators: Review, Interpretation and Analysis*, London: DTLR

Wong, C., Ravetz, J. and Turner, J. (2000) *The UK Spatial Planning Framework: A Discussion*, London: Royal Town Planning Institute

Wong, C., Baker, M., Hincks, S., Schultz-Baing, A. and Webb, B. (2012a) *A Map for England*, London: RTPI

Wong, C., Baker, M., Hincks, S., Schultz-Baing, A., and Webb, B. (2012b) 'Why we need to think strategically and spatially', *Town and Country Planning* 81(10): 437–40

Wood, C. (1991) 'Urban renewal: the British experience', in Alterman, R. and Cars, G. (eds) *Neighbourhood Regeneration: An International Evaluation*, London: Mansell

Wood, C. (1994) 'Local urban regeneration initiatives: Birmingham Heartlands', *Cities* 11(1): 48–58

Wood, C. (1996) 'Private sector housing renewal in the UK: progress under the "new regime" in Birmingham', Paper Presented to Second ACSP-AESOP Joint International Congress, Toronto, Canada, July

Wood, C. (1999) 'Environmental planning', in Cullingworth, J. B. (ed.) *British Planning: 50 Years of Urban and Regional Policy*, London: Athlone

Wood, C. (2000) 'Ten years on: an empirical analysis of UK environmental statement submissions since the implementation of the Directive 85/337/EEU', *Journal of Environmental Planning and Management* 43(5): 721–47

Wood, C. and Bellinger, C. (1999) *Directory of Environmental Impact Statements*, Working Paper 179, Oxford: Oxford Brookes School of Planning

Wood, C. and Jones, C. (1991) *Monitoring Environmental Assessment and Planning* (DoE), London: HMSO

Wood, R., Handley, J. and Bell, P. (1998) 'The character of countryside recreation and leisure appeals: an analysis of planning inspectorate records', *Journal of Planning and Environment Law* November: 1007–27

Wood, W. (1949) *Planning and the Law*, London: Percival Marshall

Woods, M., Gardner, G. and Gannon, K. (2006) *Research Study of the Quality Parish and Town Council Scheme*, London: DEFRA

World Commission on Environment and Development (WCED) (1987) *Our Common Future*, Oxford: WCED

Worpole, K. (ed.) (1999) *Richer Futures: Fashioning a New Politics*, London: Earthscan

Worpole, K. (2013) *Park Life Revisited*, London: HLF

Worskett, R. (1969) *The Character of Towns: An Approach to Conservation*, London: The Architectural Press

Worthing, D. and Bond, S. (2008) *Managing Built Heritage: The Role of Cultural Significance*, Oxford: Blackwell

Wraith, R. E. and Lamb, G. B. (1971) *Public Inquiries as an Instrument of Government*, London: Allen and Unwin

Yiftachel, O. (1989) 'Towards a new typology of urban planning theories', *Environment and Planning B* 16(1): 23–39

Young, M. and Willmott, P. (1957) *Family and Kinship in East London*, London: Routledge and Kegan Paul

Zonneveld, W. (2005) 'Expansive spatial planning: the new European transnational spatial visions', *European Planning Studies* 13(1): 1–18

Index of statutes

General index

Page numbers in **bold** indicate tables and those in *italics* indicate figures.